Cement-Based Composites
Second Edition
Materials, mechanical properties and performance

Cement-Based Composites
Second Edition

Materials, mechanical properties
and performance

Andrzej M. Brandt

Routledge
Taylor & Francis Group

LONDON AND NEW YORK

First published 1995 by E & FN Spon

2 Park Square, Milton Park, Abingdon, Oxon OX14 4RN
711 Third Avenue, New York, NY 10017, USA

Routledge is an imprint of the Taylor & Francis Group,
an informa business

First issued in paperback 2017

Copyright © 1995 Taylor & Francis
Copyright © 2009 Andrzej M. Brandt

Typeset in Sabon by Pindar NZ, Auckland, New Zealand

All rights reserved. No part of this book may be reprinted or
reproduced or utilised in any form or by any electronic, mechanical,
or other means, now known or hereafter invented, including hereafter
invented, including photocopying and recording, or in any information
storage or retrieval system, without permission in writing from
the publishers.

Notice:
Product or corporate names may be trademarks or registered
trademarks, and are used only for identification and explanation
without intent to infringe.

British Library Cataloguing in Publication Data
A catalogue record for this book is available from the British Library

Library of Congress Cataloging-in-Publication Data
Brandt, Andrzej Marek.
  Cement-based composites: materials, mechanical properties, and
  performance/Andrzej M. Brandt. — 2nd ed.
    p. cm.
  Includes bibliographical references and index.
  1. Cement composites. 2. Cement composites—Mechanical
  properties. 3. Reinforced concrete. I. Title.
  TA438.B73 2008
  620.1'35—dc22   2008022424

ISBN13: 978-0-415-40909-4 (hbk)
ISBN13: 978-1-138-11539-2 (pbk)

Contents

Preface to the first edition

The book is the result of the author's studies and research over the last few years. In that time not only has research activity created a basis of some general and specific opinions, but also it was possible to gather the views and comments of many outstanding scientists and professional engineers encountered on different occasions, including internal seminars at the Institute in Warsaw, national and international workshops and conferences, and on visits to universities in several countries. It is impossible to mention here all those who on many of these occasions and in various ways agreed to share with the author their research results, comments, doubts and opinions. Without all the friendly contacts with numerous outstanding personalities the book could never have appeared and very kind thoughts are addressed to them.

Given the rapid development of the science in the last few decades, a book covering in a limited volume a relatively large scope of knowledge cannot attempt to encapsulate all the concepts or develop them in a homogeneous way and with equal competence. Unavoidably, advanced readers will find omissions, errors and imprecision. Some chapters are written in more detail than others. With certain problems the author had less personal experience than in others. All these weaknesses of the book are due to the author's subjective viewpoint. All fruits of human activity are imperfect and that is the case with this book. It is the author's hope that the most dissatisfied readers will decide to write something better in the near future.

It is the author's pleasure to present thanks and acknowledgements to several individuals who agreed to help him in the preparation of the book.

Thanks are given to my closest co-workers in the Department of Strain Fields of the Institute of Fundamental Technological Research (IFTR), Polish Academy of Sciences in Warsaw, namely to Professor J. Kasperkiewicz, Dr A. Burakiewicz, Dr M. A. Glinicki, Dr M. Marks and Dr J. Potrzebowski. Without their cooperation in all tests and studies and their helpful discussions it would have been impossible to prepare the book. Thanks are due to Mr M. Sobczak, who prepared majority of drawings and participated in some testing. Among other members of the staff of IFTR special thanks are given to Professor W. Marks who agreed to participate in research which completed Chapter 13. Results mentioned in Sections 10.4 and 10.5 were based on experiments carried out with competence and courage by Professor G. Prokopski at the Częstochowa Technical University.

Particularly warm thanks are due to Professor L. Kucharska from the Wrocław Technical University who kindly agreed to read most of the chapters and who made many useful comments and criticisms.

Partial financial support in completing Chapter 13 was provided by the National Committee for Research in Poland (Grant No. 700409101) which is gratefully acknowledged.

The book was written while the author was a staff member of the IFTR, Centre of Mechanics, and all support from colleagues and administration is kindly acknowledged.

The kind cooperation and patience exhibited by the staff of Chapman & Hall in London and particularly by Ms Susan Hodgson and Mr Nick Clarke is acknowledged with pleasure.

Warsaw,
November 1993

Preface to the second edition

The last 10 to 20 years have been very fruitful in the development of cement-based composites from various viewpoints: the extension of applications in outstanding civil engineering structures, the increase of mass production of buildings and ordinary structures, many new materials and technologies appearing on the market, and also considerable development of new methods of testing, computation and modelling. That is why the manuals and textbooks in this field have a relatively short life: they become obsolete quickly, and new ones are needed by the same groups of readers.

There is a growing importance attached to having relevant information on various aspects of these materials, presented comprehensively and covering other related fields. The principal aim of this book is to provide an overview of the present state of knowledge and technology in this quickly developing area of cement-based materials. These materials play an increasing role in our world where a growing population and its rightful expectations of building and civil engineering facilities create new challenges to all involved in this field.

The needs in various countries and regions of the world are different because of the level of development, climatic conditions, traditions and so on, but the raw materials for concrete-like composites are similar and basic laws of physics, chemistry and economics are the same. That is why the knowledge and technology of materials are created and exploited at the world level and there is rational understanding of their importance. We should be aware of how a large portion of national and regional resources are devoted to constructing and maintaining all buildings and civil engineering infrastructure, and how much may be gained or lost through more or less appropriate design and execution of all these facilities: buildings, roads, bridges, airports, dams, harbours, and the like. The quality and longevity of infrastructure are directly related to the sustainable development that is vital for all countries. And most of the infrastructure is made with concrete.

The book is aimed at students of building faculties, at professional engineers and at researchers who want to be familiar with the modern approach to the mechanics of cement-based composites. The information presented here should also help to deal with new materials that will appear on the market or that will have some application in coming years. Without any doubt, it is not possible to include all theoretical and applied information on such a vast subject in one modest book. The author may only expect that for all interested readers the book will provide the basis by which they may begin to solve their own particular problems by further reading and developing their own research and experience.

In preparing this second edition the author has tried to create a volume similar to the first one. However, there has been great progress in the number and value of new products and technologies, testing methods and test results, many of them interesting and worth studying. Therefore, it has been necessary to carry out a severe but subjective selection process. This is well visible in the lists of references, which have been limited and so do not contain many valuable papers and books. Their authors are asked to forgive these omissions – many of the omitted works have been carefully studied and have helped significantly to create an overall picture of the present situation in the vast field of cement-based composites.

As for the first edition, the author is grateful to all mentioned previously, and particularly to Professor Michał A. Glinicki and Dr Daria Jóźwiak-Niedźwiedzka for useful discussions on a few selected items. The support from all colleagues at the Section of Strain Fields and administration of the Institute of Fundamental Technological Research, Polish Academy of Sciences, in Warsaw, is kindly acknowledged.

The kind cooperation and patience exhibited by the staff of Taylor & Francis in Milton Park, Abingdon, UK, and particularly by Mr Simon Bates, is acknowledged with pleasure. Thanks are addressed to Ms Camille Lowe from Pindar NZ for her friendly and efficient coordination of all works during the final stage of preparation of the book. The author is thankful to T & F for their suggestions related to the second edition of my book.

Warsaw,
April 2008

1 Introduction

The term 'cement-based composites' requires some explanation. Why not simply 'cements and concretes'? In fact, the proposed term covers a larger group of materials and its use is justified for various reasons.

Cement mortars and concretes have been used extensively in their present form since the beginning of the nineteenth century. There is no need to set out here their history or to describe their properties and applied technologies, because nearly every year several excellent books are published on these subjects. However, there is perhaps at present a new phase of development of Portland cement-based materials, stimulated by the following factors:

- New types of these materials have appeared and are available. They satisfy new performance exigencies but they also create new requirements for their use and demands on staff competence.
- New theoretical and experimental methods are available, derived from various technical fields, but mainly from the field of high strength advanced composites.
- The behaviour of concrete-like materials is more correctly understood and these materials are better used in practice if a modern approach is applied to them as well as to the composite materials.
- Durability is considered the main property of structures and this expectation imposes new requirements on cement-based materials.
- Concretes should satisfy various requirements that in the past century were not imposed or that did not have the importance they have now; this concerns various ecological aspects, because sustainability is a key worldwide concern.

The last two factors are of particular importance in building and civil engineering works. The corrosive environment in which many outstanding bridges, dams and marine structures are constructed, loads varying with the service life of buildings, the impacts and thermal actions imposed on industrial structures and building facades, degradation due to intensive traffic, freezing and thawing of roads, highways and runways – these are only some examples of situations which are common in all parts of the world.

Traditional concretes cannot satisfy new requirements and do not behave satisfactorily to meet present demands; it is sufficient to mention that many concrete structures have required considerable effort and money to restore and maintain their serviceability because they failed to support the conditions of use shortly after completion. The possibility of using waste materials in the production of concrete is sometimes crucial in deciding what kind of building materials should be used in a given investment.

It is the author's strong belief that, using present knowledge and without any great increase in unit cost of investment, it will be possible in the near future to avoid the necessity of restoring and rehabilitating the thousands of civil engineering and building structures which after only a few years are nearly in an unserviceable state. Durability is one of the most important aspects considered in the book, also as an element of sustainable development.

Cement-based materials correspond well to the definition of composites. Their behaviour and properties may be better understood, designed and predicted using a modern approach than was possible with traditional concrete technology. These materials belong to a larger group of brittle matrix composites which, among others, also contain ceramics. Their brittleness is the source of the majority of failures of different magnitudes. The approach based on composite materials engineering is consequently presented throughout the book.

Advanced composite materials used in vehicles, aircrafts and rockets are designed and manufactured using the results of extensive studies. Non-classical phenomena such as cracking, fracture, plastic deformations, fatigue and the like are analyzed at various levels from macro to nano to ensure adequate fracture toughness and reliability in various situations. These methods have long been available for cement-based composites too but their direct application is not always possible: if the basic principles and methods are similar, other features and all quantitative relations are completely different. That is why the mechanics of concrete-like composites require a new set of rules, methods and research results, which should bridge the gap between material scientists and civil engineers. This book is a modest attempt to collect together the available knowledge in this field.

There is another special aspect of cement-based materials: they are used in large quantities and in various conditions in the field, therefore they cannot be too expensive and their technologies cannot be too sophisticated. Local component materials should be used as far as possible and relatively simple manufacturing methods should be adopted. Consequently, the design and testing methods should account for all these circumstances. This does not mean, of course, that high performance concretes may be produced using poor technology or by non-competent staff – on the contrary!

In the book, information on classical cement-based composites, their properties and technology, is limited. On the other hand, new composite materials with special internal structures are examined in more detail and traditional materials are also studied using the composite approach. Material structure and composition, crack propagation and control, the influence of interfacial zones and of debonding processes, and new methods of testing, are described using theoretical studies, experimental results and practical examples.

One of the features of composites is the importance of their internal structure. It is well known that the same volume fractions of the main components, but arranged differently in space, may produce completely different materials. That observation leads to material design and optimization in cement-based materials. The rational composition of layers, pores, fibres and particles may respond perfectly and in the most efficient way to particular requirements of exploitation or to production technology. That way of thinking may help to tailor-make concrete and to improve the design and execution of concrete structures.

To keep the length of the book within acceptable limits, many problems which are usually included in textbooks for students at technical universities, are here omitted or presented very briefly. For some topics, which are perhaps peripheral to the book's scope, the reader is referred to selected references where basic data may be found; for example, concrete-like composites without cement, like polymer concretes (PC) and bitumen concretes, are only briefly mentioned. For similar reasons the scope of the book is limited to materials, and composite structures are not considered. Mechanical properties – strength, deformability, toughness, etc. – are more closely examined than other physical properties of cement-based composites.

The book is not a traditional collection of detailed information about one material after another. It would perhaps be inappropriate to present in such a way this field of materials science in which progress has been so fast in the last few decades. A very detailed presentation along the lines of a catalogue of new materials and techniques could become obsolete soon after its publication.

The aim of this book is to analyze, together with the reader, how to examine a material's properties and behaviour in relation to the material's structure and composition. That understanding is then applied to particular problems which arise in material design, execution, testing and prediction of properties.

In Chapter 2 brief general information is given on composite materials. It is explained how they have been developed in various fields of technical activity, and particularly in those where their advantages of high strength and stiffness, combined with low weight, are needed. These properties are often associated with special requirements, such as high resistance in corrosive environments and in elevated temperature, enhanced resistance against fatigue and impact.

In Chapter 3 the basic notions are introduced and the main groups of materials are presented which are related particularly to cement-based materials. Chapters 4 to 7 concern different components or phases, their properties and distribution, which form a material structure.

The remaining chapters deal with the main mechanical properties of the materials considered and describe their characteristic features in various loading and environmental conditions. The problems related to design and optimization are considered, together with some comments about the economics of these materials. The main problems and questions are discussed that involve an engineer attempting to apply non-classical cement-based materials or to understand better the behaviour of classical ones. One chapter (Chapter 13) is devoted to so-called 'high performance concretes' – a family of new materials that is developing quickly. In the final chapter the main directions of applications and future subjects of research are briefly reviewed.

The future will bring many new solutions to the present problems in design, testing and exploitation of cement-based composites but, as in all other fields of human knowledge, without any doubt there will appear more new questions than answers.

2 Composites and multiphase materials

2.1 Properties and requirements

A multiphase material is usually defined as a heterogeneous medium composed of two or more materials or phases, which occupy separate regions in space. If the term 'multiphase materials' is understood to include the larger group containing natural and man-made materials, the term 'composites' is more often reserved for man-made ones.

The properties of a composite or a multiphase material derive from those of its constituents but there are also synergetic properties. This means that some properties may be derived from the properties of constituents and their volume fractions using the rule of mixtures. Others, however, are different and new and are caused by interactions between the constituents. These interactions are called synergetic effects and are important for understanding the behaviour of composites; see also Section 2.5.

Two conclusions are obvious from these definitions:

1 Between separate regions occupied by component materials there are boundaries in which mutual interactions are transferred; this is a clear difference between composites and alloys.
2 Unlike in mixtures, most properties of composite materials cannot be simply deduced from their composition and should be specially determined considering their structure, in most cases by experiment.

In the above definitions, a phase is understood as a region of a material that has uniform physical and chemical properties. This means that not only are cement paste and rock aggregate grains or voids different phases in concrete when it is considered as a composite material but, for example, so also is any water or ice which fills these voids.

In various books and manuals the above definitions are formulated in different ways but the differences are in form and convention and do not deserve much attention here.

Multiphase natural materials such as bone, timber or rock have been encountered and used by humans since the beginning of their history on the globe. Composites have also been known for several thousands of years. The first reference to mud bricks reinforced by straw is to be found in Exodus, the second book of the Old Testament. Layers of mud with chopped straw or horsehair were used for roads in Babylon and in China. The Babylonians and Assyrians bonded stone and semi-baked bricks together with natural bitumens. Plaited straw with mud or clay was used to build walls independently by civilizations in different parts of the world. Gypsum and lime mortars are found in Egyptian pyramids. In the ancient Roman Empire the use of natural pozzolanic binders with sand and crushed stone helped to build the magnificent structures which are still admired in present times.

The use of composites was not limited to building materials. Medieval armour and swords in China and Japan were constructed with wrought iron and steel forged together. In Central Asia, bows were built of animal tendons, timber lamellae, and silk threads with adhesives as binding agents. The resulting laminated composites had ductility and hardness from their constituents but toughness was one of the synergetic effects of their performance.

The continuity of application of composites is reflected in modern languages and certain words in materials science have common Latin origin. For example, the word *caementum* meant finely crushed ceramics with binding properties. The word *betunium* was used to designate a mixture of stone and crushed bricks with lime or pozzolanic mortar. Also, the word *pozzolana* was taken from a locality of Pozzuoli near Naples where volcanic tuffs were found in abundance. *Concretus* meant something which had grown together.

The evolution of materials engineering over the centuries was perhaps less spectacular than the evolution of structures but certainly more important for human civilization. In archaeology some periods are named after the mastery or universality of particular materials used for weapons, tools and jewellery: the Neolithic period – the last part of the Stone Age – was followed by the Bronze Age and the Iron Age.

In modern times there are several examples when an advance in new forms of construction was not possible until an appropriate material was available. The prestressing of concrete structures was attempted in the years 1938–44 in Germany with negative results. The reason was very simple: ordinary low quality steel was used for the prestressing wires. The relaxation of the steel and creep of the concrete quickly reduced the designed prestressing force to zero. But in 1945 the series of successful tests by Gustav Magnel of the University of Ghent in Belgium and by Eugène Freyssinet, the great French engineer, were based on high strength steel with a yield point over 1000 MPa, where losses due to both above-mentioned effects were equal only to a small percentage of the imposed prestressing force.

As another example, the titanium-based alloys may be mentioned. Their application in the aircraft industry permitted the barrier imposed by elevated temperature at high speed to be overcome, which had previously ruled out the use of aluminium and magnesium alloys, and new generations of airplanes were built with technical success.

In recent times the application of composite materials has been growing rapidly. It is difficult to find a field of technical activity or objects produced for everyday life where composites are not used. The materials science dealing with composites has become an independent branch of science in itself, and not just a part of metallurgy as it was before. Faculties at universities, research institutes and international organizations are concerned with materials science and technology. Many types of production of advanced composites are considered as symbols and proof of technical development in most advanced countries.

The recent progress in composite materials is connected to new requirements imposed on modern equipment in industry, building, transportation and other fields. These requirements cannot be satisfied either by natural materials, by metallic alloys or by the composites available up to now. New composite materials are therefore indispensable with their tailored properties, obtained by optimum material design and appropriate technology. These properties are of various characters: starting from high strength and low deformability through low weight and excellent insulation properties, up to enhanced toughness and temperature resistance. Limiting the considerations to mechanical properties only, it may be concluded that in composite materials two main goals are achieved: high strength or hardness simultaneously with appropriate ductility or toughness.

There is a special group of requirements imposed on composite materials which concern the cost and production conditions. In several fields of application of composites for relatively expensive equipment only small volumes of a high performance material are needed for safety or reliability of the entire systems. For example, thin layers of high temperature resistant materials used for shields of spacecraft and for cutting edges of tools are made of particularly hard composites. In these cases the unit price of an appropriate composite may be of lesser importance in view of the overall cost of the equipment with enhanced or new performance. The performance viewpoint is vital, among others, in aircraft construction: if certain metallic parts can be replaced by composite ones with lower weight, this saving may be used to increase payload or range or both, and in that way the overall performance of that aircraft may be completely changed, largely covering the extra cost of the composite material used.

The situation is completely different for composite materials applied in building and civil engineering. Their unit price must be well adapted for the large volumes used, to ensure the right proportions between the cost of all parts of the work. But here, also, there are particular regions where a small amount of a high quality material can considerably improve the performance of the entire structure, e.g. beam-column joints in aseismic structures.

Similar differences exist in manufacturing methods. High strength composites are produced in specialized factories or laboratories with sophisticated equipment and in precisely controlled conditions. Cement-based materials are often manufactured *in situ* or in field factories where only relatively simple technologies are possible and where quite large variations in composition and component properties are unavoidable. However, when composite building materials are developed, higher education and skill of both people in design offices and *in situ* is required than it was for the application of wood and steel. All these issues are developed in more detail in the following chapters.

It may be concluded from the above that composites are strongly hetero-geneous and anisotropic materials, with random and irregular or regular internal arrangement of different phases. Their properties and behaviour result from their constituents' composition, distribution in space (structure) and also from their interaction (synergetic properties). Their behaviour is conditioned also by local effects at interfaces between matrix and inclusions. The high quality often means their resistance against various external actions, and – frequently – improved durability.

Composites may be characterized according to type of anisotropy. The structural anisotropy created by appropriate distribution of fibres or inclusions should be distinguished from anisotropy of crystals or natural organic materials like bone or wood. The anisotropy of composites is generated more or less purposefully, according to design and adequate technology. At various scales composites with random, unidirectional, bi-directional (laminates) and multidirectional anisotropy are produced.

2.2 Components of composite materials

The components of composite materials may be presented in a somewhat simplified way in three main groups: binders, fillers and reinforcements. That classification is based on functions fulfilled by the components, but all three groups are not always represented by different materials. Sometimes the roles of particular components are more complicated; for example, in polymer concretes the polymers play both roles of binder and reinforcement for the aggregate grain skeleton.

In another classification of composite components attention is paid to their characteristic form. The continuous phase (matrix) embeds the dispersed phase (inclusions). The inclusions may have the form of more or less regular particles and grains, fibres and pores or voids separated or interconnected. The terms matrix and inclusions may be used at different levels: the matrix itself may be composed of inclusions embedded in a binder.

Between matrix and inclusions there is an intermediary region called the interface. Stresses are transmitted from matrix to inclusions (particles, grains or fibres), and vice versa, through that interface. Local failures occur-ring there in the form of plastic yielding or cracks considerably modify the behaviour of the composite material under external actions.

Binders are the materials that by their binding properties and bonds to other components ensure the transmission of stresses. The chemical or mechanical bond (friction) may play a more or less important role according to the constituent properties and situation. The binders are used as matrices or they form matrices together with other constituents, e.g. smaller inclusions.

As ceramic matrices the following may be pointed out: silicon carbide (SiC), silicon nitride (Si$_3$N$_4$) and aluminium oxide (Al$_2$O$_3$). A few large groups of matrices may be mentioned: polymers, metals, asphalts and ceramics. In some classifications cement matrices belong conventionally to ceramics, and in others they form a separate group. In Table 2.1 examples of binders and matrices are listed with their mechanical characteristics. The data given in that table, as well as in the others in the book, should be considered only as general information not suitable for particular calculations, because the variety of materials used all over the world and their new types appearing each year on the market do not allow the preparation of comprehensive listings. Therefore, actual catalogues of produced materials should always be consulted.

The fillers or inclusions are used as particles or grains of various shapes and dimensions. They are introduced into the matrix for different reasons: to improve the mechanical properties by hard particles in sufficient volume fraction, to control crack propagation or, in the case of small particles, to control the movements of dislocations. Sometimes cheap and porous inclusions are introduced to decrease the material cost or to improve the thermal insulation properties. The inclusions may be made of superhard particles, crushed stones, sand, voids filled with air or specially produced gas, etc. In laminar composites, internal layers are often made of cheap and lightweight materials only to maintain the distance between external rigid and hard layers and thus to increase the overall stiffness.

The reinforcement also may have various forms: stiff and tough layers, fibres and fabrics; grains of hard materials are also used as reinforcement.

In certain cases hybrid reinforcement is used; this means that the reinforcement is composed of two or more types of fibres, layers or grains, carefully designed for their separate purposes.

In brittle matrices the reinforcement may be either in the form of ductile fibres to increase toughness or of deformable particles and pores to block the crack propagation. Stiff fibres or hard particles are introduced into ductile matrices to increase their strength. Short chopped fibres are used in many cases, but also long continuous fibres and different types of nets and fabrics. The polymer structure introduced to a porous stiff matrix of hardened concrete may be also considered as its reinforcement.

In Table 2.2, examples of fibres used as reinforcement for composite materials are presented. In Figure 2.1, examples of fibre properties are plotted in such a way that their relative tensile strength f_t/γ [10^6 mm] and relative Young's modulus E/γ [10^{6}mm] are ordinate and abscissa, respectively, of a rectangular system of coordinates. The points and regions in the diagram show schematically how strong and stiff the fibres are, both properties being related to their density.

There is a large variety of materials which are used as composite constituents (components). Their selection is based on their properties with respect to final goal, their availability and cost, and also to the applied technology. It

Table 2.1 Examples of binders and matrices of composite materials

Type of material	Density $\gamma \cdot 10^3 (kg/m^3)$	Strength		Modulus of elasticity $E(GPa)$	Specific strength $f_c/\gamma \cdot 10^6$ (mm)	Specific modulus $E/\gamma \cdot 10^6$ (mm)
		compressive f_c (MPa)	tensile f_t (MPa)			
Resins hardened by polymerization:						
epoxies	1.1–1.4	90–200	35–100	2–7	3.2–8.3	135–580
polyesters	1.1–1.5	90–250	30–90	1.5–6	2.9–7.0	140–460
acrylics	0.95–1.25	50–150	40–120	3–6	4.2–9.6	315–480
polyurethanes	1.1–1.2		20–70	1.5–2.5	1.8–5.8	135–210
styrenes	1.0–1.2	70–150	45–90	2–5	4.5–8.2	200–450
Resins hardened by polycondensation:						
furanics	1.1–1.2	80–150	9–15	1–14	0.8–1.25	90–1200
phenolics	1.15–1.2	25–70	8.5	1–9	0.7	90–750
urethanes	1.25	22–50		1.0		80
Metals:						
aluminium	2.7		70	70	2.6	2600
al. alloys 4–14% Cu			110–220			1200–1400
4–23% Si			150–220			
4–12% Mg			150–280			
copper and alloys	8.9		100–400	107–125	1.1–3.4	

Table 2.1 continued

Type of material	Density $\gamma \cdot 10^3$ (kg/m³)	Strength		Modulus of elasticity E (GPa)	Specific strength $f_c / \gamma \cdot 10^6$ (mm)	Specific modulus $E/ \gamma \cdot 10^6$ (mm)
		compressive f_c (MPa)	tensile f_t (MPa)			
Hydraulic binders						
lime paste	1.0	1–10	0.2		0.02	
gypsum paste	1.2	10–12	3.0		0.25	
Portland cement paste	2.0–2.2	10–25	2.0–9.0	10–20	0.10–0.40	500–900
Portland cement mortar	1.8–2.3	10–80	1–5	10–30	0.05–0.30	480–1300
Bitumens:						
asphalts and tars	1.2–1.4					

Table 2.2 Examples of fibres as reinforcement of composite materials (informative values from various sources)

Type of material, characteristic dimensions of cross-section d, characteristic length ℓ (mm)	Density $\gamma \cdot 10^3$ (kg/m³)	Tensile strength f_t (MPa)	Modulus of elasticity E (GPa)
Metallic fibres:			
steel $d = 0.1 - 1.0$; $\ell = 3 - 60$	7.85	500–2000	210
aluminium	2.65–2.90	29	70
tungsten $d = 10$ μm	19.2	2700	400
Glass fibres:			
alkalic glass A	2.45	3200–3450	69–72.5
low alkali glass B, E	2.5 –2.6	3450–3700	69–81
corrosion resistant glass C	2.5	3150–3300	70
high modulus glass M	2.9	3500	110
high strength glass S	2.45	4300–4500	88–90
alkali resistant glass CemFIL			
$d = 10 - 15$ μm	2.6	2500	80
Basalt fibres	2.6	480–750	70–120
Strong fibres:			
polycrystalline oxides			
Al_3; SiO_2; TiO_2	2–6	1000–2000	400–500
boron $d = 100$ μm	2.6–2.7	3000–3800	385–420
silicon carbide SiC	2.5–3.5	1000–2750	200–450
boron carbon B_4C	2.7	2400	580
PAN-based carbon fibres[1]	1.7–2.0	1900–7100	230–390
pitch-based carbon fibres			
$d = 14 - 18$ μm; $\ell = 3 - 10$ mm	1.9–2.15	1400–3000	160–700
Asbestos fibres:			
chrysotile $d = 0.02$ μm – 0.4 mm	2.4–2.6	3100–3500	60–160
crocidolite (blue) $d = 0.1$ μm	3.2–3.4	3500	180–200
Amorphous metal (metal-glass) fibres:			
thickness 20–30 μm	7.52	3500	155
width 1–2 mm, length 15–60 mm			
Natural organic fibres:			
cotton $d = 12–30$ μm	1.5	300–600	6–10
flax $d = 20–35$ μm		440–700	
wool $d = 12–35$ μm	1.3	130–200	
coconut coir $d = 0.1–0.3$ mm	1.12–1.15	40–200	1.9
$\ell = 1800–2800$ mm			

Table 2.2 continued

Type of material, characteristic dimensions of cross-section d, characteristic length ℓ (mm)	Density $\gamma \cdot 10^3$ (kg/m³)	Tensile strength f_t (MPa)	Modulus of elasticity E (GPa)
jute d = 0.1–0.4 mm	1.02–1.04	250–350	
ℓ = 50–350 mm			
Indian hemp d = 15 μm		750–850	
sisal d = 125–500 μm	1.45	400–850	
cellulose d = 13–120 μm	1.20–1.50	300–500	15–45
horse hair d = 50–400 μm	1.3	160	10–70
Polymeric fibres:			
polyethylene d = 0.025–1.3 mm	0.90–0.95	100–650	
polypropylene d = 10–200 μm	0.9	200–600	0.3–5.0
nylon d = 0.025 mm	1.15	300–900	4–8
polyester d = 0.020 mm	1.3–1.4	400–1100	1–5
polyolefin d = 0.15–0.65 mm	0.9	300	3–15
Kevlar (aramid)[2] ℓ = 6–60mm	1.44	3500	3–4
PRD 149 d = 10–12 μm	1.44	3400	185
PRD 49 d = 10–12 μm	1.58	2800	125
PRD 29 d = 10–12 μm	1.3	900	65
PVA (polyvinyl alcohol)	1.3	800–1500	29
d = 0.015–0.65 mm			30–35
Whiskers of various materials		6900–34500	
SiC d = 0.5–10 μm	3.2	21000	1260–2300
Al₂O₃ d = 0.5 μm	4.0	43000	490–880
			500

1 PAN: polyacrylonitrile
2 aramid: aromatic polyamide

is important that, before the composite constituents are selected, possible synergetic properties and local effects at stress transmission regions are also thoroughly analyzed.

2.3 General description of composite materials

2.3.1 Classifications

Composite materials may be classified according to various criteria. The most important from the physical viewpoint relates to the type of discontinuities created in the material structure. Systems of discontinuities may be introduced on the levels of macro-, meso-, micro- or nano-structure. The

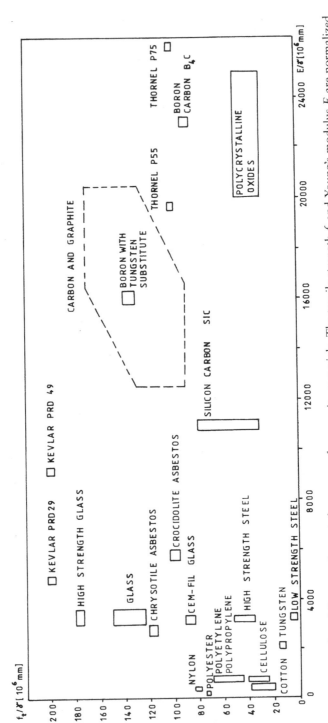

Figure 2.1 Examples of fibres used as reinforcement of composite materials. The tensile strength f_t and Young's modulus E are normalized with respect to specific density γ and put on the ordinate and abscissa, respectively.

structural elements are distributed at random or regularly, using many different techniques, which are described in specialized books and manuals. Different elements may fulfil various conditions: blocking and controlling the crack propagation, improving strength, increasing porosity, improving thermal isolation, modifying the transfer of fluids and gases across the material, and so on. Four main groups of composite materials with different types of discontinuities are shown schematically in Figure 2.2:

1 dispersion strengthening, e.g. metals with fine dispersed particles of hard material which restrain the motion of dislocations;
2 grain strengthening, e.g. Portland cement-based matrices with sand and aggregate grains randomly distributed;
3 fibre-reinforced, e.g. cement paste with asbestos fibres or polymer matrices with glass fibres;
4 laminate composites, e.g. timber plywood or laminated plastics.

As a fifth group there are perhaps functionally graded materials (FGM), which appeared around 1990. Their micro- and macrostructure is continuously changing across the thickness, allowing for different properties on both sides of the element, e.g. mechanical strength and resistance against high temperature; see Figure 2.3, (Pindera *et al.* 1998). Such materials play an important role, among others, as shields for advanced aircraft and aerospace vehicles.

The distinction between composite materials and composite structures is not sharp and laminated composites form a group that may be logically considered for different purposes either as materials or as structures. Composite materials are used to build up composite structures, but homogeneous ones may also be used. It is usual to consider concrete and fibre-reinforced concrete as composite materials. A concrete beam reinforced with steel bars or a set of steel beams connected together with a concrete steel-reinforced slab to create

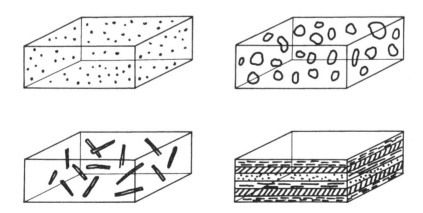

Figure 2.2 Schematic representation of four groups of composites.

a bridge deck are examples of composite structures. The distinction is one of convention and is of limited importance; the composite structures are not considered in this book.

Suggestions worth consideration for classifying composite materials have also been published by Tsai and Hahn (1980) and by Jones (1998).

2.3.2 Composites with dispersion strengthening

In this type of composite materials the main load-bearing constituent is the matrix. Small hard particles are distributed evenly in the matrix to block dislocations. The mechanism of the strengthening is described by several authors, among others several years ago by Kelly (1964) and later by Ashby and Jones (2005a). The matrices are made of metals and polymers. The particles are of different origin: silica powder or fine sand are used for polymer matrices and, for metallic matrices, second phase precipitates and particles made of oxides, nitrides, carbides and borides. Precipitates are small particles which crystallise from impurities dissolved in metal alloys. The crystallization during the cooling process results in very small and hard particles, closely distributed in the metal matrix. When ceramic particles

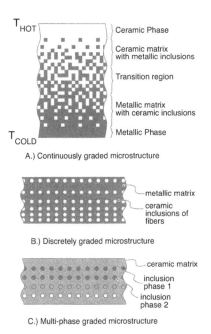

A.) Continuously graded microstructure

B.) Discretely graded microstructure

C.) Multi-phase graded microstructure

Figure 2.3 Examples of different types of functionally graded microstructures. (Reproduced with permission from Pindera J. M. *et al.*, Higher-order micro-macrostructural theory for the analysis of functionally graded materials; Kluwer Academic Publishers, 1998).

are used the resulting composites are called cermets. The size of that type of particle varies from that of atoms to micrometres for ceramic ones and their volume fraction from 1% to 15%.

When particles are bigger than 1 μm and their volume increases, for example exceeding 15%, then both the matrix and the particles share the load-bearing function in appreciable proportions and the materials obtained may be considered as belonging rather to the second group of above proposed classification. These composites maintain several properties from the matrix as their continuous phase: its ductility and toughness, heat and electric conductivity, etc.

If a uniform distribution of particles may be assumed, then the material structure is described by two values, as in Ashby and Jones (2005a): d – particle diameter or its other characteristic dimension for irregular shapes; and V_d – volume fraction of particles in the composite.

Using d and V_d, the mean distance L between particles may be calculated, which determines the mechanism of reinforcement that prevents the plastic flow; this is explained schematically in Figure 2.4. The driving force which pushes the dislocations is provided by the stress t_y. The force acting on one segment of the row of particles is equal to $t_y bL$, where b is the perpendicular dimension. The critical situation occurs when that force is balanced by the line tension T:

$$t_y bL = 2T.$$

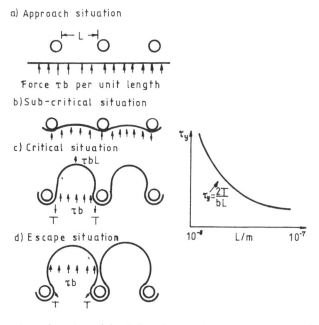

Figure 2.4 Schematic explanation of the dislocation motion across a system of small particles dispersed in a matrix, after Ashby and Jones (2005a).

The dislocations move and yielding is initiated when the equilibrium is exceeded by the growing stress t_y.

The reinforcement effect increases with d and V_d of hard particles. As a result a higher energy of external load is required to ensure the motion of dislocations.

The boundary between dispersion-strengthened and grain-reinforced composites is based on the particle size and dimension level of the processes: the grains control the crack propagation and not that of dislocations.

Composites with dispersed particles are not used only as load-bearing materials. Other uses include, for example, rocket propellants in which aluminium powder or other particles are dispersed to obtain a steady burning, and paints with silver or aluminium flakes to ensure electrical conductivity or excellent surface coverings.

2.3.3 Grain-reinforced (particulate) composites

This group of materials with so-called binary micro- or macrostructure and their non-linear behaviour is described, among others, by Axelrad and Haddad (1998). In this group different concrete-like materials are also included. They are composed of a matrix in which grains of apparent diameter from 1 μm up to 100 mm (or even more) are embedded. The volume fraction of these inclusions usually exceeds 15%.

Hard grains strengthen the matrix in ordinary concretes and in other structural composites. In materials used as insulation layers on building walls or for various non-structural purposes the lightweight and porous grains are applied or systems of pores are created.

In the range of elastic deformations hard grains reduce the matrix deformability. After crack opening, the grains control the crack propagation because each deviation of a crack to contour a hard grain requires additional input of external energy. Soft grains, beside improvement of insulation properties, also block the cracks which are passing through these grains.

The matrices are usually brittle, based in most cases on Portland cement, but also on other kinds of cement or gypsum. In the group of concrete-like materials ductile matrices are also used. These are polymers and bitumens, which behave as brittle at least in certain conditions, but in others they flow plastically.

The strength and deformations of grain-reinforced composites may be designed and forecast using various theoretical models. In the following chapters these issues are discussed in more detail.

2.3.4 Fibre-reinforced composites

This is perhaps the most important group of modern materials applied since the 1940s in various fields: from aircraft and spacecraft structures, through housings for electronic equipment up to concrete layers on runways and highways or industrial floors.

The fibre reinforcement causes a high degree of anisotropy and heterogeneity of the composites. The role of fibre reinforcement depends on the matrix that is reinforced. Two cases may be distinguished: in ductile matrices, hard and stiff fibres are introduced to increase the overall stiffness and strength, and in brittle matrices, ductile fibres control the crack opening and crack propagation, their reinforcing effects being often of secondary importance. At fracture, different phenomena appear between fibres and matrix: debonding, fibre fracture and fibre pull-out; see Figure 2.5.

The volume fraction of fibres and their distribution determine the way in which the fibres influence the behaviour of composite materials. All situations occur in practice as the fibre volume content may be as low as 0.3% in certain types of polypropylene or steel fibre-reinforced cements and may reach theoretically in unidirectional advanced composites used in construction of aircraft even 78.5% under square array and 90.7% under hexagonal array; see Dutta (1998).

The role of fibres depends also on their form, i.e. whether they are used as short chopped fibres (Figure 2.6), continuous single fibres or used in rovings, mats or fabrics, distributed at random in the matrix or arranged in a more regular way (Figure 2.7).

The bonding between the fibres and the matrix is another decisive element. It depends on the quality and processes that appear in the fibre/matrix interface (interphase). In some advanced composites, but also in glass fibre-reinforced cements, the chemical interaction between these two constituents may be destructive for the composite integrity. The fibre-matrix bond is ensured by different processes: by adhesion, mechanical anchorage and by friction, depending on the chemical and mechanical properties of both phases.

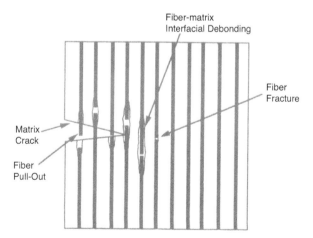

Figure 2.5 Debonding, fibre fracture, and fibre pull-out in a unidirectional continuous fibre composite, after Mallick (1997).

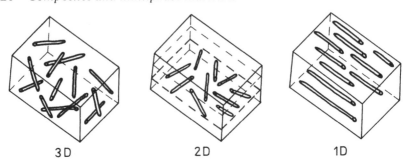

3 D   2 D   1D

Figure 2.6 Idealized orientation of short fibres in a matrix.

The quality of bonding may vary in time with the intensity of load and its duration. The role of the fibre/matrix interface is considered in more detail in Chapter 7.

A typical stress-strain diagram for a composite material, in which strong and brittle fibres are reinforcing a ductile matrix, is shown in Figure 2.8, together with separate diagrams for fibres and matrix. The failure of a composite material occurs shortly after the failure of fibres and the post-failure behaviour of the matrix is highly non-linear.

A detailed description of mechanical problems arising in fibre-reinforced composites exceeds the scope of this chapter. Below, only a few groups of problems are briefly described, and the issues relevant for brittle matrix composites are considered further in more detail in the following chapters. The interested reader may find further information concerning high strength

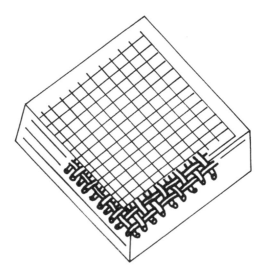

Figure 2.7 Composite material reinforced by a woven fabric.

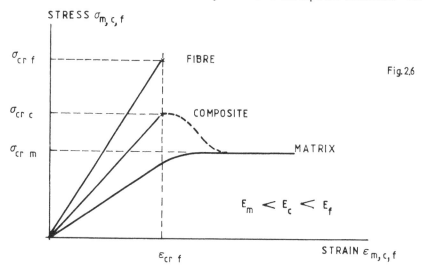

Figure 2.8 Typical stress-strain diagram for fibre, matrix and composite material, after Tsai and Hahn (1980).

fibre composites in specialized manuals such as by Harris (1999) and Ashby and Jones (2005b).

Polymer resins reinforced with glass fibres (GFRP – glass fibre reinforced plastics) were at the beginning of the high strength composites. The GFRP did not satisfy the growing requirements for stiffness in load-bearing elements of aircraft. New types of composites were introduced with resin matrices reinforced with, among others, high modulus boron or carbon fibres (CFRP).

The next generation of fibre-reinforced materials were the metal matrix composites (MMC) of ductile alloys reinforced with hard carbon or ceramic fibres and whiskers. The MMC offer serious advantages over other types of composites, mainly in the following aspects:

- better temperature capability;
- improved wetting of fibres and better load transfer by a strong matrix-fibre bond;
- higher strength and stiffness.

A particular type of fibre-reinforced composite is the directionally solidified eutectics in which matrix and reinforcement are made in one single process from one or two different materials. The reinforcement usually has the form of plates or rods and their orientation may be designed and controlled in the solidification process. Only the volume fraction of the reinforcement cannot be designed arbitrarily and is determined by the applied technological processes.

Eutectic composites have excellent mechanical properties at normal and elevated temperature: high strength and Young's modulus may be maintained after exposure during hundreds of hours at temperature over +1200°C. In several applications these properties are necessary to ensure the fulfilment of service conditions by the composite material.

2.3.5 Laminated composites

Laminated composites are constructed of at least two different layers connected together, but more often the number of layers is higher. To bond together laminae of different materials or of the same material but differently oriented is the most effective and economical way to achieve a material which satisfies simultaneously several different requirements, some of them conflicting like high strength and reduced weight, or abrasion resistance and thermal insulation. This is possible by applying different layers which have particular roles in the final laminate composite.

The layers are built of homogeneous materials or also of composite materials. Metal sheets and polymers are used as well as timber and light-weight foams. The layers are flat or curved, e.g. in pipes of different kinds. As examples of laminated composites produced in large volumes the following may be mentioned.

- Two external layers made of stiff and hard material are bonded to an internal layer of a lightweight core which maintains the external layers' distance and ensures the flexural rigidity of the composite, so-called sandwich plates (Figure 2.9a).
- Two or more layers are bonded together to satisfy conflicting requirements, such as corrosion or thermal resistance of the outer skin and strength of the next lamina, necessary for the overall bearing capacity (Figure 2.9b).
- Several layers of orthotropic composites, e.g. reinforced with parallel systems of fibres, are stacked together in such a way that adjacent laminae are oriented in different directions and thus the strength and stiffness of the composite plate may be tailored to the specific design requirements (Figure 2.9c).

The mechanical behaviour of laminated composites under external load and other actions is very complex. The overall properties should be designed and tested together with those of each single lamina. In addition, the conditions of their bonding should be verified because destruction may occur via different processes:

1   failure of external resistant layer by tension, compression or buckling;
2   failure of internal layers by shearing;
3   shearing along interfaces between layers (delamination);

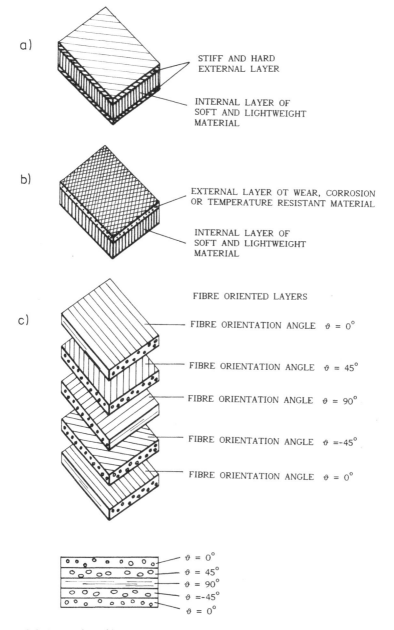

a)

STIFF AND HARD
EXTERNAL LAYER

INTERNAL LAYER OF
SOFT AND LIGHTWEIGHT
MATERIAL

b)

EXTERNAL LAYER OT WEAR, CORROSION
OR TEMPERATURE RESISTANT MATERIAL

INTERNAL LAYER OF
SOFT AND LIGHTWEIGHT
MATERIAL

c)

FIBRE ORIENTED LAYERS

FIBRE ORIENTATION ANGLE $\vartheta = 0°$

FIBRE ORIENTATION ANGLE $\vartheta = 45°$

FIBRE ORIENTATION ANGLE $\vartheta = 90°$

FIBRE ORIENTATION ANGLE $\vartheta = -45°$

FIBRE ORIENTATION ANGLE $\vartheta = 0°$

$\vartheta = 0°$
$\vartheta = 45°$
$\vartheta = 90°$
$\vartheta = -45°$
$\vartheta = 0°$

Figure 2.9 Examples of laminated composites:
  (a) sandwich panel composed of two thin strong sheets with a thick core of light and less resistant material
  (b) composition of an external layer ensuring wear, corrosion or thermal resistance with a hard internal lamina giving overall bearing capacity
  (c) composition of five fibre-reinforced unidirectional layers with different fibre orientation angle.

4   local failures and damage of various types;
5   overall buckling or rupture.

The forms of failure depend on the loading and supporting conditions but also on local properties of the constituents and their bond properties.

The mechanical properties of laminated composites as well as their fracture behaviour have been considered in many books and papers. Interested readers are referred to the classical book on anisotropic plates by Lekhnitskiy (1968), to a review of basic mechanical problems by Tsai *et al.* (1980), to design and optimization methods presented by Gűrdal *et al.* (1999) and to books by Daniel and Ishai (1994), Soutis and Beaumont (2005), and many others.

2.4 Structure of composite materials

The material structure of a composite consists of its elements (grains, particles, fibres, voids), of their volume fractions, shapes and orientations and of their distribution in space. A complete description of the material structure should contain characteristic properties of the constituents. Depending on the precision required, that information should indicate:

1   type of matrix and reinforcement with their volume or mass fractions;
2   forms and dimensions of the elements of the discontinuous phase, with the statistical distribution of these data, if needed;
3   arrangement of all these elements in the space;
4   orientation of the elements, for instance, the direction of fibres;
5   properties and dimensions of the interfacial zones between the constituents.

In particular cases other data also may be furnished.

The physical and chemical data of the constituents are given separately and are usually not included in the description of their structure.

The material structure may be considered on various levels, both by theoretical and experimental methods. These levels are defined conventionally in many ways by different authors. For the purpose of the following chapters, the levels below are distinguished.

1   Molecular level or nano-level, in which atomic bonds are considered and observations are made by various indirect methods; the characteristic dimensions are of the order of $1 \text{ Å} = 10^{-10} \text{ m} = 0.1 \text{ nm}$.
2   Structural level or micro-level, in which crystals and dislocations are observed with different types of electronic microscopes; the characteristic dimension is of the order of $1 \text{ μm} = 10^{-6} \text{ m}$.
3   Meso-level in which the main objects which measure about $1 \text{ mm} = 10^{-3} \text{ m}$ may be observed using optical microscopes; this means that the mortar situated between the biggest grains of aggregate also contains grains of sand and pores.

4   Macro-level of large grains or that of structural elements made of composites where the smallest characteristic dimension is $10^{-2} - 10^{-1}$ m.

In investigations of composite materials, usually the micro- and meso-levels are particularly considered and the macro-level is characteristic for composite structures. In the book by Ashby and Jones (2005b) a more diversified system of levels has been proposed, also characterized by dimensions of observed objects.

At the nano-level two areas may be distinguished. In nano-science knowledge is extended to understanding material behaviour such as the hydration of Portland cement, infiltration of materials of nano-dimensions into materials' structure, development of the Alkali-Aggregate Reaction (AAR) at nanoscale, using nano particles in coatings, etc. In nano-technology rather ambitious operations are considered for the modification of the structure and properties of materials, e.g. in the nanostructure of the calcium silicate hydrate CSH gel. Very active research is being conducted in this challenging field.

All materials are heterogeneous if considered at a sufficiently low level. The selection of an appropriate level for consideration and observation of the structure of a given composite is important for the results and depends considerably on the aims of the investigation. At every level different objects appear and different processes are observed while others are neglected because they are out of scale. If the macro-level is considered, the composites are treated as homogeneous materials and all processes taking place at lower levels remain unknown. Moreover, it may appear that without entering these lower levels certain phenomena cannot be understood. That is the case with several properties of cement-based composites: important factors and agents act in the lower levels and results appear in the forms of cracks or spalling on the macro-level or on that of structural elements.

At the micro- and meso-levels the stresses and strains in the fibres and in the matrix are considered separately, and such a study is considered as micromechanics. At the macro-level average values characterize the composite material's overall behaviour.

The material structure may be described qualitatively and quantitatively and represented in various ways: on radiograms and photograms without any magnification or obtained through all kinds of optical, electronic and acoustic microscopes. Other types of specialized equipment may indicate the chemical composition of particular structural elements together with their various physical properties. The precision of all these indications varies with the requirements and methods. Local data concerning certain points, i.e. sufficiently small regions of composite constituents, is required but so is global information. For instance, the porosity may be given by one number indicating the volume fraction of voids or may be described with full details of the dimensions of pores, the distance to the nearest neighbour, their statistical distribution, etc. For some overall material characteristics, such as frost resistance of concretes, global information on the porosity is imprecise and nearly useless and much more data is needed. Material structure is also often represented on schematic images.

Special computerized equipment has been in use since the 1970s to assist with the numerical description of material structure and its quantitative analysis. Machines were first produced e.g. under the commercial name Quantimet. Later, several other firms started to produce such equipment, and their performance is quickly developing together with growing applications in various fields of material science. Image analysis is used in biology, petrography and metallurgy but also for the treatment of satellite or aerial photographs for civilian and military purposes, which explains the rapid development of that technique.

Images of material structures may be analyzed at full scale as obtained from polished faces or from cross-sections of specimens. Also, images obtained by means of all types of microscopes may be analyzed. The main objective of the analysis is the quantification of images. This means that, in an observed field, particular objects like grains, fibre cross-sections, cracks and voids are distinguished and quantified as to their area fraction, shape, perimeters, distribution, and so on.

The observed fields should be prepared by different technical means to improve the visibility of the objects of interest and illuminated in an appropriate way. The image is recorded by a video camera, transformed into electronic signals and stored in a computer. With special programs the stored images are analyzed according to the aim of the research. The output is obtained in a graphical and numerical form, ready for further treatment. Using known stereological methods, image analysis gives ample information on the three-dimensional structure of examined materials. Moreover, thanks to the large storage capacity and high speed of operations of modern microcomputers, analysis and comparison of great numbers of images are possible within relatively short time frames. In this way, the results of various processes may be seen, e.g. the progress of ageing or corrosion, or the influence of different technologies, and the quantified results are obtained for further treatment. The image analysis methods are described in many books; see also Chapter 6.

In the micromechanics of material structures, the notion of the representative volume element (RVE) is important. It is the smallest region of material in which all structural elements are represented in appropriate proportions, and stresses and strains can be considered as uniform on the macroscopic level. This means that the statistical homogeneity is assumed in the RVE determination and it implies that material averages and RVE averages are the same. It is often assumed that the minimum linear dimension of the RVE is equal to four times the maximum linear dimension of an inhomogeneity such as a grain, pore or crack. For further discussion of the RVE the reader is referred to the book by Jones (1998) and papers by Kanit *et al.* (2003) and by Cheng Liu (2005). Examples of RVE for different material structures are shown in Figure 2.10. Particular problems of calculation of the RVE

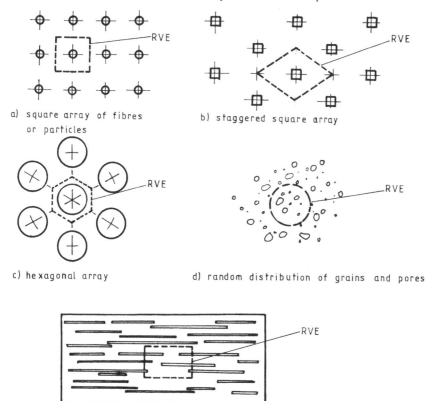

Figure 2.10 Examples of the representative volume elements in composite materials with various material structures.

dimensions for asphalt concrete and for engineering cementitious composites (ECC) are studied in papers by Romero and Masad (2001) and by Kabele (2003), respectively.

In the determination of RVE dimensions for a material structure with randomly distributed elements, e.g. short fibres, the probability should be established that the material is correctly represented by its RVE. When the required probability is increased then the RVE dimensions are also increased.

The dimensions of the RVE are not constant for a material structure, i.e. they may vary with the loading and with development of the crack pattern if the RVE is selected to represent the cracked material in its actual state. Also the RVE may be different for different material characteristics: if the focus of material representation is reduced to only one group of its characteristics, then the RVE does not take into account the others. In that sense two different groups of material properties are distinguished: structure-sensitive and structure-insensitive ones.

The structure-sensitive or constitutive properties of a material under consideration, such as strength, stiffness or water permeability, are, in a more or less direct way, related to the distribution and orientation of the structure elements and to their reciprocal interaction. Other properties, called structure-insensitive or additive, depend only on the material composition, i.e. the volume (or mass) fractions of all its constituents, and on their properties. There are few structure-insensitive properties among composite materials; as an example of one property, specific weight may be quoted.

Because the most important material properties of the composites are structure-sensitive, the design and optimization of the material structure are believed to be one of the main approaches in material science. The material structure is closely connected to the material technology and they should be designed together if an optimal structure of a composite material has to be obtained as a result of an industrial or field production.

2.5 Models and theories

When a composite material is designed for a particular application and its behaviour under load and various other actions should fulfil certain given requirements, the natural approach of a designer is to create a theoretical model of the material. A model represents a system of more or less simple rules and formulae, which describe material behaviour. That description should comprise the following elements:

1   all coefficients for stress-strain relations in linear elastic deformations and for a given system of coordinates;
2   rules for non-linear, viscous and inelastic behaviour;
3   behaviour over time, i.e. under long-term loads and imposed deformations, including the influence of the rate of loading;
4   strength criteria under various stress states and types of fracture processes.

In this way a system of constitutive equations is established, giving complete stress-strain relations, together with a description of the mode of fracture.

Other parts of the material behaviour description should consider the variation of the material properties with position. These variations are related to initial material structure or to local discontinuities and damages due to load and other actions. In composite materials the assumption of homogeneity may be justified with several restrictions only: for limited regions, under small loads and for considerations with a well-defined degree of approximation. If the relations for homogeneous media are applied, special methods of homogenization are used, e.g. RVE. It means that effective variable properties are replaced in well-defined regions by calculated constant values. Instead of a real heterogeneous material with properties varying from one point to another, a fictitious material is considered in which homogeneous regions have calculated properties; see Section 2.4.

In general, it is difficult or even impossible to express completely material behaviour by theoretical considerations. The attempts are aimed at approximate information, usually submitted to experimental verification. The simpler the model used, the easier is to get information. However, very simple models cannot describe effectively the properties of a real material.

The basic approach to material behaviour of composites is derived from Hooke's law in its generalized form, which in fact concerns only linear elastic and homogeneous media. Then additional mechanisms are used to account for the complexity of the material behaviour, ranging up to highly non-linear behaviour with inelastic deformations. For composite materials, a few simple models are used which were initially proposed for use in general solid mechanics; considerably more complex models are also proposed.

The Hookean model (Figure 2.11a) takes the form of a spring, which represents the linear stress-strain relation. In a three-dimensional orthogonal system of coordinates it is represented by the following equations:

$$\sigma^{ij} = C^{ijkl}\, \varepsilon_{kl}, \; i,j,k,l = 1,2,3 \tag{2.1}$$

where σ^{ij} is the stress tensor, ε_{kl} is the strain tensor and C^{ijkl} is the tensor of elastic constants. Taking into account the symmetry of indices, 21 independent elastic constants determine an anisotropic material. For materials with particular symmetry with respect to certain planes or axes, the number of constants is reduced, and for an isotropic material only two constants are needed, e.g. coefficient (modulus) of elasticity E and the Poisson ratio v. A complete development of that general approach may be found for example in the books by Fung (1965) and Harris (1999). Features arising from different kinds of anisotropy are discussed in Hodgkinson (2000).

In the Hookean model all supplied energy is stored and the displacement appears immediately after application of load.

The above-mentioned concept of an isotropic material described by two constants is practically never encountered when real materials are studied, and it is only used for:

- very rough approximations of examined materials;
- school exercises and training examples;
- particular cases where, at the macroscopic level, it may be admitted that in certain conditions the microscopic grains, particles or fibres are distributed at random in space and randomly oriented.

Therefore, in all other cases the composite materials are considered as having different properties in different directions. This is characteristic for most fibrous and all laminated materials where certain directions are intentionally reinforced.

When a composite is treated as an anisotropic elastic material, the stress-strain relations are based on a general approach, which is valid in a region of small deformations. These theoretical relations were developed for application in high strength composites which, under service loads, should remain elastic. That part of the composite materials theory is not examined here because there are many excellent handbooks covering that subject, which may be considered as a classical one. Moreover, for the brittle composites the zone of elastic deformations is very limited. These materials are characterized by early appearance of non-elastic behaviour in the form of microcracks of various kinds. They obey Hooke's law only when subjected to small loads and under higher loads develop more or less plastic, quasi-plastic and non-linear deformations. To express the material behaviour close to fracture as well as the inherent plasticity and viscosity of many materials the Hookean model is completed by other simple models shown in Figure 2.11.

In the Newtonian model (Figure 2.11b) the displacement is proportional to load P and to time t, but here all energy is dissipated. The model has the form of a piston in a cylinder filled with fluid (a dashpot). The displacement is inversely proportional to a coefficient of viscosity η which characterizes the motion of the piston in the cylinder.

In the de St Venant model (Figure 2.11c), a block starts to slide when the load P reaches a certain value P_o. Below that value there is no motion in the model.

Even for approximate calculations, these simple models are used in combinations. Examples of such combinations are presented in Figure 2.12. Further details on various models for composites may be found in Jones (1998) and in various handbooks on solid mechanics, e.g. Fung (1965).

The calculations of composite material properties require the application of rather complex models, composed of several simple elements. The validity

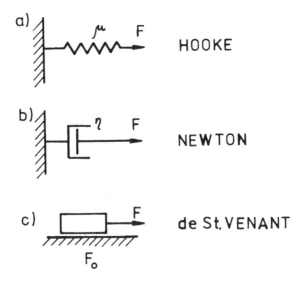

Figure 2.11 Simple theoretical models.

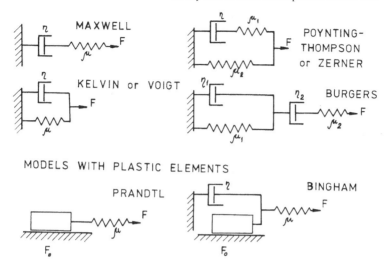

Figure 2.12 Models of viscoelastic materials.

of results depends on the type of composite material and its heterogeneity and also on the value of the load and intensity of other actions. The theoretical model may express small deformations of materials having regular structure, e.g. laminates composed of orthogonal layers, much better than ensuring satisfactory representation of a randomly reinforced and strongly heterogeneous material under intensive load, producing local plastic deformations and cracks. In the latter case, a satisfactory model probably does not exist.

An example of the application of theoretical models is the calculation of Young's modulus of a two-phase composite. Both phases behave as linear elastic and homogeneous materials according to Hooke's model, but having different Young's moduli E_1 and E_2. They are randomly distributed in the space with respective volume contents V_1 and V_2, $V_1 + V_2 = 1$. To obtain upper and lower limits for the Young's modulus of composite material under compression two extreme possibilities of internal structure are considered, as shown in Figure 2.13. In a parallel system both phases are subjected to equal strain and, in a series system, to equal stress.

It may be derived that, for a parallel model, the following formula gives the Young's modulus of the composite:

$$E_c = E_1 \, V_1 + E_2 \, V_2 = E_1 \, V_1 + E_2 \, (1 - V_1) \tag{2.2}$$

which is a simple application of the rule of mixtures.

For the series model it is:

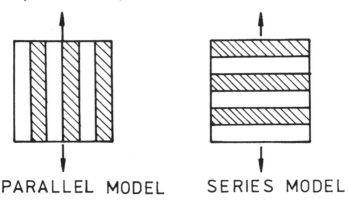

PARALLEL MODEL   SERIES MODEL

Figure 2.13 Two extreme possibilities of structure of two-phase materials.

$$E_c = \left[\frac{V_1}{E_1} + \frac{V_2}{E_2} \right]^{-1} \tag{2.3}$$

These expressions for E_c are the upper and lower limits for all possible phase arrangements, and may be used for approximate calculations, providing that they give close numerical results – if the difference is too large, the obtained answer is practically useless.

As an example, a two-phase composite material is considered, of unknown phase arrangement with equal contents $V_1 = V_2$. Two cases of Young's moduli of both phases are considered:

1   $E_1 = 2\, E_2$;
2   $E_1 = 100\, E_2$.

Using these values of elastic constants the upper and lower limits may be calculated from equations (2.2) and (2.3):

1   $E_c = 1.5\, E_2$ and $E_c = 1.33\, E_2$;
2   $E_c = 50.5\, E_2$ and $E_c = 1.98\, E_2$.

In the first case both formulae give close values which may be used for approximate calculations. However, in the second case the obtained values are so different that they cannot give any useful approximation of the range of volume fractions. These two examples are shown in Figure 2.14. Further proposals were therefore aimed at constructing a more sophisticated model; see Ashby and Jones (2005b) in which the two previous ones are combined. The phase arrangements in the models shown in Figure 2.15 are built up to obtain better results for heterogeneous bodies.

The Young's modulus represented by the Hirsch's model is expressed by the following equation:

a)

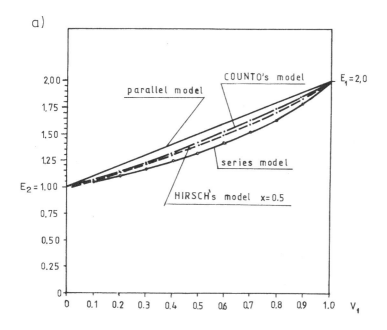

b)

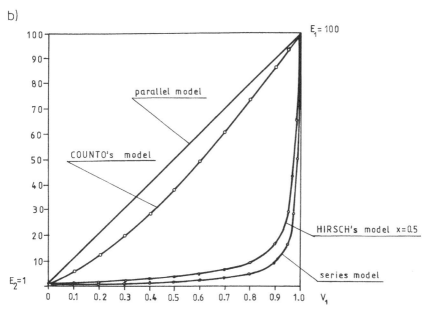

Figure 2.14 Examples of relations between the modulus of elasticity of two-phase
$(V = V_1 + V_2)$ composites and the volume fraction V_1 for four different
theoretical models:
(a) for a weakly heterogeneous material with $E_1 = 2E_2$
(b) for a strongly heterogeneous material with $E_1 = 100E_2$.

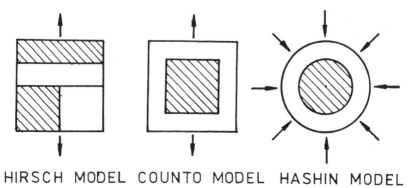

HIRSCH MODEL COUNTO MODEL HASHIN MODEL

Figure 2.15 Models for heterogeneous materials.

$$E_c = \left[\frac{x}{V_1 E_1 + V_2 E_2} + (1-x) \left(\frac{V_1}{E_1} + \frac{V_2}{E_2} \right) \right]^{-1} \tag{2.4}$$

where $x \in (0;1)$ and is equal to the relative fraction corresponding to the parallel model.

A similar equation for Counto's model has the following form:

$$E_c = \left[\frac{1-\sqrt{V_2}}{E_1} + \frac{1}{\left[\left(1-\sqrt{V_2} \right)/\sqrt{V_2} \right] E_1 + E_2} \right]^{-1} \tag{2.5}$$

The curves obtained from the above formulae as functions of V_1 with $x = 0.5$ are also shown in Figure 2.14. The conclusion is that even more complicated models do not furnish limits, which may be practically useful for strongly heterogeneous materials. Problems that concern these kinds of materials are also examined in Sections 10.5 and 11.7. Practical calculations for advanced composites are presented and discussed by Jones (1998).

The phenomenon of the fracture of composites is not included in the theoretical description given by the models discussed above and is much more difficult than the prediction of elastic properties of the materials. Fracture may have different forms and it is preceded by various effects which should be represented by other models. There are, for example:

- brittle fracture with small or even quite negligible plastic deformations;
- ductile fracture with large deformations;

- rupture when a tensile element shows necking up to the complete disappearance of its cross-sectional area.

Under increasing load or other external actions (imposed deformations, temperature) in an element built of composite material, various forms of local failure appear: cracks in the matrix, fractures of the fibres, failures of bonds between phases, plastic deformations, etc. All these effects develop in the fracture processes under the influence of increasing external actions and also with ageing, corrosion, shrinkage, fatigue and so on. To study the failure process, not only individual phenomena should be considered separately but also their interactions. In many cases, the homogenization and superposition of effects are not valid or should be used with caution within well-defined limits.

In very complicated processes of fracture all theoretical approaches and models are aimed at qualitative explanation of basic features. Quantitative predictions should be completed by extensive experimental data.

In a group of similar composites, different types of fracture may be observed in various conditions, e.g. materials considered as ductile behave as brittle at low temperature or under high rates of loading. Therefore, if a group of materials like glasses and ceramics are called brittle without particular explanations, this means that it concerns their behaviour in normal conditions. The notions brittleness and ductility have only relative meaning and do not include any quantification. It depends on various circumstances what plastic deformation accompanying the fracture is considered as small. However, brittle fracture is perhaps more often studied because its appearance in any structure without plastic deformations as prior warning is particularly dangerous. In the following chapters, brittle fracture is considered in more detail as characteristic for brittle matrix composites: fracture in such materials is initiated by pre-existing cracks, which propagate due to load or other imposed actions. Plastic or quasi-plastic deformations are usually restricted to small regions around the tip of the propagating crack. In many cases the existence of plastic deformations therefore may be neglected to simplify matters.

Fracture mechanics for brittle materials is basically derived from the theory presented by Griffith (1920). Initially it was proposed for an elliptical crack in an infinite elastic and homogeneous medium subjected to distant tensile forces and the conditions for crack propagation were formulated for that case. Later, the brittle fracture mechanics were developed for various situations in real structural elements, with non-negligible plastic deformations and with several complications necessary to account for heterogeneity of materials, time effects, etc. In that approach the crack's appearance and propagation is considered as a basic effect of loading and as phenomena directly related to final failure.

In all of the above considerations a consistent deterministic approach has been applied. This means that all variables – material properties, phase distribution, local discontinuities and their interactions – were treated as having well-determined values, positions, directions, and so on. In reality, all these parameters are subjected to random variations and detailed data on their statistical distributions are usually not available. To simplify the considerations the notion of representative volume element (RVE) has been proposed; see Section 2.4. The notion of the RVE is often applied even without detailed analysis of its validity, simply using certain statistical moments like average values or variance to characterize the composite material.

In conclusion, the following three observations may be formulated.

1   For high strength composites, the theoretical models are based on the generalised Hooke's law, with all necessary complements for local effects, time and temperature influences, non-homogeneity, etc.
2   For brittle matrix composites, in which cracks and various types of internal discontinuities are also considered in normal conditions and under service loads, the formulae for prediction of behaviour should be derived from fracture mechanics or should at least account for fracture phenomena. The analytical representation of fracture processes is, however, not entirely available because of their complexity and heterogeneity.
3   All values that characterize any composite material should be considered as random variables. If a detailed statistical analysis is impossible because of the complexity of the problem or incomplete data, an appropriate approximate procedure should be chosen, completed by experimental results.

For composite materials, due to the complexity of their structure and variability of their properties in both deterministic and probabilistic approaches, experimental results and their quality are of particular importance. They are necessary not only for verification of all theoretical calculations but also they should supply numerical data for factors, coefficients and distributions, which are used in the formulae.

Another aspect has to be considered here to round out the general information on composites. Synergetic effects mentioned above are of special importance and are introduced into the definition of a composite material: its properties and behaviour are not only determined by the properties and volume fractions of the components, but also certain new features appear due to the joint action of different components and external agents. Synergism is defined according to Aristotle as 'the whole is greater than the sum of its parts'. It may be expressed in an example of a fibre-reinforced matrix. If $f(F,M)$ is a property of this composite and $f(F)$ and $f(M)$ are these properties of the fibres and the matrix, respectively, then it follows:

$$f(F,M) = \underbrace{f(F) + f(M)}_{\substack{additive \\ mechanism}} + \underbrace{f(F) \cdot f(M)}_{\substack{synergetic \\ effects}} \tag{2.6}$$

In this example the synergetic effects are produced by action of the matrix that increases the apparent strength of the fibres with respect to the dry fibre bundle because the matrix acts as a bridge around individual fibre breaks; see Swanson (1997).

Synergetic effects may be positive or negative, meaning that they may act in line with our needs and may be created by the designer, or, by contrast, they may go in an opposite direction. Examples are strength and toughness of a composite material that are increased by joint influence of the components. Negative synergetic effects may appear, for example, when an external load and a corrosive agent produce larger or quicker damage than their additive effects.

The design of composite materials and structures is based on an appropriate application of positive synergetic effects.

There is another group of materials which are now being developed rapidly – nanoscale materials with some features on the scale of nanometres, i.e. 10^{-9} m. This group includes nanocomposites. Their components are analysed on a nanometer scale. Nanocomposites are formed to obtain new properties superior to those of composites in micro- and macroscale. The definition of nanocomposite material has been extended considerably to cover a large variety of materials. In the field of new materials the particles at nanoscale appear in zeolites, clays, metal oxides, etc. Other new and promising applications include nano-wires and nano-tubes as reinforcement. With building materials, however, the active application of nanomaterials and nanocomposites is only developing and so is not considered in more detail in this book. For further reading, see the manual by Gogotsi (2006).

References

Ashby, M. F., Jones, D. R. H. (2005a) *Engineering Materials 1*, Butterworth Heinemann, 3rd ed.

——.(2005b) *Engineering Materials 2*, Butterworth Heinemann, 3rd ed.

Axelrad, D. R., Haddad, Y. M. (1998) 'On the behaviour of materials with binary microstructures', in: *Advanced Multilayered and Fibre-Reinforced Composites*, Y. M. Haddad ed., London: Kluwer Academic Publishers, pp. 163–72.

Cheng Liu (2005) 'On the Minimum Size of Representative Volume Element: an experimental investigation', *Experimental Mechanics*, 45(3): 238–43.

Daniel, I. M., Ishai, O. (1994) *Engineering Mechanics of Composite Materials*. New York: Oxford University Press.

Dutta, P. K. (1998) 'Thermo-mechanical behaviour of polymer composites', in: *Advanced Multilayered and Fibre-Reinforced Composites*, Y. M. Haddad ed., London: Kluwer Academic Publishers, pp. 541–54.

Fung, Y. C. (1965) *Foundations of Solid Mechanics*. Englewood Cliffs, New Jersey: Prentice-Hall.

Gogotsi, Y. (2006) *Nanomaterials Handbook*. Boca Raton, Florida: CRC Press.

Griffith, A. A. (1920) 'The phenomena of rupture and flow in solids', *Phil.Trans.Roy. Soc.*, A 221, London: 163–98.

Gűrdal, Z., Haftka, R. T., Hajek, P. (1999) *Design and Optimization of Laminated Composite Materials*. New York: Wiley-Interscience.

Harris, B. (1999) *Engineering Composite Materials*, 2nd ed. Cambridge: Woodhead Publishing.

Hodgkinson, J. M., ed. (2000) *Mechanical Testing of Advanced Fibre Composites*. Cambridge: Woodhead Publishing.

Jones, R. M. (1998) *Mechanics of Composite Materials*. 2nd ed. London: Taylor & Francis.

Kabele, P. (2003) 'New development in analytical modeling of mechanical behavior of ECC', *Journal of Advanced Concrete Technology*, Japan Concrete Institute, 1(3): 253–64.

Kanit, T., Forest, S., Galliet, I., Mounoury, V., Jeulin, D. (2003) 'Determination of the size of the representative volume element for random composites, statistical and numerical approach', *International Journal of Solids and Structures*, 40: 3647–79.

Kelly, A. (1964) 'The strengthening of metals by dispersed particles', *Proc. Roy. Soc.*, A282, 63, London.

Lekhnitskiy, S. (1968) *Anisotropic Plates*, transl. from Russian, 2nd ed., New York: Gordon and Breach.

Mallick, P. K. (1997) *Composites Engineering Handbook*, New York: Marcel Dekker Inc.

Pindera, J. M., Aboudi, J., Arnold, S. M. (1998) 'Higher-order micro-macrostructural theory for the analysis of functionally graded materials', in: *Advanced Multilayered and Fibre-Reinforced Composites*, Y. M. Haddad ed. London: Kluwer Academic Publishers, pp. 111–32.

Romero, P., Masad, E. (2001) 'Relationship between the Representative Volume Element and mechanical properties of asphalt concrete', *Journal of Materials in Civil Engineering*, ASCE, 13(1): 77–84.

Soutis, C., Beaumont, P. W. R. (2005) *Multi-Scale Modelling of Composite Material Systems*. Cambridge: Woodhead Publishing.

Swanson, S. R. (1997) 'Design methodology and practices', in: P. K. Mallick ed., *Composites Engineering Handbook*. New York: Marcel Dekker Inc., pp. 1183–1206.

Tsai, S. W., Hahn, H. T. (1980) *Introduction to Composite Materials*, Westport, Connecticut: Technomic Publishing.

3   Concrete-like composites
Main kinds

3.1 Definitions and classification

The definitions and classifications of cement and concrete-like composite materials presented below are used in the chapters that follow. Various authors have put forward slightly different definitions and classifications, which can be found in Hannant (1978), Venuat (1984), Mindess *et al.* (2003), and others. This variety is not surprising, because concrete-like composites are produced in the largest mass and volume of any building materials in the world. They are applied in different structural and non-structural elements and are made with various constituents, used in different combinations. The term concrete-like composites is somewhat wider than cement-based composites. In fact, in this chapter a few materials are mentioned which do not contain any kind of cement and in which other binders are applied.

Throughout the book the terms taken from classic concrete technology are used together with those from the field of advanced composite materials. This is one of the results of application of the composite materials science to cement-based materials.

The large group of cements and concretes covers many traditionally used materials that have for some 20 years been the subject of rapid development and innovation. Also, their variety has been extended, mainly due to use of various polymer additions and chemical and mineral admixtures. In this chapter the main kinds of cements and concretes are introduced briefly.

Concrete-like composites are the composite materials and ordinary concrete is the most representative example of the group. They are composed of a matrix and inclusions, and possibly also of a system of reinforcement. This group of materials is by definition larger than the scope of this book, which is defined as cement-based composites.

In ordinary concretes, Portland cement combined with sand and water is used as the matrix or the binder. The inclusions are composed of coarse aggregate made of natural stones with specially selected grain fractions. Under the term concrete-like composites falls a large group of materials which, with respect to ordinary concretes, all constituents may be replaced by other materials. Not only are Portland cements used as binders but also for other types of hydraulic cements, bitumens, polymers and resins. The natural stone grains are replaced by various kinds of artificial lightweight aggregate, organic fillers and also by systems of separated or interconnected voids. The natural sand as the matrix component may be also replaced by other fine grain fillers.

The short dispersed or long fibres, and also polymers, which impregnate the void systems, are used as reinforcement. Hard grains of any dimension also play a role in reinforcement of the matrix. If the reinforcement has a form of regularly distributed steel bars, tendons or cables, according to the classification proposed in Section 2.3.1 it is a composite structure rather than a material.

Cement-based materials are limited to composites with cements as binder. This means Portland cements as well as all other kinds of cement.

Binders can include many kinds of materials, which provide a good bond with other components. Their role is to bind together fine and coarse aggregate grains and eventual fibres. Portland cement is a binder.

The word binder has two meanings: it may denote the Portland cement which when mixed with water binds together the sand grains, but it is also applied to cement paste and mortar which are binders for the coarse aggregate grains.

In most cases when cement is mentioned, Portland cement is what is meant. However, the word cement also means other inorganic and even organic binders. If doubt may arise, the name Portland cement is used, but in other cases the word cement is sufficient for the same meaning. In concrete-like composites two groups of organic binders are used: synthetic polymers and bitumens. The ordinary Portland cement is used together with organic binders or may be replaced entirely.

For some types of binders, such as polymers or bitumens, water is not needed in preparation of the mixes.

Concrete admixtures are special chemical compounds, natural minerals and other substances which are added to the concrete or mortar batch before or during mixing with the aim of improving certain properties of the composite material in its fresh or hardened state. Some authors use the term admixtures also to cover substances added at the cement manufacturing stage. These substances are more often termed *additives*. Traditionally, admixtures and additives were added in relatively small quantities, e.g. as small percentages of the Portland cement mass. However, at present quite often 50% or even more of cement is replaced by other materials such as fly ash or ground granulated blast-furnace slag and in such a case the words admixture or additive are less appropriate. These kinds of components of the blended cements are called fillers, microfillers or supplementary cementitious materials (SCM).

Aggregate is a system of grains of various materials, forms and dimensions. The boundary between the dimensions of so-called coarse and fine aggregates is defined conventionally; e.g. the maximum dimensions of fine aggregate is 4.75 mm according to ASTM, 9.52 mm according to BSI and 4 mm or 2 mm according to European Standard EN 12620:2002; see Section 4.2.

The definition has particular importance when different methods of the mix design are examined.

Coarse aggregate is treated as a discontinuous phase of a composite. Each particular grain is an inclusion in the matrix – the discontinuities and stress concentrations caused by the grains are often the origins of cracking.

The grains of fine aggregate are made of natural sand in most kinds of concrete-like materials and are considered as a matrix when mixed with cement and water. On a lower level they may be also considered as inclusions causing microcracks in cement paste.

The fine grain fraction may be also composed of other materials, such as non-hydrated cement particles, fly ash, and the like.

Different kinds of solid waste materials are used as concrete aggregate. These are slags from blast furnaces or metallurgical plants, crushed concrete and bricks, rubber particles and even municipal wastes. The word filler is applied in most cases to these kinds of aggregate which have low strength and are used to fill a certain volume with a cheap and lightweight material.

Voids between aggregate grains are packed with a matrix in which a system of pores exists filled with air or other gases, and partly and temporarily by water. Pores in the matrix are created in a natural way during mixing processes of concretes and mortars but may be also specially induced to increase the insulation properties, to decrease the weight or to improve resistance against freezing.

The pores may be separated or interconnected by channels or microcracks in the matrix and the overall pore fraction is an important material characteristic. However, more precise information may be needed, e.g. pore diameter distribution and average distance between pores are often required to evaluate the material's frost resistance; see. Section 6.5.2. In addition, aggregate grains themselves contain pores.

Reinforcement is the component in cement-based materials which should increase their toughness and control crack propagation. Its impact on particular kinds of strength depends on its nature, distribution and volume; see Chapter 5.

In composite materials thin dispersed fibres, either short or continuous, are applied and distributed randomly or regularly, and also in the form of mesh or fabrics.

When a system of interconnected pores in a hardened concrete matrix is filled with a polymer during the impregnation process, another type of reinforcement is created.

Ordinary concrete is a composite material made of Portland cement, sand, coarse aggregate and water, also with some admixtures. The specific weight of ordinary concrete is usually within the limits of 1900–2500 kg/m^3.

High strength and *high performance* concretes (HSC and HPC) are made basically with the same constituents as ordinary concrete. However, better quality constituents are used and in different proportions. High performance may mean various properties are present: not necessarily only strength but also higher impermeability, better durability, higher resistance against given loads and actions, and so on; see Section 13.4.

Lightweight concrete is a modification of concrete in which special lightweight aggregate or admixtures producing pore systems are used to lower its specific weight. It is usually assumed that the density of lightweight structural concrete (LSC) ranges from 1400 to 1900 kg/m^3. There are three main classes: lightweight aggregate concrete, foamed concrete and autoclaved aerated concrete; the latter two are used rather as insulation layers in building applications.

Lightweight structural concrete is made with different lightweight aggregates of natural origin or various industrial by-products that have a porous structure. The main reason for using LSC is to enable reduction of the dead load in concrete structures, particularly when long cantilever beams are built in the temporary stages of construction, e.g. bridge decks and girders. The compressive strength of LSC may reach that of ordinary concretes; even so-called high performance lightweight concretes are produced.

Heavyweight concrete is a kind of concrete obtained with special heavy aggregate to increase its specific weight over 2400 kg/m^3 and to obtain its special properties.

Portland cement mortar is a mixture of Portland cement, sand and water, without coarse aggregate.

Portland cement paste is a mixture of Portland cement and water, without any aggregate. One of the oldest research papers on cement paste published in modern times was by Powers and Brownyard (1947).

The last five terms are mentioned only briefly here, because all necessary information related to them may be found in any handbook on concrete technology or building materials and also in the chapters that follow.

The examples of classification of concrete-like composites given below were selected according to the importance of the classification criteria. For the sake of brevity it is assumed here that concrete-like composites are made with three groups of components:

- continuous phase, i.e. cement based matrix;
- discontinuous phase, i.e. grains and pores;
- micro-reinforcement, i.e. fibres or polymer systems.

All three groups may be represented in a three-dimensional system of coordinates as seen in Figure 3.1. Not all regions in this space diagram are related to real materials, but all materials may be shown in particular regions. For instance, a region is indicated that corresponds to steel-fibre reinforced concrete (SFRC) with lightweight aggregate.

The proposed classification does not cover all kinds of composite materials because only three variables are used. Therefore, several kinds of materials obtained with various types of admixtures and by using different technologies may be imagined only in a multi-dimensional space.

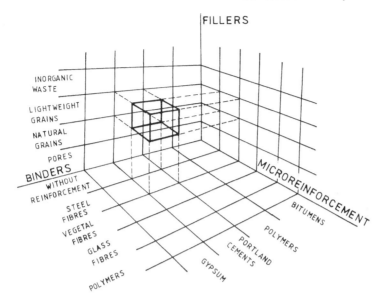

Figure 3.1 Concrete-like composites presented in a system of rectangular coordinates, after Brandt *et al.* (1983).

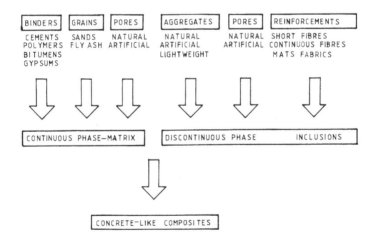

Figure 3.2 Constituents of concrete-like composites, after Brandt *et al.* (1983).

In Figure 3.2 the main constituents of concrete-like materials are shown with their relations to the material structure. According to that scheme, fine aggregate grains have a double role in composite materials: together with a binder they form a continuous phase around inclusions and reinforcement, but also these grains should be considered as inclusions with respect to cement paste. Also, single fibres should be treated as inclusions when stress concentrations are examined, for instance around the fibre ends.

The composites with brittle and ductile matrices are distinguished with respect to the matrix behaviour. The Portland cement-based materials belong to the group of brittle matrix composites as well as materials with ceramic matrices. Also, resin concretes which behave basically in a ductile manner may show some brittleness in particular conditions: in low temperature or when subjected to high rate loadings.

The brittleness of the material does not have an objective measure and it is determined in a relative way by conventional values. This issue is considered in more detail in Chapters 9 and 10.

The concrete-like composites may also be classified according to their application in construction. In such a way it is possible to distinguish between the structural and non-structural materials, the latter being used for insulation, decoration, etc. In this book attention is paid chiefly to structural materials, the strength and toughness of which are considered to be important.

3.2 Main kinds of concrete-like composites

3.2.1 *Ordinary concretes*

Ordinary Portland cement concrete (OPC) is a composite material and its constituents are cement mixed with water, fine aggregate (sand) and coarse aggregate which is available in the form of natural gravel or crushed stones; see Section 4.2.

The matrix called cement mortar is composed of cement paste with fine aggregate and the coarse aggregate grains are embedded in the matrix. Spaces between grains are filled with matrix which cements them together. There are voids between the grains and in the matrix itself. The voids may be considered also as inclusions in a continuous matrix, causing stress concentration and crack initiation.

The OPC is used as a structural material for non-reinforced elements subjected to more or less uniform compression. Because of its low tensile strength and brittleness concrete is chiefly used with passive or active reinforcement in the form of reinforced or prestressed concrete elements.

The concrete is characterized by its compressive strength at an age of 28 days after mixing. However, this one-parameter characterization of structural concretes is considered as insufficient and other properties are believed to be necessary for many purposes. This issue is examined in Chapter 8.

Starting from the OPC, several modifications may be obtained with different kinds of aggregate, by the introduction of special pore structure of appropriate parameters, by variation of quality and quantity of cement and also by application of a large variety of available admixtures and reinforcements. The application of various technologies also contributes to enlarging the field of materials based on ordinary concrete.

3.2.2 Fibre-reinforced cement matrices

This group of materials, also called fibre-reinforced concretes (FRC), contains concretes and mortars reinforced with short fibres, distributed at random or arranged in a certain way also as mats and fabrics. The fibres are of different materials: steel, glass, polypropylene, asbestos, etc., as well as natural organic fibres; see Chapter 5.

The volume of fibres is usually limited due to technological reasons, because increasing the fibre content means obtaining a homogeneous material becomes difficult or even impossible. Also the cost of fibres and the inefficiency of their use are arguments put forward to reduce the fibre volume to the lowest amount necessary.

The influence of fibres on the composite behaviour of elements subjected to loading is complex. In the limits of elastic deformations the fibres are not active and their role may be derived from the law of mixtures. When microcracks open, the fibres act as crack-arrestors and control their propagation. Total strength can be increased due to the contribution of fibres. The load corresponding to the crack appearance is the same or is increased when an especially high volume of fibres is added. The main effect of fibres is a considerable increase in deformability and in the energy of an external load which needs to be accumulated before a rupture occurs. The load-displacement diagram is completely different and quasi-ductility of the material behaviour is observed. The fibre reinforced concretes and mortars offer increased toughness and that advantage is decisive for their applications; see EN 14651:2005 and EN 14721:2000.

Fibre reinforced composites are considered in several subsequent chapters.

3.2.3 Ferrocements and textile reinforcement

(a) Components and applications

Ferrocement is the oldest kind of reinforced concrete and was invented by J. L. Lambot in 1848. If long and thin steel wires in the form of nets and grids are introduced into the cement-based matrix the composite material obtained is called ferrocement. The nets are welded in the nodes, woven or plaited, usually they are plane but three-dimensional forms are also possible. The wires are thicker than for fibre-reinforced concretes and diameters from 1 mm to 2 mm are used.

A rather broad definition proposed by the ACI 549.2R-04 for ferrocement is: 'A type of thin wall reinforced concrete constructed of hydraulic cement mortar reinforced with closely spaced layers of continuous and relatively small wire diameter mesh. The mesh may be made of metallic or other suitable materials.' In this definition all kinds of non-metallic meshes are included, as well as fabrics and mats made with non-metallic fibres that are treated as textile reinforcement.

The matrices for ferrocement consist of hydrated Portland cement paste with a filling material such as sand or micro-aggregate, similar to these used for FRC. The basic proportion of sand to cement is 3:1, by mass; slight modifications are possible. An adequate penetration of the mesh by the fresh cement mix is required and its fluidity should be adjusted by modification of the water/cement (*w/c*) ratio and by using appropriate admixtures. The concreting is executed by various methods from well instrumented automatic shotcreting to hand plastering and patching. Concreting with or without a side mould is possible, depending on the position and density of reinforcement. Construction methods can be adapted to the equipment locally available and to the more or less complicated shape of the final product: walls, roofs, boat hulls, etc. Particular advantages of ferrocement are obtained when large and curvilinear structures are built with fewer forms and scaffolding required.

The reinforcement of the ferrocement is composed usually of several layers of reinforcement of various shapes and structure. The following examples may be mentioned:

- woven and interlocking mesh;
- welded mesh of rectangular, diagonal or more often square pattern;
- expanded metal lath;
- punched or perforated metal lath;
- continuous filament irregularly assembled into a two-dimensional mat;
- rarely used three-dimensional structure of wires.

Examples of mesh for ferrocement are shown in Figure 3.3. An extensive review of various issues related to ferrocement can be found in Naaman (2000).

The application of nets has a somewhat similar influence to that of dispersed fibres, but the technology is quite different. Structural elements of various forms may be obtained at a low cost, without complete formwork, because the mortar may be placed directly onto steel nets by various procedures. Curvilinear shells, thin-walled pipes and plates are produced

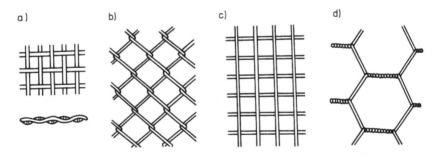

Figure 3.3 Examples of mesh for ferrocement, a) woven; b) plaited; c) welded; d) twisted. Source Brandt *et al.* (1983).

with ferrocement, as well as boat hulls and sculptures, using only simple equipment and by relatively unskilled workers.

The wires for making the reinforcement are used as received, but may be galvanically coated or covered with polymer coating to increase wire–matrix bonding and to reduce the danger of possible corrosion. Expanded metal lath and welded meshes perform better than woven ones because the latter causes discontinuities at nodes, which act as stress concentrators.

The cement-based matrices for ferrocement may be subjected to modifications, e.g. by admixtures of polymers and resins to the fresh mix or by impregnation of hardened ferrocement. The modifications are aimed at particular improvements of the final product, mainly an increase of strength or impermeability.

(b) Mechanical properties

The mechanical properties of ferrocement depend on the volume fraction and type of reinforcement as well as on the properties of the matrix. In particular:

- the ultimate tensile strength under static load depends on the efficiency of the reinforcement;
- the cracking resistance and so-called first crack load depends mostly on the matrix properties, which may be considerably improved, provided that a relatively high volume fraction of reinforcement is used;
- the compressive strength is only infrequently determined for ferrocements, because local compressions are usually supported directly by the matrix.

The volume fraction of the reinforcement is rarely lower than 5–6%, which corresponds to 400–500 kg/m^3. This may be considered as a relatively high reinforcement, compared with ordinary concrete structures. The spacing of the wires varies from 5 to 10 mm and a specific area of reinforcement S_R should satisfy the condition: $S_R = 4\ V_f/d > 0.08$ [mm^{-1}], where d is the wire diameter.

In meshes with a rectangular pattern the reinforcement area S_R is divided into transversal and longitudinal directions: $S_R = S_{RT} + S_{RL}$. The value of the ultimate matrix strain ε_{mu} is difficult to determine because it is related to the definition of a crack. It is generally admitted that for a non-reinforced matrix $\varepsilon_{mu} = 100 - 200.10^{-6}$. The first deviation from the linear behaviour as observed on the stress-strain or load-deformation curves for ferrocement elements corresponds to $\varepsilon_{mu} = 900 - 1500.10^{-6}$, but these values are strongly influenced by the volume and type of reinforcement. It has been proved that thinner wires, densely distributed, perform better than thicker ones.

The spacing between layers of reinforcement is ensured by special links; usually the spacing is uniform across the depth, but for elements subjected to bending it may be smaller in the tensioned zone in order to increase the efficiency of the reinforcement and the bearing capacity of the element. An approximate formula proposed for the number of layers N is: $N > 0.16\ t$, where t is total depth in millimetres. This means that a spacing between 4 and 6 mm is commonly used for ferrocement plates or walls.

The load is transferred from matrix to wires not only by bonding but also by nodes and transverse wires. That is the reason the crack pattern visible on ferrocement elements often reflects the distribution of wires and nodes. The nodes in nets are the origin of stress concentrations and the regular distribution of wires often corresponds to the crack pattern that appears in service. There are several techniques to improve ferrocement properties by addition of thin fibres, impregnation, etc.

The elastic modulus of ferrocement after matrix cracking depends on that of the reinforcement and is denoted E_R. It is lower than E_s for steel because the wires are not straight and, particularly in woven meshes, they stretch when subjected to tensile load. For rectangular welded mesh only, $E_R = E_s$. Diagonal wire systems are also characterized by decreased stiffness and reinforcing efficiency, e.g. for wires oriented at $45°$ the efficiency is estimated to be between 65 and 80%.

The fatigue strength of the ferrocement depends on the efficiency and type of reinforcement. It is higher for woven than for welded meshes and the failure as a result of fatigue is always initiated from the external steel mesh due to tensile fatigue. The deflections and crack width of ferrocement elements under bending may be calculated as functions of the number of cycles (Balaguru *et al.* 1979, Naaman 2000).

Durability of the ferrocement elements is a function of cover depth and cracking characteristics. When improved durability in a corrosive environment is required, a small diameter reinforcement with high specific surface should be used, and impermeability of the matrix should be ensured by additional measures. In such a case, as well as for liquid retaining structures, the admissible crack width is up to 0.02–0.05 mm, while for ordinary structures and in ordinary conditions 0.10 mm is considered as acceptable for ferrocement.

Impact strength of the ferrocement is essential for such applications as boat hulls, container walls and ballistic panels. It was tested by several authors, e.g. Grabowski (1985), with drop hammers, Charpy hammers and air gun projectiles. The results of comparative tests are expressed in crater depth or perforation characteristics and failure energy. The failure energy increases linearly with the wall depth and specific surface of the reinforcement. In Figure 3.4 tensile strength and Charpy impact strength are compared as functions of the specific surface S_R of reinforcement.

(c) Ferrocement with reinforced matrices

Crack control and resistance against local loads in ferrocement is improved considerably by the addition of fibres and polymers into the matrix com-

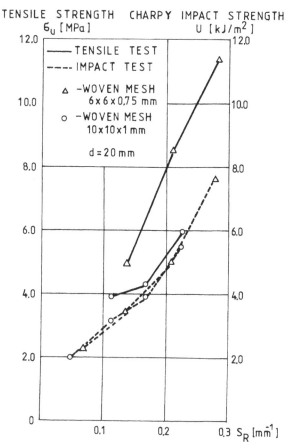

Figure 3.4 Tensile strength and Charpy impact strength of ferrocement elements versus specific surface of reinforcement, after Grabowski (1985).

position. Various kinds of fibres, such as steel, glass, low modulus carbon, vinylon and polypropylene, are used with different volume contents, usually similar to that for fibre concretes. Polymers are also used to modify the matrix, both as liquid resins as in polymer cement concretes (PCC), or by using an impregnation technique as in polymer impregnated concretes (PIC) (see next section). The main reinforcement is based on a classic steel mesh or on a hybrid combination of steel wire fabric and crimped wire cloth, as tested by Ohama and Shirai (1992). As a result, increased first crack and maximum loads were observed due to a significant improvement of matrix quality. Also, deflections at the ultimate stage were increased thanks to improved bonding. The improved resistance of such composites against impact permitted their use for shields and protective layers in industrial and military structures.

Ferrocement overlays used in the strengthening of masonry walls were successfully tested by Tan *et al.* (2007).

(d)   Cement composites with textile reinforcement

Composite elements with Portland cement matrices and textile reinforcement have been used for some time as thin roof plates, partition walls, cladding, and the like. Some discussion of textile reinforcement is presented in Section 5.9. As well as systems of plane fabrics, three-dimensional (3D) reinforcement is also manufactured and there have been tests of elements with 3D carbon fibre reinforcement (Zia *et al.* 1992).

The reinforcement of cement matrices with textiles has certain advantages, namely:

- better mechanical properties due to more regular distribution of wires than in the case of short chopped fibres;
- the possibility of more efficient orientation of wires, if the direction of principal strain components due to loading is known;
- the elimination of accidental regions of weak reinforcement;
- increased durability by application of non-corroding fibres and by ensuring at least the minimum depth of cover for all wires;
- the possibility of the application of advanced industrial techniques (e.g. lamination) and simple production of complicated shapes with plane sheets laminated as sandwiches and folded before hardening;
- the possibility of application of very simple technologies, e.g. hand-patching or laying mats and fabrics while spraying them with cement slurry and low pressure rolling.

The mechanical properties of hardened materials are similar to those of ordinary fibre-reinforced composites. They are determined by the quality of the matrix and reinforcement, and the interface layer between these two. Particularly interesting are the improved resistance and toughness against impact loading. Modelling of a reinforced concrete beam strengthened by ferrocement thin plates was presented by Elavenil and Chandrasekar (2007).

For a review of the test results of elements with textile reinforcement the reader is referred to the proceedings presented by Hamelin and Verchery (1990, 1992) and Naaman (2000).

3.2.4 Polymer modified concretes (PMC)

The modification of cement concretes and mortars by polymers and resins is aimed at improvement of their mechanical properties: compressive and tensile strengths, matrix-reinforcement bond, impermeability, corrosion resistance, etc. There are three main groups of polymer modified concretes:

1   polymer concretes (PC) composed of aggregate and polymer matrix;
2   polymer cement concretes (PCC) where Portland cement and polymer form the matrix;
3   polymer impregnated concretes (PIC) – hardened Portland cement concrete with a pore system impregnated with a polymer.

By appropriate composition and technology a large variety of different materials may be obtained. For that purpose several kinds of monomers are used, which means that low viscosity liquid organic materials can be formed from simple molecules which are capable of combining chemically into polymers, composed of very large molecules. In Section 4.1 basic information is given on applied synthetic materials used in cement composites, and in Section 13.3 the properties of Polymer Modified Concrete (PMC) are described.

3.2.5 Asbestos cements

Asbestos cements may be considered as one of the oldest man-made fibre-reinforced materials. They were already being produced in 1900 at a time when other fibre-reinforced composites were not known. Natural inorganic asbestos fibres were used only with neat cement paste and a composite material was obtained with high tensile and flexural strength and several other excellent mechanical properties; see Section 5.2.

Asbestos cements were extensively used in the construction of buildings, mostly in the form of tiles and plates for roofs and thin-walled pipes. Over the last 40 years, however, evidence has emerged of a serious hazard to the health of workers in the asbestos fibres industry, and the application of asbestos cements in building construction has decreased considerably. The danger for the inhabitants of buildings has also been taken seriously. Asbestos fibre cements are now excluded from the construction of all buildings by regulations in most countries, but their application has continued in other products. However, asbestos cements installed in earlier times frequently occurs in existing buildings and very careful procedures of repair, extraction and storage are imposed.

Among several production methods the most commonly used were the so-called Hatschek, Magnani and Manville processes. The asbestos fibre cement elements were also produced by shotcreting. More details on the production methods may be found, among other sources, in Hannant (1978) and in Bentur and Mindess (1990).

The mechanical properties of asbestos fibre cements depend on the quality, amount and orientation of the fibres; the data given in Table 3.1 should be considered as relating to orientation only. The asbestos fibre cements were considered as cheap and durable, with high corrosion and abrasion resistance, good performance in elevated temperature and good resistance against freeze/thaw cycles. Usually the volume fraction of fibres in composites was 6%, sometimes increased up to 20%. Asbestos fibres (mixed

Table 3.1 Mechanical properties of asbestos cement composites, after Hannant
(1978) and other sources

Properties	Values	Units
Density	1.800–2.100	kg/m³
Young's modulus	24	GPa
Tensile strength	12–25	MPa
Tensile strength under bending	30–60	MPa
Compressive strength	50–200	MPa

with glass fibres) were also used extensively as reinforcement for resins and
plastics, improving considerably their mechanical and thermal properties.
When cement paste in a 'wet-mix' process was used, there was less danger
for workers' health.

The important strengthening effect of asbestos fibres is due to their high
strength and Young's modulus and excellent bonding to cement paste. The
fibres control cracks in the brittle matrix and the rupture of the composite
is accompanied by the pull-out of fibres. The linear elastic behaviour is
observed nearly to the maximum load and then the shape of the descending
line is determined by the efficiency of the fibres; see Figure 3.5.

The pull-out mechanism was examined by Akers and Garrett (1983) and

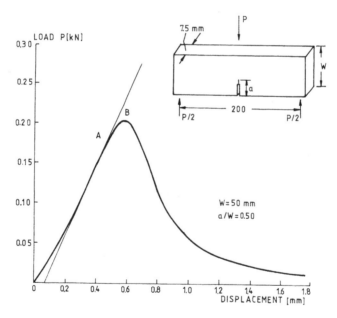

Figure 3.5 Characteristic load-displacement curve for the notched specimen made of
asbestos cement, after Mai *et al.* (1980).

it appears that the load-displacement curve is similar to that for steel fibres. An appreciable difference of shear strength at the interface was observed for two main kinds of asbestos fibres: for chrysotile fibres τ_{max} = 1.8 MPa; and for crocidolite fibres τ_{max} = 3.1 MPa. Chrysotile fibres are flexible and crocidolite fibres are stiff. The results obtained cannot be explained solely by that difference and further reasons should be looked for in the inter-fibre bond within the fibre bundles. In the manufacturing process of composite elements the crocidolite fibres are subjected to more damage than chrysotile ones. The mechanical behaviour of asbestos cement elements is not directly related to the results of the pull-out tests of single fibres or their bundles.

The mechanical properties of asbestos fibre cements may be calculated from the law of mixtures or by using the fracture mechanics formulae from which can be seen the specific work of fracture and R-curve. Mai, *et al.* (1980) observed also that crack initiation was close to the bending strength, which was related to a quasi elastic and brittle behaviour. For specimens with a depth greater than 50 mm the size effect on mechanical behaviour was negligible. For smaller specimens the pull-out fibres across cracks could not be developed before quick crack propagation took place followed by the failure of the specimen.

In extensive studies with a view to replacement of asbestos fibre by other kinds of fibres, it has been shown that, for hybrid asbestos-cellulose fibre reinforcement the modified law of mixtures provided good agreement with experimental results and that linear elastic fracture mechanics formulae may be applied to these composites. Even with fibres of relatively short length the aspect ratio ℓ/d is high enough so that the fracture energy is provided mostly by the fibre pull-out processes in 95% of cases. As replacements for asbestos fibres, glass fibres, polypropylene and vegetal fibres are used (Krenchel and Hejgaard 1975); see Sections 5.2 and 5.8.

3.2.6 Bitumen concretes

In bitumen concretes two kinds of organic binders are used: asphalts (asphalt cements) and tars. Various kinds of bitumen binders are found directly in nature and are obtained from natural oil processing as residue. Their properties are described in more detail in Section 4.1.4.

Asphalt mixes are obtained by mixing asphalt or tar with stone aggregate and sometimes up to 5% of rubber admixture may be added, e.g. styrene-butadiene rubber (SBR). The aggregate fraction may vary in large limits from 5% up to 85% of the total volume. The grains should not exceed 25 mm in diameter, with sufficient volume of fine grains and dense grading to avoid segregation during preparation and placing. Because the asphalt mixes are used mostly on roads and runways a good skid resistance is required as well as resistance of the aggregate to freeze-thaw cycles. Angular grains of strong and tough coarse aggregate are preferable, because interlocking of grains and internal friction in aggregate structure are increased and the stability of the road pavement under heavy traffic and temperature variation is better ensured. For the same reason, natural fine aggregate may be replaced by industrial ash. These measures also prevent the ravelling caused by rolling of layers.

Three main types of aggregate structures may be distinguished:

1   Dense graded structures with low total volume of voids and increased number of intergranular contacts are appropriate when a hot mix with bituminous matrix is prepared.
2   Open-graded structures that do not contain the lowest sizes are used for cold mixes.
3   One-size structures (cf. Figure 6.6a) are appropriate when spraying with bituminous matrix is executed, after spreading and compacting the layer of the aggregate (Young *et al.* 1998).

To improve crack control and tensile strength various kinds of continuous fibre reinforcement are used for road and runway overlays with different fibres: polystyrene, nylon, glass or steel.

The structure of aggregate is usually designed in such a way that upper layers are composed with small grain sizes only, mostly with fine sand and stone flour as a filler, lower layers with continuous grading and bottom layers with increased volume of bigger grain sizes.

The aggregate structure is designed for bituminous mixes in such a way that the total volume of voids that are filled with bitumen can be maintained as low as possible, because with the excess of the matrix the viscous deformations enhanced by long-term loading and higher temperature in summer may compromise the stability of the aggregate structure, leading to its segregation. Long-term properties should also be checked: low creep deformations and adequate durability of matrix-aggregate bond are needed. Another group of practical requirements concerns the fresh mix workability.

Asphalt concretes are basically outside the scope of this book and more information may be found in publications on the materials for road construction, e.g. Domone and Illstone (2001), Piłat and Radziszewski (2004).

3.2.7 Soil cements

(a) General information

Soil cements are defined by the ACI 230 1R-90 (reapproved 1997) as 'a mixture of soil and measured amounts of Portland cement and water compacted to a high density' and the main difference with the concrete is that 'individual particles are not completely coated with cement paste.' Application of other binding materials, like natural or artificial pozzolans, is also possible. The main reason for its application is the relatively low cement content

(6–8% mass) and the lower cost than for ordinary concretes because in all cases local soil is used.

Soil cement may form layers under bituminous or concrete pavements in road structures. It is also used in construction of slope protection, liners for channels or reservoirs and various kinds of stabilization layers under structural foundations. The application of soil cements for construction of local roads in rural regions, particularly in developing countries, is spreading. Usually, the soil cement layers are maintained permanently in high-moisture conditions during service.

Nearly all kinds of soils may be used as a component for soil cements; however, granular soils are considered to be better than clays, but often it is a mixture of both. The volume fraction of organic materials should preferably be lower than 2%. Aggregate grains are restricted to 50 mm and fine fractions are limited because their excess may cause increased cement consumption. In certain regions of the world, rice husk (10–15%) is added. The best composition is closely related to local conditions and requirements imposed on the final product – price, exposition to external actions, required durability, etc.

Adobe, which is a material made with soil (clay or sand), stabilizer, binder and water mixed and moulded together to form sundried blocks, is not considered here.

(b) Mechanical properties

The density of soil cement is dependent upon the moisture content and increases with the cement content up to a certain optimum value and then it decreases. The compressive strength of the soil cement f_c may be characterized by an increase in density compared to that of a wet soil without cement or any other binder f_{co} and that increase is represented by a coefficient a, $a = (f_c - f_{co})/ f_{co}$. The coefficient a for Portland cement admixture of 8% per volume may vary from 4.5 after 7 days to 12.6 at an age of 180 days. The improvement of strength depends considerably on the soil properties and ageing conditions and is increasing linearly with the Portland cement content up to 15% (Ambroise *et al.* 1987). From experimental data it appears that values up to $f_c = 15$ MPa may be reached.

The compressive strength of the product should be determined experimentally because any prediction is not reliable enough as it concerns its increase with ageing. The flexural strength (modulus of rupture) may be approximately calculated from the formula proposed in FHA (1979): $f_r = 0.51 f_c^{0.88}$ (in psi) or $f_r = 0.27 f_c^{0.88}$ (in MPa).

High shrinkage should be avoided and reduced by moist curing and limited cement content. Shrinkage cracking may considerably increase permeability, and the durability of the soil cement layers may then be reduced. Experimental tests (compression, splitting, Young's modulus, shrinkage, water permeability) have proved influence of stabilization by the addition of cement, mechanically by static, dynamic and vibro-compaction and both, i.e. by cement and mechanical means (Bahar *et al.* 2004). Specimens have been made with sandy clay soil (typical in that region of Algeria) with grains below 0.63 mm, and their properties depended on cement content, quality of the initial curing and later moisture conditions. The best compressive strength has been obtained in dry state with cement content over 8% by mass. The appropriate water content was determined experimentally according to other conditions. Examples of strength versus age curves are shown in Figure 3.6.

(c) Granular soil with fibre reinforcement

Granular soil with fibre reinforcement is a material composed of granular natural soil (i.e. sand) and thin continuous fibres distributed in the material volume with or without cement or any other binding agent. The synthetic (polyester, polypropylene or polyamide) fibres are very thin and their volume fraction is low: starting from 0.1% or 0.3%; also natural fibres may be used (e.g. sisal, coir). The fibres are distributed in a uniform and isotropic way (3D) or are parallel to a selected plane (2D). The main influence of the reinforcement is the cohesion of sand, which is enhanced with the fibre content, while the angle of the internal friction remains constant.

The specimens of soil with fibres and cement or cactus pulp as a binder were subjected to bending, abrasion, erosion and water absorption. Sisal fibres 40–50 mm long and polypropylene fibrillated fibres 20 mm long were

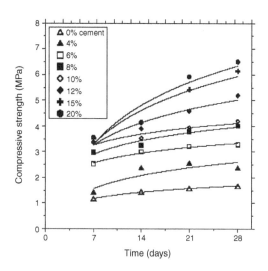

Figure 3.6 Effect of age of curing on compressive strength at different cement contents of soil-cement specimens. (Reproduced with permission from Bahar R. *et al.* Performance of compacted cement-stabilized soil; published by Elsevier Limited, 2004.)

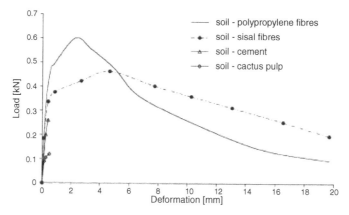

Figure 3.7 Behaviour of different reinforced soil-cement specimens under bending. (Reprinted from *Cement and Concrete Composites*, 27, Mattone, R. pp 6, Copyright 2005, with permission from Elsevier.)

used with V_f. = 1%. The matrix was composed with sand 42%, silt 44% and clay 14%. In Figure 3.7 the load-deflection curves are shown for plates 300×400 mm and 40 mm thick under four-point bending. The influence of fibre reinforcement is significant and the brittleness of soil cement is considerably reduced; the composite behaves similar to fibre-reinforced concrete elements (Mattone 2005).

The tests of the specimens under compressive load proved a considerable increase of bearing capacity due to fibres. The descending branch of the load-deformation curve is long and its shape is strongly related to the kind of fibres used. In the case of elastic fibres a quasi-linear behaviour is observed, but for ordinary polymeric fibres the elastic quasi-plastic behaviour should be expected. Thanks to the artificial cohesion, the reinforced sand is used for construction of walls and slopes with steep gradients between 60° and 90°; also in various geotechnical works (Aldea 2007).

3.2.8 *Roller compacted concrete (RCC)*

Roller compacted concrete (RCC) is a dry concrete consolidated by the use of external vibrating rollers. Its initial consistency makes the main difference in comparison with ordinary concretes – its conventional slump value is equal to zero. The moisture content should be high enough that mixing is possible and the appropriate distribution of cement is feasible, but it should be sufficiently low to support heavy rolling and vibrating equipment. The aim of rolling is to decrease the porosity that should not exceed 3%, particularly when frost resistance must be ensured.

The most frequent application of RCC is in concrete dams, both for construction and rehabilitation. Concrete is placed *in situ* in layers of 300–400 mm and then compacted. An example of the mixture composition was given by Abdel-Halim *et al.* (1999): coarse aggregate (max. grains 38 mm) and fine aggregate (10–15% of fines below 0.075 mm), cement 96 kg/m³, natural pozzolan 85 kg, water 90 l, sand 613 kg, coarse aggregate 1,386 kg and admixture 0.9 l. Average density was 2,263 kg/m³ and compressive strength 9.4 MPa at an age of 93 days. In contrast to the soil cement, coarse aggregate is used and high dosage (over 30%) of fly ash may be applied.

Extensive experimental studies on RCC with various mixture proportions have been performed by Kokubu *et al.* (1996) and optimal proportions of components were proposed.

3.2.9 Fibre reinforced gypsum

Gypsum is extensively applied inside buildings for plasters, decorations, claddings and partition walls. It is considered a non-water-resistant material and is brittle if not reinforced (cf. Section 4.1.2).

Since the late 1960s, α-gypsum with some addition of polymer has been used as a matrix for glass fibres as dispersed reinforcement. Ordinary E-glass fibres are used and this low cost composite is suitable for various indoor applications, having excellent fire resistance thanks to crack control by fibres. E-glass fibres do not corrode in gypsum matrices as they do in all kinds of Portland cement matrices. These fibres are particularly suitable for massive applications because of their low price and their content is usually 5–6% by weight (cf. EN 15283).

In several applications, the Glass Fibre Reinforced Gypsum (GFRG) may replace the glass fibre cement composites, which, in spite of many attempts and techniques, have more or less limited durability (cf. Section 5.3). It is lightweight with a density between 1600 and 1800 kg/m³, may be easily adapted to various shapes, is inexpensive and has flexural strength between 20 and 30 MPa and tensile strength 8–10 MPa.

The example of test results is shown in Figure 3.8. Curves show relations between modulus of rupture and fibre content for two different matrices. Specimens were cut out of sheets 10–13 mm thick, and subjected to bending. The slurry of water/plaster ratio between 0.4 and 0.6 was sprayed and reinforced with glass fibres 50 mm long, distributed by a chopper mounted on a spray gun; the excess water was extracted by suction (Ali and Grimer 1969).

The specimens of glass fibre gypsum with and without the addition of polymer emulsion have been tested by Bijen and Van den Plas (1992); their compositions are presented in Table 3.2. The test results proved better mechanical properties of these composites than did Portland cement matrices reinforced with glass fibres (GRC). Also, the durability of composites with polymer admixture, when tested according to various accelerated ageing procedures, proved to be quite good in high moisture, corresponding to outdoor applications and no decrease of strength and durability was observed.

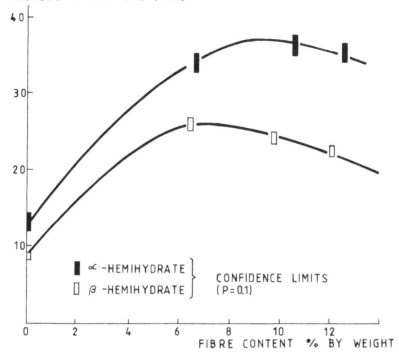

Figure 3.8 Modulus of rupture for glass fibre reinforced gypsum specimens as a function of fibre content, after Ali and Grimer (1969).

Table 3.2 Glass fibre reinforced gypsum compositions

Components	Plain gypsum (GRG) % mass	Polymer modified gypsum (PGRG) % mass
α-hemihydrate gypsum	73.7	54.5
Polymer emulsion	–	19.0
Melamine resin	–	5.40
Total water	26.5	19.50
Antifoaming agent	–	1.20
Catalyst	–	0.27
Retarder	0.01	0.01
Glass fibres	13.00	13.00

Source: Bijen, J. and Van den Plas, C. (1992) *Polymer Modified Glass fibre Reinforced Gypsum,* London: Chapman and Hall/Spon.

3.2.10 *Self-compacting concretes (SCC) and self-levelling concretes (SLC)*

SCC (sometimes deciphered also as self-consolidating concrete) is defined as a concrete that can flow under its own weight and completely fill the formwork without any vibration, even when access is hindered by dense reinforcing bars, whilst homogeneity is maintained. It is obvious that such behaviour of the fresh concrete mix is of a great advantage from the viewpoint of execution *in situ* regarding the economies in construction time, manpower and energy. Self-compacting fresh mix is considered as the basic remedy against heterogeneity of properties of ordinary concrete in regions with congested reinforcement, which is frequent in many structural elements of bridges and industrial buildings. The first investigations on this kind of concrete were initiated in Japan in late 1980s and soon SCC was applied in bridge structures. Since that time the knowledge and technology has allowed the use of SCC with good results. This innovation in the domain of concrete technology is considered to be of the highest importance over the past few decades.

As a basic parameter of the material's composition, the ratio of water to total powder w/p is considered, where total powder means a sum of concrete and additional powder, which is often limestone powder. The proportion between   and the dosage of superplasticizer decides on the behaviour of the mix. The mixture proportions vary in large limits, depending on the quality of components, but typically are a cement content of 350–400 kg/m^3, 180–350 l/m^3 of water, a maximum amount of aggregate of 60% of total volume, filler or additive between 25% and 100% of cement weight and a superplasticizer between 1–2% of cement mass. Sometimes a viscosity admixture is required to avoid segregation if a filler or additives are less than 50% of the cement mass. Maximum aggregate size is usually limited to 20 mm and the use of fly ash is frequent. Several examples of mix design proportions are published by Ouchi *et al.* (2003), which are based on separate compositions of fine aggregate and of coarse aggregate. This method may be called 'Japanese.' In contrast, Su *et al.* (2001) proposed to compose the self-compacted concretes using a different method (Chinese), in which all aggregate is considered simultaneously, but the voids between coarse grains should be filled by cement mortar. According to the authors, this method leads to better compositions with less cement and is therefore less expensive. These two different approaches (Japanese and Chinese) to mix design of SCC are discussed and three example proportions are designed, apparently according to the Chinese approach, and tested by Brouwers and Radix (2005). As shown in Table 3.3, the differences are only in the selection of sand gradation. The results both in workability and strength properties of the hardened concretes are similar.

Due to the high fines content, SCC may have larger early age shrinkage and creep than ordinary concrete, although some results proved a similar behaviour and even a lower drying shrinkage at early ages of SCC with regard to ordinary concrete with similar properties after hardening. These

Table 3.3 Examples of mix proportions [kg/m³] of SCC

	Mix A	Mix B	Mix C
Components			
cement	310	315	320
limestone powder	189	164	153
sand 0–1 mm	–	–	388
sand 0–2 mm	–	306	–
sand 0–4 mm	1018	719	628
gravel 4–16 mm	667	673	687
water	170	173	174
superplasticizer	6.0	5.51	5.21
water/cement ratio	0.55	0.55	0.55
water/powder ratio	0.34	0.36	0.37
Strength [MPa] at 28 d			
compressive	51.2	50.7	53.6
tensile splitting	4.2	4.1	4.7

Source: Reprinted from *Cement and Concrete Research*, 35, Brouwers, J.H. and Radix, H.J., pp. 21, Copyright 2005, with permission from Elsevier.

discrepancies are justified by different fines and other components used in various cases.

The mechanical properties of SCC, as stiffness and strength, do not differ significantly from the corresponding properties of standard concrete, particularly as it concerns concretes of high strength of 90–100 MPa (Domone 2007). In most cases, SCC are produced in the range between medium (50 MPa) and high (100 MPa) compressive strength at an age of 28 days. Therefore, basic rules and recommendations adopted for standard concretes may be successfully applied to the structures made with SCC. Certain doubts in the applicability of SCC still exist with respect to structures in seismic regions.

The higher amount of fine particles increases compacity and reduces porosity; therefore lower permeability and improved durability of SCC may be expected.

Early age shrinkage of SCC appears, as for ordinary concretes, and produces cracking due to restraints. The inclusion of a low amount of glass fibres does not significantly modify the SCC performance, its flowability in the fresh state or its mechanical properties in the hardened state. The cracking is partly reduced by AR (Alkali Resistant) glass fibres in two different modes: (i) the total cracked area may be reduced; and (ii) the maximum length of cracks may be reduced (Barluenga and Hernández-Olivares 2007).

In the tests performed by Aydin (2007), a relatively high volume fraction $V_f = 2\%$ of hybrid fibres (pitch-based carbon fibres and steel fibres) was used.

For all proportions between these two kinds of fibres there were no difficulties in mixing and no decrease of workability. As a result, the correct distribution of fibres was maintained and mechanical properties of hardened composite were improved in a similar degree to that of ordinary fibre concrete.

A series of experimental verifications carried on in situ by Zhu *et al.* (2001) proved that homogeneity of self-compacted concrete placed in tall columns and walls was entirely comparable with that of ordinary vibrated concretes. Test were carried out using drilled cores at different levels, as well as using various non-destructive methods, and have shown that the results for SCC were even slightly more uniform along the height of tested elements of natural scale.

When SCC is designed for a floor, for example an industrial floor in a store or for parking, then its property to get a flat surface (e.g. with tolerance of 1 mm over a length of 4 m) without additional operations is very attractive and the mix composition is modified into SLC. An example of such material was presented and discussed by Rols *et al.* (1999). The compressive strength over 40 MPa was sufficient and it was obtained with 260 kg/m³ of cement and a *w/c* ratio equal to 0.68. To reduce bleeding and segregation of the fresh mix, three different agents stabilising viscosity were tested and two were selected as acceptable: starch solution 1.3 kg/m³ and precipitated silica solution 3.9 kg/m³. The excessive drying shrinkage, which was greater by 50% in comparison to ordinary concrete of similar cement content, required an especially intensive cure.

3.2.11 Pervious concretes

The concept of porous concrete (PoC), also called enhanced porosity concrete (EPC), is based on an assumption that air and water permeate freely through systems of interconnected voids and pores and in this sense it is a kind of 'green concrete.' Such material is used for various kinds of pavements, parking lots, river banks, etc., where different aims may be reached:

- limitation of interference into natural water balance;
- easy drainage of floods in the case of thunderstorms (Beeldens and Brichant 2004);
- construction of pavements for limited or local traffic where different kinds of plants may be grown, thus improving the aesthetics of the place (Matsukawa and Tamai 2004);
- reduction of noise from the interaction between tyre and road pavement (Naithalath *et al.* 2006).

Porous concrete should maintain sufficient strength and durability under the conditions of root penetration into voids, the influence of freezing-thawing cycles, the development of microorganisms and bacteria, the influence of contaminated ground water, etc. In the case of road overlays, resistance against abrasion is required and only the top overlay should be built with EPC. Because of its porous structure the compressive and tensile strength is lower than normal concretes, but values of 50 MPa and 5 MPa, respectively, may be reached with small size aggregate (below 30 mm or even 10 mm) and using appropriate composition with silica fume and superplasticizers will ensure the strength of the aggregate-cement interface (ACI 522.1-08).

The porous concretes are preferably composed with blast-furnace slag cements, gap graded crushed aggregate with grain size above 5 mm and without fines or minimizing sand volume fraction with superplasticizers and fly ash. Pavements with increased porosity are built in a few layers with different tasks. A minimum required permeability between $7 \cdot 10^{-5}$ m/s and $3 \cdot 10^{-4}$ m/s was proposed by Beeldens and Brishant (2004).

The values that characterize pervious concrete should be close to the permeability of soil, which is estimated as between approximately $1.2 \cdot 10^{-5}$ and $10.5 \cdot 10^{-4}$ m/s for sandy soils and between $5.2 \cdot 10^{-6}$ and $3.1 \cdot 10^{-7}$ m/s for loam. To compare, the respective value of permeability of ordinary and high performance concretes is usually in the range of $10^{-12} - 10^{-14}$ m/s.

Depending on its role – water evacuation or noise reduction – the material composition and structure are purposefully selected.

As an alternative to the pervious concrete, pavements built with blocks may be considered, as the filtration through the joints between blocks is potentially equivalent.

The parameters of a concrete that is efficient for evacuation of water are not necessarily the same as for a concrete that is acoustically effective. In the former case, open porosity with large pore sizes and reduced pore constriction is looked for. In contrast, for the latter case, smaller pore size and high constriction are better at fulfilling the task. In certain cases, a material answering both kinds of requirements is necessary (Naithalath *et al.* 2006).

References

Abdel-Halim, M. A. H., Al-Omari, M. A., Iskender, M. M. (1999) 'Rehabilitation of the spillway of Sama El-Serhan dam in Jordan, using roller compacted concrete', *Engineering Structures*, 21: 497–506.

Akers, S. A. S., Garret, G. G. (1983) 'The relevance of simple fibre models to the industrial behaviour of asbestos cement composites', *International Journal of Cement Composites and Lightweight Concrete*, 5(3): 173–9.

Aldea, C. M. ed. (2008) *Design & Applications of Textile-Reinforced Concrete*. ACI SP 251(CD-ROM).

Ali, M. A., Grimer, F. J. (1969) 'Mechanical properties of glass-fibre reinforced gypsum', *Journal of Materials Science*, 4(5): 389–395.

Ambroise, J., Pera, J., Taibi, H., Tuset, J., Accetta, A. (1987) 'Amélioration de la tenue à l'eau de la terre stabilisée par ajout d'une pouzzolane de synthèse', in: Proc. 1st Int. Congress of RILEM, Versailles 1987, vol. 2, *Combining Materials: Design, Production and Properties*, London: Chapman and Hall, pp. 575–82.

Aydin, A. C. (2007) 'Self compactability of high volume hybrid fiber reinforced concrete', *Construction and Building Materials*, 21: 1149–54.

Bahar, R., Benazzoug, M., Kenai, S. (2004) 'Performance of compacted cement-stabilized soil', *Cement and Concrete Composites*, 26: 811–20.

Balaguru, P. N., Naaman, A. E., Shah, S. P. (1979) 'Fatigue behaviour and design of ferrocement beams', *Proc. ASCE*, 105, ST7: 1334–46.

Barluenga, G., Hemández-Olivares, F. (2007) 'Cracking control of concretes modified with short AR-glassfibers at early age. Experimental results on standard concrete and SCC', *Cement and Concrete Research*, 37: 1624–38.

Beeldens, A., Brichant, P. P. (2004) 'Structural development of water permeable pavements', in: *Proc. RILEM Int. Symp.* College of Eng. Nihon University, Koriyama, Japan, pp. 303–10.

Bentur, A., Mindess, S. (2006) *Fibre Reinforced Cementitious Composites*, 2nd ed., London: Taylor & Francis.

——.(1990) *Fibre Reinforced Cementitious Composites*, London and New York, Elsevier Applied Science.

Bijen, J., Van den Plas, C. (1992) 'Polymer modified glass fibre reinforced gypsum', in *Proc. Int. Workshop on HPFRCC*, H. W. Reinhardt and A. E. Naaman eds, RILEM/ACI, Mainz, 23–26 June 1991, London: Chapman and Hall/Spon, pp. 100–15.

Brandt, A. M., Czarnecki, L., Kajfasz, S., Kasperkiewicz, J. (1983) *Bases for Application of Concrete-like Composites*, (in Polish), Warsaw: Centre for Building Information.

Brouwers, H. J. H., Radix, H. J. (2005) 'Self-compacting concrete: theoretical and experimental study', *Cement and Concrete Research*, 35(11): 2116–36.

Domone, P. L. (2007) 'A review of the hardened mechanical properties of self-compacting concrete', *Cement and Concrete Composites*, 29: 1–12.

Domone, P. L. J., Illston, J. M. eds (2001) *Construction Materials*, London: Taylor & Francis.

Elavenil, S., Chandrasekar, V. (2007) 'Analysis of reinforced concrete beams strengthened with ferrocement', *International Journal of Applied Engineering Research*, 2(3): 431–40.

Grabowski, J. (1985) 'Ferrocement under impact loads', *Journal of Ferrocement*, 15(4): 331–41.

Hannant, D. J. (1978) *Fibre Cements and Fibre Concretes*. Chichester: J. Wiley & Sons.

Hamelin, P., Verchery, G., eds (1990) *Textile Composites in Building Construction*, Proc. of Int. Symp. in Lyon, Part 1, Pluralis.

——.(1992) *Textile Composites in Building Construction*, Proc. Int. Symp. in Lyon, Parts 1 and 2, Pluralis.

Kokubu, K., Cabrera, J. G., Ueno, A. (1996) 'Compaction properties of Roller Compacted Concrete', *Cement and Concrete Composites*, 18: 109–17.

Krenchel, H. and Hejgaard, O. (1975) 'Can asbestos be completely replaced one day?' in *Proc. RILEM Symp. Fibre-reinforced Cement and Concrete*, Sheffield, pp. 335–46.

Mai, Y. W., Foote, R. M. L., Cotterell, B. (1980) 'Size effects and scalling laws of fracture in asbestos cement', *International Journal of Cement Composites* 2(1): 23–34.

Matsukawa, T., Tamai, M. (2004) 'Fundamental study on porous concrete and application for planting material', in: *Proc. RILEM Int. Symp.* College of Engineering, Nihon University, Koriyama, Japan, pp. 295–301.

Mattone, R. (2005) 'Sisal fibre reinforced soil with cement or cactus pulp in bahareque technique', *Cement and Concrete Composites*, 27: 611–16.

Mindess, S., Young, J. F., Darwin, D. (2003) *Concrete*, 2nd ed. New Jersey: Prentice-Hall, Pearson Ed.

Naaman, A. E., (ed.) (2000) *Ferrocement & Laminated Cementitious Composites*. Proc. of 6th Int. Lambot Symp. on Ferrocement. University of Michigan, Techno Press.

Naithalath, N., Weiss, J., Olek, J. (2006) 'Characterizing enhanced porosity concrete using electrical impedance to predict acoustic and hydraulic performance', *Cement and Concrete Research*, 26: 2074–85.

Ohama, Y., Shirai, A. (1992) Development of polymer-ferrocement, in: *Proc. Int. Workshop on HPFRCC*, H. W. Reinhardt and A. E. Naaman eds, RILEM/ACI, Mainz, 10–13 July 1992, London: Chapman and Hall/Spon, pp. 164–73.

Ouchi, M., Nakamura, S., Osterson, T., Hallberg, S. E., Lwin, M. (2003) 'Applications of self-compacting concrete in Japan, Europe and the United States'. ISHPC. Available at: http//.www.fhwa.dot.gov/bridge/scc.pdf

Piłat, J. and Radziszewski, P. (2004) *Asphalt Pavements* (in Polish), Warsaw: WKŁ.

Powers, T. C., Brownyard, T. L. (1947) 'Studies of the physical properties of hardened cement paste', *Journal of American Concrete Institute*, 43.

Rols, S., Ambroise, J., Pera, J. (1999) 'Effects of different viscosity agents on the properties of self-levelling concrete', *Cement and Concrete Research*, 29: 261–66.

Su, N., Hsu, K. C., Chai, H. W. (2001) 'A simple mix design method for self-compacting concrete', *Cement and Concrete Research*, 31(12): 1799–1807.

Tan, K. H., Hendra, S. S., Chen, S. P. (2007) 'Out-of-plane behavior of masonry walls strengthened with fiber-reinforced materials: a comparative study', in: *Proc. Int. Workshop on HPFRCC5*, H. W. Reinhardt and A. E. Naaman eds, Mainz, 10–13 July 2007, RILEM PRO53, pp. 419–26.

Venuat, M. (1984) *Adjuvants et Traitements*, Paris: published by the author.

Young, J. F., Mindess, S., Gray, R. J., Bentur, A. (1998) *The Science and Technology of Civil Engineering Materials*. New Jersey: Prentice-Hall.

Zhu, W., Gibbs, J. C., Bartos, P. J. M. (2001) 'Uniformity of in situ properties of self-compacting concrete in full-scale structural elements', *Cement and Concrete Composites*, 23: 57–64.

Zia, P., Ahmad, S. H., Garg, R. K., Hanes, K. M. (1992) 'Flexural and shear behaviour of concrete beams reinforced with 3D continuous carbon fibres', *Concrete International*, 14(12): 48–52.

Standards

Note: Only basic international standards are listed, published by three organizations:

- European Committee for Standardization (CEN);
- American Concrete Institute (ACI);
- American Society for Testing and Materials (ASTM).

Interested readers are referred also to other particular standards, guides and recommendations of these and other organizations as well as to national standardization institutions. It is important to always consult the recent issues of all these documents.

ACI 522.1-08 *Specification for Pervious Concrete.*

ACI 230.1R-90 (reapproved 1997) *Report on Soil Cement.*

ACI 549.2R-04 *Report on Thin Reinforced Cementitious Products.*

EN 14651:2005 *Test Method for Metallic Fibre Concrete – Measuring the flexural tensile strength (limit of proportionality (LOP), residual).*

EN 14721:2000 *Test Method for Metallic Fibre Concrete – Measuring the fibre content in fresh and hardened concrete.*

EN 15283-1; 2:2008 *Gypsum boards with fibrous reinforcement – Definitions, requirements and test methods.* Part 1: Gypsum board with mat reinforcement. Part 2: Gypsum fibre board

FHA (Federal Highway Administration) (1979) *Soil Stabilization in Pavement Structures. A User's Manual,* 2.

4 Components of matrices of cement-based composites

4.1 Cements and other binders

4.1.1 Cements

Cements and other binders, by their compatibility to other composite constituents, bind together and ensure transmission of forces and displacements. The following are the main groups of inorganic binders:

1   hydraulic cements (Portland cements), also blended with various secondary cementing materials, like pozzolans, slag cement, fly ash, etc.;
2   high alumina cements, natural cements;
3   non-hydraulic cements (gypsum, lime, magnesium cement);
4   hydrothermal cements;
5   sulphur cement.

The term 'hydraulic' means that the product is water-resistant when hardened. Non-hydraulic materials are those that decompose when subjected to moisture and cannot harden under water. Among non-hydraulic cements, lime may be quoted as an exception, because its hardening is caused by carbon dioxide from atmospheric air; however formed $CaCO_3$ (calcium carbonate) is water resistant.

Portland cements are by far the most important binder for concrete-like composites. They are obtained by grinding and mixing raw materials that contain argil and lime (i.e. clays, shales and slates) together with limestones and marls. The ground materials are fused in a temperature of about 1,400°C into clinker, which is ground again into a powder of particles below 100 μm and at that stage gypsum is added to control the rate of hydration process of certain cement compounds. The resultant product is completed by various additions to obtain the final composition, which therefore may be well adapted to designed properties: strength after hardening, hydration rate and hardening time, colour, resistance against chemical agents, etc. Average Blaine fineness of Portland cement ranges from 300 m²/kg to 500 m²/kg.

Raw materials to produce various types of Portland cement are exploited in nearly all regions of the world and this is perhaps one of the reasons for its universal application and importance.

Not only the composition but also the fineness of cement has a considerable influence on the hydration process and properties of the hardened product.

The fineness is expressed as a ratio between surface and mass of the grains and is usually kept within the limits from 200 m^2/kg to 900 m^2/kg. The fine cements offer a higher hydration rate and higher final strength. On the other hand, the higher water requirement of fine cement increases its shrinkage and the rapid hydration is accompanied by an increased rate of heat generation.

The main types of Portland cement are:

1   *standard* for ordinary concrete structures without special features;
2   *rapid setting and high early strength* for winter concreting and repair work;
3   *low hydration heat* for massive structures like dams, large foundations, etc.;
4   *high strength* for high performance concretes in outstanding structures, prestressed elements, etc.;
5   *sulphate-resistant* for elements exposed on organic sewage;
6   *various kinds of blended cements* with properties adapted to particular purposes.

In various countries, several types of Portland cements are produced as a result of the different composition of basic raw materials, different finenesses and different additives. In Tables 4.1 and 4.2 mineralogical and chemical compositions of Portland cement are shown after data published by several authors. The composition and properties of various kinds of cements is beyond the scope of the book and the reader is referred to special manuals, e.g. Kurdowski (1991), Popovics (1992), Neville (1997), Bensted and Barnes (2001), Mindess, *et al.* (2003).

The variety of cements is quickly developing and interested readers should consult recent editions of international recommendations: ACI 225R-99,

Table 4.1 Mineralogical composition of Portland cements

Chemical name	Abbreviated name	Chemical notation	Abbreviated notation	Mass contents %
Tricalcium silicate	Alite	$3CaO \cdot SiO_2$	C_3S	38–60
Dicalcium silicate	Belite	$2CaO \cdot SiO_2$	C_2S	15–38
Tricalcium aluminate		$3CaO \cdot Al_2O_3$	C_3A	7–15
Tetracalcium aluminoferite	Celite	$4CaO \cdot Al_2O_3 \cdot Fe_2O_3$	C_4AF	10–18
Pentacalcium trialuminate		$5CaO \cdot 3Al_2O_3$	C_5A_3	1–2
Calcium sulfate dihydrate	Gypsum	$CaSO_4 \cdot 2H_2O$	CSH_2	2–5

Table 4.2 Chemical composition of Portland cements

Chemical name	Abbreviated name	Chemical notation	Abbreviated notation	Mass contents %
Calcium oxide	Lime	CaO	C	58–66
Silicon dioxide	Silica	SiO_2	S	18–26
Aluminium oxide	Alumina	Al_2O_3	A	4–12
Ferric oxides	Iron	$Fe_2O_3 + FeO$	F	1–6
Magnesium oxide	Magnesia	MgO	M	1–3
Sulphur trioxide	Sulphuric anhydrite	SO_3	S	0.5–5
Alkaline oxides	Alkalis	K_2O and Na_2O	K+N	1.0

ASTM 150-07 and EN 197:2004, which provide information on cement classes and their characteristics. National guides and information published by cement producers also should be consulted for recent developments. These types of documents define the chemical composition of several kinds of cement as well as their required physical properties.

Further progress in the variety of Portland cements is directed on special types, such as:

- very low heat cements for massive structures;
- cements with spherical particles for high strength and for flowing concretes;
- rapid set and rapid hardening cements for repair works;
- cements with particles of small and very small dimensions for high strength and improved tightness.

Ultrafine cement is obtained by grinding (high energy milling (HEM)) and is available from different producers. The average particle size is 2.5 μm, with approximately 90% finer particles than 5 μm, or even a great particle size reduction may reach 50% of the powder below 1 μm and 100% below 6 μm for so-called nano/micro cements. Cement is considered as ultrafine if its Blaine fineness exceeds 700 m²/kg; some kinds may even reach 1,700 m²/kg. Its characteristics are a very short setting time and possible retrogression of strength after a few days. Special pastes are produced with ultrafine cement with some retarders and superplasticizers (HRWRA); also ordinary cement and bentonite may be admixed. Because of its excellent penetration capacity, such grouts are used for the repair and strengthening of cracked dams, wells, containers and other concrete structures (Sarker and Wheeler 2001).

The progress in the cement industry is directed towards very low exhaustion of deleterious gases and powders, reduction of energy required and use of supplementary fuel materials, such as used tyres, etc. The majority of cement production in the world concerns blended Portland cement with various kinds of pozzolans and microfillers, called supplementary cementitious materials (SCM) (cf. Section 4.1.2).

Another direction in the development of cement production is the control of dispersion of its properties by using real-time control of production parameters (temperature, duration, etc.) in the function of varying properties of the raw materials.

Natural cements are produced from the same raw materials as Portland cements but are in natural composition. Therefore they are less homogeneous, are cheaper and offer lower strength. Its local use is decreasing with large availability and low cost of ordinary Portland cement.

High alumina cements (also called calcium aluminate cements) are obtained from bauxite Al_2O_3 fused together with limestone at a temperature of 1,600 °C into clinker, which is then ground into powder. The chemical composition of high alumina cement is given below after Neville (1997) and Bukowski (1963):

lime	CaO	35–44%	magnesia	MgO	1%
alumina	Al_2O_3	33–44%	titanium dioxide	TiO_2	1%
ferric oxide	Fe_2O_3	4–12%	sulfur trioxide	SO_3	0.5%
silica	SiO_2	3–11%	alkalia	K_2O and Na_2O	0.5%
ferrous oxide	FeO	0–10%			

High alumina cements are about three times more expensive than Portland cement, mainly because of the high consumption of energy necessary for grinding hard clinkers.

Hydration of high alumina cement requires more water, which results in mixes of better workability. The hydration process starts later, after mixing with water, but develops more quickly with high rate of liberated heat. Also, hardening is quick and usually within 24 hours about 80% of final resistance is obtained. The hardened cement has both lower porosity and permeability.

Mortars and concretes made with high alumina cement are resistant against sulphate attack and also are better at resisting CO_2 from ordinary drinking or mineral water. In comparison, the resistance against alkalis is lower than that of Portland cements.

As a result of exposure to humidity and even a slightly elevated temperature, a conversion reaction may start in hardened cement paste. It develops in the calcium aluminate hydrates where hexagonal crystals are transformed into cubic ones, which have smaller volume. This causes an increase of porosity and considerable decrease of strength. It has been proved that even a temperature over 20°C can initiate the reaction for which the remaining amount of mixing water may be sufficient. The conversion reaction may occur at any time during the life of a structure, without prior warning, when favourable conditions appear, namely when temperature and humidity increase. This was the origin of serious failures of roofing structures in Great Britain and discussions on the applicability of high alumina cement still appear in the journals (Neville 2003). The enhanced possibility of corrosion of prestressing reinforcement also reduces the application of high alumina cement.

High alumina cement is used in mortars for refractory brick walls and in refractory concretes when mixed with special aggregate, like corundum or fire clays. At present the application of high alumina cement is, in general, very restrained; it is limited to non-structural elements and to refractory structures where there is no risk of high humidity. Other applications are in repair where rapid hardening is required (Barnes and Bensted 2001).

Expansive cements are kinds of Portland cements, which contain various special constituents that increase their volume during hydration and hardening. By appropriate composition and careful cure, the shrinkage of Portland cement may be compensated (so-called shrinkage-compensating cements), or even the final increase of total volume may be obtained (Popovics 1992). The control of swelling and shrinkage of concrete with expansive cements is difficult and not always perfectly reliable. The expansive cements are used at a limited extent and only for special purposes, e.g. for repairs of old concrete or for structures in which shrinkage cracking should be excluded and where water tightness should be perfect.

4.1.2 Other binders

Gypsum is a kind of non-hydraulic binder obtained from a mineral of the same name, composed mainly with $CaSO_4 \cdot 2H_2O$ and certain amounts of different impurities. The production is based on crushing the raw material and heating it to a temperature between 130°C and 170°C for dehydration. There are two main kinds of gypsum: β-hemihydrate, also called plaster of Paris and α-hemihydrate. The latter requires lower water content, resulting in a higher strength of hardened product, which also occurs when burnt at a higher temperature. The last stage of production is grinding into a fine powder and mixing with additives, which delay and control the setting time to enable effective mixing with water and easy placing in forms.

Gypsum is used as a paste with water for indoor decorations, partitions and plasters for the surface finish of walls and ceilings. When mixed with sand and lime, various kinds of mortar are obtained for non-structural and even structural elements. The compressive strength varies within large limits from 1.0 to 7.0 MPa for the plaster of Paris type and from 12 to 15 MPa for α-hemihydrated gypsum. Fibre reinforced gypsum is briefly described in Section 3.2.9.

Lime as quicklime CaO is obtained from natural limestone or dolomite burnt in temperature between 950°C and 1,100°C. The slacked lime obtained in the form of porous grains is mixed with water using about 2–3 m³ of water for 1,000 kg of lime. The dense paste should be stored in conditions preventing drying and away from the influence of atmospheric factors, e.g. in a hole in the ground and covered by a layer of soil. The duration of maturing is decisive for the product quality: from 3 weeks up to 6 months – or even longer. The density of the lime paste is about 1,400 kg/m³.

The hardening of lime is a slow process called carbonation, which is based on the absorption of CO_2 from the atmosphere when the lime hardens in the open air. The lime is then transformed again into limestone $CaCO_3$ and free water is evaporated. In thick brick walls the joints that are filled with lime mortar harden over several years.

Lime mortar is obtained by mixing lime with sand in proportions from 1:2 up to 1:4.5, adding water according to the required consistency. Portland cement is also added for increased strength. The compressive strength f_c = 0.6 MPa of lime mortar specimens has been obtained on cubes of 70 mm at an age of 28 days, and tensile strength was f_t = 0.2–0.3 MPa (Żenczykowski 1938). Nowadays, the application of lime mortar is limited to minor jobs and reconstructions of traditional buildings.

Magnesium oxychloride is produced from magnesium oxide MgO (*Sorel cement*) by burning magnesite MgCO in temperature between 600°C and 800°C. The MgO is mixed with magnesium chloride MgCl and not with water. The hardening process is quick and even after just 28 days the compressive strength of about 100 MPa may be reached. The application of magnesium oxychloride is limited to interiors of buildings because its durability is insufficient when exposed to moisture. Its main use is as a binder for sawdust to produce a composite material for the top layers of floors.

Hydrothermal cements are produced from finely ground lime and silicates mixed together and cured in high-pressure steam (autoclaving). The obtained material is used mostly to produce bricks. As silicates, the fly ash may be used, but its addition increases the hydration time (Venuat 1984).

Sulphur cement has been used as a molten bonding agent for centuries. At present, sulphur cement is used to make sulphur cement concrete for repairs and chemically resistant applications. It is modified with carbon and hydrogen in various proportions (not less than 80% of sulphur). Sulphur cement does not contain Portland or hydraulic cement. For mixing with aggregate a temperature between 130°C and 140°C (approximately) is required. The gradation of aggregate should ensure the appropriate volume of voids and minimum cement consumption, usually between 10% and 20% mass.

Sulphur cement concrete offers over 70% of its final strength already after 24 hours when cooled at 20°C. The hardened concrete has considerable resistance against the highly corrosive influence of acids and salts (ACI 548.2R-93). The use of sulphur cement requires special care and equipment *in situ*. The disadvantages of sulphur cement are the brittleness of hardened elements, low frost resistance and a low melting point of 113°C; these limit any larger use than as a molten bonding agent.

Supersulphated cement is obtained from blast-furnace slag (80–85%), calcium sulphate (10–15%) and lime or Portland cement klinker (approximately 5%). After hardening, the strength comparable to that of ordinary Portland cement may be obtained with a considerably lower heat of hydration. Supersulphated cement may be used in various special concrete structures, particularly in situations where the action of acid fluids, sea water and oils should be expected; e.g. for foundations and harbour structures. Mixing supersulphated cement with Portland cements and special treatment, such as accelerated hardening, is not possible.

4.1.3 Supplementary cementitious materials (SCM)

Portland cement is often blended with other cementing materials, called also mineral admixtures. Ternary and quaternary cement blends are those mixtures containing two or three supplementary cementing materials, such as various kinds of fly ash, grounded slag and silica fume (SF), in addition to Portland cement. For various reasons they are commonly applied for ordinary concretes and also for high strength and high performance concretes. Supplementary cementitious materials not only allow a decrease in Portland cement production with its all negative ecological effects, but also several properties of hardened concrete may be improved. These supplementary materials have limited cementitious properties, but in a finely divided form with water the pozzolanic reaction is initiated in which calcium hydroxide (CH) is transformed to CSH. This reaction, being even slower than the hydration of plain Portland cement, improves the binding properties of blended cements.

SCM have different functions: they modify the properties of the binder, some are made of waste materials and their use helps to solve ecological problems, others are used as a partial replacement of Portland cement to decrease the total cost. Also, the small particles eliminate pores and increase the density of the hardened composite, so they play a role of active fillers and microfillers. The non-hydrated particles of Portland cement also act as microfillers in the concrete mix. In fact, all of them play these various roles simultaneously and currently, in almost all kinds of concretes, supplementary binding materials are used. Their important effect is that they fill the voids between the fine cement grains that otherwise would be occupied by water, and this improves particle packing. These chemical and physical effects result in a denser and stronger hardened cement paste that is particularly important in the interfacial transition zones (ITZ) around aggregate grains.

Various blended cements may give lower concrete strength at early ages. However, after longer periods of curing, such as 90 or 180 days, with sufficient moisture, the strength is usually not lower than for plain Portland cement.

In different standards, there are various categories of blended cements that are recognized with respect to the specified mass fraction of supplementary materials (EN 197:2004; ASTM C595-08; ACI 232.2R-03). Also, natural pozzolans are used where are available. The worldwide requirement of sustainability is translated in the cement industry by several regulations and in general the use of SCM is estimated to be 20% of the total production of clinker.

Examples of the physical characteristics of a few SCM are shown in Table 4.3 (Mindess 2004).

Natural and artificial pozzolans

Application of pozzolanic mineral admixtures is considered beneficial for ordinary concrete and indispensable for high strength concrete.

Pozzolans were obtained initially from volcanic rocks and natural volcanic ashes in Pozzuoli in Italy, for example. After grinding, a kind of cement was produced, but at the moment that production has only local importance. Artificial materials with 'pozzolanic' properties are fly ash, ground slag or clays and shales calcinated at 650–1,000°C, SF, rice husk ash and other. When these materials are added to Portland cements they play the role of supplementary and less active binders and also of very fine fillers.

Pozzolanic materials are added to cement mix for following practical reasons:

1   to increase concrete density, impermeability and strength;
2   to improve the workability without increasing cement content;
3   to decrease risk of segregation during transport and handling;
4   to decrease cement content for lower heat generation and cost;
5   to improve resistance of concrete in environments with various chemicals.

Table 4.3 Physical characteristics of selected supplementary cementitious materials (SCM)

Material	Mean size of a particle [μm]	Surface area [m²/kg]	Particle shape	Specific gravity
Portland cement	20–30	300–400	Angular, irregular	3.1–3.2
Natural pozzolans	10–15*	< 1,000	Angular, irregular	< 2.5
Fly ash (F and C)	10–15	1,000–2,000	Mostly spherical	2.2–2.4
Silica fume	0.1–0.3	15,000–25,000	Spherical	2.2
Rice husk ash	10–75*	< 50,000	Cellular, very irregular	< 2.0
Calcined clay (metakaolin)	1–2	~ 15,000	Platey	2.4

Source: after Mindess (2004) and other sources.
*   depends on grinding

An example of curves showing the concrete strength development in time for various blended cements are shown in Figure 4.1 (Swamy and Laiw 1995). It has been confirmed that fly ash concrete needs a longer curing period than ground granulated blast-furnace slag (GGBS) and silica fume (in this figure silica fume is indicated as MS – microsilica) before reaching an equal or higher compressive strength than reference ordinary Portland cement concrete.

Fly ash (known as pulverized fly ash [PFA] in the UK) is an inexpensive waste material exhausted with gases by furnaces at coal power stations and is trapped by special electric devices, which are installed to reduce air pollution. It is accumulated in industrial zones and its utilization is beneficial for ecological reasons. Obtained as a finely divided product, fly ash does not require grinding; the particles are spherical and are of comparable dimensions to the cement grains. It is a kind of artificial pozzolan, which is used with Portland cement as an addition or partial replacement. The main reasons for its large application are (ACI 232.2R-03, ASTM C311-07, EN450:2005):

1   fly ash is cheaper than cement and its utilization is a solution for ecological requirements in industrialized regions;
2   the addition of fly ash improves workability of fresh mix;
3   partial replacement of cement decelerates the hydration processes and reduces the rate of hydration heat;

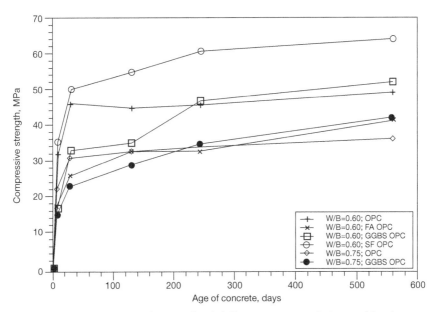

Figure 4.1 Development of strength of different cement-admixture blends: pure ordinary Portland cement (OPC), with 10% silica fume (MS), 30% fly ash (FA) and 65% ground granulated blast-furnace slag (GGBS), with different values of w/b, after Swamy and Laiw (1995).

4   the addition of about 25% of fly ash with respect to cement mass decreases the risk of alkali-aggregate reaction;
5   concrete resistance against chemical attack, e.g. sulphates, is increased;
6   it improves the impermeability of hardened composite by the inclusion of additional fine grains in the mix.

Fly ash is composed of silica, alumina and ferric oxides (at least 70%), a small amount of MgO, SO_3 and Na_2O. The quality of fly ash depends on its fineness and high silica content and on reduced carbon volume. As for a waste material, there are large variations of its composition and properties, which are functions of many factors: quality of coal, parameters of the combustion process, method of accumulation, etc. Fly ash is classified into two categories: class F (low calcium) produced from combustion of bituminous or anthracite coal and class C generated from burning of lignite and sub-bituminous coals. In fly ash class F, the lime is limited to 10%. High-calcium class C fly ash has more than 10% of lime (CaO) and should be used with caution for concrete because of possible negative effects. Only 10–30% of particles for both classes F and C are larger than 45 μm with specific surface between 250 and 400 m^2/kg; most particles are spheroidal with an average diameter of 20 μm or below. Fly ash class F has been used in concrete for more than 60 years in the USA, and class C has long been considered as less useful, particularly with regard to the durability of concrete. Since the 1980s, due to modification of combustion installations, its production increased and various investigations proved its applicability for high performance concretes also (Naik and Singh 1995). For some time, the amount of fly ash available in the market is decreasing for two reasons: (i) modification of the industrial processes in which less waste is produced; and (ii) an increasing demand from concrete producers.

Fly ash may be added directly to concrete mix as an additional admixture or as a cement replacement. It may be also added to the cement clinker before grinding. In both cases, usually the mass fraction of fly ash is limited to 30%, but additions over 50% are also reported with good results. The initial strength (at an early age) of concretes blended with fly ash is lower, up to approximately 28 days, and later recovery of strength depends on the intensity of pozzolanic reaction. Strength increase, due to active fly ashes, may be expected at the age of 90 days.

The replacement of a part of Portland cement by fly ash or crushed limestone flour improves the rheology of concrete mix and allows reduction of the amount of superplasticizers necessary to obtain the required workability.

Fly ash from fluidized bed combustion is obtained as a kind of by-product from new combustion installations that are rapidly built in many countries. In these boilers, the process called 'circulating fluidized bed combustion' (CFBC) with a lower temperature of combustion (approximately 800–850 °C in place of traditionally 1,150–1,750 °C) allows the use of less fuel and the reduction of CO_2 and NO_x emission into the atmosphere. Because of the application of various kinds of fuel, for example, brown coal and coal with high-sulphur petroleum coke, (Sheng *et al.* 2007), and the of use of limestone or dolomite as a sorbent for sulphur dioxide that appears in the combustion, the obtained fly ash is characterized by different chemical composition than the conventional fly ash. Other combustion processes known as PFBC or AFBC (where P denotes pressurized, and A means atmospheric) also lead to new kinds of fly ash. They are all outside the present recommendations for additives to cement concrete because of too high a content of unburnt carbon (loss on ignition), high content of SO_3 and of unreacted (free) lime (CaO) in a highly reactive form. Direct application of that kind of fly ash is limited by the present standard requirements in most countries. In Table 4.4 after Rajczyk *et al.* (2004) an example of the chemical composition of these kinds of fly ash is shown to compare with that of conventional fly ash class F. Unlike conventional fly ash, they do not contain a glass phase or mullite and grains have non-spherical shape. The particles are mostly below 50 μm in size, and approximately 50% are below 10 μm.

Furthermore, there are significant variations in the properties of fly ash due to different conditions of combustion: temperature, duration, kinds of fuel and the type of fluidal boiler. The pozzolanic activity of fly ash may be increased by different processing: heat (autoclaving with vapour cure), chemical (adding alkalies) or mechanical (grinding) (Fu *et al.* 2008).

Recently published results showed that with certain precautions this kind of fly ash may be used for structural concretes, though limits of 15–20% of cement replacement should not be normally exceeded (e.g. Glinicki and Zieliński 2008).

Before this kind of fly ash is applied in concrete, several concrete properties should be verified with respect to the soundness and sensibility to sulphuric corrosion; also a possibility of formation of etringitte has to be tested and the properties of the fly ash itself should be verified with respect to their variability in time. For normal use, certain modifications in standards are needed, because at present there are serious legal obstacles in using these

Table 4.4 Chemical composition of fluidized bed combustion fly ashes as compared to that of class F fly ash from conventional combustion installations

Component	Fly ash from black coal combustion in fluidized bed %	Fly ash from brown coal combustion in fluidized bed %	Fly ash from black coal combustion in conventional furnace %
L.o.i.	11.1	4.1	1.8
SiO_2	33.6	31.1	52.3
CaO	16.4	26.8	4.1
Al_2O_3	18.1	22.6	28.5
Fe_2O_3	6.9	3.1	6.4
MgO	2.7	1.8	2.4
SO_3	6.5	4.9	0.4

Source: after Rajczyk et al. (2004).

kinds of fly ash as concrete components. However some kinds of fly ash may appear to be unacceptable for structural concretes, e.g. because of too high a content of MgO or residual coal (Havlica *et al.* 1998).

Silica fume (SF) was initially considered as a waste material, which was obtained during the production of ferrosilicon and silicon metal. It started to be applied in 1969 in Norway as a valuable admixture for high strength cement composites when its pozzolanic properties were discovered. Already in the 1980s in North America and Europe the use of SF increased considerably. SF has a form of very small spherical particles of average diameter between 0.01 and 0.2 μm, which may fill capillary pores in cement paste, thereby increasing the concrete density and impermeability. The average particle size is 100 times smaller than concrete particles and Blaine surface may reach 20,000 m^2/kg. Because of high content of amorphous SiO$_2$ (usually above 80% and sometimes up to 96% by mass) its reaction with lime liberated from cement hydration develops appreciable pozzolanic effects and increases the strength of hardened cement paste. It contains only small amount of alkalis (Na$_2$O and K$_2$O) and is white due to its low content of carbon.

SF is supplied on the market mostly in two forms: as slurry and as dry densified material – the latter is by far the most common form. In densified SF, single spheres are agglomerated into clusters of approximately 20–100 spheres, and 3D clusters may have dimensions of a few millimetres (Diamond and Sahu 2006). The agglomeration is necessary for technological reasons: very low apparent density and safety of workers make it difficult to handle originally obtained SF and it is not available on the market as a dust. Strongly linked spheres do not disperse easily and completely during ordinary mixing of concrete, which is possible only by special ultrasonic treatment in laboratory conditions. Therefore their role in filling the pores in hardened materials may be only partly accomplished because SF clusters are larger than cement grains. The properties and dimensions of densified SF may vary widely from one supplier to another.

SF is used in concretes mostly as a partial replacement for Portland cement. Its influence on the compressive strength of concrete was first observed and was partly attributed to a reduction of the aggregate-matrix transition zone thickness. That reduction from about 50 μm to less than 8 μm has been observed by Mehta and Monteiro (1988). Reinforcing that zone by both reaction products of SF with Portland cement and the presence of hard silica particles is also suggested by Bentur *et al.* (1988).

The influence of fly ash and SF was tested by many researchers, and since the 1990s these microfillers have been considered as necessary constituents of high performance concretes (Malier 1992). As an example, tests carried out according to Spanish standards by Sánchez de Rojas and Frias (1995) gave interesting results, which are shown in Figure 4.2. The total amount of heat of OPC mortar is compared with that liberated by mortars with 30% cement replacement with fly ash and SF. The replacement with fly ash decreased the hydration heat due to its slower pozzolanic activity while that

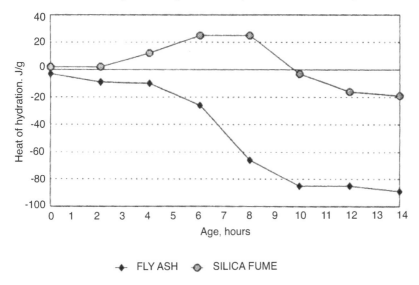

Figure 4.2 Increase and decrease of the total heat of hydration of mortars with 30% SF and 30% FA, respectively, compared to 100% OPC mortar with heat generation considered as zero; development in time of hydration, Sánchez de Rojas and Frias, (1995).

with SF increased the heat due to greater activity at an early age. After the first 10 hours the decrease of heat was observed for both admixtures.

SF acts also as a superplasticizer and the improved workability of fresh mix is obtained with lower water volume. However, it is always used together with a superplasticizer because of its high water demand. Mixing time is increased to allow a good distribution of SF with all the other components of the mix; nonetheless, a complete dispersion of the particle clusters is achieved with difficulty.

The content of SF is usually between 7% and 15% of cement mass as an addition or replacement and in most cases 8% is considered as optimum. A higher amount may adversely decrease the hydrated cement paste alkalinity below normal value of pH = 12.5 which is considered as necessary as protection against corrosion of steel reinforcement. This limitation does not concern special concretes like Reactive Powder Concrete (RPC) where even 30% of SF is used, e.g. in DUCTAL® (Jovanovic *et al.* 2002), cf. Section 13.4.6). Indications for the use of SF in concrete may be found in the ACI 234R-06, ASTM C1240-05 and EN 13263:2005 recommendations.

The influence of the alkali content of the SF on its pozzolanic properties has been studied by de Larrard *et al.* (1990) using 20 different SF compositions with various kinds of cements and plasticisers. The authors have proposed certain empirical relations to forecast the compressive strength of very high performance mortars and concretes.

The polymer mortars with SF were tested by Ohama *et al.* (1989) in view of their application for industrial floors, decorative coatings and repair of concrete structures. The results were interesting and have shown different influence of SF, that is, an increase of compressive strength and resistance to water permeation and penetration of chloride ions, but no improvements have been observed of resistance to water absorption, carbonation and drying shrinkage. Also, adhesion of the composite to covered concrete surfaces was not increased.

The addition of SF has a positive influence on wet mix shotcrete by improving the following characteristics (e.g. FIP 1988):

1 extremely low rebound < 5%;
2 high early and final strength;
3 high resistance against chemical attack;
4 improved density thanks to additional formation of CSH (calcium silicate hydrate) during hardening of cement paste;
5 reduced carbonation due to increased density.

Examples of SF properties are given in Table 4.5. With its increasing application, SF is no longer considered to be a waste material but rather a useful admixture for high quality concrete in so-called ternary blends: Portland cement, SF and various kinds of fly ashes. The price of SF equals approximately three times that of Portland cement and is increasing; therefore its application influences the total material cost.

SF is sometimes erroneously called microsilica (crushed crystalline silica), which is also a microfiller of similar chemical composition but with larger particles ranging from 4 to 40 μm.

Rice Husk Ash (RHA) is obtained from an agricultural waste. Over 100 countries produce rice and there are millions of tons of rice husks received as a by-product. After combustion at properly controlled conditions to reduce its carbon content, a very highly pozzolanic material is obtained with properties comparable to those of SF (Bui 2001). However, it has other properties that require special treatment. Due to its very porous structure, a

Table 4.5 Chemical composition and physical properties of silica fume, after various sources

a) *(in %)*

ig.loss *)	SiO_2	Al_2O_3	Fe_2O_3	CaO	MgO	Na_2O	K_2O	SO_3
0.7– 3.0	75–96	0.25–1.50	0.2–5.0	0.2–7.0	0.1–3.5	0.4–0.9	0.8–3.2	0.3–1.2

b)	*Specific gravity kg/dm³*	*average particle dimension (μm)*	*BET N_2 surface area (m²/g)*
	1.9–2.3	0.1–0.3	14–26, av. 20

* Loss on ignition: loss of mass of a sample after heating to 1000°C.

considerably large amount of water is necessary when mixing with other concrete constituents. Another problem is the reduction of a relatively high content of unburnt carbon, ranging from 5% to 15%. Both of these differences with SF may have a negative influence on various properties of hardened concrete; different response of the fresh mix to air-entraining agents, reduced corrosion resistance of reinforcement, etc. Besides, RHA is normally ground to a fineness of 10–75 μm, which is approximately 100 times coarser than SF. Therefore, the possibility of filling the voids in the ITZ is negligible.

With all these properties, RHA provides a valuable admixture, which is particularly important in developing countries where the cost of SF is relatively high and the agglomeration of the rice husk as a waste produces ecological problems.

Ground Granulated Blast-Furnace Slag (GGBS) is obtained from rapidly cooled blast-furnace slag. It has long been known that granulated blast-furnace slag (GBS) possessed a latent hydraulic property when subjected to processes of grinding and mixing with lime and Portland cement in different proportions. At the same composition of GBS, grinding techniques are a chief factor that affects particle geometric characteristics of GGBS. As an active additive, GGBS is widely applied in cement and concrete, especially in high performance concrete. The specific surface of ordinary GGBS ranges from 300 m^2/kg to 400 m^2/kg, while ultrafine GGBS may have over 600 m^2/kg.

The replacement of cement clinker by GGBS may vary from 10% up to 70% or even more for pure slag cements, and different kinds of blended cements are available. The use of GGBS as a partial replacement of Portland cement has a significant ecological effect. The deposits of industrial waste are transformed into a constituent of concrete. The emissions of dusts and gases into the atmosphere from cement production are limited and the need of natural raw materials is proportionally reduced (Bijen 1996).

The influence of GGBS is the reduction of mobility of ions and water in the pore system and as a consequence there is:

- a reduction of diffusion of chlorides (up to 50 times) and therefore, better resistance against corrosion of reinforcement;
- a lower probability of the alkali-aggregate reaction;
- an improvement of the resistance against sulphate corrosion of mortars and concretes.

The last property depends mainly on chemical composition and fineness.

Also, thinner ITZ around aggregate grains and lower free lime CaO content are observed. The experimental results by Gao *et al.* (2005) demonstrated that GGBS significantly decreases both the quantity and the orientating arrangement of CH crystals at the ITZ and the CH crystal size becomes smaller. The weak ITZ between aggregate and cement paste is strengthened as a result of the pozzolanic reaction of GGBS. The above improvements become much more significant with the decrease of particle sizes of GGBS.

GGBS considerably decreases the content of $Ca(OH)_2$ crystals in the aggregate-mortar ITZ, as determined from XRD (X-ray Diffraction) and SEM (Scanning Electron Microscopy) analysis. Moreover, it reduces the mean size of $Ca(OH)_2$ crystals, which make the microstructure of ITZ more dense and strong. The more the mortar contains fine particles below 3 μm of GGBS, the higher its early strength is and particles 3–20 μm increase its long-term strength.

On the other hand, the rate of hydration at an early age is reduced and long and careful curing of the external surface in moisture is advised. The strength development is considerably slower under standard 20°C curing conditions than that of Portland cement concrete, although according to certain studies, the ultimate strength is higher for the same water-binder ratio. Therefore, GGBS is not used in those applications where high early age strength is required. At higher temperature strength gain is much more rapid and the improvement in early age strength is more significant at higher dosage of GGBS. Even a 10°C increase above standard curing temperature considerably accelerates the strength development of mortars containing high levels of GGBS and at 40°C and 50°C, the strength of GGBS mortars is more or less equivalent to that of Portland cement mortar after 3 days (Barnett *et al.* 2006). However, the results obtained among the others by Tomisawa and Fujii (1995) have shown that for contents of GGBS over 50% the compressive strength decreased also after 28 and 90 days, while at early age that decrease was observed for any content. With higher Blaine fineness better strength was obtained (Figure 4.3). The compressive strength of blended cement paste (85% of GGBS) increased with higher Blaine fineness (Figure 4.4).

In the large structural concrete elements where heat dissipation is slow, there can be a significant rise in temperature within the first few days after casting due to the exothermic reaction of the binder. This leads to higher early age strengths, which can only be determined by temperature matched curing (Barnett *et al.* 2006).

The addition of GGBS modifies significantly the rate of heat liberation as it is shown in Figure 4.5 with the heat evolution curves. Concrete specimens made with OPC and with 50% replacement with GGBS of different fineness have been compared. In all mixes water content was constant at 160 kg/m^3. The presence of GGBS of a different fineness had beneficial effect in reducing the peak heat evolution and in extending the time of these peaks. Addition of HRWR had set retarding influence. The control of heat liberation, and particularly its rate, is of primary importance for execution of large concrete elements (Swamy 2004).

According to test results by Khatib and Hibbert (2005) the long-term strength and durability of concretes containing GGBS exceed that of neat PC concrete, provided that a proper curing is ensured, depending upon the GGBS content. In addition, the presence of GGBS increases the workability of concrete and therefore an equal 28-day strength to the control can be

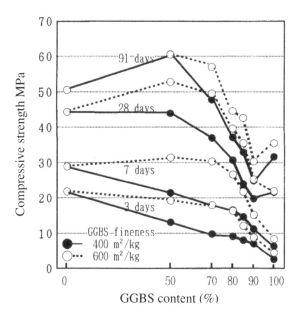

Figure 4.3 Relationship between compressive strength of cement paste specimens for various contents and two kinds of GGBFS, Tomisawa and Fujii (1995).

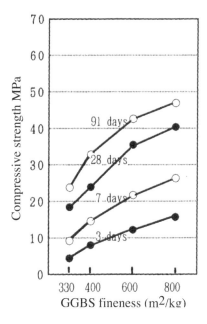

Figure 4.4 Compressive strength of cement paste specimens as functions of the GGBFS fineness for blended cement with 85% of GGBFS, Tomisawa and Fujii (1995).

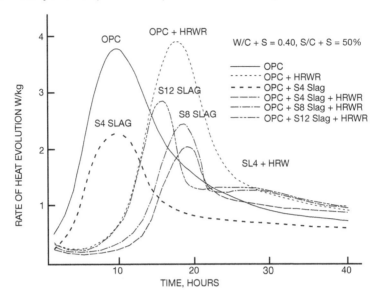

Figure 4.5 Rate of hydration heat as a function of time for various concrete compositions, fineness of OPC 323 m²/kg, slag S4 453 m²/kg, slag S8 786 m²/kg, slag S12 1260 m²/kg; Swamy (2004).

achieved by lowering the water to cementitious materials ratio in the mix. The incorporation of up to 60% GGBS to partially replace PC in concrete causes an increase in long-term compressive strength and E_d.

The density of mortar increased with additions of GGBS. Tests carried on by Ohama *et al.* (1995) have shown that for 40% replacement of Portland cement by GGBS (with fineness of 400, 800 and 1,500m²/kg) the pore-size distribution was considerably modified in a positive sense (Figure 4.6). However, the results obtained by Nakamoto and Togawa (1995) proved that the rate of carbonation increased by the addition of GGBS above 50%. Relationships in Figure 4.7 show that the carbonation remained linear with respect to square root of number of days, as for plain Portland cement. The carbonation is higher than in the ordinary Portland cement concrete because the carbonation products do not fill the pores in concrete when GGBS is used.

For high slag contents with respect to neat Portland cement concrete the carbonation was significantly enhanced and these results confirmed earlier publications, e.g. Portland slag cements with substantial fraction (25–85%) of ground blast-furnace slag show increased resistance against chemical corrosion, but durability may be decreased when exposed to outdoor conditions.

The lower cost of GGBS and its reduced ecological impact compared with Portland cement are also advantages of this supplementary binding material. Useful details on properties and application of GGBS for concretes may be found in ACI 232.2R-03, ACI 233-95 and ASTM 989-06.

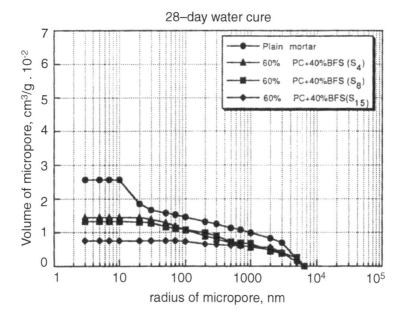

Figure 4.6 Pore-size distribution of mortars cured 28 days in water for various fineness of GGBFS compared with plain Portland cement. Reprinted from *Cement and Concrete Composite*, Ohama, Yoshihiko, Vol. 20, 'Polymer-based admixtures', pp. 24., Copyright 1998, with permission from Elsevier.

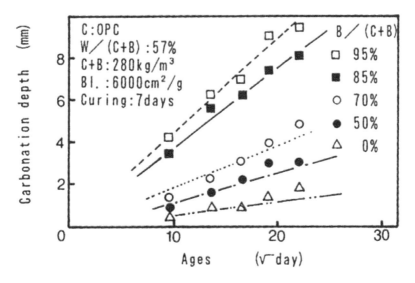

Figure 4.7 Progress of carbonation with time in the field tests for high ratios of B/C+B (GGBS to Portland cement+GGBS), Nakamoto and Togawa (1995).

Metakaolin (calcined clay) is an ultra fine pozzolana, produced from the mineral kaolin at temperature between 700°C and 900°C; in that temperature kaolinite loses water. This thermal activation is also referred to as calcining. The obtained product is highly pozzolanic and in recent years there has been an increasing interest in the utilisation of metakaolin used as partial replacement of Portland cement (up to 30%), often as an alternative to the use of SF.

Metakaolin reacts rapidly with the calcium hydroxide in the cement paste, converting it into stable cementitious compounds, thus refining the microstructure of concrete and thereby reducing its permeation properties. Whilst a limited amount of research data exists, all sources indicate that the use of metakaolin provides better durability of concrete, particularly with respect to various effects of chloride penetration. Chloride penetration may be reduced by 50–60% for concretes with 8–12% replacement of cement with metakaolin. Also it was applied with success to control alkali-silica reactivity of aggregate (Gruber *et al.* 2001). Tests with 15% replacement of Portland cement and with combined replacement with fly ash proved that the strength development was quicker during the first days and final strength at an age of 28 days was not disturbed (Jalali *et al.* 2006). The increase in the compressive strength of blended concrete is thought to be due to the filling effect where metakaolin particles fill the space between cement particles, to the acceleration of cement hydration and pozzolanic reaction of metakaolin (Khatib and Hibbert 2005).

The general conclusion is that various kinds of supplementary binding materials are used to produce different blended cements, but this rapidly developing practice should be accompanied by deep knowledge and care to obtain results corresponding well with the expectations.

4.1.4 Bitumens

Bitumens belong to a larger group under the general name of 'organic cements,' which also covers polymers and resins. Bitumens are described briefly below, with emphasis on their properties as the bonding components in concretes and without considering their application as insulation layers for roofing.

Bitumens are composed of two groups: asphalts and tars, this generic name covers also pitches. All are hydrocarbons with small fractions of sulphur, oxygen, nitrogen and traces of metals. The typical composition is given by various authors: Piłat and Radziszewski (2004), Domone and Illston (2001):

- carbon 80–87%
- hydrogen 9–15%
- oxygen 2–8%
- nitrogen 0.1–1%
- sulphur 0.2–7%
- metals: iron, nickel, vanadium 0–0.5%

Asphalts are obtained mostly by different refining processes after distillation of crude oil as a residue left at the bottom of the still. Natural asphalts are also exploited from surface deposits (Trinidad, Syria) or from pores filled by asphalt in sandstones and limestones where asphalt content may reach 25% of the total volume of the rock. Natural asphalts are added to those obtained from oil to increase their hardness.

Asphalts may be improved by addition of natural or synthetic caoutchouc or polyethylene. In modern road construction, modifications with synthetic rubber are extensively used, e.g. with SBR. In modified asphalts the bonds to stone aggregate is increased as well as their resistance against atmospheric agents (cf. Section 3.2.6, Young *et al.* 1998).

Tars are by-products from the distillation of coal, brown coal, timber and peat. Synthetic resins are used as additions to modify properties of tars for application as matrix for road overlays.

The principal difference between asphalts and tars relate to temperature sensitivity and wetting ability for aggregate grains. Physical properties of tars are more dependent than asphalts on temperature: they become softer in high temperature and brittle in lower temperature and such effects may be observed on road pavements in summer and winter. On the other hand, tars offer higher skid resistance and wet the aggregate grains better; good bond is less related to the quality of grain surface: its cleanness and roughness. Ageing processes accelerated by influence of oxygen from the atmosphere and by sunlight are quicker in tars and they lead to increased brittleness.

Mechanical behaviour of asphalts is basically viscoelastic. Under high loading rates they are elastic and brittle. When the load is imposed over a long time their deformability is similar to viscous materials. With respect to temperature variations, tars behave as thermoplastics; this means that with increasing temperature they are transformed gradually from brittleness to fluidity, with simultaneous decrease of material adhesion and cohesion when softening temperature is attained. That transformation is entirely reversible within a certain range of temperature.

To characterize asphalts several physical values should be measured, e.g.:

1   dynamic viscosity, expressed in Pa.s;
2   softening point which is the temperature in °C mentioned above;
3   penetration grade as the distance expressed in 0.1 mm units that a standard needle penetrates in 5 seconds under 100 g load and in temperature of 20°C;
4   ductility expressed by the distance in 10 mm units to which a 10 mm square cross-section briquet can be deformed by tensile load applied at 20°C and with displacement rate equal to 50 mm/min.

These are examples of basic properties that are determined according to various standards (cf. ASTM D1074-02, EN 13108 (series) 2005 and 2006), but in different countries other properties are also verified.

Asphalts are used as insulation layers and coatings, but their use as binders in mixes with natural stone aggregate for road and runway pavements is the main application. The addition of sulphur (up to 50%) improves the penetration index and matrix-aggregate bond. In special cases, like pavement repair, asphalts may also be used in the form of emulsions and cutbacks. Tars are usually more expensive than asphalts and their application for road pavements may be considered as non-rational.

4.1.5 Polymeric components

The synthetic polymers and resins considered here are those that are used for polymer cement concretes (PCC). Properties of ordinary cement mortars and concretes are significantly improved by polymer-based admixtures, which are classified into four main types (Ohama 1998):

- polymer latex or polymer dispersion;
- redispersible polymer powders;
- water-soluble polymers;
- liquid polymers.

Polymers in the form of latexes are applied for modification of concrete properties and three main groups of these modifications may be distinguished: (i) as binders for aggregate systems together with Portland cement (PCC); (ii) as a sole binder for aggregate grains (PC); and (iii) as impregnants for hardened concretes and mortars (PIC).

Two groups of synthetic materials are used.

1   Thermoplastics become ductile when heated but within certain range of temperature there are no permanent chemical transformations. Here polyvinyls, acrylics, polyamides, polymethyl methacrylates and polystyrenes may be mentioned as examples.
2   Thermosets are stiff and rigid in normal temperature, but when heated the irreversible modifications appear. Here the examples are polyepoxies, polyesters, aminoplastics, polyurethanes, silicons, etc.

Properties of ordinary cement mortars and concretes are significantly improved by polymer latex modifications. Already the fluidity of the fresh mix is generally improved and that is exploited in so-called self-consolidated concretes (cf. Section 3.2.10.) In the hardened composites, the microcracks are bridged by the polymeric films or membranes, and crack propagation is controlled and matrix-aggregate bond is increased. By improved water tightness the resistance against chemical agents is enhanced. Hardened polymers considerably increase the strength, toughness and durability of the final products. Moreover, the composite properties may be tailored by selection of monomer composition and by appropriate technology.

The volume fraction of polymeric admixtures should be always adapted to the purpose and its results carefully checked from various viewpoints: possible appearance of negative effects, material incompatibility, etc. Independently of the kind of admixture, some modifications in the cement hydration may be expected, because chemical and physical interactions occur.

The monomers which are used for polymer modified concretes should satisfy, to a larger or smaller extent, the following conditions:

- low viscosity and high wetting ability for aggregate grains;
- good bond to aggregate and reinforcement;
- controlled hardening allowing the mix to be cast in prepared forms;
- low toxicity, particularly after hardening;
- appropriate durability in a humid environment and – as far as possible – resistance in elevated temperature.

Viscosity is the internal resistance to flow exhibited by a fluid and to suspended particles; their size, volume content, etc. Low viscosity is one of the main characteristics required for monomers used in PIC, because it is essential to obtain the required depth of impregnation within a reasonable time of a few hours.

The monomers should also have capability of quick polymerization by heating or radiation, or preferably by chemical means without additional expenditure of energy. The following may be mentioned as examples:

- methyl metacrylate (MMA);
- styrene (S);
- trimethylolpropane trimethacrylate (TMPTMA),
- and prepolymers: polyester styrene (PE/S).

The basic description of PCC and PIC composites is given in Section 13.3.

Polymeric admixtures are covered by several standards and recommendations (e.g. ACI 548.3R-03).

4.2 Aggregates and fillers

4.2.1 *Types of aggregates*

The amount of aggregate or filler is usually between 60% and 85% of volume fraction in all kinds of concrete-like composites. Therefore their influence on mechanical and other properties is important. Also the unit cost of composite material is closely related to the kind of aggregate used.

Aggregates may be classified according to their density into the following categories:

- heavy-weight aggregate with density above 2,000 kg/m^3;

- aggregate of density between 1,500 and 1,700 kg/m³, applied for ordinary and high performance concretes;
- light-weight aggregate of density below 1,120 kg/m³.

Here the density is understood as granular bulk density, which means the voids between grains and the inherent porosity of the material is taken into account; the latter varies from a few percent for hard rocks to much larger values for porous natural or artificial aggregates.

Aggregates may be also classified according to the grain dimensions of coarse and fine (sands), according to the origin: natural and artificial, from deposits in dunes, river or sea banks and pits. Other criteria for classification are certainly possible.

In concrete-like composites natural stone aggregate is used in the form of gravel or crushed rock and sand. Aggregates are obtained from all kinds of rocks as a result of the natural processes of abrasion and weathering. Natural aggregates are also produced in quarries where rock is crushed into applicable sizes. The main types of rocks are basalts, granites, sandstones and limestones, but other rocks are also used according to their availability in proximity to the place of concrete production. For ordinary applications it is usually meaningless to distinguish other types of rocks according to their origin and mineralogical composition. Only for high performance concretes or when danger of alkali-aggregate reaction exists, is there a need for deep petrographical and mineralogical studies (cf. Section 13.4.2).

Artificial aggregate may be obtained from various kinds of slags, fly ashes, burnt clays and shales, crushed bricks and recycled concrete, etc. Also other by-products, mineral and organic wastes, are used as fillers for ordinary concrete.

As a special heavy-weight aggregate, barite ($BaSO_4$), and other heavy-weight minerals as well as iron ores or scrap iron are used for concrete shields against radiation and for other structures where increased weight is needed. Barite, together with coarse marble aggregate, is used for industrial anti-static floors.

Examples of aggregates and fillers are listed in Table 4.6. An excellent manual on aggregates was published by Alexander and Mindess (2005) and guides for use of various kinds of aggregate are available from ACI 221R-96, ASTM 33-07 and EN 12620:2002.

4.2.2 *Properties of aggregates*

The most important characteristics of aggregate grains are:

1   shape and texture;
2   compressive strength;
3   other mechanical properties;
4   distribution of grain sizes.

Table 4.6 Examples of aggregates and fillers for concrete-like composites

Types of materials	Characteristic dimension	Density ρ	Compressive strength f_c	Modulus of elasticity E
	(mm)	(10^3 kg/m³)	(MPa)	(GPa)
Natural aggregates:				
Pit and river sand	0.05–2.0	2.6–3.0	150–350	40–90
Grave and crushed aggregates:				
Basalt, gabbro	2–300	2.9–3.0	200–400	80–160
Granite	2–300	2.5–3.0	160–330	40–50
Sandstone	2–300	2.4–2.7	50–250	60–80
Limestone	2–300	1.7–2.8	100–240	60–80
Porphyrite	2–300	2.6–2.8	160–270	
Quartz	2–300	2.5–2.6		50–60
Schist	2–300	1.5–3.2	60–170	
Natural lightweight aggregates:				
Pumice, scoria, tuff	5–50			
Volcanic cinders	5–80	0.5–1.0	8–12	
Sawdust, wood chips	1–20	0.3–0.8		
Cork	5–20	0.06–0.12		
Lightweight porous limestones:	5–50	1.2–1.6		
Artificial lightweight aggregates:				
Burnt porous clays and shales	3–25	0.6–1.8		
Burnt porous schists	3–25	0.8–1.9		5–15
Perlite, vermiculite	1–5	0.03–0.40	0.8–1.2	
Polystyrene beads	2–5	0.01–0.05		
Glass beads	0.5–5.0	2.1–3.0	50–120	40–80
Glass bubbles	10–30			
Blast-furnace slag (lightwt.)	5–30	1.0–1.1	80–100	
Blast-furnace slag (heavywt.)	5–30	2.7–2.9	120–240	
Expanded blast-furnace slag	5–30	0.65–0.9		
Crushed bricks	2–120	1.6–1.8	5–35	10–30
Crushed concrete	10–120	1.9–2.3		
Sintered fly ash	2–300	1.3–2.1		
Heavyweight natural aggregates:				
Barites and iron ore		3.4–6.3	50–80	
Heavyweight artificial aggregates:				
Iron beads, scrap iron	2–20	7.8	300–400	210

Other properties which should also be considered are cleanness, humidity, thermal expansion coefficient, chemical relation to Portland cement, durability in general, etc. Frost resistance of aggregate grains is required for structures that are exposed to climatic actions, e.g. road pavements and bridge elements. This is the most important quality to look for when sourcing aggregate.

According to their shape, rounded grains of natural gravel and sand are distinguished from angular grains obtained from crushed rocks. The grains may be more or less close to spheres and cubes or very elongated and needle-like, which are considered inadequate for high performance concretes, though it is still a disputable opinion. The particle shape also influences the workability of the fresh mix.

In concrete-like composites, the aggregate grains participate in load bearing together with the cement matrix, relative to the stiffness of both these phases. However, the necessary condition is sufficient bond strength along the aggregate-matrix interface. The bond depends not only on the roughness of grain surface, but also on its cleanness, lack of very fine clay particles and other impurities. Also, in some cases, a chemical bond between grains and cement paste appears, but the aggregate-matrix bond is mostly of a mechanical character. The bond is never ideal and continuous; local defects and discontinuities always exist in the interface.

The grain surface texture is very important. To improve the cement-aggregate bond, it is better when the aggregate surfaces are rather rough and not smoothly polished. For instance, the surface of basalt grains is much less developed than that of limestone. That is one of the reasons why the basalt grain-matrix bond is usually weak and cracks are more susceptible to develop and propagate in the interface than in the case of a limestone. Another reason is the chemical affinity that may improve the bond. The problems related to aggregate-matrix bonding are presented more in detail in Section 8.4.

Hard and strong aggregate is a necessary constituent of a good quality structural concrete, particularly when high strength is required. For ordinary concretes, the aggregate strength is assumed to be higher than the cement paste matrix. A number of different rocks (granite, basalt, limestone, diabase, etc.) are the origin of aggregate suitable for concrete, developed in a natural way (river or pit gravel) or produced by crushing stones. Less strong kinds of aggregate are more acceptable when lower concrete strength is required. However, the relation between the strength of aggregate and that of hardened composite is not linear, mostly because of the influence of interface, as mentioned above. When strong aggregate is used, then the matrix is weaker and cracks are propagating across the matrix. In contrast, in concretes with lightweight aggregate, the matrix is stronger and cracks cross the aggregate grains (cf. Section 9.2). However, it is generally admitted that for high performance concretes both characteristics, grain strength and aggregate-matrix bond, should be considered. Furthermore, it is often required that the coarse aggregate should have a relatively small maximum size, limited to 16 mm,

8 mm or even lower; this is not valid for particularly large structures, such as foundations, concrete dams, etc.

The shock and abrasion resistance of aggregate is important for structures where actions of that kind are foreseen. The porosity of grains, even as it increases their bond, may cause excessive variability of moisture content and adversely affecting the frost resistance of outdoor structures. In many regulations and standards concerning concretes, the aggregate properties and percentage of weak grains, impurities etc. are limited.

In several countries, the deposits of hard natural rocks or gravels, which may be used for concretes, are already exploited or are distant from regions of mass concrete production. Therefore, lower quality aggregates are used, which often implies serious difficulties in maintaining the concrete quality. Lower concrete strength and inadequate durability, lower workability or higher cement consumption, poor aspect after few years and other negative outcomes are observed as resulting from the use of low quality aggregate.

The chemical properties of aggregate should be carefully considered, because certain kinds of rocks are susceptible to reacting to moisture and highly alkaline Portland cement paste where pH is equal approximately to 13.5.

The main type of alkali-aggregate reaction (AAR) is observed when aggregates with a reactive form of silica are used, such as opal, chalcedony and tridymite; this kind of reaction is also called alkali-silica reaction. The origin of alkalis is in Portland cement, in de-icing agents or in other concrete constituents. For the first time, Stanton (1940) showed that chemical reactions may develop between certain kinds of aggregate and the alkaline environment of Portland cement paste, which induce swelling due to the accumulation of the reaction products. Because of his tests, the traditionally adopted Féret's thesis about the inert character of aggregate in concrete lost its general meaning.

In all cases, the products of reaction in the form of gels $nNa_2O \cdot mSiO_2 \cdot pH_2O$ increase their volume and swell, which may have disastrous consequences on the integrity of the hardened concrete elements. Even after several years the large cracks appear on external faces of concrete elements and serious repairs are necessary. Among numerous descriptions, a study case is described in the paper by Kucharska and Brandt (1998), as an example, together with an attempt of simple calculation of stress distribution as a function of the aggregate grain diameter, its swelling and position in concrete element.

In many countries and regions, the types of rocks mentioned above are encountered rarely and are avoided without serious difficulties. In other countries, like Japan, where volcanic rocks are prevailing, their use should be made possible with Portland cements blended with fly ash or SF to decrease their alkalinity (Okada *et al.* 1989). SF binds alkalis from the Portland cement into stable and non-expansive products, thus decreasing alkalinity approximately from a pH of 13.6 to 12.5. There also exist possibilities for the use of these rocks with other binders, in asphalt concretes, for example. The methods of checking the aggregate for that kind of risk are long, expensive and not perfectly reliable and for important works the use of aggregate tested by previous experience is advised. It is sometimes not clear why, in many structures made with potentially reactive aggregate, no deleterious results were observed, while in other cases large cracks and even disasters occurred. This different behaviour is probably related to different moisture conditions and temperature: reaction is quickest when temperature is close to +40°C.

In the last decades many accidents prove that also in the countries apparently free from any volcanic activity, like Poland, certain categories of post-glacial aggregates may produce dangerous effects of AAR, encountered, for example, in concrete structures of buildings a few months or even years after construction.

The principal measures aimed at preventing the destructions due to AAR may be summarized as:

- the use of low alkali Portland cements – meaning below 0.6% of equivalent of Na_2O;
- the limitation of the Portland cement content to keep alkali content below 3 kg/m³;
- blending the reactive aggregate with non-reactive ones;
- the application of mineral admixtures and fillers like condensed SF, fly ash, stone powder, which together may replace up to 20% of Portland cement.

The influence of fly ash on the development of AAR was experimentally examined by Alsali and Malhotra (1991) and later by many other researchers, and it was proved efficient in inhibiting the expansion due to AAR products from potentially reactive aggregate.

Beside the alkali-silica reaction described above, similar phenomena may occur in the case of reactive dolomites and limestones. The so-called alkali-carbonate reaction (ACR) is less frequent and not completely understood. When the effects of ACR are observed, two similar remedies are also necessary: either to keep the content of alkalis in concrete as low as possible, or to decrease the percentage of deleterious aggregate in the concrete mix.

All these measures for both ASR and ACR should be considered when, for economical reasons, total elimination of reactive aggregate is not possible.

Extensive research results on AAR are reported by Okada *et al.* (1989) with several papers on test results; the basic description of the phenomena and their effects may be found in Mindess *et al.* (2003). For further details on the aggregate properties the reader is also referred to handbooks for concrete technology, e.g. Popovics (1992) and to special recommendations on that subject published systematically by ACI 221.1R-98, ASTM C289-07 and ASTM C1260-07.

4.2.3 Lightweight aggregates

Natural lightweight aggregates (LWA) are mostly obtained from volcanic rocks and their application is of local character. Artificial aggregates are produced in several countries for three main reasons:

1   to satisfy the increasing demand for aggregate, if possible, close to mass concrete production;
2   to exploit large volumes of industrial wastes, accumulated in the neighbourhood of metallurgical and other plants;
3   to obtain various kinds of lightweight concrete for nonstructural elements and insulation layers;
4   to produce lightweight high performance concrete for structures in which reduction of own weight is of primordial importance, e.g. long-span concrete bridges.

The production of artificial aggregates requires several procedures, according to particular conditions and requirements.

Blast-furnace slags are crushed into grains. Clays and shales are burnt in temperature between 1000°C and 1200°C to generate gases and to obtain porous structure of final material after cooling and crushing. Expanded slag is produced by cooling molten slag with water to entrap steam and to obtain porous material.

The application of non-ground-granulated blast-furnace slag (NGGBFS) was proposed by Yüksel *et al.* (2006) as partial replacement of natural sand. The use of this material is not only environmentally beneficial, but also does not require additional energy for grinding. Test results show that while strength is slightly decreasing, this replacement improves durability and the unit weight of concrete may be reduced by 10%. Replacement percentage of natural sand should be maintained between 25% and 50%.

Expanded glass and polystyrene beads are of lowest density and they create a system of pores in the structure of hardened concrete or mortar.

The bulk density of lightweight aggregate may vary between 30 kg/m^3 and 1800 kg/m^3 (cf. Table 4.6). Because of the rough surface the adhesion of cement mortar to the grains is usually excellent and this is one of the reasons explaining that in high performance concretes lightweight aggregate may be used. More information on lightweight aggregates may be found out in CEB/FIP Manual (1977) and in EN 13055:2002.

Selected fractions of lightweight aggregate are pre-wetted before mixing and used as partly replacement of fine fractions of aggregate. That is a kind of internal curing in order to supply water for cement hydration in the bulk concrete. Besides soaked, lightweight aggregate (pumice, perlite sand, etc.), super-absorbent polymers (SAP), which are much more efficient, are also used. Internal curing improves various properties of hardened concrete, but the main reasons for its application is the reduction of autogeneous shrinkage, enhanced strength development and improved durability (Bentz *et al.* 2005). By a higher degree of hydration, lower porosity, densification of ITZ, reduced internal stresses and micro-cracking enhanced durability is expected (Lura *et al.* 2007).

In the case when damage of the concrete surface is expected, due to cycles of freezing and thawing in the presence of de-icing salts, the internal curing also is beneficial. Investigations have proved that resistance against scaling may be considerably enhanced and that this way may be cheaper and easier than the traditional application of air-entraining agents (Jóźwiak-Niedźwiedzka 2005). Extensive research and test results are collected in the report by the RILEM Technical Committee 196-ICC (2008).

4.2.4 Recycled aggregate

Natural aggregate may be replaced with recycled aggregate from old crushed concrete in various proportions, or even only recycled aggregate may be used.

Durability of concrete elements made with recycled aggregate may be lower when exposed to various aggressive media and should be carefully checked. In such a case, application of natural fine aggregate (natural sand) is preferable (Brozovsky *et al.* 2006). The use of recycled aggregate is very important for ecological reasons, but should be preceded by careful testing because with old concrete various deleterious influences may be introduced to new structures. The bulk application of recycled aggregate is limited to secondary structures, e.g. as a sub-base for roads and pavements.

Valuable recommendations are published in ACI collected volume by Liu and Meyer (2004) and in ACI 555R-01.

4.2.5 Fillers and microfillers

Fillers and microfillers are terms used for the kinds of aggregates, which, because of very low Young's modulus, do not participate in load bearing. Their strength is low or quite negligible. The role of the fillers is to:

- improve the insulation properties of concretes;
- decrease the weight of these concrete elements and layers, where high strength is not needed;
- fill a part of volume with relatively lightweight and inexpensive material;
- decrease the amount of cement, which is expensive and is a source of shrinkage.

Different kinds of materials are used as fillers: sawdust and timber chips, cork, rubber waste and other types of wastes.

The most important kind of fillers are pores obtained in fresh mix by different procedures and admixtures, generating foam in cement paste. Systems of pores are specially created in the matrix to increase the freeze-thaw resistance. Air-entrained mixes also behave better during handling and placing in forms. In the so-called cellular concrete a perfect control of final material density is achieved by selection of pore volume and dimensions. The structures formed with pores are described in more detail in Section 6.5.

The following are used as fillers:

- fly ashes, which do not always improve the material properties and sometimes are considered as decreasing the durability of concrete for outdoor structures;
- SF which is of particular large application in last decades (cf. Section 4.1.2);
- calcareous powders with specific area around 0.6 m³/g;
- metakaolin;
- glass beads.

Intergranular spaces and micropores may be partly blocked by microfillers acting as inert components; this effect is enhanced by intrinsic pozzolanic activity.

4.2.6 Aggregate grading

The grain size distribution of aggregate should be carefully designed and checked if concrete of good quality is to be prepared. The grading is established after sieve analysis using sieves in accordance with local standards; usually separate sieve systems are used for coarse and fine aggregates.

There are various national and international standards, e.g. ACI 221R-96, ASTM 33-07 and EN 12620:2002. The older standards indicated areas of the sieve diagrams of recommended aggregate grading and examples of such 'good advice' are shown in Figure 4.8, after past Polish Standard and ASTM specification.

The aggregate grading may be represented not only by sieve curves, but also by triangular diagrams. An example of such a diagram is shown in Figure 4.9 in which limit sieve curves from Figure 4.8 are reproduced. In that representation only three aggregate components are taken into account and the corresponding admissible region is determined.

The aggregate grading may be described not only graphically, but also in a numerical way; several commercial programs are available for such tasks, (cf. Section 12.5).

In continuous grading, all grain sizes are represented and theoretically, the volume is best packed by grains. The highest strength may be then expected, but the workability of such a mix risks being unsatisfactory; therefore the mixing and placing in forms may be expensive and ineffective. For such a mix an additional amount of cement paste is necessary.

When the grading is discontinuous, called gap grading, it means that some grain sizes are absent and certain segments of a corresponding curve obtained after sieve separation are horizontal, even if that curve is situated

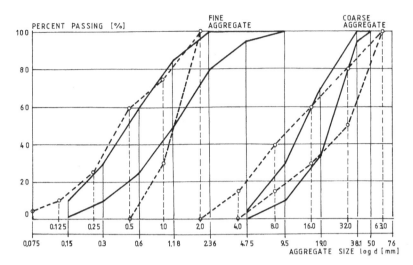

Figure 4.8 Examples of limit grading curves for fine and coarse aggregates, according to old ASTM recommendation (continuous line) and to past Polish Standard for Ordinary Concrete (broken line). Admissible regions are situated between the respective curves.

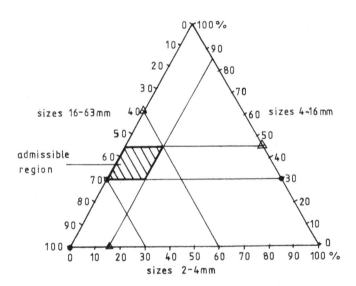

Figure 4.9 Limit curves from Polish Standard as indicated in Figure 4.8 represented in the form of a triangular diagram.

within the prescribed limits. In that case, usually less sand and cement may be used for the same concrete strength, but the danger of segregation arises during transport and handling of the fresh mix.

In the design of grading, several factors should be analysed and it is generally assumed that there is no 'ideal' grading suitable for all occasions. For concretes of controlled quality it is usually necessary to separate available aggregate into sizes and to recombine it according to desired proportions (cf. Sections 12.4 and 13.4.2). The design of concrete composition, including the selection of optimum aggregate grading, considering both mechanical and economical factors is a typical problem for mathematical optimization (cf. Popovics 1982). The criteria of best packing and appropriate workability should be completed by minimum cost, taking into account available sources of aggregate and the composition of fractions in these sources. In any case, every designed grading of aggregate should be checked by a trial mix to determine its effective workability, eventual segregation and obtained properties of concrete.

For further information concerning selection and properties of aggregate the reader is referred to above mentioned sources and other text books for concrete technology, e.g. Mindess *et al.* (2003), and basic European Standard for concrete EN 206-1:2000.

4.3 Chemical admixtures

4.3.1 Kinds of admixtures

The admixtures are added to the batch before or during mixing (cf. Section 3.1). The objective of admixtures is to modify and improve certain properties of the fresh mix or of the hardened composite; though a well-designed and correctly executed concrete may already have these properties to a greater or lesser degree. Use of various admixtures should never replace the proper selection of concrete constituents of good quality, adequate mixture proportions and appropriate mixing and curing techniques.

There are many different kinds of admixtures and it would exceed the scope of this chapter to describe them all in detail, and new products appear frequently on the market. The main groups of admixtures are:

• plasticizers and superplasticizers, which increase the fluidity of the fresh mix or reduce the volume of mixing water for the same fluidity;
• accelerators or retarders, which influence the process of cement hydration to help in application of special technologies;
• air-entraining agents, which introduce a system of well-sized and correctly distributed pores to the matrix;
• other admixtures, which improve the resistance against certain external actions, reduce the rate and amount of hydration heat of Portland cement, decrease the permeability, modify the colour of hardened concrete, etc.

New generations of admixtures present very advanced properties, for example enabling considerable water reduction up to 30%, obtained together with air-entrainment.

The main kinds of admixtures were described already by Venuat (1984) and later in the basic monograph by Mailvaganam and Rixom (1999). Also in several handbooks for concrete technology and in ACI 212.3R-04 and ASTM C494/C494M-08 important data are published.

The admixtures are manufactured by various specialised firms as ready-to-use products with guaranteed properties. Their producers are obliged to give all technical data together with detailed instructions for use. Also, the main and additional effects of admixtures should be clearly stated. However, it is advisable to verify every admixture in local conditions before serious application on a larger scale controlling the results obtained and compatibility with other components of the mix.

4.3.2 Accelerators

The acceleration of hydration and hardening is needed for different purposes, e.g. concreting in low temperature, repair works, etc. The most commonly known accelerator is calcium chloride ($CaCl_2$); its amount is limited to up to 2% of the cement mass. By accelerating the hydration of Portland cement, the rate of heat is increased, which helps to maintain acceptable temperature in cold weather. The temperature should not fall lower than a few degrees below 0°C. The amount of heat liberated with and without the admixture for concreting at a temperature of +5°C is shown in Figure 4.10. For higher temperature the admixture is less effective.

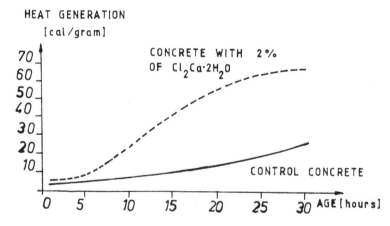

Figure 4.10 Heat generated during cement hydration – influence of an accelerating admixture, after Venuat (1984).

It is considered as inadmissible to increase the amount of $CaCl_2$ because several negative effects may be expected: increase of drying shrinkage, lower resistance against frost and sulphate agression and higher risk of aggregate alkali reaction. However, the most important effect is accelerated corrosion of reinforcement and that is why the application of $CaCl_2$ is, in many countries, admitted only for plain concrete elements.

There are accelerating admixtures based on other compounds, but these are more expensive, e.g. $Ca(NO_3)_2$.

Another kind of admixture is used when an acceleration of normal processes of cement hydration is not needed, but a very rapid set in a few minutes is necessary to execute particular technological operations, such as repair or shotcrete of wet surfaces. Such admixtures are based on aluminium chloride, potassium carbonate, ferric salts and others.

4.3.3 Set-retarders

These admixtures are used in hot climates to slow down the cement hydration and to reduce the rate of heat liberation. A similar effect is needed in the execution of external surfaces, which are finished by special procedures, like exposing coarse aggregates. In other cases these admixtures are used when more time is needed for casting fresh concrete before cement hydration takes place.

Various compounds are used as set-retarders based on sugar, lignosulphonic or hydroxy-carboxylic acids and salts, and also inorganic salts. For example, sugar added as 0.05% of cement mass may delay the set by approximately four hours. The long-term concrete strength is not modified by set-retarders.

4.3.4 Plasticizers

The use of plasticizers and superplasticizers for concretes initiated on a large scale in the 1980s was a considerable step forward in the development of concrete as a structural material. These admixtures enabled production of high strength and high performance concrete due to reduced volume of water necessary for required workability of the fresh mix. Together with other innovations, this was the basis for a new and fascinating period of development of concretes, which extends to the present time (Brandt and Kucharska 1999).

Plasticizers are applied for the following reasons:

1   to obtain higher strength of hardened concrete by decreasing the water/cement ratio with the same workability of fresh mix;
2   to obtain the same workability with smaller cement content and reduced generation of internal heat during cement hydration;
3   to increase the workability and to reduce the risk of segregation, which results in cheaper and more effective handling of fresh mix;

4   to improve impermeability and frost resistance together with other properties related to reduced water/cement ratio.

Certain plasticizers may also accelerate or delay the hydration processes; these effects should be indicated by the producer.

The action of plasticizers consists of increasing the mobility of cement particles by reduction of the interfacial tension and increasing electrokinetic potential. That way the flocculation tendency of cement particles, and also of other fine particles in the fresh mix, is considerably reduced and thin water cover appears around these particles.

The reduction of water by 5–10% is normally achieved by use of plasticizers.

Because the active surface of cement is increased, higher early and final strength of concrete is obtained as well as better durability.

The plasticizers are produced on the basis of several organic compounds: lignosulphonic and hydroxylated carboxylic acids and salts, polyols and others. Usually between 0.2% and 0.5% of cement mass is added in an appropriate way, more often directly to the mixing water. The increased amount of a plasticizer may cause retardation effects and longer setting times.

4.3.5 Superplasticizers

Superplasticizers (high-range water reducers HRWR, ACI 212.4R-04) are used for similar purposes as plasticizers, but with increased effects, and water may be reduced by 15–30%. Such results cannot normally be obtained by increasing the amount of a plasticizer without a risk of large side effects: considerable slow down of the rate of cement hydration or excessive porosity.

The result of a superplasticizer is shown schematically in Figure 4.11, in the form of a relation between the water/cement ratio and the workability. By using superplasticizers all modifications indicated by arrows 1, 2 and 3 are possible. The plasticizers act principally as water reducers for constant workability, which is indicated by arrow 3.

The improvement of workability is limited to a short time and usually after 30–45 minutes it returns to normal. Therefore, the handling of concrete and placing it in forms should be appropriately scheduled.

Different compounds are used as superplasticizers, such as:

* sulphonated melamine formaldehyde condensates;
* sulphonated naphthalene formaldehyde condensates;
* and next generation: polycarboxylate ether-based (PCE).

The hydration and hardening processes in concrete were not modified by superplasticizers of earlier generations, and only a slight increase in porosity

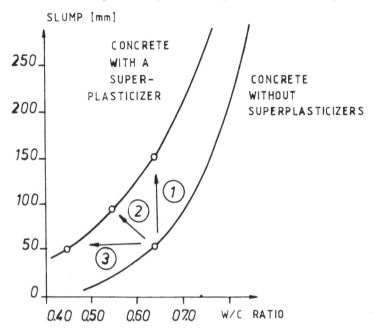

Figure 4.11 Influence of a superplasticizer on the workability and *w/c* ratio of concrete; 1 increase of workability, 2 increase of workability together with decrease of *w/c*, 3 decrease of *w/c* ratio, after Venuat (1984).

was sometimes observed. If the water/cement ratio is decreased, then a considerable increase in strength may be expected – even 50%. Additional increase in strength occurs because of better packing of particles in the fresh mix. Various kinds of high strength and high performance concretes are designed and executed with low and very low values of *w/c*, which is possible by appropriate application of superplasticizers. On the other hand, there are flowing concretes that may be pumped without difficulties.

The application of superplasticizers may also facilitate the achievement of good workability with composition of aggregate grains, which is far from perfect. In such a case, by avoiding improvements of aggregate considerable economies may be obtained. Superplasticizers containing reactive polymeric dispersants allow workability to be maintained during a 60-minute period or longer, which helps considerably in all on-site operations. These are used for special mixture compositions and allow the production of SCC, which are described in more detail in Section 3.2.10. Superplasticizers may also be added in two portions if necessary; e.g. before and after transportation of the fresh mix.

New kinds of superplasticizers allow the modification of not only workability, but also rate of strength increase. The market is varied and interested readers should refer to on-line information.

4.3.6 Air entrainers

The purpose of air-entrainment in cement mortar and concrete is to improve their frost resistance. Only uniformly distributed air pores of appropriate dimensions can increase the resistance against freeze-thaw cycles. Concrete structures exposed directly to natural weather conditions in regions of cold and moderate climate are in danger of surface destructions of various kinds. Other kinds of large pores and air voids caused by an excess of water and incomplete compaction of the fresh mix have no effect on that property, which is very important for the durability of outdoor structures. The necessary amount of air-entraining agent is determined as a function of the volume of water in the capillary pores where it might be transformed into ice. An efficient system of air-entrained pores should reply to several requirements (cf. Section 6.5.)

In very high performance concretes with low water/cement ratio the air-entraining agents may be considered unnecessary (cf. Section 13.4.2). In many countries, application of air-entrainers for such structures is imposed by standards.

An effective system of air pores may be obtained only by application of special admixtures that are manufactured from various compounds, e.g. salts of wood resins, sulphonated lignins and petroleum acids, and synthetic detergents, producing a foam of dense air bubbles in the cement based matrix, which can survive the handling and mixing operations. The application of air entrainers also has other results: the workability of fresh mix is increased and the impermeability is improved, therefore better resistance against de-icing agents is achieved.

Air-detraining admixtures also exists, e.g. silanes in an emulsified mineral oil base or tributyl phosphate (Manning and Northwood 1980).

4.3.7 Water repellents

These admixtures decrease water penetration across a hardened porous cement matrix, but cannot be used to eliminate water migration and the term 'damp-proofing' is not correctly applied. They are used on concrete surfaces against rain and ground water penetration. The water repellents are produced on the basis of silicones, acrylic resins or petroleum products.

4.3.8 Admixtures reducing permeability

These are composed of fine mineral particles (bentonites, ground limestone, ground dolomite or lime, rock dust) and chemical compounds (e.g. sulpho-nates) in order to limit water migration when added as 1–2% of mass of Portland cement. Other effects are the workability improvement and decrease of mixing water by 5–10%.

Inert admixtures and secondary binding materials, like fly ash and SF, may enhance concrete impermeability. Though they are basically inert, finely ground mineral materials have some bonding effect and contribute to strength by pozzolanic reaction.

Finely ground blast-furnace slag and natural pozzolanic materials of volcanic origin in blended cements are cheaper than pure Portland cement and are used in large dosage, e.g. the amount of ground slag may reach 90%, but should be clearly indicated in delivery. These kinds of blended cements have increased resistance against chemical corrosion, particularly against sulphate attack.

4.3.9 *Other admixtures*

There are other admixtures not described here, which improve some particular properties of concrete composites. These are, (for example):

1 admixtures to obtain volume expansion during hydration and hardening in order to compensate natural drying shrinkage of Portland cement, and are used in repair works or in grouting of prestressing reinforcement ducts;
2 gas-forming admixtures to produce lightweight concretes;
3 admixtures increasing bonding, usually polymer emulsions (latexes), are used most often in repair works when fresh concrete is placed on old, also concrete bonding to steel reinforcement is increased;
4 colouring admixtures for decoration elements of building facades.

References

Alexander, M., Mindess, S. (2005) *Aggregates in Concrete*. London; Taylor & Francis.

Alsali, M. M., Malhotra, V. M. (1991) 'Role of concrete incorporating high volumes of fly ash in controlling expansion due to alkali-aggregate reaction,' *American Concrete Institute Materials Journal*, 88(2): 159–63.

Barnett, S. J., Soutsos, M. N., Millard, S. G., Bungey, J. H. (2006) 'Strength development of mortars containing ground granulated blast-furnace slag: effect of curing temperature and determination of apparent activation energies', *Cement and Concrete Research*, 36(3): 434–40.

Bensted, J., Barnes, P. (2001) *Structure and Performance of Cements*, 2nd ed. London; Taylor & Francis.

Bentur, A., Goldman, A., Cohen, M. D. (1988) 'The contribution of the transition zone to the strength of high quality silica fume concretes,' in: *Proc. Int.* Symp. *Bonding in Cementitious Composites*, S. Mindess, S. P. Shah eds, Boston, 2–4 December 1987, Pittsburgh; Materials Research Society: pp. 97–103.

Bentz, D. P., Lura, P., Roberts, J. W. (2005) 'Mixture proportioning for internal curing,' *Concrete International*, 27(2): 35–40.

Bijen, J. (1996) *Blast Furnace Slag Cement for Durable Marine Structures*. Association of the Netherlands Cement Industry.

Brandt, A. M., Kucharska, L. (1999) 'Developments in cement-based materials,' in: *Proc. Int. Sem. Extending Performance of Concrete Structures*, R. K. Dhir and P. A. J. Tittle eds, Dundee; Thomas Telford: pp. 17–32.

Brozovsky, J., Zach, J., Brozovsky, J. Jr. (2006) 'Durability of concrete made from recycled aggregates,' in: *Proc. Int. RILEM-JCI Sem. Concrete Life '06*, Israel; Ein-Bokek: pp. 436–43.

Bui, D. D. (2001) *Rice Husk Ash as a Mineral Admixture for High Performance Concrete*. Delft, Netherlands; Delft University Press.

Bukowski, B. (1963) *Concrete Technology*, parts 1 and 2, (in Polish), Arkady, Warsaw.

De Larrard, F., Gorse, J. F., Puch, C. (1990) 'Efficacités comparées de diverses fumées de silice comme additif dans les bétons à hautes performances,' *Bulletin de Liaison des Laboratoires des Ponts et Chaussées*, 168: 97–105.

Diamond, S., Sahu, S. (2006) 'Densified silica fume: particle sizes and dispersion in concrete,' *Materials and Structures*, 39: 849–59.

Domone, P. L. J., Illston, J. M. eds (2001) *Construction Materials*, London; Taylor & Francis.

Fu, X., Li, Q., Zhai, J., Sheng, G., Li, F. (2008) 'The physical-chemical characterization of mechanically-treated CFBC fly ash', *Cement and Concrete Composites*, 30, 3: 220–6.

Gao, J. M., Qian, C. X., Liu, H. F., Wang, B., Li, L. (2005) 'ITZ microstructure of concrete containing GGBS,' *Cement and Concrete Research*, 35(7): 1299–1304.

Glinicki, M. A., Zieliński, M. (2008) 'The influence of CFBC fly ash addition on phase composition of air-entrained concrete,' *Bulletin of the Polish Academy of Sciences*, 56(1): 45–52.

Gruber, K. A., Ramlochan, T., Boddy, A., Hooton, R. D., Thomas, M. D. A. (2001) 'Increased concrete durability with high-reactivity metakaolin,' *Cement and Concrete Composites*, 23(6): 479–84.

Havlica, J., Brandstetr, J., Odler, I. (1998) 'Possibilities of utilizing solid residues from Pressured Fluidized Bed Coal Combustion (PSBC) for the production of blended cements,' *Cement and Concrete Research*, 28(2): 299–307.

Jalali, S., Peyroteo, A., Ferreira, M. (2006), 'Metakaolin in concrete – beneficial impact on performance parameters' (abstract), in: *Proc. Int. RILEM Workshop on Performance Base Evaluation and Indicators for Concrete Durability*. Madrid, p. 49.

Jovanovic, I., Paatsch, A., Durukal, A. (2002) 'Ductal: a new generation of ultra high performance fiber reinforced concrete,' in: *Proc. of 6th Int. Symp. Utilization of High Strength/High Performance Concrete*, Leipzig, pp. 1089–95.

Jóźwiak-Niedźwiedzka, D. (2005) 'Scaling resistance of high performance concretes containing a small portion of pre-wetted lightweight fine aggregate,' *Cement and Concrete Composites*, 27: 709–15.

Khatib, J. M., Hibbert, J. J. (2005) Selected engineering properties of concrete incorporating slag and metakaolin,' *Construction and Building Materials*, 19(6): 460–72.

Kucharska, L., Brandt, A. M. (1998) 'Local stresses in concrete structures due to expanding aggregate grains,' in *Proc. of 5th Int. Workshop Durable Reinforced Concrete Structures*, Weimar: Aedifitio Publ., pp. 249–63.

Kurdowski, W. (1991) *Chemistry of Portland Cement* (in Polish), Warszawa: PWN.

Liu, T., Meyer, C. eds (2004) *Recycling Concrete and Other Materials for Sustainable Development*, ACI SP219.

Lura, P., Jensen, O. M., Igarashi, S-I. (2007) 'Experimental observation of internal water curing of concrete,' *Materials and Structures*, 40(2): 211–20.

Mailvaganam, N. P., Rixom, M. R. (1999) *Chemical Admixtures for Concrete*, 3rd ed., London; Taylor & Francis.

Malier, Y. (1992) *Les bétons à hautes performances. Caractérisation, durabilité, applications*. Presses de l'Ecole Nat.des P. et Ch.

Manning, D. G., Northwood, R. P. (1980) 'The rehabilitation of structures on an urban freeway: a case study,' in: *Proc. Conf. Bridge Maintenance and Rehabilitation*, IABSE and ASCE, West Virginia, 12–16 August, pp. 644–68.

Mehta, P. K., Monteiro, P. J. M. (1988) 'Effect of aggregate, cement and mineral admixtures on the microstructure of the transition zone,' in: *Proc. Int. Symp. Bonding in Cementitious Composites*, S. Mindess, S. P. Shah eds, Boston, 2–4 December 1987, Materials Research Soc., Pittsburgh, Pennsylvania, pp. 65–75.

Mindess, S., Young, J. F. and Darwin, D. (2003) *Concrete*, 2nd edition. New Jersey; Prentice-Hall, Pearson Ed.

Mindess, S. (2004) 'High performance concrete: recent developments in material design,' in Lecture Notes 18: *Some Aspects of Design and Application of High Performance Cement Based Materials*, A. M. Brandt, ed., Warsaw; IFTR: pp. 9–36.

Naik, T. R., Singh, S. S. (1995) 'Use of high-calcium fly ash in cement-based construction materials,' in: *Proc. 5th CANMET/ACI Int. Conf. on Fly Ash, Silica Fume, Slag and Natural Pozzolans in Concrete*. Milwaukee WI; Suppl. Papers: pp. 1–44.

Nakamoto, J., Togawa, K. (1995) 'A study of strength development and carbonation of concrete incorporating high volume of blast furnace slag,' in: *Proc. 5th CANMET/ ACI Int. Conf. Fly Ash, Silica Fume and Natural Pozzolans in Concrete*, V. M. Malhotra ed., Milwaukee, WI: pp. 1121–39.

Neville, A. M. (1997) *Properties of Concrete*, 4th edition. New York; John Wiley.

Neville, A. (2003) 'Should high alumina cement be re-introduced into design codes?' *Structural Engineering*, 81(24): 35–40.

Ohama, Y. (1998) 'Polymer-based admixtures,' *Cement and Concrete Composites*, 20(2): 189–212.

Ohama, Y., Demura, K., Morikawa, M., Ogi, T. (1989) 'Properties of polymer-modified mortars containing silica fume,' in: *Proc. Int. Symp. Brittle Matrix Composites 2*, Cedzyna 20–22 September 1988, A. M. Brandt and I. H. Marshall eds, London; Elsevier Applied Science: pp. 648–57.

Ohama, Y., Madej, J., Demura, K. (1995) 'Efficiency of finely ground blast furnace slags in high-strength mortars,' in: *Proc. 5th CANMET/ACI Int. Conf. Fly Ash, Silica Fume and Natural Pozzolans in Concrete*, V. M. Malhotra ed., Milwaukee, WI: pp. 1031–50.

Okada, K., Nishibayashi, S., Kawamura, M. eds (1989) *Proc. 8th Int. Con. Alkali-Aggregate Reaction*, Kyoto July 1989, London; Elsevier Applied Science.

Piłat, J. and Radziszewski, P. (2004) *Asphalt Pavements* (in Polish), Warsaw; WKŁ.

Popovics, S. (1982) 'Production schedule of concrete for maximum profit,' *Materials and Structures*, *RILEM*, 15(87): 199–204.

——.(1992) *Concrete Materials. Properties, Specifications, and Testing*, 2nd edition, New Jersey; Noyes Publications.

Rajczyk, K., Giergiczny, E., Glinicki, M. A., (2004) 'Use of DTA in the investigations of fly ashes from fluidized bed boilers,' *Journal of Thermal Analysis and Calorimetry*, 77: 165–70.

Sánchez de Rojas, M. I., Frias, M. (1995) 'The influence of silica fume on the heat of hydration of Portland cement,' in: *Proc. 5th CANMET/ACI Int. Conf. Fly Ash, Silica Fume and Natural Pozzolans in Concrete*, V. M. Malhotra ed., Milwaukee, WI; vol. 2, pp. 829–43.

Sarkar, S. L., Wheeler, J. (2001) 'Important properties of an ultrafine cement,' *Cement and Concrete Research*, 31(1): 119–23.

Sheng, G., Li, Q., Zhai, J., Li, F. (2007) 'Self-cementitious properties of fly ashes from CFBC boilers co-firing coal and high-sulphur petroleum coke,' *Cement and Concrete Research*, 37(6): 871–6.

Stanton, T. F. (1940) 'Expansion of Concrete through Reaction between Cement and Aggregate,' *Proc. Amer. Soc. of Civ. Engrs.*, 66: 1781–95.

Swamy, R. N. (2004) 'High performance cement based materials holistic design for sustainability,' in Lecture Notes 18: *Some Aspects of Design and Application of High Performance Cement Based Materials*, A. M. Brandt, ed., Warsaw; IFTR: 37–81.

Swamy, R. N., Laiw, J. C. (1995) 'Effectiveness of supplementary cementing materials in controlling chloride penetration into concrete,' in: *Proc. 5th CANMET/ACI Int. Conf. Fly Ash, Silica Fume and Natural Pozzolans in Concrete*, V. M. Malhotra ed., Milwaukee, WI; 2: 657–74.

Tomisawa, T., Fujii, M. (1995) 'Effects of high fineness and large amounts of GGBFS on properties and microstructure of slag cement,' in: *Proc. 5th CANMET/ACI Int. Conf. Fly Ash, Silica Fume and Natural Pozzolans in Concrete*, V. M. Malhotra ed., Milwaukee, WI: 951–73.

Venuat, M. (1984) *Adjuvants et Traitements*, Paris; published by the author.

Young, J. F., Mindess, S., Gray, R. J. and Bentur, A. (1998) *The Science and Technology of Civil Engineering Materials*, New Jersey; Prentice-Hall.

Yüksel, I., Özkan, Ö., Bilir, T. (2006) 'Use of granulated blast-furnace slag in concrete as fine aggregate,' *American Concrete Institute Materials Journal*, 103(3): 203–8.

Żenczykowski, W. (1938) 'Mortars for Building,' in: *Calendar of Building Review* (in Polish), I. Luft, ed., pp. 80–94.

Standards

ACI 212.3R-04 *Chemical Admixtures for Concrete*.

ACI 212.4R-04, *Guide for the Use of High-Range Water-Reducing Admixtures (Superplasticizers) in Concrete*.

ACI 221R.-96 (reapproved 2001), *Guide for Use of Normal and Heavyweight Aggregates in Concrete*.

ACI 221.1R-98, *Report on Alkali-Aggregate Reactivity*.

ACI 225R-99 (reapproved 2005), *Guide to the Selection and Use of Hydraulic Cements*.

ACI 232.2R-03, *Use of Fly Ash in Concrete*.

ACI 233.2R-03 *Slag Cement in Concrete and Mortar*.

ACI 233-95, *Ground Granulated Blast-Furnace Slag as a Cementitious Constituent in Concrete*.

ACI 234R-06 *Guide for the Use of Silica Fume in Concrete*.

ACI 548.2R-93 (reapproved 1998) *Guide for Mixing and Placing Sulphur Concrete in Construction*.

ACI 548.3R-03 *Polymer-Modified Concrete*.

ACI 555R-01 *Removal and Reuse of Hardened Concrete*.

ASTM 33-07 *Standard Specification for Concrete Aggregates*.

ASTM C150-07 *Standard Specification for Portland Cement*.

ASTM C289-07 *Standard Test Method for Potential Alkali Reactivity of Aggregates (Chemical Method)*.

ASTM C311-07 *Standard Test Methods for Sampling and Testing Fly Ash or Natural Pozzolans for Use in Portland-Cement Concrete*.

ASTM C494/C494M-08 *Standard Specifications for Chemical Admixtures for Concrete*.

ASTM C595-08 *Standard Specification for Blended Hydraulic Cements*.

ASTM 989-06 *Standard Specification for Ground Granulated Blast-Furnace Slag for Use in Concretes and Mortars*.

ASTM C1240-05 *Standard Specification for Silica Fume Used in Cementitious Mixtures*.

ASTM C 1260-07 *Standard Test Method for Potential Alkali Reactivity of Aggregates (Mortar Bar Method)*.

ASTM DI074-02 Standard Test Method for Compressive Strength of Bituminous Materials.

CEB/FIP (Comité Européen du Béton/Fédération Internationale de la Précontrainte) Manual (1977) *Lightweight Aggregate Concrete. Design and Technology*, The Construction Press.

EN 13108 (series): 2005 and 2006. Bituminous mixtures. Material specifications.

EN 197:2004 *Cement*, Part 1: *Composition, specifications and conformity criteria for common cements*, Parts 2–4.

EN 206-1:2000 *Concrete – Specification, Performance, Production and Conformity.*

EN 450:2005 *Fly ash for Concrete*, Parts 1 and 2.

EN 12620:2002 *Aggregate for Concrete.*

EN 13055:2002 *Lightweight Aggregates*, Parts 1 and 2.

EN 13263:2005 *Silica Fume for Concrete*, Parts 1 and 2.

FIP (1988), Fédération Internationale de la Précontrainte. *Condensed Silica Fume in Concrete*. State of art report, London: Th. Telford.

RILEM Report of TC 196-ICC, 2008.

5 Reinforcement of cement-based composites

5.1 Purpose of reinforcement

The purpose of reinforcement in the cement-based composites is to increase their fracture toughness, which means to improve resistance against cracking by control of crack opening and propagation. When the fibres bridge opening cracks, then several phenomena appear (cf. Figure 2.5) and by that way pseudo post-cracking ductility is exhibited.

The reinforcement increases the composite material tensile strength if fibre reinforcement is sufficiently effective. The tensile strength of the matrix itself is low; for instance, that of Portland cement mortars and concretes is approximately equal to 10–12% of their compressive strength. Much higher composite tensile strength is obtained thanks to various systems of reinforcement, including systems with two or more different fibres (hybrid reinforcement).

Traditional reinforcement of concrete elements in the form of steel bars and prestressing cables or tendons is not studied here as it is mentioned in Section 2.3.1 and only dispersed fibre reinforcement is considered. Macro-fibres are usually of 10–60 mm in length and 0.1–1.0 mm in the least dimension. Micro-fibres are of 10–30 μm in diameter and below 10 mm in length. There are several kinds of microfibres: non-metallic, including asbestos, polypropylene, mica, wollastonite and xonotlite, and steel fibres. In the following chapter both these groups of fibres are considered; made with different materials (cf. ACI 544.1R-96).

5.2 Natural mineral fibres

Asbestos fibres are extracted from natural deposits in several regions of the world. The best known are in Quebec province in Canada, near Sverdlovsk in Russia, in Cyprus, near Turin in Italy, in Zimbabwe and in the Republic of South Africa. The rock containing asbestos is crushed to separate fibres by mechanical methods.

The main kinds of asbestos fibre are:

- chrysotile asbestos, 93.5% of world production
- crocidolite (blue) 4.0%

- amosite 2.2%
- anthophyllite 0.3%
- tremolite and actinolite, which have no industrial importance

The most frequently available chrysotile asbestos is the fibrous form of serpentine $3MgO \cdot 2SiO_2 \cdot 2H_2O$ – hydrated magnesium silicate. The fibres are thin; their diameter varies from 0.012 up to 0.03 µm. The fibre length is normally 5 mm, rarely reaching 40 or even 100 mm and there is an empty channel inside a fibre (Hannant 1978). The asbestos fibres are considered as microfibres.

The tensile strength of asbestos fibres is assumed within the limits of 550 and 750 MPa. However, single fibres may have much higher strength: up to 1000 MPa for chrysotile asbestos and to 3500 MPa for crocidolite asbestos. The fibres are lightweight and have high resistance against most chemical agents, low electric conductivity and maintain strength at high temperature. The bond is excellent due to their affinity to cement paste and a large surface area. The strong fibre/matrix interface does not contain increased amounts of CH, which is formed around other kinds of fibres. Because of these outstanding properties, asbestos fibres have been used extensively since the beginning of the twentieth century as reinforcement for thin plates and pipes made of neat cement paste; the commercially available asbestos fibres are considered too short to reinforce cement mortar and concrete. Usually the fibre content reaches 10–15% of the cement mass.

For more than 40 years it has been known that asbestos dust, which is produced at all stages of asbestos fibre extraction and handling, is very dangerous to human health. The risk for occupants of buildings is also considered, but it may occur only when careless repair work or demolition is carried out. In most countries, the use of asbestos fibres in building and civil engineering is forbidden; in others it is strictly limited.

The imposed restrictions on the application of asbestos fibres in buildings as reinforcement of cement-based composites stimulated extensive research directed at other kinds of fibres. Because of the excellent mechanical properties, good durability and relatively low cost of asbestos cements, finding an appropriate replacement has been difficult. The research was mainly concerned with cellulose, polypropylene and carbon fibres and positive results have been achieved (Krenchel and Hejgaard 1975; Mai 1979; Krenchel and Shah 1985). The properties of asbestos cements are briefly described in Section 3.2.5.

Basalt fibres are made of basalt rock after melting at high temperature of 1300–1700°C and spinning it into fine filaments. As to the composition, basalt fibres may be compared to E-glass fibres, with some minor differences. Basalt fibres have higher Young's modulus and lower tensile strength than E-glass fibres. Basalt fibres as reinforcement of cement composites have similar advantages and limitations and their application is restricted.

Wollastonite fibres are obtained from natural mineral deposits found in large quantities in China, Australia, USA, Canada, Greece and in smaller amounts in other countries. Wollastonite has two main components: 49–52% of SiO_2 and 44–49% of CaO, with traces of different metal ions: aluminium, iron, magnesium, potassium and sodium. By crushing the rock fibres are readily obtained with diameters from 10 to 40 μm and aspect ratios (length to diameter ℓ/d) from 15 to 200. Wollastonite fibres are relatively inexpensive, are not harmful to human health and may be considered partly as a substitute for asbestos fibres.

Tests on wollastonite fibres as micro-reinforcement for cement pastes and mortars have proved their applicability (Low and Beaudoin 1993; Kucharska and Logoń 2005). Rheological properties of the fresh mix are not significantly modified (more viscosity than the yield value), but the latter authors obtained an appreciable increase of compressive strength and bending strength when the fibres were used as 14% replacement of Portland cement, with silica fume as an addition. Multiple cracking and high fracture toughness was observed. They also found a considerable increase in resistivity and impermeability, with a decrease in sorptivity. Under load the fibres were either broken or pulled out from the matrix.

Wollastonite fibres are used to reinforce plastics and to stabilize ceramics; their extensive application to cement-based composites may be forecast.

5.3 Glass fibres

There are several kinds of glass used to produce fibres and a few examples are presented in Table 5.1; their variety is large and potential users should always verify their actual availability, properties and prices. The fibres are produced in various forms:

- single fibres (filaments), continuous or chopped, of circular cross-section and diameter 5–10 μm; there are also tubular and rectangular cross-section fibres (USA);
- strands, bundles of single filaments;
- rovings, which are 1–60 bunches of single fibres; rovings may be coated with various polymeric materials;
- yarn of single fibres, continuous or discontinuous, also remnant shorter fibres may be used;
- woven fabrics, made of continuous or discontinuous fibres, also of rovings;
- non-woven mats of fibres, bound together mechanically or by gluing.

For reinforcement of concrete-like composites the short fibres are mostly used in lengths of about 25 mm chopped from rovings.

The first applications of glass fibres in concrete-like composites were published by Biryukovitch *et al.* (1965), who carried out tests on elements with high alumina cement. The paste made with that cement is characterized by pH from 11.8 up to 12.05 with total amount of alkalis from only 0.15% to 0.20% and during hydration calcium hydroxide does not appear. In the

Table 5.1 Main components of glass for glass fibres

Types of glass	Components							
	SiO_2	Al_2O_3	B_2O	CaO	Mg_2O	K_2O	Na_2O	ZrO_2
High alkali A glass, used for thermal and acoustic insulation in the form of wool, d = 15–30 μm	72.0	1.3	–	10.0	2.5	–	14.2	–
Low alkaline glass, used for filtration layers, as fibre reinforcement for plastics, and in textiles, d = 5–10 μm	65.0	4.0	5.5	14.0	3.0	8.0	0.5	–
Nonalkaline E-glass, used as electric insulation, as fibre reinforcement for plastics and gypsum	54.5–58.0	14.0–14.5	6.5–8.0	17.5–21.0	4.5	0.5	–	–
Zirkonian G 20 glass, obtained by Majumdar et al. in Build. Research Establ. in 1968	70.27	0.24	–	0.04	0.04	11.84	0.04	16.05
CemFIL glass, Pilkington, UK, 1970	63.0	0.8	–	5.6	–	–	14.8	16.7
CemFIL 2 glass, Pilkington, UK, 1979	60.0	0.7	–	4.7	–	0.3	14.2	18.0
NEG, Nippon Electric Glass	61.6	0.2	–	0.4	–	2.3	15.8	19.1
AR Fibre-Super (with 10% of oxides of rare earths)	56.4	–	–	–	–	0.9	15.3	16.9
Zhongyan ARG (with 5% of TiO_2)	60.0	–	–	–	–	–	–	15.0

Portland cement paste the pH is higher from 12.5 to 13.0 with a large amount of crystallized calcium hydroxide. These are the reasons why the glass fibres are not subjected to corrosion in high alumina cement paste. But the elements made of that kind of cement are exposed to dangerous conversion when in moist conditions and at higher temperature (cf. Section 4.1) and as a result they may lose a considerable part of their bearing capacity at any time during exploitation.

According to tests published by Biryukovitch *et al.* (1965) the composite elements made with a mixture of high alumina cement, granulated blast-furnace slag and gypsum, and reinforced with ordinary glass fibres, had invariable strength over a few years. These results were not confirmed by Majumdar and Nurse (1975) and they observed a decrease of strength in similar elements with fibres made of E-glass. That decrease was more important the more humid the storage conditions were for tested elements.

The apparent divergences in these two series of tests can be attributed to possible substantial differences in properties of fibres and matrices, and also in other conditions of testing.

According to a generally accepted conclusion the application of E-glass fibres is not advised in elements made of high alumina cement if they may be exposed to high humidity. That condition does not concern non-structural elements, whose strength is needed during transportation and construction only, for example, lost forms incorporated into structural elements and cladding plates.

In matrices based on Portland cement, the glass fibres are subjected to three different actions that appear in various proportions and with various importance. These are:

- alkaline aggression by hydroxyl ions from cement paste during its hydration;
- agglomeration of hydration products (CH) around filaments, decreasing their flexibility and increasing bond strength;
- densification of the interfaces around the filaments (Purnell *et al.* 2000, Bentur and Mindess 2006).

These processes gradually decrease the strength and fracture toughness of composite elements. The rate of the processes is a function of environmental conditions: humidity, temperature, and fibre and matrix properties; that is, fibre sensitivity to corrosion and pH of the matrix. The main problem consists of identifying appropriate ways of prevention, which are:

- modification of matrix by decreasing its alkalinity;
- covering the fibres with various kinds of protective coatings;
- increasing the fibre resistance against alkaline corrosion.

Sometimes different measures are used simultaneously.

After numerous tests by Majumdar (1975), Cohen and Diamond (1975) and other authors, it has been shown that fibres made of ordinary E-glass are subjected to corrosion in Portland cement matrices in the first weeks, particularly when stored in high humidity. In similar elements stored in dry air no appreciable decrease of strength was observed even after a few years.

It was expected that a considerable improvement may be obtained by the introduction of special alkali resistant (AR) fibres called CemFIL. The fibres were produced with large addition of zircon oxide ZrO_2 by Pilkington Brothers in the UK after a concept published by Majumdar and Ryder (1968). The fibres were chopped from roving into lengths of about 40 mm and their diameters varied from 12 µm to 14 µm.

The test results of elements reinforced with CemFIL fibres did not fully confirm the expectations. Some types of the fibres made of alkali-resistant (AR) glass also showed slow corrosion when elements were stored in water or in natural outdoor conditions. Thin mortar plates were tested in different conditions over the course of ten years and it appeared that with access of humidity only after six months, their strength decreased considerably. At the end of the ten-year period the strength of reinforced plates was already equal to that of non-reinforced mortar, which was attributed to advanced corrosion of fibres (Figure 5.1).

The mechanism of corrosion of fibres and some aspects of its influence on composite strength are complex and not entirely explained. The introduction of ZrO_2 has two different effects: alkali-resistance of fibres is increased, but their bond to the matrix is reduced. Low alkali-resistance of fibres and high environmental humidity increase bonding, but also accelerate their corrosion.

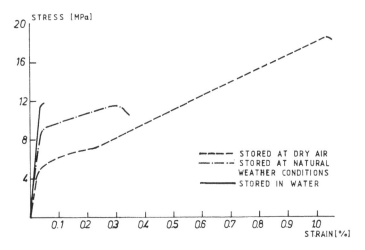

Figure 5.1 Tensile stress-strain curves for CemFIL glass fibre reinforced mortar elements stored during 5 years at different conditions. Fibre length 40 mm, 8.2% vol., after Majumdar (1980).

When fibres are more alkali-resistant and elements are stored in dry air, then a risk of poor bond strength appears and the elements may behave as a brittle non-reinforced matrix.

The quality of bonding was determined not only indirectly after composite strength, but also the surface of fibres taken out of elements was closely examined using SEM, and serious destruction was discovered.

Research led by Pilkington Brothers was directed at providing an additional coating on AR glass fibres and at the beginning of the 1980s a new kind of fibre called CemFIL 2 appeared. An extensive program of tests in various storage conditions was undertaken to verify their durability in long periods.

Tests reported by Majumdar (1980) were carried on specimens with blended cement composed of 80–85% of granulated blast-furnace slag, 10–15% of gypsum and a small amount of lime or Portland cement. That kind of cement ensured high resistance against most chemical agents and lower alkalinity, which enabled the application of glass fibres as reinforcement, but the mechanical strength at an age of 28 days was significantly lower than that of Portland cement specimens. Final results of the long-term tests with this cement were promising (Majumdar and Laws 1991).

In production of these fibres in the laboratories of Centre de Recherches of St. Gobain in Pont-à-Mousson (France), the detrimental influence of Portland cement paste on the durability of glass fibre reinforced composites was decreased. Using metakaolin as an admixture to the matrix and AR fibres in a composite material called 'Cemfilstar,' both the above-mentioned processes were significantly slowed down or even stopped (François-Brazier *et al.* 1991). Tests performed on specimens made of 'Cemfilstar' and subjected to accelerated ageing have shown that long-term performance may be expected (Glinicki *et al.* 1993). The producers of CemFIL® fibres consider their product as durable. Fibres are produced in rovings and strands, and also chopped into short sections.

The problem of the acceleration of normal ageing and therefore the ability to predict the long-term strength of GRC was discussed. The model proposed by Proctor was presented in terms of an acceleration factor; for example, that one day in water at 50°C is equivalent to approximately 100 days of weathering in UK climatic conditions. That approach was questioned by Purnell and Beddows (2005), who observed that using hot water may lead to severe mistakes in the estimation of the product's durability, because other strength loss mechanisms that are not significant at service temperature may be introduced.

Nippon Electric Glass (NEG) produces AR fibres also with a high percentage of zirconia (ZrO_2) and thus with enhanced resistance to alkalis. Two kinds of strands are offered: high integrity strands for conventional premix technology and water dispersible strands that are used in technologies developed for asbestos fibres (the Hatschek and Magnani process) ensuring the dispersion of single filaments in cement slurry. The properties of fibres are shown in Table 5.2. There are also other producers of AR glass fibres

Table 5.2 Dimensions and properties of selected AR glass fibres

Glass fibres	Diameter	Length	Tensile strength	Young's modulus	Density	Critical strain	Water uptake	Soften. temper.
	(µm)	*(mm)*	*(MPa)*	*(GPa)*	*(−)*	*(−)*	*%*	*°C*
CemFIL®	14	6,12,24	1700	72	2.68	0.024	< 0.1	860
NEG	13.5	6–25	1400	74	2.7	0.02		830
Zhongyan ARG	13	6–30	1400–1600	72.5	2.78			

and new types are still appearing on the market.

The attempts were also continued to decrease the alkalinity of cement matrices by various admixtures. Among others the fly ash decreases the 28 days strength of mortar and concrete because of lower filtration of water in the pore system and slower hydration of cement, but in longer periods that difference is gradually disappearing. The admixtures of polymer dispersions were not helpful in that respect, because their influence is weak in high humidity when alkali corrosion of fibres is the most dangerous.

The E-glass fibres in gypsum matrices are not subjected to alkaline corrosion and are applied in partition walls, internal cladding, etc., (cf. Sections 3.2 and 14.5).

Discussion on the efficiency of various methods was continued with an analysis of images from thin-sections and it has been suggested that the densification of the interfacial space around glass fibre strands is not the main reason for the loss of mechanical properties of glass fibre reinforced cements. Also, the fibre corrosion was not observed by Purnell *et al.* (2000), who concluded that the enlargement of subcritical flaws in the filament, induced during composite manufacture, should be responsible for the decrease in composite strength. These conclusions are in some contradiction with general opinion and should be confirmed.

In the present state of technology the use of AR glass fibres still requires some caution. Probably only combinations of AR glass fibres with special modifications of Portland cement matrices should be used for the elements if their long-term serviceability is required. Ordinary E-glass fibres are limited to the elements where reinforcement is needed only during the first short period of construction (Glinicki 1999). Nevertheless, the application of glass fibres of various kinds in concrete-like composites is very large and is still developing in all non-structural elements and with matrices other than Portland cement (ENV 1170-8:1996).

5.4 Steel fibres

Steel fibre reinforced concretes (SFRC) were originally developed in the USA, invented by J.P. Romualdi as US Patent 3 429 094 in 1969 and produced

Table 5.3 Examples of steel fibres for reinforcement of cement-based matrices

No.	Description	Dimensions (mm)		Tensile strength (MPa)
		diameter	length	
1	Straight, plain, round ITFR (Poland)	0.40	40	600–700
2	Straight, plain, round Type 3	0.30	25	960
	Trefil-Arbed Type 2	0.38	25	
3	Straight, plain, round stainless steel	0.25–0.64	13–60	
	National Standard (USA)	0.25	13, 25, 60	
		0.30	25	
		0.38	13, 25, 60	
		0.40	25	
		0.64	13, 25, 60	
4	Straight, plain, round Svabet kamma	0.25–0.50	25	780
	Scanovator A/B (Sweden)	0.60	60	
5	Straight, plain fibres glued together in bundles Dramix OL (Belgium)	0.25–0.50	6–30	
6	Straight, plain, square or rectangular cross-section Sumitomo Metal Ind (Japan)	0.80	30	
7	Straight, indented in two perpendicular	0.25	13, 25, 30	1000
	planes Duoform National Standard (USA)	0.38	13, 25, 60	
		0.64	13, 25, 60	
8	Straight, plain, with enlarged ends EE, Australian Wire Ind.		18	
9	Straight, plain, with enlarged ends Tibo	0.80	50	1150
10	Waved, made with plain wire Johnson and Nephew (UK)	0.40–0.60	20–40	
11	Waved on the ends Trefil-Arbed (7+11+7) = 25 mm	0.25–0.50	25	
12	Dramix ZL Bekaert Wire Corp. (Belgium) Plain with hooks at the ends, stainless			
	steel glued together in bundles,	0.38	30	1400
	separated, normal and higher quality	0.38	25	1400
	separated, normal and higher quality	0.38	30	1400
13	Melt extracted fibres, also stainless steel	$0.3{\times}1.0^{1}$	10–50	
	irregular shape Battelle Corp. Ohio (USA)	$0.4^{2}; 0.58^{2}$	35;38	
14	Melt overflow stainless fibres Microtex	0.025x	10–50	1500–
	E= 150 GPa, ε_u = 0.15, Fibre Technology	0.150^{2}		2000
	Ltd (UK)			
15	Straight, rough, machined, irregular,	$0.8{\times}2.0^{1}$	16–32	1630–
	Harex (Japan)	$0.15{\times}4.0^{1}$		2100
16	Fibercon	$0.25{\times}0.56$	60	

1 Only the smallest and biggest dimensions of a series of fibres produced are given here.
2 Characteristic dimensions of irregular cross-sections.

initially by the Batelle Development Corporation of Columbus, Ohio, followed by several other producers of chopped steel fibres.

There are now many different forms of short steel fibres used as reinforcement of concrete-like composites. Several types are presented in Table 5.3 as examples of different shapes, dimensions and producing companies. Some fibres have already disappeared from practice and have been replaced by new ones. It is always necessary, as for glass fibres, to verify the actual situation (price and availability) on the market. A few examples of steel-fibres are shown in Figure 5.2.

The fibres are produced by various methods and only a few are mentioned here:

a   chopped from cold drawn wire of circular cross-section, mostly indented (Duoform), waved (Johnson and Nephew), with hooks (Bekaert) or enlargements at the ends (Tibo);
b   cut out from stripes of thin plates, of square or rectangular cross-section, often twisted along their longitudinal axes during cutting, some kinds also with enlargements at the ends (EE);
c   machined, of rough surface and varied cross-section related to the technology of machining (Harex);
d   obtained from molten metal, with rough surface (Johnson and Nephew).

Fibres of type (a) were applied since the beginning of SFRC, but their unit cost is relatively high due to complicated technology of cold drawing and cutting. Additional operations are required to increase their bonding by surface indentation or wavings. Plain, straight fibres are rarely used because of their low efficiency due to poor bonding. Fibres with hooks that are glued together in bundles to facilitate dispersion are probably the most universally used.

Fibres of type (b) are less expensive and have better bond strength without special operations.

Fibres of type (c) were introduced later and their application quickly developed. They are less expensive than fibres chopped from wires and their uniform distribution in the fresh matrix is easier. Their shape ensures better adherence to the matrix.

Fibres of type (d) are also inexpensive and have increased bond strength. However, there are few results available of tests or practical applications of these fibres and at present they have apparently disappeared from the market.

In comparison to short fibres, fibres of 100–200 mm in length are rarely used. A few experimental applications are known, together with the technologies called SIFCON and SIMCON (cf. Section 13.5).

The surface of steel fibres may be covered by a thin coating of other metals like copper or of epoxy resin to increase their resistance against corrosion – if that feature is needed.

The main characteristics of fibres which determine their efficiency as reinforcement for brittle matrices are:

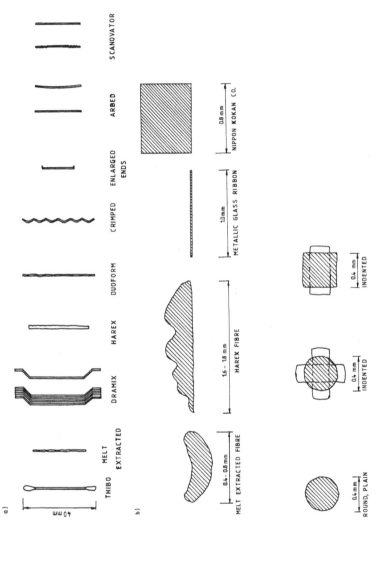

Figure 5.2 Examples of a few kinds of short steel fibres:
(*a*) shapes
(*b*) cross-sections.

- aspect ratio, which means length to diameter ratio ℓ/d;
- quality of surface and eventual anchorages at the ends;
- mechanical properties of the material, which means tensile strength and ductility (EN 14651:2005).

The efficiency of fibre reinforcement is dependent very much upon distribution and orientation of fibres in the matrix. The fibres form an internal structure and various kinds of these structures are described in Section 6.6.

The anchorage of steel fibres in the matrix is indispensable for their efficiency. Because of weak chemical adherence, the load is transferred from matrix to fibres mostly by mechanical bond. The shape of the fibre surface and its cleanness as well as the shape of a fibre itself and the quality of the matrix are decisive. The fibre-matrix bond was tested by many authors (e.g. Maage 1977; Pinchin and Tabor 1978; Burakiewicz 1978; Potrzebowski 1986; Baggott and Abdel-Monem 1992; Banthia *et al.* 1992).

If the bond along a fibre is not sufficient, and usually for plane straight fibres a correct anchorage would require increased length, then hooks or enlarged ends are necessary. The length of a fibre should approximately satisfy a condition $\ell/d \leq 100$, otherwise the distribution of fibres is difficult. The fibres with improved bond may be much shorter and then their appropriate distribution in the fresh mix is easier.

The unit price of steel fibres is an important factor deciding upon their application, because they are always the most expensive component of the mix. That is why many producers are looking for new kinds of cheaper fibres, and melt extract or machined fibres were proposed.

The fibres are produced of various kinds of steel, but usually they are of low carbon and medium strength steel. Their tensile strength is equal between 500 and 1000 MPa and only rarely higher strength steel up to 2400 MPa is used. For refractory elements, such as walls in blast furnaces, the fibres made of stainless steel ensure better durability. Young's modulus of all kinds of steel is close to 210 GPa and good ductility is an advantage.

The forces appearing in a fibre under load are low because of their short length and relatively weak bond to Portland cement paste. As a result, the fibres are never ruptured when cracks are propagating and elements are subjected to failure. Fibres which cross a crack are pulled out of the matrix and by that way the crack opening and its further propagation are controlled. Therefore, the use of high strength steel for fibres is usually not justified. The situation may be different in polymer-modified cement-based composites or in other kinds of matrices where bonding is considerably improved.

The total amount of fibres in a composite material is defined as a fraction of its volume rather than mass. It is believed that the upper limit for fibre fraction is 3% for the normal technology of premix for ordinary steel-fibre cements. Generally, the applied volumes vary between 0.5% and 2.0% and these limits are based on the following arguments:

- The workability of fresh mix decreases rapidly when higher fractions of fibres are added and the porosity, due to entrapped air voids, is increased. Even with intensive vibration it is difficult to place and to compact correctly the fresh mix in the forms when too many fibres are added; only external vibration is advised.
- The fibres have a tendency to form 'balls' in the fresh mix and with a higher fraction it is very difficult to distribute them properly.
- The total price of composite increases considerably with high fibre fractions.
- The optimal efficiency of fibres depends on several factors, but in most cases it corresponds to volumes between 1% and 2%.

In certain applications, for reasons of economy, lower volumes of steel fibres are sometimes used; for example, 0.2% or 0.3% in industrial floors, but frequent cases occurred when such reinforcement was not sufficient and cracks appeared. In such cases expensive repair works are necessary and litigations between interested parties usually follow (EN 14721:2000, ASTM A820).

The corrosion of steel fibres in cement matrix was a subject of concern in the first period of application of SFRC, but later it was proved that only short sections 2–3 mm long that stretch out of the concrete surface corrode and disappear in a short time. Cracked specimens maintained during 18 months in four different environments proved that no decrease of strength could be found in the non-cracked concrete and rusty fibres were found only in the immediate surface in the carbonated zone (Nemeeger *et al.* 2003). The test carried on by Granju and Balouch (2005) have confirmed that even fibres which bridged open cracks of 0.5 mm wide and maintained a year in a marine environment exhibited only minor traces of corrosion that did not progress inside the concrete matrix.

Special technologies were tested to introduce many more fibres, even up to 20% by volume. Fibres were put into the forms and then soaked with fluid cement slurry. The technology, called SIFCON (slurry infiltrated concrete), was applied in the repair works of concrete structures (Lankard 1985; Naaman and Homrich 1989; Reinhardt and Fritz 1989). Steel-wool mats composed of thin continuous wires (SIMCON) are also used as cement paste reinforcement. Wires with irregular and high aspect ratio up to 500 are produced by shaving or slitting and shredding, from thin sheets or foils of low carbon steel. Mats about 6 mm thick of variable density from 0.85 to 1.75 kg/m^3 are used, keeping the volume content of fibres below 4% (cf. Section 13.5). Composite plates are produced by the infiltration of mats with cement slurry and hardening under constant pressure. Obtained plates have high toughness and strength comparable to those of asbestos cement products (Bentur 1989).

5.5 Metallic Glass Ribbons

In 1986, a special type of reinforcement for concrete-like composites was proposed in France in the form of thin ribbons made of metallic glass. The ribbons are produced by rapid quenching of a liquid alloy, which is a composition of $Fe_{75}Cr_5P_8C_{10}Si_2$. The ribbons are 20–30 µm thick, 1–2 mm wide and are produced in variable lengths from 15 mm to 60 mm. Under direct tension a ribbon behaves as nearly linear elastic material without plastic deformations.

Because of the amorphous structure of the material, the ribbons have excellent corrosion resistance. The pull-out tests proved their good adhesion to cement-based matrices. When crossing a crack, the ribbons are gradually pulled out like ordinary metal fibres. The tests of composite elements reinforced with ribbons have shown their higher efficiency than other kinds of steel fibres, with relatively low volume fraction required (De Guillebon and Sohm 1986; Rossi *et al.* 1989; Kasperkiewicz and Skarendahl 1988), but their development in practical application has not been largely developed.

5.6 Synthetic (polymeric) fibres

In this section, man-made fibres, other than metallic, glass and natural vegetal, are considered. Carbon fibres are described in Section 5.7.

Polymeric fibres are produced and supplied in various forms of separate monofilaments, bundled filaments or multifilament fibrillated strands (cf. Table 2.2).

Polypropylene fibres have been produced since 1954 and since 1965 were applied as concrete reinforcement. Initially, the plain and straight polypropylene fibres were used, but their bond to a cement matrix was insufficient because mechanical interlocking was weak and the chemical bond did not exist at all. This is the reason that single filaments with fibrillated surface and twisted strands are now used.

Considerable progress has been achieved and large research programmes were started in the 1970s directed at replacement of asbestos fibres (Krenchel and Jensen 1980; Naaman *et al.* 1984). Special types of fibres called 'krenit' invented by H. Krenchel were produced by splitting extruded tapes, which were prestretched and heat treated. The Young's modulus of these fibres is equal E =18 GPa and tensile strength f_t = 600 MPa (Krenchel and Shah 1985). Ultimate elongation at rupture may reach 5–8%. The bond strength to cement matrix is increased by special electrical surface treatment and thanks to fine fibrils protruding from fibre edges after cutting. The fibres have rectangular cross-section 20×100 µm or 30×200 µm and are chopped into lengths from 3 mm to 20 mm or longer. The cement mortar reinforced with these fibres may reach maximum elongation equal to $0.4\% = 4000.10^{-6}$ as compared to approximately $0.02\% = 200.10^{-6}$ for plain mortar.

There are several kinds of polypropylene fibres, for example, very thin of 18 µm in diameter and 18 mm long known as 'crackstop' and 'cemfiber.'

Other trademarks of polypropylene fibres – fibermesh, naplon, forta and others – are available on the markets and new proposals appear every year. Twisted polypropylene fibres and 3D fabric are available with the mechanical bond considerably increased (Ramakrishnan 1987).

Polypropylene fibres have a low melting point (at approximately 200°C) and lose strength at higher temperature, which is practically without importance for their main application against plastic shrinkage cracking. Chopped fibrillated fibres of length 25–75 mm are used to reinforce mortar and concrete. They are traditionally mixed with fresh mortar and used for various thin-walled, pre-cast elements for buildings and other purposes (flower boxes, tanks, boats, etc.). These fibres are also used in shotcrete, mainly for repair works, and their different applications are quickly developing (Banthia and Gupta 2006).

The fibres of low Young's modulus do not represent a valuable reinforcement in comparison with steel and glass fibres as it concerns the increase of composite strength of hardened matrix. Their role is to control initial cracking in early-age concrete due to plastic shrinkage and thermal deformations of cement paste. However, certain positive influence is also observed on the final strength of hardened concrete because of the control of initial cracking. This is particularly important for elements subjected to impact loading.

It has been proved by many applications that even quite a low percentage addition of $0.6–0.9$ kg/m^3 of polypropylene fibres, when distributed evenly in the bulk of the fresh mix, has a significant effect on early-age cracking and the application of polypropylene fibres is very large; for example, in industrial floors made with steel fibres.

Polyacrylonitrile (PAN) fibres under various patented names are of diameter of 13–100 μm with length 2–60 mm and are used up to 3.5% vol, with the best results for 30 35 kg/m^3. Fibre/matrix bond is good and is estimated equal approximately 4 N/mm2. Reinforcement with PAN fibres reduces shrinkage cracking, increases ductility of the matrix and Modulus of Rupture (MOR) for elements subjected to bending. Its application ranges from non-structural elements like tunnel and building cladding to main reinforcement of edge beams of bridge decks and secondary reinforcement of industrial floors (Wörner and Miller 1992).

Polyethylene and polyolefin fibres are used mostly as short single filaments, with deformed surface to increase bond. Because of higher Young's modulus, the scope of application is not only control of shrinkage cracking (Banthia and Yan 2000), but also reinforcement of composite elements (Ramakrishnan 1996).

Polyvinyl alcohol (PVA) fibres have higher Young's modulus (cf. Table 2.2). As a replacement for asbestos fibres, they are used in single filaments with diameter of 14 μm and length of 4–6 mm. PVA fibres are also available in diameters of 0.2; 0.4 and 0.6 mm, and 30 mm long, also in bundles.

Acrylic and aramid fibres are rarely applied as dispersed reinforcement. The latter, in the form of continuous multifilament cables and rods are used as non-metallic components of Fibre Reinforced Plastics (FRP).

5.7 Carbon fibres

Carbon fibres are obtained from different organic fibres (precursors) by pyrolysis, which consists of decomposition into smaller molecules at high temperature. The process of fabrication of carbon fibres from special PAN fibres includes two steps: oxidative stabilization at low temperature and carbonization at high temperature in an inert atmosphere. Due to the high cost of raw materials (e.g. PAN fibres) and of this production process, carbon fibres are still expensive. Carbon fibres may be also produced from crude oil deposits like pitches or asphalts. Three main groups of carbon fibres are considered as possible composite materials reinforcement:

1   Graphite fibres with 99% carbon content after additional treatment, with high strength f_t = 2300–3000 MPa and E = 400–700 GPa; for example, Thornel produced by Union Carbide and PAN fibres produced from precursors of polyacrylonitrile.
2   Carbon fibres produced with the same precursors, but without complete graphitisation; their properties are f_t = 900–1200 MPa, E = 30–50 GPa.
3   Carbon fibres produced from less expensive materials like pitch with much lower properties: f_t = 600–900 MPa, E = 30–40 GPa, (cf. Table 2.2).

The last group of carbon fibres is considered as a possible reinforcement for cement-based composites for general application because of their reasonable price (Ohama *et al.* 1985). More expensive carbon fibres are used only for particular purposes like reinforcement in a very corrosive environment or in structures in which the application of any metallic elements is excluded. Because of their dimensions they are considered to be microfibres.

Carbon fibres have high chemical resistance and may be applied in elements exposed to elevated temperature and mechanical wear. Their small dimensions and relatively high stiffness enable the control of early microcracking, as shown by Brandt and Glinicki (1992). Extensive tests of composite specimens with pitch-based carbon fibres used as single filaments 3 mm long have shown high strength and fracture toughness (Kucharska and Brandt 1995; Brandt and Kucharska 1996). Acoustic emission records were used to distinguish the influence of carbon and ordinary steel fibres: the first, reduced microcracking at the initial stage of deformation, and the second controlled larger cracks and increased overall strength of composite elements subjected to bending. Because of low density, the correct distribution of carbon fibres in a fresh cement matrix requires special attention; for example, the application of silica fume and superplasticizers (Katz and Bentur 1992). Carbon fibres are used together with other kinds of fibres as so-called hybrid reinforcement (Kucharska *et al.* 2002).

5.8 Natural vegetal fibres

The classification of organic fibres was proposed by Gram *et al.* (1984) as shown in Figure 5.3. Here, only natural vegetal fibres are considered in more detail because of their importance and application as reinforcement for brittle matrices. Beside their low cost and availability for local applications, natural fibres are also efficient as reinforcement. Elements produced show increased tensile strength and post-cracking resistance, sufficient fatigue strength and energy absorption at fracture. Examples of stress-strain relation are shown in Figures 5.4 and 5.5; the influence of volume fraction on composite strength is also presented.

Unskilled workmen using simple equipment may produce various structural elements for roofing of low-cost houses and building non-structural elements with concrete reinforced with natural fibres.

Several problems arise connected to the fibre corrosion in highly alkaline cement matrix and the influence of high outdoor humidity. The durability of the fibres and of composite materials is also endangered by biological attack (e.g. bacteria, fungus). The fibres' instability appears in high humidity and flow of moisture. All these factors are particularly dangerous in tropical climates. Because of the alternations of wet and dry periods in tropical and subtropical conditions, cracking of elements may cause rapid degradation by corrosion. The process of embrittlement due to fibre corrosion is dangerous for the strength of elements. A short review of the application of vegetal fibres in cement composites is given by Brandt (1987). The main countermeasures against poor durability are impregnation of fibres by various chemical agents, resin coatings of fibres, reduction of the matrix alkalinity and improvement of impermeability by sealing the internal pore systems in the matrix. Efficient impregnation agents are mentioned in Andonian *et al.* (1979): photochemicals and calcium stearate, boric acid and polyvinyl dichloride, sodium nitrite and chromium stearate. Formine and stearic acid are considered as the best pore blocking and water repellent agents. The alkalinity of the environment may

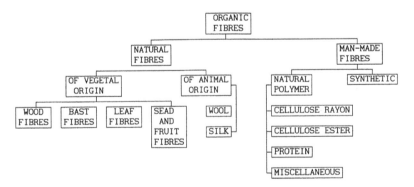

Figure 5.3 Classification of organic fibres, after Gram *et al.* (1984).

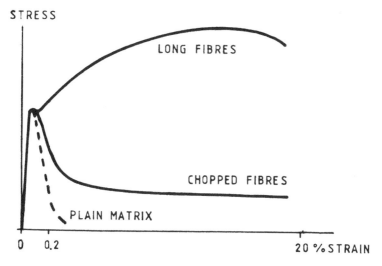

Figure 5.4 Examples of stress-strain curves for plain mortar elements and reinforced with sisal fibres, after Gram *et al.* (1984).

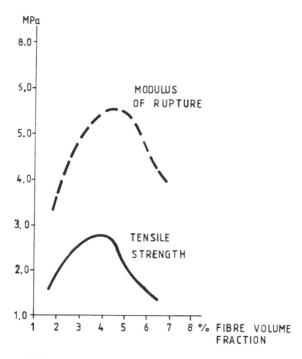

Figure 5.5 Effect of fibre content on tensile strength and modulus of rupture of composite elements, after Das Gupta *et al.* (1978).

be reduced by replacing 30–50% of Portland cement with fly ash or other pozzolan. All these measures may be used conjointly and considerable research effort is spent to ensure that elements reinforced with vegetal fibres are durable in local conditions.

Another problem is the fibre-matrix bond, which is based mostly on mechanical interlocking. For example, the surface of bamboo fibres should be subjected to special treatment like sand blow to increase its roughness.

The amount of fibres in a cement matrix may vary from 0.1% to 0.9% of mass of cement (Rafiqul Islam and Khorshed Alam 1987; Singh 1987). The dispersion of fibre properties is usually rather large because of natural variations in the plant population and simple production techniques also increase the variability of composite material properties.

Vegetable fibres are also used to replace asbestos fibres, which are expensive and dangerous to health. Coconut fibres were tested for that purpose and their strength and deformability, as well as thermal and acoustic properties, and were proved comparable with those of asbestos fibres (Paramasivam *et al.* 1984). Similar tests on specimens reinforced with flax fibres from New Zealand and Australia also showed their ability to replace asbestos in thin cement sheets (Coutts 1983).

Vegetable fibres are used mostly in developing subtropical and tropical countries in Africa and South-East Asia as reinforcement for concrete elements for housing. The application of cheap and locally available fibres may help considerably in the building of low-cost houses (Nilsson 1975).

Wood fibres are produced in the form of chips, which is usually a waste material in the wood industry. Wood chips mixed with cement paste have been used since the 1920s for the production of sheets applied for thermal insulation in housing. The chips are subjected to chemical pre-treatment to avoid any disturbance of cement hydration by organic acids. The application of wood-origin fibres as a reinforcement for minor structural elements has been developing at a local level.

Bast fibres are obtained from a few kinds of plants, for example, bamboo, hemp, flax, jute and ramie. The fibres are longer, stronger and stiffer than other vegetal fibres; jute fibres, for example, may be 3.0 m in length. For reinforcement of brittle matrices the jute fibres are chopped for sections of 12–50 mm. Bamboo fibres have low Young's modulus and tend to be used in the form of woven meshes.

Leaf fibres are mainly obtained from agave plants and are called sizal. Sizal is planted on an industrial scale in a few countries – the most important producers are Indonesia, Tanzania and Haiti. Sizal fibres are chopped or used as continuous fibres up to 1.5 m in length for making non-woven mats. Their maximum strain is 2–4%. Sizal fibres are also used as twine and spun with short chopped fibres and a small amount of steel fibres. Such a hybrid reinforcement has proved to be cheap and efficient (Mwamila 1985).

Seed and fruit fibres are limited mostly to coconut coir if it concerns application as reinforcement. Fibres are usually considered as waste in the production of copra from the coconuts. The fibres are extracted from the space between the external shell and the seed inside. Maximum length of the fibres is about 300 mm, with a maximum strain of around 30%. The fibres are also used to produce ropes and mats.

Other plants with fibres applicable for reinforcement are sugarcane bagasse, akwara, elephant grass, water reed, plantain and musamba (Aziz *et al.* 1981).

Cellulose fibres are obtained from softwood and hardwood. Softwood fibres are 30–45 μm in diameter and 3–7 mm long. Hardwood fibres are thinner (10–20 μm) and shorter (1–2 mm). Their tensile strength is by an order greater than that of the wood; for example, lumber wood without macro-defects may have tensile strength equal to 70 MPa and a single fibre, 700 MPa. The fibres are extracted from wooden pulp by different chemical and mechanical processes.

Cellulose fibres are used in the production of flat and corrugated thin cement sheets, pipes and other elements. Fibre volume content varies between 6% and 10%. Production is similar to the classic Hatschek process developed for asbestos fibre cement elements. The final properties of the product depend upon composition and production technique. At the fibre content of about 8% by volume, the maximum flexural strength was obtained according to Soroushian and Marikunte (1992), but up to 14% was required for maximum flexural toughness. The addition of fly ash and SF increases the flexural strength. Thin sheet elements reinforced with cellulose fibres exhibit considerably increased strength with respect to non-reinforced matrix and great flexural toughness. Flexural behaviour is compared with other composites in Figure 5.6.

Following tests in Australia, it has been found that at fracture in tension the cellulose fibres are both pulled-out and broken and the relation between specific work of fracture and fibre content has been established (Figure 5.7). In Nordic, European and American countries with large timber industries the application of cellulose fibres to reinforce cement mortar or paste may be interesting, but has not been developed yet, probably because other kinds of fibres are sufficiently inexpensive.

5.9 Textile reinforcement

Textile reinforcement is not defined in an unambiguous way, but it is usually admitted that all non-metallic fibres, fabrics and mats are considered as such. The general classification of textile fibres is shown in Figure 5.8 and some of them are used for cement-based materials. Several kinds of textile fibres, short chopped or long continuous ones, are described in more detail in other chapters and only their common features are considered here.

Besides fibres, reinforcement with textiles can take the form of platelets, mats and woven or non-woven fabrics. Several interesting data on textile reinforcement and its application may be found in the *Proceedings of the Symposia* in Lyon (Hamelin and Verchery 1990, 1992).

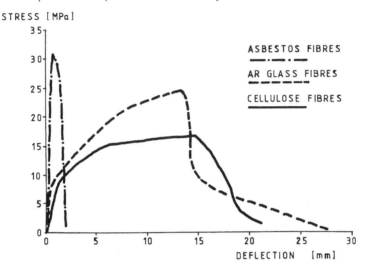

Figure 5.6 Comparison of flexural behaviour of cellulose fibre cements with glass fibre cements and asbestos cements, after Soroushian and Marikunte (1992).

The main mechanical problems in the brittle matrix composites with textile reinforcement concern the bond between cement paste and fibres and the durability of the fibres (Brandt 1990). These problems are interrelated. If the alkalinity of the cement paste is not corrosive for the material of the fibres, then the chemical bond does not exist and adherence should be ensured other ways; for example by:

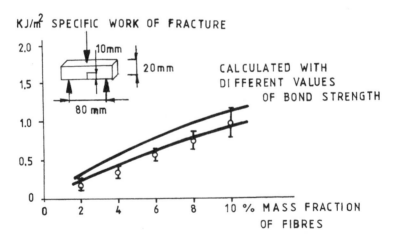

Figure 5.7 Variation of specific work of fracture with cellulose fibre mass fraction in composite elements subjected to bending, after Andonian *et al.* (1979).

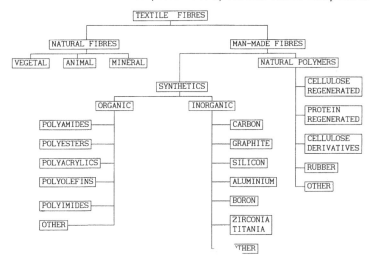

Figure 5.8 Classification of fibres used in textile reinforcement, Kelly (1989).

- increasing the length of a single fibre;
- using the fabrics, mats or meshes, where purely mechanical bonding is concentrated at the nodes and enhanced by entanglements;
- using fibres with additional hairs (fibrillated fibres).

The nodes not only provide a better bond, but may also induce appreciable stress concentrations and act as crack initiators; it is observed that the crack pattern reflects the distribution of fibres.

Textile reinforcement with natural vegetal fibres or with glass fibres is particularly vulnerable to alkaline corrosion (cf. Sections 5.3 and 5.8).

Textile fibres are useful as hybrid reinforcement, which means two kinds of fibres; for example, steel and polymeric ones, when mixed together, can control cracking at different stages: hardening with early shrinkage, drying shrinkage, excessive strain due to external loading, etc.

The design of composite materials with textile reinforcement is executed mostly by trial and error and using previous experience or test results. Sound and reliable methods for design and optimization are not available, partly because of highly non-linear behaviour.

Textile reinforcement is applied for several types of cement mortar or concrete products:

- shingles for roofing with cement blended with fly ash, characterized by good durability when exposed to climatic actions;
- inflated shell roofs or entire buildings;
- folded core sandwich panels, sometimes of complicated shapes, used for roofing;

- liners for lightweight precast concrete elements or thin plates for building facades.

The cost of composite materials with textile reinforcement depends not only on the cost of reinforcement itself, but also on the selection of adequate technology to a large measure. The overall cost should be calculated taking into account the durability of the product during exploitation. Some textile reinforcements, like mineral wool, are particularly cheap and are used to reinforce structural elements for low-cost buildings. However, even in such a case, the cost of reinforcement is comparable with that of the matrix. The production costs depend upon whether the reinforcement is carefully positioned in the forms or situated in a random way without much care. In the former case, considerable gains can be obtained by optimum positioning and orientation of reinforcement. The latter case is more appropriate for simple production of building elements by non-trained workers and without the use of expensive equipment.

Textile reinforcement, called geotextiles, are used extensively in geotechnics for earth constructions: walls, footings, road and runway foundations, etc. Asphalt matrices are also reinforced with polymer fabrics.

References

Andonian, R., Mai, Y. W., Cotterell, B. (1979) 'Strength and fracture properties of cellulose fibre reinforced cement composites', *International Journal of Cement Composites*, 1(4): 151–8.

Aziz, M.A., Paramasivam, P. and Lee, S.L. (1981) 'Prospects for natural fibre reinforced concrete in construction', *International Journal of Cement Composites and Lightweight Concrete*, 3(2): pp. 123–132

Baggott, R., Abdel-Monem, A. E. S. (1992) 'Aspects of bond in high volume fraction steel fibre reinforced calcium silicates', in: Proc. RILEM Int. Workshop *High Performance Fiber Reinforced Cement Composites*, H. W. Reinhardt and A. E. Naaman eds, Mainz, 23–26 June 1991, Chapman and Hall/Spon, pp. 444–55.

Banthia, N., Trottier, J. F., Pigeon, M., Krishnadev, M. R. (1992) 'Deformed steel fiber pull-out: material characteristics and metallurgical processes', in: Proc. RILEM Int. Workshop *'High Performance Fiber Reinforced Cement Composites'*, H. W. Reinhardt and A. E. Naaman eds, Mainz, 23–26 June 1991, Chapman and Hall/Spon, pp. 456–66.

Banthia, N., Gupta, R. (2006) 'Influence of polypropylene fiber geometry on plastic shrinkage cracking in concrete', *Cement and Concrete Research*, 36: 1263–7.

Banthia, N., Yan, C. (2000) 'Shrinkage cracking in polyolefin fiber reinforced concrete', *American Concrete Institute Materials Journal*, 97(4): 432–7.

Bentur, A. (1989) 'Properties and reinforcing mechanisms in steel wool reinforced cement', in Int. Symp. *Fibre Reinforced Cements and Concretes. Recent Developments*, Cardiff, 18–20 September 1989, R. N. Swamy and B. Barr eds, London: Elsevier Applied Science.

Bentur, A., Mindess, S. (2006) *Fibre Reinforced Cementitious Composites*, 2nd ed. London: Taylor & Francis.

Biryukovich, K. L, Biryukovich, Yu. L., Biryukovich, D. L. (1965) *Glass-fibre-Reinforced Cement*. Budivelnik, Kiev, CERA Translation, no 12.

Brandt, A. M. (1987) 'Present trends in the mechanics of cement based fibre reinforced composites', *Construction and Building Materials*, 1(1): 28–39.
——.(1990) 'Cement-based composite materials with textile reinforcement', in: Proc. Int. Symp. *Textile Composites in Building Construction*, Lyon, Ed. Pluralis, part 1, pp. 37–43.
Brandt, A. M., Glinicki, M. A. (1992) 'Flexural behaviour of concrete elements reinforced with carbon fibres', in: Proc. RILEM Int. Workshop *High Performance Fiber Reinforced Cement Composites*, H. W. Reinhardt and A. E. Naaman eds, Mainz, 23–26 June 1991, Chapman and Hall/Spon, pp. 288–99.
Brandt, A. M., Kucharska, L. (1996) 'Pitch-based carbon fibre reinforced concretes', in: *Proc. 4th Materials Engineering Conf.*, ASCE, Washington, 10–14 November, pp. 271–80.
Burakiewicz, A. (1978) 'Testing of fibre bond strength in cement matrix', in: Proc. RILEM Symp. *Testing and Test Methods of Fibre Cement Composites*, R. N. Swamy ed., Sheffield: The Construction Press, Lancaster, pp. 355–65.
Cohen, E. B., Diamond, S. (1975) 'Validity of flexural strength reduction as an indication of alkali attack on glass in fibre-reinforced cement composites', Proc. RILEM Symp. *Fibre-reinforced Cement and Concrete*, Sheffield, pp. 315–25.
Coutts, R. S. P. (1983) 'Flax fibres as a reinforcement in cement mortars', *International Journal of Cement Composites and Lightweight Concrete*, 5(4): 257–62.
Das Gupta, N. C., Paramasivam, P., Lee, S. L. (1978) 'Mechanical properties of coir reinforced cement paste composites', *Housing Science*, 2(5): 391–406 (London: Pergamon Press).
De Guillebon, B., Sohm, J. M. (1986) 'Metallic glass ribbons – a new fibre for concrete reinforcement', Proc. RILEM Symp. *Developments in Fibre Reinforced Cement and Concrete*, University of Sheffield.
François-Brazier, J. Soukatchoff, P., Thiery, J. and Vautrin, A. (1991) 'Comparative study of the mechanical damage and durability of glass cement composites', in Proc. Int. Symp. *Brittle Matrix Composites 3*, Warsaw, September 1991, A. M. Brandt and I. H. Marshall eds, London and New York: Elsevier Applied Science, pp. 278–89.
Gram, H. E., Persson, H., Skarendahl, Å. (1984) *Natural Fibre Concrete*, Stockholm: Swedish Agency for Research Cooperation with Developing Countries.
Granju, J.-L., Balouch, S. U. (2005) 'Corrosion of steel fibre reinforced concrete from the cracks', *Cement and Concrete Research*, 35: 572–7.
Glinicki, M. A. (1999) *Brittleness Mechanisms and the Durability of Glass Fibre Reinforced Cement Composites* (in Polish). Inst. of Fund. Techn. Res., Report 11.
Glinicki, M. A., Vautrin, A., Soukatchoff, P., François-Brazier, J. (1993) 'Impact performance of glass fibre reinforced cement plates subjected to accelerated ageing', in *Proc. of 9th Biennial Congress of The Glass fibre Reinforced Cement Association*, Copenhagen, pp. 1/1/I–1/1/X.
Hamelin, P., Verchery, G. eds (1990) Proc. Int. Symp. *Textile Composites in Building Construction*, Lyon, 16–18 July, Editions Pluralis, parts 1, 2 and 3.
——.(1992) Proc. 2nd Int. Symp. *Textile Composites in Building Construction*, Lyon, 23–25 June, Editions Pluralis, parts 1 and 2.
Hannant, D. J. (1978) *Fibre Cements and Fibre Concretes*, Chichester: J. Wiley.
Kasperkiewicz, J., Skarendahl, Å. (1988) 'Fracture resistance evaluation of steel fibre concrete', in: Proc. Int. Symp. *Brittle Matrix Composites 2*, A. M. Brandt, I. H. Marshall eds, Jabłonna, London: Elsevier Applied Science, pp. 619–28.
Katz, A., Bentur, A. (1992) 'High performance fibres in high strength cementitious matrices', in: Proc. RILEM Int. Workshop *High Performance Fiber Reinforced Cement Composites*, H. W. Reinhardt and A. E. Naaman eds, Mainz, 23–26 June 1991, Spon/Chapman and Hall, pp. 237–47.

Kelly, A. ed. (1989) 'Fibers and Textiles: An overview', in: *Concise Encyclopedia of Composite Materials*, Oxford: Pergamon Press, pp. 92–8.

Krenchel, H., Hejgaard, O. (1975) 'Can asbestos be completely replaced one day?' in: Proc. RILEM Symp. *Fibre-reinforced Cement and Concrete*, Sheffield, pp. 335–46.

Krenchel, H., Jensen, H. W. (1980) 'Organic reinforcing fibres for cement and concrete', in: Proc. Symp. *Fibrous Concrete CI 80*, Lancaster: The Construction Press, pp. 87–94.

Krenchel, H., Shah, S. P. (1985) 'Applications of polypropylene fibres in Scandinavia', *Concrete International*, 7(3): 32–4.

Kucharska, L. and Brandt, A. M. (1995) 'High performance cement mortars with and without silica fume reinforced with low amount of carbon fibres', in: *Proc. 5th CANMET/ACI Int. Conf. on Fly Ash, Silica Fume, Slag and Natural Pozzolans in Concrete.* Suppl. Papers, pp. 445–59.

Kucharska, L., Brandt, A. M., Logoń, D. (2002) 'Hybrid Fibre Reinforcement – is superposition of effects of different fibres always valid?' in: Proc. Int. Symp. *Non-Traditional Cement & Concrete*, Brno, pp. 376–86.

Kucharska, L., Logoń, D. (2005) 'Wollastonite fibres in cementitious composites' (in Polish). *51st Conf. KILiW and KNPZITB*, Krynica, 12–17 September, pp. 163–70.

Lankard, D. R. (1985) 'Preparation, properties and application of cement-based composites containing 5 to 20% steel fibres', in: *Steel Fiber Concrete*, US–Sweden Joint Seminar, S. P. Shah and Å. Skarendahl eds, CBI Stockholm, pp. 189–217.

Low, N. M. P., Beaudoin, J. J. (1993) 'Flexural strength and microstructure of cement binders reinforced with wollastonite micro-fibres', *Cement and Concrete Research*, 23(4): 905–16.

Maage, M. (1977) 'Interaction between steel fibres and cement based matrices', *Materials and Structures*, RILEM, 10(59): 297–301.

Mai, Y. W. (1979) 'Strength and fracture properties of asbestos cement mortar composites', *Journal of Material Sciences*, 14: 2091–102.

Majumdar, A. J. (1975) 'The role of the interface in glass-fibre-reinforced cement', *Composites*, January: 7–16.

——.(1980) 'Properties of GRC', Symp. Concrete International 80, Fibrous Concrete, April, London: The Construction Press, pp. 48–68.

Majumdar, A. J., Nurse, R. W. (1975) 'Glass-fibre-reinforced cement', *Building Research Establishment Current Papers*, CP 65/75, July.

Majumdar, A. J., Ryder, J. R. (1968) 'Glass fibre reinforcement of cement products', *Glass Technology*, 9(3): 78–84.

Majumdar A. J., Laws V. (1991) *Glass Fibre Reinforced Cement*, Oxford: BSP Professional Books.

Mwamila, B. L. M. (1985) 'Natural twines as main reinforcement in concrete beams', *International Journal of Cement Composites and Lightweight Concrete*, 7(1): 11–19.

Naaman, A. E., Homrich, J. R. (1989) 'Tensile stress-strain properties of SIFCON', *American Concrete Institute Materials Journal*, 86(3): 244–51.

Naaman, A. E., Shah, S. P., Throne, J. L. (1984) 'Some developments in polypropylene fibers for concrete', in: Proc. Int. Symp. *Fibre Reinforced Concrete*, SP-81, Detroit: ACI, pp. 375–96.

Nemeeger, D., Vanbrabant, J., Stang, H. (2003) 'Brite Euram program on steel fibre concrete subtask durability: corrosion resistance of cracked fibre reinforced concrete', in: Proc. Int. RILEM Workshop *Test and Design Methods for Steelfibre Reinforced Concrete*, RILEM Publ. SARL, pp. 47–66.

Nilsson, L. (1975) 'Reinforcement of concrete with sisal and other vegetal fibres', *Swedish Council for Building Research*, Doc. D14.

Ohama, Y., Amano, M., Endo, M. (1985) 'Properties of carbon fibre reinforced cement with silica fume', *Concrete International*, 7(3): 58–62.

Paramasivam, P., Nathan, G. K., Das Gupta, N. C. (1984) 'Coconut fibre reinforced corrugated slabs', *International Journal of Cement Composites and Lightweight Concrete*, 6(1): 151–8.

Pinchin, D. J., Tabor, D. J. (1978) 'Interfacial contact pressure and frictional stress transfer in steel fibre cement', in: Proc. RILEM Symp. *Testing and Test Methods of Fibre Cement Composites*, R. N. Swamy ed., Sheffield, Lancaster: The Construction Press, pp. 337–44.

Potrzebowski, J. (1986) 'Behaviour of the fibre/matrix interface in SFRC during loading', in: Proc. Int. Symp. *Brittle Matrix Composites 1*, A. M. Brandt, I. H. Marshall eds, Jabłonna, September 1985, London: Elsevier Applied Science, pp. 455–69.

Purnell, P., Beddows, J. (2005) 'Durability and simulated ageing of new matrix glass fibre reinforced concrete', *Cement and Concrete Composites* 27: 875–84.

Purnell, P., Short N. R., Page C. L., Majumdar A. J. (2000) 'Microstructural observations in new matrix glass fibre reinforced cement', *Cement and Concrete Research*, 30: 1747–53.

Rafiqul Islam, M. D., Khorshed Alam, A. K. M. (1987) 'Study of fibre reinforced concrete with natural fibres', in: Proc. of Int. Symp. *Fibre Reinforced Concrete*, Madras, 16–19 December, pp. 3.41–3.53.

Ramakrishnan, V. (1987) 'Materials and properties of fibre reinforced concrete', in: Proc. Int. Symp. *Fibre Reinforced Concrete*, Madras, 16–19 December, pp. 2.3–2.23.

——.(1996) 'Performance characteristics of polyolefin fiber reinforced concrete', *Proc. 4th Materials Engineering Conf.*, ASCE, Washington, 10–14 November, pp. 93–102.

Reinhardt, H. W., Fritz, C. (1989) 'Optimization of SIFCON mix', in: Proc. Int. Symp. *Fibre Reinforced Cements and Concretes. Recent Developments*, Cardiff, 18–20 September, R. N. Swamy and B. Barr eds, London: Elsevier Applied Science.

Rossi, P., Harrouche, N., de Larrard, F. (1989) 'Method for optimizing the composition of metal-fibre-reinforced concretes', in: Proc. Int. Symp. *Fibre Reinforced Cements and Concretes. Recent Developments*, Cardiff, 18–20 September, R. N. Swamy and B. Barr eds, London: Elsevier Applied Science, pp. 1–10.

Singh, R. N. (1987) 'Flexure behaviour of notched coir reinforced concrete beams under cycle loading', in: Proc. of Int. Symp. *Fibre Reinforced Concrete*, Madras, 16–19 December, pp. 3.55–3.66.

Soroushian, P., Marikunte, S. (1992) 'High performance cellulose fiber reinforced cement composites', in: Proc. Int. Workshop *High Performance Fiber Reinforced Cement Composites*, Mainz, 23–26 June, pp. 44–59.

Wörner, J. D., Müller, M. (1992) 'Behaviour, design and application of polyacrylonitrile fibre concrete', in: Proc. Int. Workshop *High Performance Fiber Reinforced Cement Composites*, Mainz, 23–26 June, pp. 115–26.

Standards

ACI 544.1R-96 (reapproved 2002) *Report on Fiber Reinforced Concrete.*

ASTM A820 *Specification for Steel Fibers for Fiber Reinforced Concrete.*

ASTM A820/A820M – 06 *Standard Specifications for Steel Fibers for Fiber Reinforced Concrete.*

EN 14651: 2005 *Test method for metallic fibre concrete. Measuring the flexural tensile strength (limit of proportionality(LOP), residual).*

EN 14721: 2000 *Test method for metallic fibre concrete. Measuring the fibre content in fresh and hardened concrete.*

EN 14845-2: 2006 *Test methods for fibres in concrete – Part 2: Effect on concrete.*

EN 14889:2006 *Fibres for Concrete. Part 1: Steel fibres – Definitions, specifications and conformity. Part 2: Polymer fibres – Definitions, specifications and conformity.*

ENV 1170-8:1996 *Test Methods for Glass-Fibre Reinforced Cement – Part 8: Cyclic weathering type test.*

6   Structure of cement composites

6.1 Elements and types of structure

The term 'structure of a material' covers the distribution of its components in space and the set of relations between them which are characteristic for the material. The characterization of a material structure also comprises data on the properties of the components.

The description of a material structure may be based either on assumptions related to an idealized perfect structure or on an analysis of a real structure. In most cases, both approaches are considered and their results are combined. The differences between these two approaches are caused by natural deviations of real structures from ideal models, and by local imperfections of different kinds. For example, the homogeneity of a material is often assumed as an approximation at a given level of analysis and when a material at lower level is considered then its heterogeneous nature becomes obvious. When the models of materials at various levels are built, they are interrelated in a systematic way, or more precisely, the models on a considered level are deduced from the results observed at a lower level (cf. Chapter 2).

Cement based composites are highly heterogeneous and their structures are composed of different elements (Brandt 2003):

- grains of fine and coarse aggregate;
- nonhydrated cement grains;
- hardened binder – cement paste with its own internal structure;
- voids, various kinds of pores and cracks;
- reinforcement, fibres and wires, steel meshes, bars and cables, polymer films and particles, etc.;
- water and air or other gases, which partly fill all voids.

A special category of components are also interface contact zones and various kinds of interfaces (ITZ – interface transition zone):

- between aggregate grain and cement paste;
- between reinforcement (fibres, mats, etc.) and cement paste;
- between old and new concrete (repair and retrofit).

The interface is understood as the region of direct contact between the two adjacent materials or material phases. The properties of interfaces are related mainly to the roughness of material surfaces, their purity, the ability of the wetting of one material by the other and to particular conditions in which thin layers of binders are hardening.

The interface also appears in composites with other kinds of binders. Further remarks on the interface types and properties in concrete-like materials are given in Chapter 7.

Several parameters are used to describe the structure: volume fractions of all components, ratio of matrix to inclusion stiffness, type of aggregate, maximum and minimum dimension of aggregate grains, their size distribution, texture and roughness of the surface aggregate grains, etc.

The grains and eventual reinforcement are bonded together in a more or less continuous way by a binding material, which fills up the inter-granular voids and also covers the grain surface with a thin film. The mechanical adhesion of the binders to surfaces of all kinds of structural elements and the chemical reactions with binders are the main factors ensuring bond; each of different importance in each case. However, it is assumed, on the basis of experimental evidence, that mechanical roughness of aggregate grains and of reinforcement surfaces decides on the quality of bonding. In the case of Portland cement paste, its shrinkage plays an important role in increasing mechanical adhesion.

The structure of concrete-like materials may be considered at different levels (cf. Section 2.4). The lowest level, where the required resolution of observation is of the order of a few Angstroms, is used mostly for examination of the microstructure of cement paste. The main interest is concentrated between so-called micro- and macro-levels, which means between the size of Portland cement grains or of small pores of a few micrometers in diameter, up to tens of millimetres for maximum grains of coarse aggregate and the diameters of steel bars.

The material's structure changes with time, mainly because during the hydration and hardening processes the fresh cement paste is becoming a hard and brittle material. This strong time-dependency is perhaps one of the characteristic features of concrete-like composites and their structures.

The structure is composed of randomly distributed and disoriented elements, but certain elements may be ordered. For research purposes, artificial materials may be constructed with regularly distributed elements (Holliday 1966; McGreath *et al.* 1969) or with selected one-size grains of aggregate (Stroeven 1973). Also, in real materials, reinforcing fibres may be aligned (Hannant 1978; Kasperkiewicz 1979). Reinforcing steel bars and meshes built of continuous wires are also regularly distributed. However, in general, the characteristics of the material's structures of concrete-like composites are related to their randomness and high heterogeneity on all levels of observation. That is the reason why there is considerable difficulty in providing any quantitative description.

The principal mechanical and other properties of concrete-like composites are related to the material's structure and only very few are structure-independent and related only to the material's composition (cf. Section 2.4). That obvious conclusion leads to a modern approach in material design: it is not sufficient to determine proportions of main constituents, but it also becomes necessary to select appropriate grain fractions and their distribution in space to find out the optimal fibre orientation and to create an efficient system of pores, as well as to decide upon several other parameters for obtaining the best structure of the material for fulfilling imposed requirements.

The importance of appropriate technologies grows in line with any complications of designed material structure, because only by adequate methods of execution can the designed parameters be obtained (Brandt and Kasperkiewicz 2003).

For theoretical studies, the structure composed of aggregate grains is simulated using various structural models. That type of approach is discussed more in detail in Section 2.5.

6.2 Methods of observation and representation of the structure of materials

A material's structure is observed on polished cross-sections or lateral faces of elements and specimens. Thin layers sawn out of specimens may be subjected to x-ray analysis and very thin layers are transparent enough to enable observation and structural analysis. The data on volume fractions of particular constituents, their spatial distribution and orientation, may be deduced from 2D images using stereological methods.

The observations are made directly on the prepared surfaces or on photograms and therefore the specific characteristics of these techniques determine the quality of obtained images, such as possible colour modifications and the sensitivity or resolution. The images should be easy for handling and diffusion, particularly by electronic means, relatively cheap and give the possibility of large magnification.

The surface of specimens should be carefully prepared for observation. Preparation is necessary to provide a finely ground and polished surface with sharp edges of all observed objects: grains, particles, pores, voids, cracks and others. Also a clear contrast between these objects and the cement paste should be produced. These requirements are similar at various levels of observation and should be rationally realized. The preparation and impregnation technique is adapted to the objects to be distinguished on the surface of a specimen and to the kind of the observation tool – a microscope. Details of preparation of specimens are described and presented on various images in many papers, e.g. Soroushian *et al.* (2003) and Brandt and Kasperkiewicz (2003).

Observations with the naked eye, possibly assisted by a magnifying glass or the application of low power microscopes, are usually sufficient for the analysis of the structure in macro-level. At that level the main elements of the material structure are clearly visible – grain of fine aggregate, fibres, large pores and cracks. The observed surface should be well polished and correctly illuminated. The observations are often executed at different ages of concrete or after various external actions in order to follow the modification of observed structures under load or during ageing processes. Objects of diameter equal to 0.5 mm may be identified without any magnification.

The observation is more difficult when certain elements are not clearly different from others. On the images of the concrete the varying intensity of grey colour of different phases may allow to distinguish them without great experience. In other cases, additional procedures are necessary – special illumination in which a particular phase will reflect the light, selective coloration, etc.

Several methods are used for enhancing the visibility of crack patterns in concretes and mortars. Cracks are more visible when soaked with coloured ink (penetrant) or with a water solution of special compounds; for example, citron acid and potassium permanganate. Fluorescent ink may be sprayed onto the surface of a specimen to facilitate observation of narrow cracks in UV light. The ink is a dispersion in light petroleum oil of fine organic-metallic particles, which form a deposit on the crack edges. Impregnation of preliminary dried specimens allows excellent visualization of cracks and interconnected voids or pores on ground and polished surfaces of cross-sections.

Figures 6.1, 6.2 and 6.3 are examples of images where structures of different kinds of concrete are shown. The images are obtained at life size with a slight magnification made possible by the reproduction of photograms.

Larger magnifications are required for micro-level where the average size of objects is approximately 1.0 micrometer. At that level, for example, the distribution of small pores induced to increase frost resistance may be observed and microcracks may be detected in the specimens cut out from the neighbourhood of a crack tip.

X-radiography is extensively used to obtain images of internal structural features, like cracks and steel fibres, but also other elements – aggregate grains and voids. The images of fibre structures are discussed in Section 6.6.

Beside the abovementioned direct methods, in which material structure and its elements may be observed, various indirect methods are applied, also giving information on material structure, for example:

- overall porosity and pore-size, distribution of pore structure (suction porosimetry), Fagerlund (1973a), mercury intrusion porosimetry, physical adsorption of gases;
- extent of cracking or of other cavities and irregularities of the material structure (ultrasonic pulse velocity across examined elements, Eddy current method);
- composition of the material, degree of cement hydration, etc. (x-ray diffraction).

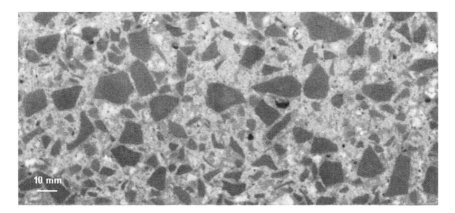

Figure 6.1 Structure of basalt concrete.

Figure 6.2 Example of concrete structure with gravel aggregate.

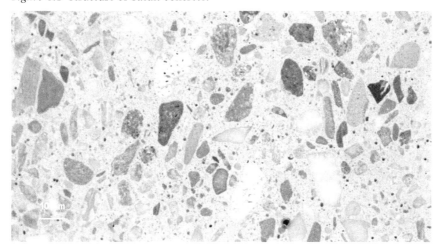

Figure 6.3 Example of concrete structure with crushed granite aggregate.

Acoustic emission events under loads and other actions are counted and recorded and may be analyzed as a result of cracking, debonding and other fracture phenomena. Simple counting of events gives information on overall material behaviour, from which conclusions on material structure may be drawn in an indirect way. The intensity of cracking in a brittle matrix varies with the intensity of loading or other action and may indicate different phases of the fracture processes (cf. Section 9.4). The more advanced level of application of the acoustic emission method is based on signal transformation into various parameters extracted from the time and frequency domains. Pattern recognition techniques and statistical analysis allow the identification of microfailures and enable their origin to be deduced – cracks in the matrix, fibre damage, delaminations, debondings, etc.

A few important groups of experimental observation methods are shown in Table 6.1. In diffractive methods, electrons and x-rays, which better penetrate the specimens, are scattered as an effect of collisions with electrons in the material. The diffraction pattern obtained is recorded and compared with reference ones. The specimens are prepared as powders of 10 μm being the maximum grain size.

For spectrographic methods infra-red spectrography is used to identify organic materials, minerals and their molecular structures. These methods

Table 6.1 Methods for identification of the microstructure of hydrated cement paste

Type	Microstructure	Methods
Indirect	Porosity	Mercury intrusion, porosimetry (MIP), gaseous adsorption (BET)
	Pore size distribution	Differential scanning calorimetry (DSC)
	Free and bound water, degree of hydration	Thermogravimetric analysis (TGA), differential thermal analysis (DTA), x-ray diffraction (XRD)
	Surface analysis	Secondary ions mass spectroscopy (SIMS)
Direct	Porosity	Back scatter electron imaging (BSE), energy dispersive x-ray analysis (EDXA)
	Morphology, intergrowth	Scanning electron microscopy (SEM)
	Inner structure, phase distribution	High voltage electron microscopy (HVEM), scanning transmission electron microscopy (STEM)
	Hydration characteristic morphology	SIMS, NMR spectroscopy
	Cement type, admixtures, water/cement ratio, compaction, alkali-silica reaction	High resolution scanning acoustic microscopy (SAM)

are based on the analysis of interaction between tested material and electromagnetic radiation. The spectrum of the amount of radiation absorbed for variable wavelengths is matched with a reference sample.

Differential thermal analysis (DTA) is based on heat transfer between the matter (absorption or liberation of heat) and the environment. The heat transfer is measured at a different temperature with respect to an inert material, for example, calcined alumina. The thermographs obtained are compared with model ones of known materials and processes such as dehydration, decarbonation, oxidation, crystalline transition, decomposition or lattice destruction.

The observation of images of composites may lead to qualitative conclusions only, but it does mean that the existence of certain objects and their reciprocal relations may be confirmed. The quantitative data and their analysis is necessary for rational design of composite materials and for effective determination of the relation between the materials' structures and properties. Computer image analysis has been developed over the last 20 years to enable quantitative analysis on the basis of images of any kind. Using a basically similar approach to an image as in manual or semi-automatic methods, the fully automatic approach offers much greater possibilities of quantitative determination of various parameters that characterize the structures of the materials.

Every image obtained by any method on a micro- or macro-scale may be subjected to computerized image analysis providing quantitative results, if particular elements in that image can be distinguished from the others. Several systems are known and commercialised by specialized companies, which allows different features of selected elements in an image to be recorded and calculated – area and perimeter, distribution, shape, etc. The possibilities of such a system depend not only upon the technical performance, but mainly on available software which allows more or less advanced quantitative analysis (cf. also Section 6.6.2).

The identification of different objects on analyzed images is carried on by several procedures and for cement-based materials the following six parameters are of primary importance:

- angle, 0–180°, between vertical axis of analyzed image (axis of the specimen) and the principal axis of the object;
- area of every analyzed object;
- dendritic length of every object reduced to the width of one pixel;
- per area is a coefficient equal to ratio between sum of areas of all objects of that kind and total area of the image;
- radius ratio between maximum and minimum distances from the gravity centre of the object to its border;
- roundness describes the shape of the object and is equal to: $(perimeter)^2/4\pi$ (area); for a circle this coefficient is equal to 1, for other shapes it is < 1.

These parameters are used for quantitative description of macrostructures composed of grains of aggregate, voids (pores), cracks, fibres, and also other objects on other levels: particles of cement, fly ash (FA), silica fume (SF), etc.

Because the analysis concerns plane images, it is necessary to apply stereological methods to obtain information on the 3D structure of a material. The elements of the structure are:

- 3D objects, like grains, particles and pores;
- 2D surfaces, e.g. faces of objects, interfacial transition zones, and other objects for which their third dimension may be neglected;
- 1D objects, like fibres;
- objects without dimensions – points, e.g. cross-sections of the fibres that are visible on cross-sections of specimens.

According to Cavalieri's law, the volume fraction of an object may be deduced analyzing cross-sections, lengths or points:

$$(V_V) = (A_A) = (L_L) = (P_P),$$

where V_V, A_A, L_L and P_P are volume, area, length and point fractions, respectively, corresponding to the analyzed phase and total value.

This equation is applied to the results of a computer image analysis, where plain images are analyzed. This is the case for pore systems, structures of cracks and other defects, distribution of fibres, etc.

The presentation of stereological methods, with examples of their application, may be found in papers by Stroeven (1973, 2003) and Hu (2004).

6.3 Microstructure of cement paste

The structure of hardened cement paste is created after the transition from a fluid to a rigid material due to progressive hydration of cement components. It is a continuous process that starts after mixing cement with water and lasts a long time with a decreasing rate. Hydrated and hardened cement paste is that continuous element of the concrete-like composites which fills the voids between fine aggregate grains and binds all elements of the material structure together. To obtain images of cement paste with its constituents, the maximum possible magnification is necessary. The micro-structure is built from following main elements:

- Calcium silicate hydrate $C_3S_2H_3$, designed also as CSH, has the form of very small and weakly crystallized particles which by their dimensions and irregularity resemble clay. CSH covers from 50% to 70 % of total volume of the cement paste and its properties and behaviour are complicated because they depend on several interrelated processes. Its representation by models and theoretical relations is not yet completely satisfactory.

- Calcium hydroxide Ca(OH)$_2$, designed as CH, is well crystallized in the form of hexagonal prisms and covers 20–25% of the total volume. Its large crystals measure from 0.01 mm to 1.0 mm.
- Calcium sulfoaluminates hydrates C$_5$AS$_3$H$_{32}$, ettringite, has the form of small crystals (1–10 μm) with long needles or plates and stripes, its volume corresponds to 10–15%.
- Capillary porosity of approximately 15% volume, depending on amount of water.

Other components of the cement paste structure are magnesium hydroxide MgH, which may occupy up to 5% of the total volume and capillary pores with volume depending mainly on the *w/c* ratio. Examples of images of the hydrated Portland cement paste are shown in Figure 6.4.

In each of these groups of solid constituents, different compounds exist, but their chemical formulae are not discussed here in detail. Interested readers are referred to handbooks on concrete technology, for example, Mindess *et al.* (2003).

The dimensions of the abovementioned constituents are from 0.1 μm for single crystal depth up to 100 μm for crystal needle length. These constituents are mixed together forming a dense and very irregular structure, partly crystallized and interlocked, which also contains non-hydrated cement grains and water in different forms. In Figure 6.5 the microstructure of hardened cement paste is shown schematically in successive phases of hydration.

In the initial stage (Figure 6.5a), thin concentric layers of crystallized needles and plates are developed around unhydrated cement grains. These crystal forms are becoming thicker because additional hydration products are growing, (Figure 6.5b). The products of hydration are called cement gel, that is, an aggregation of colloidal material. After progression of hydration, the neighbouring grains get in contact and a dense gel structure fills available spaces. This leaves thin gel pores in the gel structure and capillary pores between gel layers, (Figure 6.5c). Unhydrated parts of cement grains remain embedded in gel and their further hydration proceeds slowly, provided that two conditions are satisfied: water is available for chemical processes and there is enough space for hydration products.

Chemical processes involved in the hydration of cement depend on several factors for their development. The constituents' composition (properties of cement and admixtures) and proportions, the amount of water available for hydration and ambient temperature, may be quoted.

The admixture of gypsum in production of cement is necessary to decrease the early hydration rate and to adapt it to the technology for handling and placing the fresh mix. The rate of hydration is indicated in regulations for different kinds of cement and on average it may be assumed that hydration is achieved six hours after addition of water to cement. In that period, only about 15% of cement volume is effectively hydrated and these processes

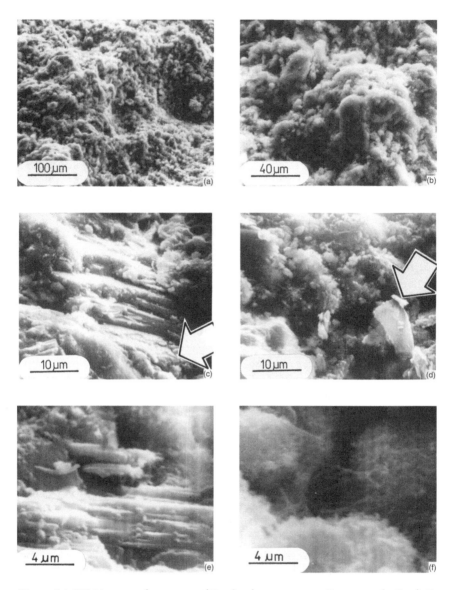

Figure 6.4 SEM images of structure of Portland cement paste. From tests by Prof. G. Prokopski (1989):

(a) Cement paste, w/c = 0.6, magnification ×200
(b) Cement paste, w/c = 0.6, magnification ×500
(c) Cement paste in gravel concrete. Transcrystalline crack across layers of crystals Ca(OH), w/c = 0.7, magnification ×2000
(d) Large crystal Ca(OH) is visible and several pores, w/c = 0.5, magnification ×2000
(e) Crystalline platelets of Ca(OH), w/c = 0.4, magnification ×5000
(f) Very thin fibre crystals of CSH (calcium silicates) are formed in pores, w/c = 0.6, magnification ×5000.

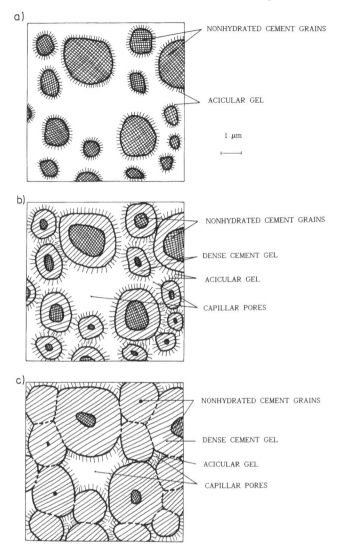

a)

NONHYDRATED CEMENT GRAINS

ACICULAR GEL

1 μm

b)

NONHYDRATED CEMENT GRAINS

DENSE CEMENT GEL

ACICULAR GEL

CAPILLAR PORES

c)

NONHYDRATED CEMENT GRAINS

DENSE CEMENT GEL

ACICULAR GEL

CAPILLAR PORES

Figure 6.5 Schematic representation of cement paste structure: sequence of hydration in time.

develop further by different mechanisms and at a rate decreasing with time during the next period of hardening. The deceleration of hardening is explained by the fact that the amount of available water is decreasing, and also its access is reduced by the hydration products, which double the initial volume of cement and gradually fill up all accessible voids and channels. It is usually admitted that for cement hydration only 22% (approximately) of water by cement mass is required. The hydration processes are described in several books; for example, Mindess *et al.* (2003), Neville and Brooks (1987),

Popovics (1979) and many others.

Finally, the cement paste becomes a hard and brittle material. Its deformability is characterized by linear elasticity and brittle fracture under a short time load and by its shrinkage and creep under varying humidity and long-term loading. The behaviour of cement paste is very brittle, that is, failure under load takes the form of a rapid progression of internal microcracks, accompanied with small deformation only. The crack propagation is controlled to some extent by inclusions – aggregate grains, pores and eventual reinforcement. These questions are dealt with in Chapters 9 and 10.

The porosity is usually considered in meso- and macro-level and is described in Section 6.5.

6.4 Structures built of aggregate grains

The aggregate grains are often called 'concrete skeleton'. In fact, in all ordinary concretes with hard stone grains they form a kind of rigid skeleton, where voids are filled with mortar and cement paste. A variety of lightweight grains are used in lightweight concretes. Examples of different structures built of aggregate are shown in Figure 6.6 and are discussed below.

The grains of similar size and in contact with each other are shown in Figure 6.6a. The voids are of large volume fraction and can only be partly filled with matrix. This structure is rarely applied for cement concretes, and is more often used for insulation granular materials or for bitumen concretes used as structural layers in road pavements.

A similar structure, but without direct contact between grains, is obtained with an additional amount of matrix. The structure as in Figure 6.6b may be obtained by mixing in an ordinary concrete mixer. However, because of equal size of grains, a large volume of the matrix is needed.

In Figure 6.6c a continuous structure of grains is schematically represented in which a large range of sizes is applied and minimum volume may be obtained. If in such a structure the biggest grains are replaced by smaller ones, the matrix volume will obviously increase (Figure 6.6d).

In Figure 6.6e the grain structure is composed of big and small grains, the intermediary ones have been removed. Gap-grading aggregates are particularly sensitive for segregation during transportation and handling. An unfavourable example is shown in Figure 6.6f; the grains are segregated and the upper part of the element looks rather like a mortar. Many books of concrete technology, cited in Sections 6.2 and 6.3, discuss the examples of grain structures.

The design of aggregate structure should begin by determining the maximum grain size, according to the density of reinforcement and characteristic dimensions of the element to be cast. In several countries for different structural elements, the maximum grain diameter is limited by prescription to 20 mm, 40 mm and 80 mm – the last value corresponding to large foundations, concrete dam walls and similar structures.

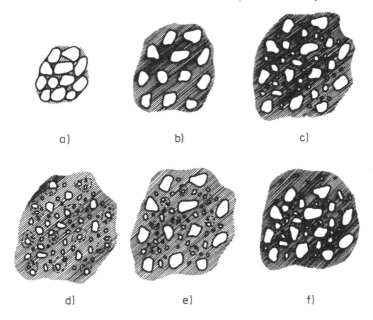

Figure 6.6 Examples of the aggregate grain structures in concrete.

A lower limit for grain size is imposed in cement paste having in mind their negative influence on crystallization processes. They may also increase water requirement and shrinkage, reduce the entrained air content and diminish durability. However, it has been proved that lowest sizes of inert aggregate grains, called stone flour, improve the workability of the fresh mix, reduce bleeding and considerably increase the impermeability of hardened composite. Fine grains of silica fume are used frequently and particularly for high quality concrete (cf. Section 4.3). The volume fraction of particles smaller than 75 µm should be limited and their influence controlled in every case, because it is closely related to the particle nature (clay, silt or harsh stone dust) and form (free particles, lumps or coating on larger grains of aggregate). A few kinds of small particles are presented in Table 6.2 showing how many single particles may be packed in 1 cm³.

For a general control of the grading one can use an approximate formula giving appropriate percentage p of material passing a given sieve of opening d, which is proposed in the following form:

$$p = 100 \ (d/D)^h,$$

where D is the maximum grain diameter and h is a coefficient which should be close to 0.5. When h is smaller then more fine grading is obtained, and when h is greater more coarse grading is obtained (cf. Figure 6.7). Examples of sieve diagrams are given in Section 4.2, where it is also mentioned that

Table 6.2 Small particles in concrete, after Brandt (2003)

Component	Average diameter (μm)	diameter (μm)	density (t/m³)	Specific surface (m²/kg)	Volume of one particle (mm³)	Mass of one particle (g)	Numbers of particles in 1 cm³ of concrete for following mass of the component		
							400kg/m³	120kg/m³	40kg/m³
Portland cement	50		3.1	250–450	6.5E-05	2.03 E-07	1970,000E+06	NA	NA
		1–100							
Fly ash	45		2.3	130–230	4.8E-05	1.10 E-07	NA	1090,000E+06	360,000 E+06
		1–150							
Silica fume	0.2		2.2	18,000–24,000	4.2E-12	9.30 E-15	NA	12.9E+12	4.3E+12
		0.1–0.3							

NA – not applicable

the choice of aggregate concerns not only its grading, but also the grain properties: their strength, porosity, surface roughness, etc.

The shape of singular grains which build the aggregate structure is of certain importance. Flat or elongated grains require more matrix to fill the voids than spherical or cubical ones (Popovics 1979), but for certain stress states such grains increase toughness (Brzezicki and Kasperkiewicz 1997). Singular grains are origins of stress concentrations, which may decide upon the apparent strength of the composite material; this problem is discussed in more detail in Section 8.1.

The structure composed of grains is always modified at the stage of a fluid mix by impenetrable boundaries, like sides of a formwork, reinforcing bars or large grains. This phenomenon occurs at various levels: structure of coarse aggregate, sand grains and other small particles when encountering larger ones. Such layers, neighbouring with various kinds of 'walls,' are weaker and more porous.

The structure of grains embedded in the mortar matrix may be quantitatively analyzed on images obtained from cross-sections as shown in Figures 6.1, 6.2 and 6.3 using the linear transverse method. It consists of superimposing a regular grid several times in a random way on the examined image and the length of lines between intercepts with aggregate grains are measured, summed up and averaged. The number obtained is called average mortar or grain intercept and its value may be considered as characteristic for the

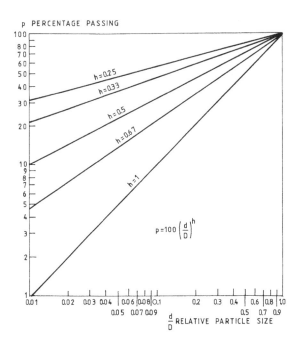

Figure 6.7 Fuller curves of various degrees, after Popovics (1979).

amount of that component in a given structure of the examined composite. The calculation procedure is explained in Figure 6.8 as one of the simplest applications of the stereological methods for quantitative analysis of the structures of materials. The manual measurements and calculations are at present usually replaced by computerized equipment; this procedure is described in Section 6.5 for analysis of the systems of pores.

According to several authors, the value of the average mortar intercept (AMI) should be equal to 3.5 mm for composites of continuous grading and manual compaction to ensure appropriate workability. When vibration of the fresh mix is applied, which is common practice at present, that value may be lower. Similar modification of that limit value is needed when, very fine sand is used. A larger value of the AMI may indicate that the composite structure is not correct, but an additional amount of coarse aggregate would probably improve it.

The increase of lower fractions will necessarily increase the amount of water and, consequently, of Portland cement. Inversely, the carefully designed gap-grading may require less matrix. In that case, the workability of fresh mix is more dependent on precise proportioning and mixing to avoid segregation. Therefore, gap-graded aggregates are not recommended when segregation is possible and a particular care in handling of the fresh mix cannot be guaranteed.

In a gap-graded aggregate mix a phenomenon of excessive bleeding may appear in cement-based materials, that is, a part of mix water is not retained by small grains and flows upward. The upper layer of hardened material becomes weak and porous due to a local increase of *w/c* ratio, with all negative consequences for local concrete strength, durability, aspect, etc. In the case of bleeding water, this also accumulates under larger aggregate grains, creating local cavities. That problem is discussed also in Section 9.2.

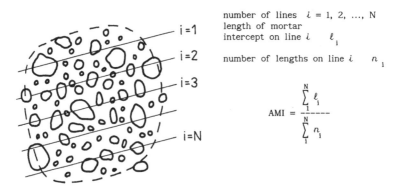

Figure 6.8 Example of application of the linear traverse method to calculate average mortar intercept.

There are several aspects that should be considered when the structure composed of aggregate grains is designed. In general, properties that have to be achieved are completely different in their nature and may be summarized as follows:

1   facility and efficiency of execution, low energy consumption, limitation of aggregate segregation, reduction of care during the hardening period;
2   high strength of hardened composite, low shrinkage, adequate fracture toughness and resistance to external actions which are to be foreseen in that particular case;
3   adequate durability in local conditions, frost and corrosion resistance.

These properties are interrelated and often depend, in a conflicting way, on the composition and structure of the aggregate. The properties may be translated into unit cost, including the cost of maintenance during service life (cf. Popovics 1982). Correctly determining the cost of composite material is a complex problem that may be included as one of several criteria in a multicriteria optimization problem. Methods of the mix design are discussed in Chapter 12 and only the questions related to the aggregate structure are discussed in this section.

The condition of minimum voids, which is often imposed for aggregate design, is derived from conditions of low permeability and high strength, which are connected to resistance against corrosion and frost destruction. A minimum volume of voids is obtained in an aggregate composition in which smaller ones adequately fill voids between big grains. However, such a composition may be very expensive and thus practically impossible. A very elaborate grading requires much work and energy consumption, because local aggregate deposits are rarely correct and require careful segregation, completion and transportation of particular fractions on long distances. Certain compromise solutions are often necessary between high quality and low cost.

For good workability the matrix fraction should be increased, which may have a negative effect on strength and durability.

For high freezing and thawing resistance, a very dense structure of material is required with an appropriate system of pores obtained by application of air-entraining admixtures. Also aggregate grains should exhibit adequate frost resistance.

The choice between crushed stones and gravel as coarse aggregate has its effects on strength and workability. For Portland cement mixes, crushed aggregate requires more cement paste to fill the void and more energy for mixing and compacting, but higher strength may be obtained than with gravel, thanks to better bonding to cement paste.

The behaviour of aggregate grains in the matrix under load depends on the ratio between their Young's moduli $n = E_m/E_a$. In structural concretes usually $n < 1$ and hard grains reinforce the matrix. The cracks propagate through matrix and aggregate grains control that propagation. In the concretes with lightweight aggregate $n > 1$ and the grain resistance is lower than that of the cement matrix. The examples of crack pattern in composites characterized by both these relations are shown in Figure 9.2.

Application of local aggregate, to reduce costs of transportation, and also application of secondary materials for ecological reasons, should be considered. For minimum unit cost of final composite material the cement volume and energy consumption should be minimized. All these factors are closely related to the aggregate system and the local conditions in situ decide often on selection of the aggregate. A theoretical solution of such a problem should always be checked in a laboratory and later, in natural scale, with actual constituents and technologies, also taking into account the qualifications of personnel and availability of technical control required for high quality (Alexander and Mindess 2005).

6.5 Structure of pores and voids

6.5.1 Porosity of cement paste

The porosity of concrete-like composites is an important characteristic which determines, to a large extent, their mechanical properties. The pores are of different shapes and dimensions and their classification is not strictly established and recognized. These pores, distributed at random, contribute to the heterogeneity of the composite material.

High porosity is strongly detrimental to the strength and permeability of cement-based composite materials, particularly if the pores are of large diameter. The permeability to gases is much higher than to fluids across the pore structure because the viscosity of the former is about 100 times lower than that of the latter. Increased permeability adversely affects durability, mainly for materials in the structures exposed to outdoor conditions. The structure of the external layers of concretes and their porosity are different from the core of the elements. The cement hydration is less advanced due to quicker desiccation, the size distribution of aggregate grains is shifted to smaller diameters and higher porosity may be observed. Due to all these effects, the strength and resistance against corrosive factors of these parts of concrete elements are usually lower.

Porosity cannot be determined by one parameter, that is, by ratio of pore volume to total volume, because for most purposes it is necessary to separately consider pores of different kinds and dimensions in relation to mechanical properties of composite materials, and pores in cement paste may be classified according to various criteria.

Using the definitions proposed by the International Union of Pure and Applied Chemistry (IUPAC) three groups are distinguished according to their characteristic dimension d:

- micropores $d < 2$ nm

- meso-pores 2 nm $< d <$ 50 nm
- macropores $d >$ 50 nm

The lower limit is not defined because it depends upon the method of pore detection and examination. The upper limit is related to capillary effects and usually is assumed as equal about 0.5–1.0 mm.

The porosity and pore systems in cement-based composites may be one example of fractality, which is characteristic for these materials. Fractal quantity depends on the scale used to measure it; for example, the fracture area of a concrete element cannot be determined in an unambiguous way if the method and scale of its determination are not given (cf. Chapter 10.5). The classification and analysis of a pore system and all quantitative results derived depend among other things on the method of observation and magnification of microscopic images (Mandelbrot 1982; Guyon 1988).

Other notions that are very important in consideration of pore systems are percolation and percolation threshold. The permeability of the system composed of strongly disordered pores is non-linearly related to pore number and total volume, and that relation may be characterized by a threshold – as long as a certain pore density is not reached, the system remains impermeable. When the density becomes higher than that critical value, then permeability appears and increases rapidly (Stauffer 1986). As this concerns the connections between pores, these may be distinguished: interconnected pores, closed pores and pores closed at one end. The roles of these kinds of pores in the flow of fluids and gases across material are different.

As it concerns their origin and role in cement paste, four different types of pores in cement paste are proposed:

- gel pores are of characteristic dimensions from 0.0005 µm up to 0.01 µm (0.5–10 nm);
- capillary pores of 0.01 to 10 µm, mostly between 0.02 and 0.03 µm of average diameter;
- intentionally introduced pores of spherical shape and defined diameters as a result of so-called air entrainment, usually between 0.05 and 1.25 mm;
- air voids larger than capillary pores, which are entrapped inevitably during mixing and compaction of fresh mix – bug holes; their total volume depends on quality of mixing and casting of concrete.

Pore distribution is schematically represented in Figure 6.9 where the four main types mentioned above are shown and their reciprocal relations and possible overlaps are explained.

Pore dimensions may be compared to other elements of the material structure:

- fine aggregate grains (sand) $>$ 100 µm;

- unhydrated cement grains 1–100 μm;
- cement needle-like crystals length 1 μm, width 0.05 μm;
- silica fume particles 0.1–0.3 μm,

and it appears that pores are distributed at various levels of the structure.

The gel pores form a part of CSH (calcium silicate hydrate), and may be classified as micro pores or meso pores. The principal difference between gel and capillary pores is that the former are too small to be filled by the hydration products and for capillary effects, it means that no menisci are formed. The gel pores occupy between 40% and 55 % of total pore volume, but they are not active in water permeability through cement paste and they do not influence the composite strength. Water in the gel pores is physically bonded. It is believed that gel pores are directly related to shrinkage and creep properties of the cement paste.

The capillary pores and air voids are partially or completely filled with water, which depends on the environmental hygrometry and takes part in continuous hydration of cement grains. The total volume of these two categories of pores is related to the decrease in strength of the hardened paste. The capillary pores are caused by that part of the mixing water which is not absorbed by cement grains during hydration. With the development of the hydration and hardening processes the capillary pores are filled with the hydration products and the possibility of water flow is gradually reduced. The w/c ratio has to be equal to 0.35–0.40 in the cement paste in order to ensure full hydration of cement (Powers 1964).

When the initial value of the w/c ratio > 0.5, there is an excess of water and then the capillary pores are large and probably cannot be filled completely.

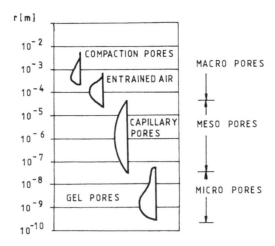

Figure 6.9 Pore-size distribution in Portland cement concrete, after CEB (1989).

Otherwise, for a *w/c* ratio < 0.5 only reduced water flow is possible. Then there is enough water for hydration and also enough space for hydration products, which swell during the process so that the capillary pores may have reduced permeability. These values for mortars and concretes depend on the mixture proportions and quality of components, and also on special admixtures added to modify the fluidity of the fresh mix.

So-called structural entrapped pores appear when the volume of water is insufficient and workability is poor or when vibration of the fresh mix is insufficient. The difference between the total volume of the fresh mix components and the final volume of this mix, (the latter is smaller due to chemical reactions of hydration) is that if they are not filled with hydration products, they appear as voids called 'contraction pores.'

Air entrained pores are described more in detail in Section 6.5.2.

Gel pores constitute approximately 28% of Portland cement paste and capillary pores between 0% and 40%, depending on the *w/c* ratio and progress of hydration processes. The voids of entrapped air should not exceed a small proportion; that is, 1% of the total volume, otherwise the concrete must be considered as inadequately executed.

The duration and quality of cure of the fresh mix determine the hydration progress and the volume of hydration products, which may eventually fill up the capillary pores and reduce their permeability. For reduced *w/c* and prolonged cure the pore size distribution diagrams are shifted towards smaller sizes, with decreasing total volume. The saturation of capillary pores plays an important role in the composite durability if the material is exposed to freeze-thaw cycles. Damage is more probable when the critical value of the saturation is attained and the increasing volume of ice cannot be accommodated without additional stresses. The hardened paste density is also important to ensure resistance against all other external attacks.

Figure 6.10 shows how the permeability of cement paste increases with the *w/c* ratio. At first it increases slowly and beyond *w/c* = 0.5 it increases rapidly. The permeability may therefore be regarded as an indirect and partial measure of porosity. Considering the components of cement-based composites it may be assumed that the permeability is determined more by matrix properties than by aggregate ones. Particularly, the matrix/aggregate interface may have the largest content of pores and microcracks which affect the overall permeability (cf. Chapter 7).

6.5.2 Entrained air-void systems in hardened concrete

The frost resistance of concrete is considerably improved when an appropriate structure of air voids (air bubbles) is formed during mixing, and is uniformly distributed and maintained in the hardened state. Special admixtures called air-entrainers are necessary for that aim.

The mechanism of preventing damage to a concrete structure by entrained air pores was explained first by Powers (1945, 1964). The conversion of water to ice causes an increase of volume of 9.03%. That increase produces hydraulic pressure of unfrozen water in capillary pores and additional tensile stress in concrete structures. The intensity of the hydraulic pressure is proportional to the length of the capillary pores. A densely distributed system of air bubbles reduces considerably the length of the capillary pores and the bubbles act in a similar manner to dilatation reservoirs in a central heating system of a house. Therefore, the air voids are helpful only when they are sufficiently small and densely distributed.

The influence of freezing water in capillary pores is a very complex phenomenon and various processes accompany simple hydraulic pressure of ice on hardened cement paste, but the model proposed by Powers was universally accepted as explaining the destructive influence on concrete elements with acceptable agreement with experimental results. That model is considered as a basis for standardized requirements for the design of structures. Furthermore, there is strong experimental evidence that the distribution of the voids is a key factor governing the frost resistance of concrete. This statement is supported by experience and investigations in several countries where the temperature in winter falls below zero. An example of such

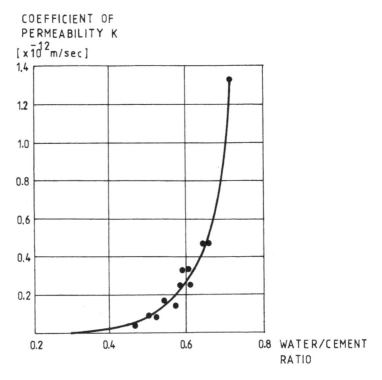

Figure 6.10 The effect of water/cement ratio on the permeability of mature cement paste, Browne and Baker (1978).

investigations is a review published by Pleau *et al.* (2001) after more than 600 laboratory and field concrete mixes were tested.

There are different recommendations for the parameters that characterize a pore system. Spacing factor determined according to CEB (1989) should not exceed 0.2 mm and that recommendation is based on results obtained by several authors who also published indications for the pore diameter. For example, Klieger (1978) proposed that for maximum aggregate grains of 63.5, 38, 19 and 9.5 mm the optimum spacing factor should be equal to 0.18, 0.20, 0.23 and 0.28 mm, respectively; for mortars, a spacing factor of 0.30 mm was indicated. Consequently, the air bubble mean diameter should be between 0.05 and 1.25 mm, according to Neville and Brooks (1987). These values are confirmed and are valid also now.

After various test results and recommendations it may be concluded that the effective air void structure in the hardened cement matrix is described by four parameters, contained in respective limits:

- A – total air content, usually between 4 and 7%, depending on the aggregate grain diameter;
- $\bar{L}$ – spacing factor that expresses the mean distance from a given point in the cement paste to the nearest free surface of an air void, below 0.2 or 0.22 mm;
- α – specific surface of air voids, over 15 or 20 mm^{-1};
- A_{300} – percentage of pores smaller than 0.3 mm, over 1.5% or 1.8 %.

The spacing factor is low when the total air content is sufficiently high and single air bubbles are small and uniformly dispersed in the matrix. Pores $\geq$ 5 µm are recognized on specially prepared cross-sections, then counted and recorded manually or using computer image analysis – in both cases based on ASTM C 457 (1991). The quality of the surface treatment of concrete specimens prior to the microscopic examination is decisive.

Automated computer analysis of images representing air void systems is developed and is now universally applied for testing specimens taken from structures. The results of a European round-robin test, analysed by Elsen (2001), proved the reliability of this method, provided that a laboratory executing such investigations has enough experience to carefully perform all stages of such a work. There are many sources of possible errors, from preparation of specimens to correct illumination of polished surfaces. Results of computer image analysis of air void systems should be carefully checked, including comparisons between the values obtained from image analysis and those from visual examination according to ASTM C 457-06 (2006) and taking into consideration various possible sources of errors (Pleau *et al.* 2001; Załocha 2003).

For different conditions of exploitation of tested concrete structures, different limits are imposed on these parameters. The technique of measurement and discussion of results are described elsewhere. In Figure 6.11, a schematic diagram, shows the relation between spacing and the durability factor, which is understood as a percentage of initial value of E_{dyn} after 300 cycles of freezing and thawing, ASTM C 666/C666M-03 (2003).

Automated image analysis may be successfully applied using only a flatbed scanner. This opens possibilities to perform quantitative image analysis of pore structures (and also of aggregate grain distribution) also by small and less equipped laboratories. Even a high-resolution scanner is much less expensive than a classic microscope with all the additional equipment, and the parameters of the air-void system may be obtained with the same precision (Załocha and Kasperkiewicz 2005).

The testing of specimens subjected to freeze-thaw cycles does not reflect very precisely the material's actual resistance in real structure. One of the reasons is that delays between particular cycles of freezing and thawing are of great importance and the possibility of the pores partly drying out are different in a laboratory test than in natural varying conditions. Nevertheless, air entrainment is required by many standards for open air concrete structures.

The formation of air bubbles in cement paste depends upon many circumstances, primarily:

- the kind and amount of air-entraining agent;
- the composition of concrete, and the eventual application of other admixtures;
- the technique of transportation, mixing and compaction of the fresh mix, and temperature during these operations.

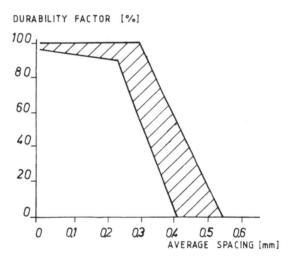

Figure 6.11 Relative durability of concrete as a function of average spacing of air entrained pores, after Mindess *et al.* (2003).

In general, the total volume of entrained air increases with a decrease of cement and increase of fine aggregate content, with lower temperature and with an increased content of angular aggregate grains.

Only the pores obtained by air-entrainment are of regular spherical shape and distribution; others are very irregular because these are created between interrelated needle- and plate-like crystals of hydrated cement. Examples of correctly distributed small air pores are shown in Figure 6.12, together with the values of all four basic parameters. In Figure 6.13 pores and voids are shown that are not created by appropriate air entraining and have no beneficial influence of the frost resistance of that concrete.

As examples, two different concretes (A and B) were analysed as to the distribution of the entrained pores diameters. The results are shown in Table 6.3 and in Figures 6.14 and 6.15. In Table 6.3 symbols + and – indicate that the respective criterion of correct air entraining is or is not satisfied. After these results it may be concluded that the concrete A is frost-resistant, according the criteria presented above, and that concrete B does not satisfy respective requirements.

A fourth category of air voids is caused by air entrapped during compaction. These voids, called bug holes, are usually larger and have no beneficial influence on freeze-thaw resistance. An excessive volume of entrapped air considerably increases the concrete permeability and decreases its strength; therefore it should be considered as inadmissible.

The relationship between pore system characteristics, its capillarity and permeability can be presented schematically (Venuat 1984; Figure 6.16). Because the pore structures are usually composed of pores of various dimensions, their capillarity and permeability reflect the pore size distribution and also existing interconnections between the pores. Therefore, the description of the pore structure only as a system of spheres is nothing more than a simplification as it is assumed that pores have regular shape.

The determination of the pore structure in hardened cement matrices is based on direct and indirect approaches.

In direct measurement methods the microscopic examination by optical or scanning electron microscopes is eventually completed by computer image analysis. Stereology formulae are used to deduce three-dimensional structure from plane images. However, the complex shapes of pores in cement paste do not facilitate the task.

In indirect methods other processes are applied and there are two main groups of methods for studying the micro-structure of hardened cement-based composites (Scrivener and Pratt 1987):

1   Bulk techniques like thermogravimetric analysis (TGA) and quantitative x-ray diffraction (QXRD) give information on the average volume fraction of particular phases, e.g. pores.
2   Selective techniques provide data on the size distribution of certain components, e.g. pore size distribution, and there are mercury intrusion

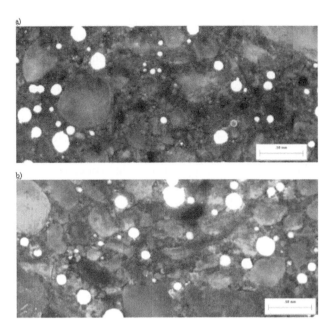

Figure 6.12 Examples of distribution of entrained air pores observed on ground and polished face of concrete specimens:
(a) α = 30,4 mm-1, A=3,99%, A_{300}=1,97%, $\overline{L}$ = 0,17 mm
(b) α = 32,5 mm-1, A=4,61%, A_{300}=2,99%, $\overline{L}$ = 0,15 mm.
(from the tests by Professor M. A. Glinicki and Dr M. Zieliński in 2006)

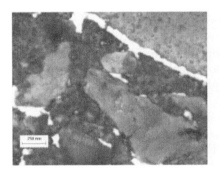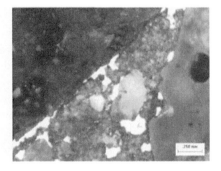

Figure 6.13 Examples of incorrect distribution of pores in air-entrained concretes:
(a) α = 28,0 mm-1, A=3,06%, A_{300}=1,10%
(b) α = 38,4 mm-1, A=2,54%, A_{300}=1,17%.
Because of interconnected space, factor $\overline{L}$ cannot be determined.
(from tests by Professor M. A. Glinicki and Dr M. Zieliński in 2006)

Table 6.3 Parameters of air pore structures in concretes A and B

Parameters			Concrete A		Concrete B	
Total air content A	%		4.6	+	5.7	+
Spacing factor $\overline{L}$	mm		0.32	–	0.15	+
Specific surface α	mm^{-1}		16.0	–	30.0	+
Percentage of micropores A_{300}	%		0.8	–	2.8	+

porosimetry (MIP), nitrogen adsorption and methanol adsorption techniques, and low temperature calorimetry.

The results obtained from the indirect methods are often controversial, because actually it is not a pore system that is examined but rather the processes applied in these methods; the results reflect only the pore size distribution response. Any established value of pore diameter has only conventional meaning and may be different than diameters obtained from other methods. The indirect methods more or less influence the object of observation and measurements because the interventions disrupt material structure. Determining of distribution of pore diameters in cement paste is performed by the mercury porosimetry method and the results are partly confirmed by observations and counting the pores by computer image analysis, but mercury intrusion may damage and alter the material microstructure. Furthermore, the intrusion of mercury into a pore is related to the orifice of the pore rather than to its real dimension (Diamond 2000). Other methods, like capillary condensation, give considerably different values.

Interesting results concerning the pore diameter distribution were published by Verbeck and Helmuth (1969) and an example from their paper is shown in Figure 6.17. The pore-size distribution for three different values of *w/c* ratio obtained by MIP is given. It appears from these diagrams that observed maxima correspond well to two categories of pores: gel pores and capillary pores. These maxima move in the direction of larger or smaller diameters when *w/c* ratio increases and decreases, respectively.

6.5.3 *Porosity and strength*

The strength of compressive composite material is reduced by the existence of a pore structure and that reduction depends upon the characteristics of the pore system and the type of loading. The problem of how to determine the influence of a pore system on the strength of concretes was considered on the basis of a simple application of the law of mixtures:

$$f_c = f_o(1 - P), \tag{6.1}$$

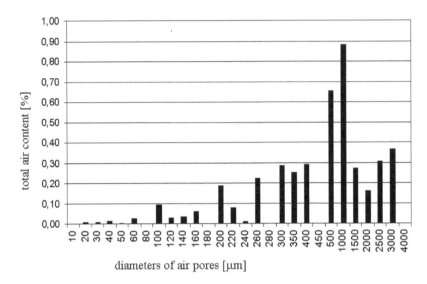

Figure 6.14 Distribution of air pore diameters in concrete A.
(from tests by Professor M. A.Glinicki and Dr. M. Zieliński in 2006)

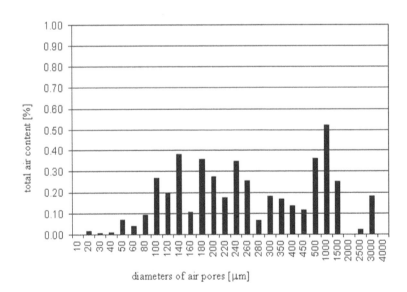

Figure 6.15 Distribution of air pore diameters in concrete B.
(from tests by Professor M. A.Glinicki and Dr. M. Zieliński in 2006)

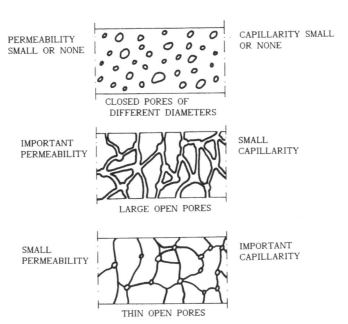

PERMEABILITY
SMALL OR NONE

CAPILLARITY SMALL
OR NONE

CLOSED PORES OF
DIFFERENT DIAMETERS

IMPORTANT
PERMEABILITY

SMALL
CAPILLARITY

LARGE OPEN PORES

SMALL
PERMEABILITY

IMPORTANT
CAPILLARITY

THIN OPEN PORES

Figure 6.16 Example of pore systems, their permeability and capillarity, after Venuat (1984).

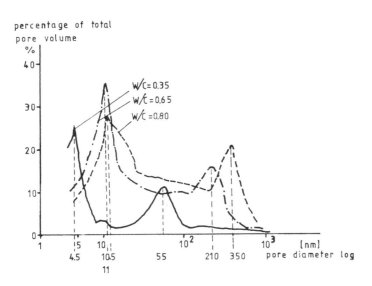

Figure 6.17 Pore-size distribution in a moist cured cement paste at an age of 11 years, after Verbeck and Helmuth (1969).

where f_c is the composite strength of a material with a system of pores, f_o is the strength of solid material and P is the porosity. The formula (6.1) may therefore be considered as the upper bound and was corrected by numerical coefficients after experimental data from tests of cement based materials. These are examples of the equations, discussed, among others, by Fagerlund (1973b):

Balshin (1949) $f_c = f_o(1 - P)^C$, where $f_o = 68.74$; $C = 8.15$ .... (6.2)

Ryshkevich (1953) $f_c = f_o \exp(-C_1 P)$, where $f_o = 74.4$; $C_1 = 8.96$ .. (6.3)

Both these formulae were initially proposed for ceramics and in equation (6.3) the value of C_1 was estimated for tensile strength to be between 4 and 7. It has been observed that for values of porosity P below a certain value P_o equations (6.2) and (6.3) are in close agreement with experimental data. However, at higher porosities for $P > P_o$ it is advisable to introduce the notion of critical porosity P_{cr} corresponding to the strength f_c approaching zero and to use the following equations proposed by Schiller (1958):

$$f_c = f_0\left[1 - \left(\frac{P}{P_{cr}}\right)C_2\right], \text{ or}$$ .... (6.4)

$$f_c = C_3 \ln\frac{P_{cr}}{C_4}$$ .... (6.5)

here C_2, C_3, C_4 are empirical constants.

In Hansen (1966) an approximate formula was proposed to relate porosity and strength, assuming that the compressive strength is proportional to the cross-sectional area of the solid material:

$$f_c = f_o(1 - 1.2^{P2/3})$$ .... (6.6)

In equation (6.6) the following factors are only approximately covered by numerical coefficients:

- the pores may act as stress concentrators, depending on the state of stress (compression, tension, shearing) and their shape;
- with increasing porosity P several phenomena may be expected, which do not interfere when the porosity is low, e.g. rupture of walls between neighbouring pores, non-homogeneous distribution of internal stresses.

The porosity P_o which corresponds to a clear variation of curve $f_c(P)$ may be considered as a kind of percolation threshold, proposed by Fagerlund (1979) and shown in Figure 6.18. The experimental relationship between concrete strength and capillary porosity is shown in Figure 6.19 and it is well confirmed by extensive experimental results published by Uchikawa (1988). Relations between volume of capillary pores of different dimensions and the strength of cement paste, mortar and concrete are presented in Figure 6.20. Large sections of the curves may be assumed as corresponding to linear relations.

The influence of porosity on the Young's modulus E may be represented by similar empirical formulae, e.g. for hardened cement paste:

$$E = E_o (1 - P_o)^3,\qquad\qquad(6.7)$$

where E_o is the value of Young's modulus for capillary porosity equal to zero ($P = 0$). After tests outlined in Te'eni (1971) equation (6.7) has been confirmed for both compressive strength and Young's modulus.

All the above formulae should be considered as only approximate, because obviously the influence of porosity on the mechanical behaviour of a com-

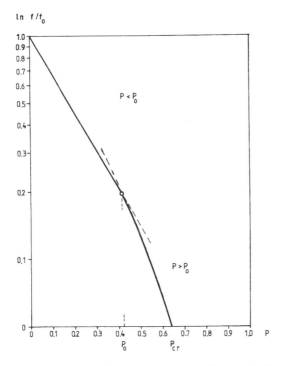

Figure 6.18 General relationship between porosity and strength of porous materials, after Fagerlund (1979).

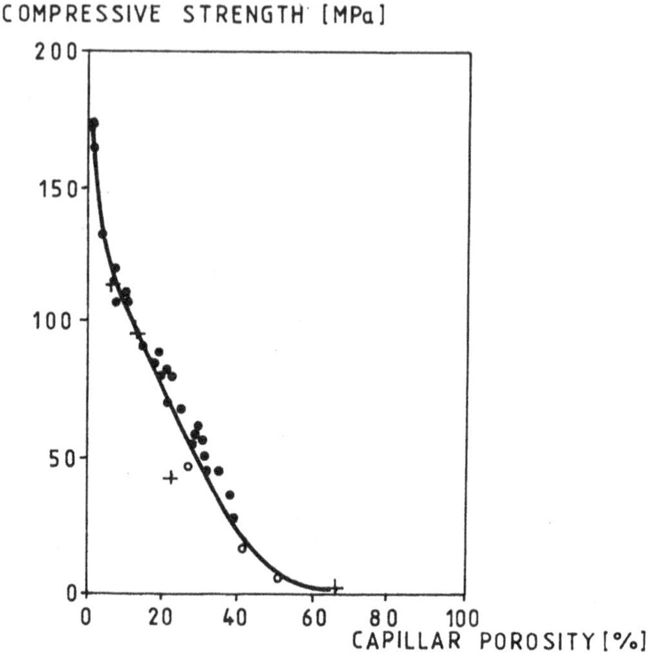

COMPRESSIVE STRENGTH [MPa]

Figure 6.19 Experimental relationship between capillar porosity and compressive strength of various concretes, after Verbeck and Helmuth (1969).

posite material cannot be reflected solely by one-parameter relations. The two- and multi-phase models mentioned in Section 2.5 may be also applied to composite materials in which a pore structure represents one phase. As shown in Figure 2.14, the theoretical models are inadequate for pores, which have no stiffness. In certain cases, spherical pores even represent obstacles for crack propagation. In the others, sharp pores in the material's structure stimulate crack opening and propagation.

The above experimental observations are reflected in Kendall *et al.* (1983) who proposed to clearly distinguish the influence of two populations of pores – small gel or colloidal pores entrapped between hydrated particles of about 2 nm in size, and larger crack-type pores of length equal to about 1 mm. The influence of the first population on the concrete strength is proportional to its total volume and is reflected by the law of mixtures or classic Féret's equation where volume of air is taken into account (cf. Section 8.2).

The colloidal pores do not start cracks as their dimensions are extremely small and this fact is correctly represented by the Griffith's equation:

$$f_t = \sqrt{\frac{E\gamma}{\pi c}} \tag{6.8}$$

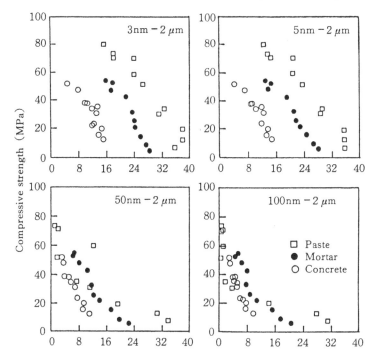

Figure 6.20 Relationships between volume of capillary pore space and compressive strength of hardened cement paste, mortar and concrete for different pore distributions, after Uchikawa (1988).

where f_t – tensile strength, E – Young's modulus, γ – fracture energy, c – characteristic dimension of a notch or a pore.

This theory combines the influence on strength of both groups of voids considered independently: the first one characterized by its total volume P, but does not contain long cracks, and the second one without appreciable volume but characterized by length c of a crack. The criterion for crack extension according to Kendall *et al.* (1983) is:

$$f_t = \left[\frac{E_o \gamma_o (1 - P)^3 \exp(- kP)}{\pi c} \right]^{1/2} \tag{6.9}$$

where E_o and γ_o correspond to non-porous body and k is a constant. When the strength f_t is plotted against porosity P it is assumed that pore length c remains constant.

The relevant curves corresponding to experimental verification are shown in Section 13.2 and Figure 13.2.

The pores and their influence on the material strength should be considered as randomly distributed structural imperfections in the approach of fracture mechanics. In such a formulation their influence depends on the parameters of pore structure and of pore shape, or eventually the empirical formulae with approximately calibrated coefficients may be applied. Odler and Rößler (1985) investigated strength as function of porosity for cement pastes and proposed to differentiate the influence of pores according to their sizes. There were several other proposals to construct a general model taking into account various processes in hardened concrete-like materials that have been published by Parrot (1985) and Luping (1986), among others. These attempts were continued by Atzeni *et al.* (1987) who characterized the pore structure using Griffith's theory and using average radius of pores r_m; by this approach following expression for tensile strength was obtained:

$$f_t = \sqrt{\frac{2E_oT_o(1-P)}{\pi\sqrt{r_m}}} \qquad (6.10)$$

where E_o and T_o are the Young's modulus and specific fracture energy of the matrix without pores, respectively.

Probably only large pores have an important detrimental influence on composite strength. Based on tests on high alumina cement specimens published by Murat and Bachiorrini (1987), it has been shown that a clear correlation existed only when porosity with pores larger than 1 μm was taken into account.

Kumar and Bhattacharjee (2003) proposed another model, taking into account the age of the concrete, exposure conditions and aggregate type. Acceptable correlation with tested specimens taken out from in situ structures was obtained.

It may be concluded that low porosity is necessary for high strength of concrete-like composites. To achieve that aim the material components must be carefully selected for best packing and a low *w/c* ratio. The optimum packing is obtained with appropriate composition of all aggregate fractions by application of superplasticizers acting as deflocculants, but also the fine voids between aggregate grains should be filled up with microfillers; for example, SF. Low water content is possible when excellent compaction is ensured by superplasticizers: the *w/c* ratio may be as low as 0.3 without decreasing the workability of the fresh mix. When a system of entrained air voids is necessary to obtain required frost resistance, then unavoidable reduction of strength should be accepted. For more detailed description of materials with low *w/c* ratio, cf. Section 13.4.

6.6 Structure of fibres

6.6.1 *Types of structures*

The structure composed of reinforcing fibres or wires in the matrix may be characterized by the following groups of parameters:

- material of fibres (steel, glass, polypropylene, etc.);
- shape of a fibre (short chopped or continuous, single fibres or mats, plain fibres or indented; meshes, fabrics);
- distribution of fibres in the matrix (random, linearized, regular);
- amount of fibres in the matrix (volume or mass fraction).

The number of above-listed parameters indicates how many different types of reinforcing fibre structures may be used in concrete-like composites.

The first two groups of parameters are considered by the designer who decides on the material's structure and composition. As described in Section 5.4, steel fibres are used as short chopped lengths, as continuous woven wires, non-woven or welded at the nodes (ferrocement) and as a kind of wool composed of thin and long entangled wires. Short fibres are, in most cases, randomly distributed, but their alignment is also possible. Glass and other kinds of fibres may be also chopped or used in the form of mats and fabrics.

The reinforcement in the form of fibres or wires is, by its nature, subject to randomness, in contrast to classical reinforcement by bars or prestressing cables that are very precisely positioned in the reinforced elements. Fibres are described by their average volume fraction and average direction; usually the precise determination of these parameters is neither possible nor needed. However, these average values of characteristic parameters are used in design and control because the required distribution and orientation is often disturbed by various factors related to execution techniques, wall and gravity effects during vibration, etc. When these influences act in a systematic way, the randomness of distribution, or its regularity, may be considerably modified.

The execution of operations aimed at the realization of designed fibre structure requires special care and control. The effective fibre distribution and direction may be determined after the analysis of images obtained on cross-sections or on radiograms.

6.6.2 *Structures composed of short fibres*

Three types of idealized structures of short chopped fibres may be distinguished:

1 linearized fibres (1D);
2 random distribution in parallel planes (2D);
3 random distribution in space (3D).

Examples of these three ideal structures are shown in Figure 2.6 in Chapter 2.

All real structures are subjected to more or less important deviations; sometimes they are considered as combinations of the idealized ones. These deviations may be introduced accidentally and some directions or regions may be more heavily reinforced. Aggregate grains influence the distribution of fibres and that effect increases with the grain size (Figure 6.21). It is interesting to learn how many single fibres are in 1cm³ of paste; this explains their efficiency in controlling the microcracks. Data on selected fibres and microfibres are given in Table 6.4.

In most cases of practical application, fibres are randomly distributed; that is, according to 2D and 3D schemes. For certain purposes in the fibre structure it may be interesting to adapt fibre direction to the fields of stress and strain, provided that it is known. Steel fibres may be linearized when a form with fresh mix is subjected to a magnetic field of appropriate intensity (Brandt 1985); also special techniques of shotcreting allow the alignment of fibres (Johnston 2001). In situ, these methods of arranging the fibres are effective within a certain acceptable tolerance; that is, most of fibres are only slightly deviated from the selected direction. When thin elements are reinforced with fibres then again the majority of fibres are situated in parallel planes and arrangement close to 2D is obtained; sometimes it may be considered as intermediary 2D to 3D distribution. The simplest way to increase the density of steel fibres toward the bottom of the element is by a vibration of fresh cement mix over a long period of time; this may be interesting for elements subjected to one sign bending moments.

The correct execution of the designed distribution requires special care and experienced workers and in most cases may be only partly achieved. Recommendations for execution of fibre reinforced concretes may be found for example in ACI 544.1R-96 (2002) and RILEM TC 162-TDF (2003).

Examples of fibre distribution are shown in radiograms in Figure 6.22.

In any type of fibre distribution the balling of fibres and their agglomerations

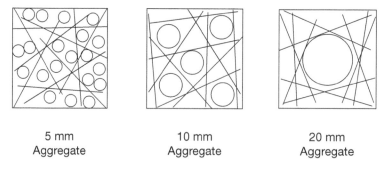

5 mm
Aggregate

10 mm
Aggregate

20 mm
Aggregate

Figure 6.21 Effect of aggregate size on the fibre structures and distribution within a square of side length equal to fibre length (40 mm), after Hannant (1978).

Table 6.4 Fibres and microfibres in concretes and mortars (3D), after Brandt (2003)

Fibres	Average dimensions	Numbers of fibres in 1 cm³			
	volume of a single fibre	0.5%	1.0%	2.0%	3.0%
Asbestos fibres	0.1μm × 4mm 3.14E-08 mm³	1.59 E+08	3.18 E+08	6.36 E+08	9.55 E+08
PAN carbon fibres	6 μm × 3 mm 8.5 E-05 mm³	59,000	118,000	236,000	354,000
Pitch carbon fibres	14.5 μm × 3 mm 0.000495 mm³	10,100	20,200	40,400	60,600
PVA fibres	24 μm × 7 mm 0.00317 mm³	1,580	3,160	6,320	9,480
Steel microfibres	0.15 mm × 6 mm 0.106 mm³	49	94	189	283
Steel fibres	0.4 mm × 30 mm 3.77 mm³	1.33	2.65	5.31	7.96

in a kind of cluster should be avoided because fibres interconnected with others are not efficient. Their bonding to the matrix is weak and reduced to local contacts and large air voids in the matrix are created. When an appropriate technology is applied to a correctly designed mix (cf. Sections 4.3 and 12.2) then the fibre clusters do not appear.

The description of idealized and real structures of fibres may be based on several different approaches. Already in 1964 a formula was proposed by Romualdi and Mandel (1964) in which the average distance between the centroids of fibres, called spacing s, is related to diameter d and fibre volume fraction V_f as demonstrated by the following equation:

$$s = 1.38d\sqrt{\frac{1}{V_f}} \tag{6.11}$$

To derive that formula it was assumed that the fibres were uniformly oriented in space and that the average length of the fibres in a direction perpendicular to the crack opening is 0.41 ℓ, where ℓ is the single fibre length. This reflects an assumption, that the directional efficiency is assumed equal to 0.41.

In that formulation the spacing s has no direct physical meaning and the proposed equation (6.11), mentioned in many books, was strongly criticized. The fibre efficiency assumed here is based on non-verifiable assumptions and has not been generally accepted. Inaccuracies in experimental verifications

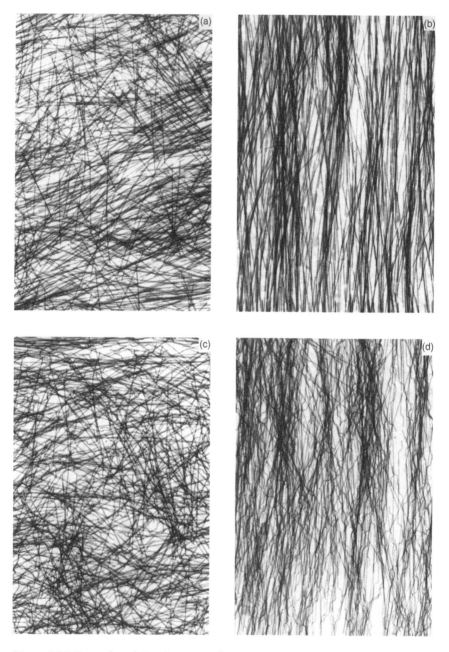

Figure 6.22 Examples of X-radiograms of cement mortar plates reinforced with steel fibres V_f = 2 %, plate depth 20 mm, direct exposition 15 min:
  (a) Plain straight fibres 0.38×25 mm, random distribution of fibres 2D
  (b) Plain straight fibres 0.38×25 mm, linearized 1D
  (c) Bekaert fibres with hooks 0.4×40 mm, random distribution 2D
  (d) Bekaert fibres with hooks 0.4×40 mm. linearized 1D.

of the equation (6.11) were indicated by Shah and Rangan (1971) and it has been proved that the coefficient 1.38 is too high for all types of ideal fibre distribution, Krenchel (1975).

The formulae proposed by Aveston and Kelly (1973) give the numbers of fibres for three ideal structures. The number of aligned (1D) fibres crossing a plane of unit area and perpendicular to their direction is:

$$N^{1D} = 4V_f / \pi \, d^2. \tag{6.12}$$

For fibres randomly oriented in a plane (2D), the corresponding equation is:

$$N^{2D} = 8V_f / \pi^2 2d^2, \tag{6.13}$$

and for random distribution in three dimensions (3D):

$$N^{3D} = 2V_f / \pi d^2. \tag{6.14}$$

Later these formulae were derived by Kasperkiewicz (1979) and presented in the form of cross-sectional areas corresponding to a single fibre:

$$\alpha^{iD} = 1/N^{iD}, \; i = 1,2,3. \tag{6.15}$$

So-called effective fibre spacing has been proposed by Swamy and Mangat (1975) in the form:

$$s_{eff} = A\sqrt{d}/\ell V_f \tag{6.16}$$

where the constant $A = 2.5$ for ultimate state of loading and $A = 2.7$ for the first crack opening. Here the fibre distribution is not analyzed and assumed as 3D.

On the basis of geometric probability theory, the efficiency and spacing concepts (meaning free spacing and average nearest neighbour distance) for idealized fibre structures have been developed by Stroeven (1978 and 1979). Using the stereological approach, a method has been determined to assess, after a cross-section or a projection analysis, the deviations from idealized structures in the form of segregation and anisometry.

Kasperkiewicz (1978a, 1978b) has proposed a coherent system of characteristics for fibre structures with two interdependent parameters: α^{iD} – the area corresponding to a cross-section in one fibre and given by equation (6.15) together with equations (6.12)–(6.14), and s_{app} – apparent spacing between fibre intercepts on a basing measuring line l_b. Both these parameters are explained in Figure 6.23. On a fracture surface or on a cross-section the fibres may be counted to establish the value of α, meaning the area corresponding to one single fibre.

Values of α^{iD} calculated on a few fracture or cross-sectional areas of examined elements may be compared with theoretical ones calculated from equations (6.12)– (6.15). Conclusions may be obtained as a result of such comparisons about the effective fibre structure and its relation to the idealized ones. In many cases the effective structures are situated somewhere in between the ideal ones and may be described as 1D to 2D or 2D to 3D.

Apparent spacing s_{app} is defined as a mean distance between intercepts of fibre projections on a basic line l_b (Figure 6.23). If w denotes the depth of the examined layer, the following equation is valid:

$$w s_{app}^{iD} = \alpha^{iD}, \text{ i} = 1,2,3, \qquad (6.17)$$

It may be observed that both parameters α^{iD} and s_{app} do not depend on the fibre length ℓ, like spacing s proposed by Romualdi and Mandel (1964).

In Figure 6.24 a few curves are shown representing α^{iD} plotted against fibre volume fraction V_f for two fibre diameters d = 40 mm and d = 25 mm and for three basic types of distribution. These curves may be used to determine the values of $\alpha^{iD}(V_f)$ or of $V_f(\alpha^{iD})$. For arbitrary values of volume fraction V_f introduced to the mix, the value of α^{iD} may be found, provided that the kind of distribution is assumed. Or, knowing the kind of distribution, values of α^{iD} from the curves may be compared to these from experimental calculations. Proposed equations were verified experimentally by Kasperkiewicz (1978a), and a satisfactory agreement has been found out. An example is shown in Figure 6.25 where $V_f(\alpha)$ is calculated from the radiograms and V_{fexp} is obtained after crushing the same specimen and washing out the fibres.

Theoretical calculations of the influence of efficiency of the fibre reinforcement on cracking stress were proposed by Romualdi and Batson as early as 1963 and examples of such curves are shown in Figure 6.26.

Experimentally the number of fibres in a cross-section is exposed by appropriate illumination and may be counted on an image as shown in Figure 6.27. Because of random orientation of fibres their cross-sections have various forms from circular to prolonged ellipses. Traces of steel fibres are not always clearly visible between aggregate grains made of natural rock and sometimes it is necessary for counting to replace a natural image by an equivalent system of points (Figure 6.28), which is then easily discerned by computerized image analysis. Counting fibres is a reliable way to check effective reinforcement in structural elements by analysis of images taken from sawn out cores. This may serve to compare number of counted fibres with assumed design volume of reinforcement or to evaluate influence of different casting methods.

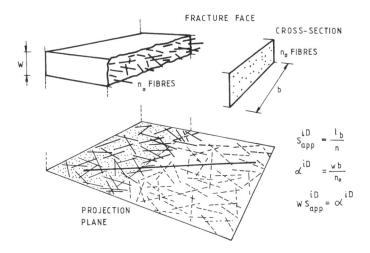

$$S_{app}^{iD} = \frac{l\,b}{n}$$

$$\alpha^{iD} = \frac{w\,b}{n_o}$$

$$W\,S_{app}^{iD} = \alpha^{iD}$$

Figure 6.23 Parameters describing fibre distribution, after Kasperkiewicz (1983).

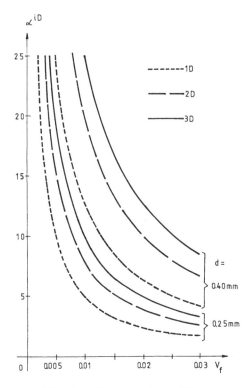

Figure 6.24 Parameters α^{iD} (i = 1, 2, 3) as functions of fibre content V_f for two fibre diameters d = 25 mm and 40 mm, after Kasperkiewicz (1978a).

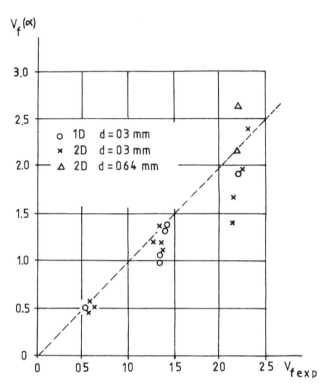

Figure 6.25 Verification of fibre content V_f (α) obtained analytically from radio-gram and curves in Figure 6.23 and V_{fexp} obtained directly by counting fibres washed out from crushed elements, after Kasperkiewicz (1978a).

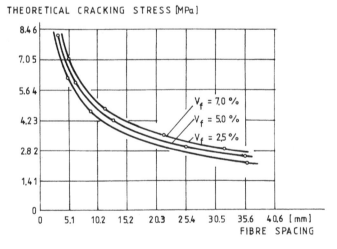

Figure 6.26 Cracking stress as a function of fibre spacing, assuming G_c = 3.5 N/m, after Romualdi and Batson (1963).

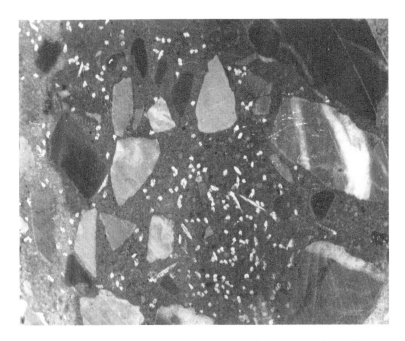

Figure 6.27 Cross-section of a steel fibre reinforced element with visible structure of fibres.

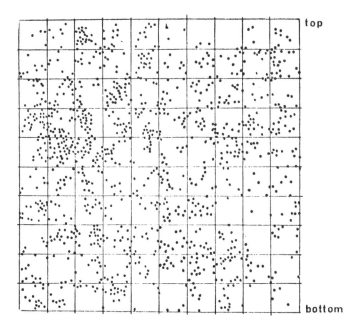

Figure 6.28 System of points representing traces of fibres in a cross-section, prepared for analysis in a computerised image analyzer.

References

Alexander, M., Mindess, S. (2005) *Aggregates in Concrete*. London: Taylor & Francis.

Atzeni, C., Massida, L., Sanna, V. (1987) 'Effect of pore distribution on strength of hardened cement pastes', in: *Proc. of 1st Congress of RILEM*, Versailles, September 1987, vol. 1, London: Chapman and Hall, pp. 195–202.

Aveston, K., Kelly, A. (1973) 'Theory of multiple fracture of fibrous composites', *Journal of Material Science*, 8(3): 152–62.

Balshin, M. Y. (1949) 'Relation of mechanical properties of powder metals and their porosity and the ultimate properties of porous metal ceramic materials', *Dokl. Akad.Nauk USSR*, 67(5): 831–34.

Brandt, A. M. (1985) 'Influence of the fibre orientation on the energy absorption at fracture of SFRC specimens', in: *Proc. Int. Symp. on Brittle Matrix Composites 1*, A. M. Brandt and I. H. Marshall eds, London and New York: Elsevier Applied Sciences, pp. 403–20.

——.(2003) 'Material structures of cement-based composites', in: Structural Image Analysis in Investigation of Concrete, Proc. *AMAS Workshop*, Inst. of Fund. Techn. Res., vol. 3, pp. 149–74.

Brandt, A. M., Kasperkiewicz, J., eds (2003) *Diagnosis of Concretes and High Performance Concretes by Structural Analysis* (in Polish) IFTR, Warsaw.

Brzezicki, J., Kasperkiewicz, J. (1997) 'Estimation of the structure of air entrained concrete using a flatbed scanner', *Cement and Concrete Research*, 35: 2041–46.

Browne, R. D., Baker, A. F. (1978) 'The performance of concrete in a marine environment', in: *Developments in Concrete Technology 1*, F. D. Lydon ed., London: Applied Sciences Publishers.

Diamond, S. (2000) 'Mercury porosimetry; an inappropriate method for the measurement of pore size distribution in cement-based materials', *Cement and Concrete Research*, 30(10): 1517–25.

Elsen, J. (2001) 'Automated air void analysis on hardened concrete: Results of a European intercomparison testing program', *Cement and Concrete Research*, 31(7): 1027–31.

Fagerlund, G. (1973a) 'Methods of characterisation of pore structure', *Lund Institute of Technology*, Lund.

——.(1973b) 'Influence of pore structure on shrinkage, strength and elastic moduli', Rep. 44, *Institute of Technology*, Division of Building Materials, Lund.

——.(1979) 'Samband mellan porositet och materials mekaniska egenskpaper' (in Swedish), Rep. 26, *Institute of Technology*, Division of Building Technology, Lund.

Guyon, E. (1988) 'Matériaux fortement hétérogènes: effects d'échelle et lois de comportement', *Materials and Structures, RILEM*, 21(122): 97–105.

Hannant, D. (1978) *Fibre Cements and Fibre Concretes*, Chichester: J. Wiley & Sons.

Hansen, T. C. (1966) 'Notes from a Seminar on structure and properties of concrete', *Stanford University, Civil Engineering Dept. Technical Report*, No. 71.

Holliday, L. (1966) 'Geometrical considerations and phase relationships, in: *Composite Materials*, L. Holliday ed., Elsevier, pp. 1–27.

Hu, Jing (2004) 'Porosity of concrete. Morphological study of model concrete', PhD Thesis, Delft University of Technology, *Optima Grafische Communicatie*.

Johnston, C. D. (2001) *Fiber-Reinforced Cements and Concretes*, London: Taylor & Francis.

Kasperkiewicz, J. (1978a) 'Apparent spacing in fibre reinforced composites', *Bulletin of the Academy of Polish Sciences, ser.sc.techn.*, 26(1): 1–9.

——.(1978b) 'Reinforcement parameter for fibre concrete', *Bulletin of the Academy of Polish Sciences, ser.sc.techn.*, 26(1): 11–18.

——.(1979) 'Analysis of idealized distributions of short fibres in composite materials', *Bulletin of the Academy of Polish Sciences, ser.sc.techn.*, 27(7): 601–9.

——.(1983) 'Internal structure and cracking processes in brittle matrix composites', (in Polish), *IFTR Reports*, 39, Warsaw.

Kasperkiewicz, J., Malmberg, B., Skarendahl, Å. (1978) 'Determination of fibre content, distribution and orientation in steel fibre concrete by X-ray technique', in: *Proc. RILEM Symp. Testing and Test Methods of Fibre Cement Composites*, R. N. Swamy ed., Lancaster: The Construction Press, pp. 297–305.

Kendall, K., Howard, A. J., Birchall, J. D. (1983) 'The relation between porosity, microstructure and strength, and the approach to advanced cement-based materials', *Phil. Trans. Roy. Soc.* A310: 139–53.

Klieger, P. (1978) 'Significance of tests and properties of concrete and concrete-making materials', *ASTM STP 169B*, Philadelphia, pp. 787–803.

Krenchel, H. (1975) 'Fiber spacing and specific fiber surface', in: Proc. RILEM. Symp. *Fibre Reinforced Cement and Concrete*, London: The Construction Press, pp. 69–79.

Kumar, R., Bhattacharjee, B. (2003) 'Porosity, pore size distribution and *in situ* strength of concrete', *Cement and Concrete Research*, 33(1): 155–64.

Luping, T. (1986) 'A study of the quantitative relationship between strength and pore size distribution of the porous materials', *Cement and Concrete Research*, 16(1): 87–96.

McGreath, D. R., Newman, J. B., Newman, K. (1969) 'The influence of aggregate particles on the local strain distribution and fracture mechanism of cement paste during drying shrinkage and loading to failure', *Bull. RILEM*, 2(7).

Mandelbrot, B. (1982) *The Fractal Geometry of Nature*, New York: Freeman.

Mindess, S., Young, J. F., Darwin, D. (2003) *Concrete*, 2nd ed. New Jersey: Prentice-Hall, Pearson Ed.

Murat, M., Bachiorrini, A. (1987) 'Résistance mécanique et notion de porosité critique: Application aux composites à base de ciment alumineux', *Proc. of 1st Congress of RILEM*, Versailles, September 1987, vol. 1, London: Chapman and Hall, pp. 203–9.

Neville, A. M., Brooks, J. F. (1987) *Concrete Technology*, London: Longman Scientific & Technical.

Odler, I., Rößler, M. (1985) 'Investigations on the relationship between porosity, structure and strength of hydrated Portland cement pastes. II. Effect of pore structure and of degree of hydration', *Cement and Concrete Research*, 15: 401–10.

Parrot, L. (1985) 'Mathematical modelling of microstructure and properties of hydrated cement', *NATO ASI Series E: Applied Science*, 95: 213–28.

Pleau, R., Pigeon, M., Laurencot, J.-L. (2001) 'Some findings on the usefulness of image analysis for determining the characteristics of the air-void systems in hardened concrete', *Cement and Concrete Composites*, 23: 237–46.

Popovics, S. (1979) *Concrete-Making Materials*, New York/Washington: Hemisphere Publishing Company and McGraw-Hill.

——.(1982) 'Production schedule of concrete for maximum profit', *Materials and Structure, RILEM*, 15(87): 199–204.

Powers, T. C. (1945) 'A working hypothesis for the further studies of frost resistance of concrete', *American Concrete Institute Journal Proc.* 41, February: 245–72.

——.(1964) 'The physical structure of Portland cement paste', in: *The Chemistry of Cements*, H. F. W. Taylor ed., vol. 1, London: Academic Press, pp. 391–416.

Romualdi J. P., Batson, G. B. (1963) 'Mechanics of crack arrest in concrete', *Journal of Engineering, Mechanical Division, ASCE Proc.*, 89(EM3): 147–68.

Romualdi, J. P., Mandel, J. A. (1964) 'Tensile strength of concrete affected by uniformly distributed and closely spaced short lengths of wire reinforcement', *Journal of the American Concrete Institute*, June: 657–70.

Rößler, M., Odler, I. (1985) 'Investigations on the relationship between porosity, structure and strength of hydrated Portland cement pastes. I. Effect of porosity', *Cement and Concrete Research*, 15: 320–30.

Ryshkevitch, E. (1953) 'Compression strength of porous sintered alumina and zirkonia', *Journal of the American Ceramic Society*, 36: 65–8.

Schiller, K. K. (1958) 'Porosity and strength of brittle solids, in: *Mechanical Properties of Non-Metallic Brittle Materials*, London: Butterworth, pp. 35–49.

Scrivener, K. L., Pratt, P. L. (1987) 'The characterization and quantification of cement and concrete microstructures', *Proc. of 1st Congress of RILEM*, Versailles, September 1987, vol. 1, London: Chapman and Hall, pp. 61–8.

Shah, S. P., Rangan, B. V. (1971) 'Fibre reinforced concrete properties', *Journal of the American Concrete Institute*, 68(2): 126–37.

Soroushian, P. Elzafraney, M., Nossoni, A. (2003) 'Specimen preparation and image processing and analysis techniques for automated quantification of concrete microcracks and voids', *Cement and Concrete Research*, 33(12): 1949–62.

Stauffer, D. (1986) *Introduction to Percolation Theory*, London: Taylor & Francis.

Stroeven, P. (1973) *Some Aspects of the Micromechanics of Concrete*, Ph.D Thesis, Stevin Laboratory, Technological University of Delft.

——.(1978 and 1979) 'Morphometry of fibre reinforced cementitious materials', *Materials and Structures, RILEM*, part I, 11(61): 31–8; part II, 12(67): 9–20.

——.(2003) 'Quantitative damage analysis of concrete', in: *Proc. Int. Symp. on Brittle Matrix Composites 7*, A.M. Brandt, V. C. Li and I. H. Marshall eds, Warsaw: ZTurek and Woodhead Publishing, pp. 121–8.

Swamy, R. N., Mangat, P. S. (1975) 'The onset of cracking and ductility of steel fiber concrete', *Cement and Concrete Research*, 5: 37–53.

Te'eni, M. (1971) 'Deformational modes and structural parameters in cemented granular systems', in: *Structure, Solid Mechanics and Engineering Design*, M. Te'eni ed., Southampton 1969, London: Wiley Interscience, pp. 621–42.

Uchikawa, H. (1988) 'Similarities and discrepancies of hardened cement paste, mortar and concrete from the standpoints of composition and structure', *Journal of the Research of the Onoda Cement Company*, 40(19), 24 pp.

Venuat, M. (1984) *Adjuvants et Traitements*, Paris: published by the author.

Verbeck, G. J., Helmuth, R. H. (1969) 'Structures and physical properties of cement pastes', in: *Proc. of the 5th Int. Symp. on the Chemistry of Cement*, part III, Tokyo, pp. 1–32.

Załocha, D. (2003) 'Image analysis as a tool for estimation of air void characteristics in hardened concrete: example of application and accuracy studies', in: Proc. Int. Workshop *Structural Image Analysis in Investigation of Concrete*, J. Kasperkiewicz, A.M. Brandt eds, Warsaw: IFTR, pp. 239–57.

Załocha, D., Kasperkiewicz, J. (2005) 'Estimation of the structure of air-entrained concrete using a flatbed scanner', *Cement and Concrete Research*, 35: 2041–6.

Standards

ACI 544.1R-96 (2002) 'State-of-the-Art Report on Fiber Reinforced Concrete', *ACI Committee 544*.

ASTM C 457-06 (2006) *Standard Test Method for Microscopical Determination of Parameters of the Air-Void System in Hardened Concrete*.

ASTM C 666/C666M –03 (2003) 'Standard Test Method for Resistance of Concrete to Rapid Freezing and Thawing', *Annual Book of ASTM Standards 2004*, 04(02), ASTM International.

CEB (1989) 'Durable Concrete Structures', *Comité Euro-International du Béton Design Guide*, Bulletin Nos. 182 and 183, Lausanne.

RILEM TC 162-TDF (2003) 'Test and design methods for steel fibre reinforced concrete $\sigma - \varepsilon$ design', *Materials and Structures*, 36(262): 560–7.

7 Interfaces in cement composites

7.1 Kinds of interfaces

The interface is a layer between two different phases of a composite material. The structure and composition of that particular layer, known as the interfacial transition zone (ITZ), depend on the properties of both neighbouring phases and also on conditions of mixing, hydration, curing and ageing of the materials.

In the concrete-like composites, the stresses are transferred from one phase to another through the interface and it should be accepted that the ITZ may be weaker or stronger, and that its structure is always different from that of the bulk matrix. The flow along and across an interface is more intensive than in other phases, because the interface layers have a lower density and may be more penetrable by fluids and gases. It is believed that the ITZ participates in the determination of the overall permeability of the material. Therefore, the influence of the ITZ properties on the mechanical behaviour of the material and on the transport of fluids and gases was studied by many researchers, even though its importance was estimated differently. The problem is, to what extent is the ITZ different from the bulk matrix and how large it is.

The quantitative determination of the influence of the ITZ is essential for mechanical properties (strength and Young's modulus) and durability (porosity) of cement-based composites, but it is difficult to determine the values of bond between two adjacent materials in concrete. Moreover, all measures aimed at modification of the ITZ, like the application of some kinds of microfillers, have an impact on the properties of the composite material itself, and these effects make analysis of test results more complex. The measurements of the ITZ properties in artificial specimens do not allow all conditions to reproduce correctly and tests performed on specimens built from concrete and stone parts have not supplied particularly useful findings. Tests performed by several researchers, for example, Roy and Jiang (1995) or Mindess and Rieder (1998) on specimens in which the ITZ was artificially created between concrete and stone parts did not supply results that may be directly used for analysis of the interface between matrix and aggregate grains in concrete.

The interface in concrete is formed partly by two neighbouring phases or, more often, predominantly by one of them, but in this region strong modifications with respect to both phases are observed. Its structure is often composed of several layers, with the inclusion of grains and pores. The main reason why the ITZ appears between two adjacent components is the so-called 'wall effect,' which means that the particles are differently packed against an impenetrable surface of an aggregate grain or of a steel bar. Moreover, as suggested by Maso (1980), after adding water to the dry mix all solid particles are covered with a water film and the network of fine ettringite crystals.

In most cases the concrete mix is subjected to vibrations that are transferred to the aggregate grains. Both these facts considerably influence the ITZ that is created in the processes of cement hydration and hardening. The structure of the interface may be very complex and it appears to be the weakest region of the composite material when exposed to external actions and loads. Interfacial regions usually occupy only a part of the material volume, but their properties considerably influence its behaviour and quality; this part is more important when the distances between neighbouring aggregate grains are small.

Several important test results and conclusions concerning the interfaces, their structure, properties and influence on overall material properties have been published in many conference proceedings, namely by Mindess and Shah (1988), and later by Maso (1993) and by Katz *et al.* (1998).

There are different types of interfaces:

- between small grains of aggregate and cement paste, which means between matrix components;
- between the matrix itself and grains of coarse aggregate from various materials;
- between matrix and reinforcement (various kinds of fibres, steel bars, cables, tendons, nonmetallic reinforcement, etc.);
- between the similar composite materials (concretes, mortars), but of different age or quality.

The transfer of forces between two adjacent phases by bonding is by far the most important factor for the mechanical behaviour of a composite material. The bond, in any of its forms ensured in the interface, is necessary to transform a mixture of various phases into a composite material. The bond itself is a combination of three phenomena, which occur simultaneously, but at different levels:

- mechanical interlocking of the cement paste and the surface of grains, bars or fibres;
- physical bonds between molecules;
- chemical reactions producing new compounds which are attached to both phases.

Even though the nature of these phenomena is basically known, their proportions in each particular case are different and were not fully understood until now. Furthermore, in many investigations related to the ITZ and carried out in the past, the newer kinds of concrete like FRC, HPC, SCC and other advanced materials have not been considered – even as their application has increased considerably over the last 20 years.

The problems of interface in concrete elements, with classic reinforcement in the form of steel bars and prestressing tendons, are not considered here because they exceed the scope of this book and are widely examined in the manuals on concrete technology and concrete structures.

7.2 Aggregate-cement paste interface

This interface was studied by many investigators who examined its morphology and nature. The present state of knowledge is far from complete, though the evidence has been obtained for the main qualitative characteristics of the interface. However, the quantitative estimation of these characteristics is still discussed in view of new tests and analyses. The different conclusions from experimental results are due to the variety of methods applied and to the lack of any standardized recommendations for test methods (Mindess 1996).

From experience in building practice it is known that for high strength concrete, material with a good bond should be used. For example, the rather weak bond of basalt grains is the reason for opinions that a high strength concrete is rarely obtained with basalt aggregate. Without a proper bond, even hard grains are partly inert fillers and not strongly strengthening inclusions. However, there is neither a universally adopted method to qualify the aggregate surface in this respect, nor is there a direct relation established between the bond quality and strength of these cement-based composites.

Some of the authors of early-published contributions to the problem of the ITZ are Farran (1956) and Luybimova and Pinus (1962). Later, a general image was proposed by Mehta and Monteiro (1988) and by Odler and Zurz (1988), among others, who presented experimental results which suggested that the transition zone was composed of a thin layer of 'duplex film' and a zone of acicular ettringite and larger portlandite crystals, which passes continuously into the bulk cement paste. The 'duplex film' consists of two different layers, portlandite crystals and CSH (calcium silicate hydrate) phase, with the latter not always being present or detectable. Duplex film is crystallized to a small extent only; its thickness is in the order of 1 μm and is usually well adhered to the grain surface. In the zone next to the duplex film, the orientation of crystals is, to some extent, influenced by the vicinity of the aggregate grain surface. The sequence of the processes in interface layers at the beginning of cement hydration was considered by many authors, without achieving conclusive results.

An example of the interface is shown schematically in Figure 7.1. However, in some cases the absence of the duplex film has been reported. Moreover,

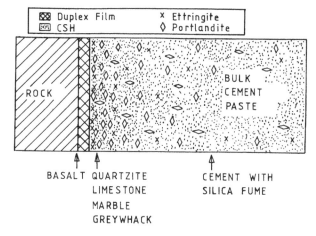

Figure 7.1 Scheme of aggregate-cement-paste interface. Possible positions of debonding fracture in composites with different components are indicated, after Odler and Zurz (1988).

because of bleeding in concrete, some kind of lens of water is often trapped under aggregate grains and, as a result, the ITZ may be different below and above these grains. The mechanical properties of the ITZ result in failure either right at the interface or at a certain distance from it. Mehta (1986) proposed another schematic presentation of the transition zone, which was frequently reproduced in subsequent books (Figure 7.2).

The structure of the interface varies as a function of the nature of aggregate and the grain surface texture. The quality of the cement paste, the mixing process and conditions of cement hydration are also of great importance. Close to the aggregate grains, ineffective packing of cement particles and different conditions of their hydration – the wall effect – were the reasons that the ITZ was found to be different from the bulk material to some extent. The influence of the aggregate size was proved by Elsharief *et al.* (2003).

One of the possible measures of the quality of the interface, presumably related to its strength, is its micro-hardness, which is understood to be hardness tested with very small forces producing microindentations. The results of measurements published by Lyubimova and Pinus (1962) are shown in Figure 7.3. It appears from these tests that depending on various circumstances, weak and strong layers may be produced in both aggregate grains and cement matrix. These authors proposed to distinguish two classes of interface according to the chemical affinity of the grains, the texture of their surface and the conditions of hydration processes in the cement matrix.

Aggregate grains obtained, for example, from basalt rock are chemically inert and hydration products exhibit weak adhesion to their surface. In that case even a local increase of strength of the matrix close to the aggregate, as shown in Figure 7.3a, has no strengthening effect on composite material,

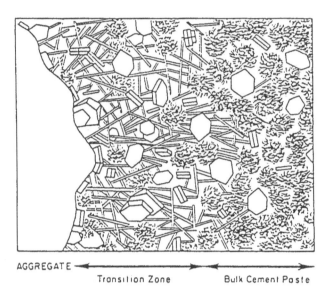

AGGREGATE ← — — — — — — → ← — — — — — — →
Transition Zone   Bulk Cement Paste

Figure 7.2 Schematic image of the transition zone and bulk cement paste in concrete, after Mehta (1986).

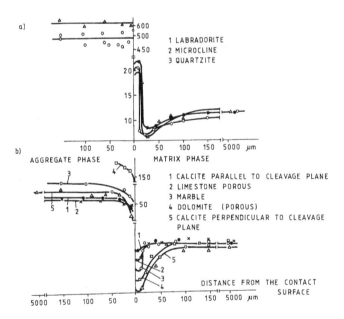

Figure 7.3 Variation of micro-hardness in the interface between aggregate grains and cement paste for various kinds of aggregate, after Lyubimova and Pinus (1962).

or only under compression stress. That increase in matrix strength may be explained by different hydration conditions in the interface layer than in the bulk matrix.

The interface between ground quartzite sand grains and cement paste creates conditions for a strong bond, which is partly of a chemical nature and from the beginning of the hydration process the sand grains behave as active elements of the material structure. In the case of the porous carbonate aggregate the bond is related to the highly developed surface of the grains. As a result, there is no clear distinction between grain and cement matrix, but the interface represents a kind of continuous passage from one to another. The strong interface of that type is not a source of cracking – on the contrary – it is a strengthening element of the composite's structure.

According to various authors, the reaction zone extends from 0 µm to 100 µm in aggregate grains, depending on their nature, and from 25 µm to 200 µm in the hardened cement paste. Other authors expressed opinions that the interface layer of higher porosity matrix around aggregate grains may even be 1–2 mm thick. In concretes with the admixture of silica fume the difference between interface and bulk material is smaller or even non-existent. Where hard rock grains are used the strong contact zone in the matrix is observed, but a layer weaker than the rock is formed on the grain surface. In comparison, the grains of carbonate rocks have weaker zones on both sides of the contact face, and this is shown in Figure 7.3b. However, there are some kinds of carbonate rocks that can develop high bond strength to the cement paste and as a result high strength concrete can be obtained. Similar effects were observed when blast-furnace slag aggregate grains were used.

Saito and Kawamura (1986) tested the micro-hardness of the interfacial zone for aggregate grains made of different rocks – limestone, andesite and granite. Their conclusions, namely that the interactions between cement paste and aggregate do not seem to contribute to the bond strength, are not fully supported by other authors; for example, Asbridge *et al.* (2002) observed that micro-hardness of the ITZ was lower than for bulk regions of mortar specimens. However, the ease of crack propagation along the interface has been shown by the majority of researchers. These discrepancies in conclusions show the inadequate knowledge that was available as to the role and importance of the interface in the fracture processes.

In Figure 7.4 the diagrams show variations in the amount of unhydrated material and porosity across the interface in concrete with an aggregate to cement ratio equal to 4:1, *w/c* = 0.4, concrete age 28 days (Scrivener *et al.* 1988). From these measurements, the thickness of the interface may be evaluated as 30–40 µm. The variation of observed properties was linear for the amount of anhydrous material and clearly non-linear for porosity. These properties of the interface explain why its micro-hardness and strength were also variable.

There are two main factors determining the strength of the adhesion of cement paste to the aggregates: chemical affinity and mechanical bond.

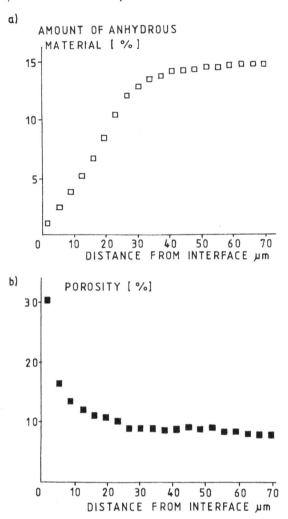

Figure 7.4 Microstructural gradients in the interfacial region of concrete:
    (a) amount of anhydrous material
    (b) porosity, after Scrivener *et al.* (1988).

Chemical reactions have been observed at the surface of siliceous grains and limestone grains, but of a different origin in both cases. Due to the reactions, an interface layer is created in most cases and its properties are different from neighbouring phases. Besides moderate chemical reactions that improve bonding and help to increase the density of the interface, the alkali-aggregate reaction (AAR) may appear in certain conditions and in an extreme case it may ruin the material's structure. The nature and effects of AAR are described in Section 4.2.2.

A mechanical bond depends on the texture of the aggregate grains. There are different possibilities, from perfectly smooth grains, which do not allow an appreciable mechanical bond, up to porous rock grains in which the abovementioned transition zone is created. A few authors found epitaxial growth of calcium hydroxide on the grain surfaces and thus mechanical interlocking was enhanced. The orientation of calcium hydroxide was measured by Grandet and Ollivier (1980a, 1980b), and it increased in the interface, which helped to explain the easier propagation of microcracks in this region. It has also been observed that ettringite was concentrated in the interface, which was probably another reason for its lower strength. Interesting comments on that subject are given in the paper by Struble (1988) who observed in 7-day-old specimens the approximately 50 μm large zone along the interface with a low proportion of large unhydrated clinker grains and a high proportion of voids. In limestone concretes a low proportion of CH has been found, which confirmed suggestions of chemical reactions between these two components. In these investigations no duplex films were found.

It has been suggested that the fracture processes along the interface were related to an increased amount of pores, voids and large crystals (Zimbelman 1985). The properties of the interface were studied by Aquino *et al.* (1995) on a system composed of a cylindrical stone specimen that was pushed out from the surrounding annulus of cement paste. The composition of the ITZ was different (fine and coarse aggregate, silica fume and latex additions) and its quality was determined from the load-slip curves. The general conclusion was that the lower porosity, the higher the bond strength.

The low strength of the transition zone is attributed to higher porosity and to the orientation of large CH crystals with weak intercrystalline bonds. Hadley (1972) has shown that hollow grains composed of Portland cement hydration products, which increased porosity, appeared more often in the interface than in bulk cement paste. These are known as 'Hadley grains,' where the pores and voids may appear in the form of clusters, which create critical flaws.

The technological reasons for the weakness of the interfacial zone may be a *w/c* ratio that is too high, incorrect packing of grains in the mix due to wrong design of the aggregate size distribution, and a defective mixing technique in which a large amount of air is entrapped in the fresh mix. The influence of the interface porosity on the durability of the concrete has been suggested by many authors.

Microcracks also appear in the interface prior to the application of any external load. They are initiated by bleeding and shrinkage of the cement paste in regions close to aggregate grain faces, which play a restraining role and cause stresses that are higher around larger grains of coarse aggregate than around sand grains. The bond to quartzite grains is also stronger than to most kinds of coarse aggregate grains. Therefore, the interface layers between the aggregate and cement matrix are the origin of microcracks. These microcracks and flaws propagate and form larger cracks under stresses and displacements caused by external loading, leading gradually to fracture. Experimental evidence of high local strain in the interface is shown in Figure 7.5 after Lusche (1974). Akçaoglu *et al.* (2004) observed that the interfacial bond was a determining factor for the concrete tensile strength while its influence on the compressive strength was negligible.

The cement-paste-aggregate grains zone was studied using computer simulation by Garboczi and Bentz (1996) who replaced this zone with properties varying along its depth by a simple shell and tried to propose the limits of such simplification. A tri-dimensional micro-structural model representing the interfacial zone with mineral micro-fillers was applied by Bentz *et al.* (1993). The influence of different admixtures was simulated using a digital image-based model. A considerable improvement was observed in the integrity of the interface due to the pozzolanic reaction of admixtures. Yang (1998) applied the Hori and Nemat-Nasser model and Mori-Tanaka theory to evaluate the equivalent Young's modulus of cement mortar as related to thickness of the ITZ. Using specimens with different values of sand to mortar volume ratio from 0 up to 50% established that the elastic modulus of the ITZ was between 20% and 40% of the matrix modulus for the ITZ thickness of 20 μm, while it was between 50% and 70% of the matrix modulus for the ITZ 40 μm thick.

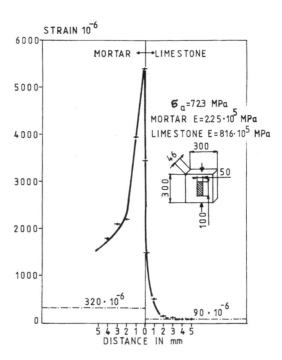

Figure 7.5 Results of the strain measurements at the border region between mortar and limestone inclusion, after Lusche (1974).

The relation between compressive strength of concrete and quality of ITZ was studied by Darwin (1995) on the basis of experimental results and FEM calculations. His conclusions were that the ITZ has an appreciable influence, but the role of the properties of the cement matrix and aggregate and non-homogeneous structure of concrete are much more important. Goldman and Bentur (1993) suggested, after tests with SF and carbon black as an inert microfiller, that the increase of concrete compressive strength was due to increased density of the ITZ by the addition of SF.

Important input to the understanding of the role of the ITZ was made by Diamond and Huang (1998). Their findings partly contested previous opinions that the composition, structure and properties of the ITZ are considerably different from bulk material, but they agree with Darwin's conclusions. After quantitative analyses of successive stripes (10 µm wide) from the aggregate surface outward they found that the hardened paste near the aggregate grains had statistically a lower content of unhydrated cement particles and a higher content of pores and of calcium hydroxide. An example of the analysis on 10 µm stripes is shown in Figure 7.6. The local variations of structure and composition of successive layers of the ITZ around the grains is a reflection of the similar patchy character of the bulk cement paste. These variations depend, to a large degree, on the duration of the concrete mixing. Also the width of the interfacial zone is very different within the same concrete and the random variations in a single layer around a cement grain were found

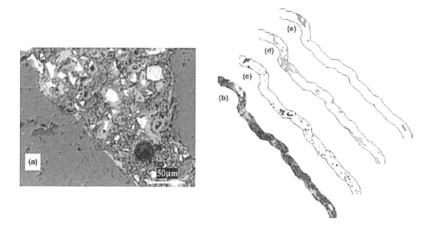

Figure 7.6 Example of the analysis of the 10 µm stripe close to the sand grain surface:
   (a) area around a sand grain to be analyzed
   (b) the first 10 µm stripe adjacent to the sand grain
   (c), *(d)* and *(e)* locations of the pixels corresponding to pores, calcium hydroxide and unhydrated cement grains, respectively. Reprinted from *Cement and Concrete Composites*, Vol 23, Diamond, S., Huang, J., 'The ITZ in concrete: a different view based on image analysis and SEM observations, pp. 10., Copyright (2001), with permission from Elsevier.

to be often greater than variations with distance from the aggregate. The general conclusion was that the mechanical effects of difference between the ITZ and bulk material are often overestimated. In the later published papers these authors even considered that influence as marginal in ordinary concretes (Diamond and Huang 2001). Also the percolation effects through the ITZ were contested by Diamond (2003) who suggested that it was due to the highly porous patches of bulk paste. The technique for specimen preparation and computer image analysis of the successive narrow stripes of the ITZ was described in detail by Diamond (2001).

In the paper by Elsharief *et al.* (2003) the percentages of porosity were determined among the other properties as a function of the *w/c* ratio, using a similar technique of narrow stripes. For *w/c* = 0.55 more porous micro-structure of the ITZ has been found, and for *w/c* = 0.4 less porous than the bulk paste. Scrivener *et al.* (2004) also have shown variations in the following characteristics (Figures 7.7 and 7.8):

- percentage of unhydrated cement grains;
- local value of *w/c* ratio;
- distribution of CH and C-S-H;
- porosity.

These results have proved the significant influence of concrete age: already after 28 days, differences between the thin zone (approximately 10 µm) close to the aggregate and bulk cement paste have decreased considerably

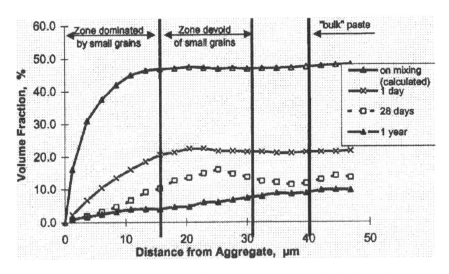

Figure 7.7 Distribution of unhydrated cement in concrete (*w/c* = 0.4) at various ages, Scrivener *et al.* (2004).

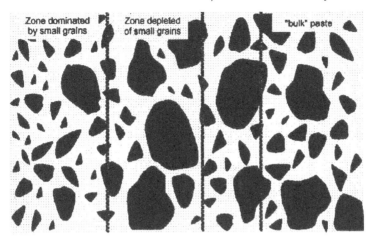

Figure 7.8 Schematic representation of grading of cement grains in ITZ, Scrivener *et al.* (2004).

and after one year have reduced to just a few percent. However, because the more porous region was 10–20 μm wide, the significant influence of the ITZ on mechanical properties was confirmed.

Kucharska (1998) confirmed, on the basis of different test results, that the importance of the ITZ is basically related to the *w/c* ratio, which means that the ITZ is of reduced width and less different from bulk material in high performance concretes than in ordinary concretes.

Even with certain contradicting opinions, the ITZ is a region where the structure is different from the bulk material due to the ineffective packing and different hydration conditions. It has been also demonstrated how the ITZ significantly depends on the composition of the binder, e.g. Kobayashi *et al.* (1998) have shown that in concrete with GGBS and limestone powder used as admixtures the ITZ had increased hardness and smaller width. Similar conclusions are proposed by Gao *et al.* (2005) who tested concrete specimens with the substitution of GGBS for Portland cement ranging from 20% to 60%. That addition considerably decreased the size and content of $Ca(OH0_2)$ crystals in the ITZ and as a result the hardness of this zone was increased. The influence was optimal for GGBS with higher specific area (600m²/kg) and for 20% replacement of the Portland cement. In such a case the ITZ as a weak zone disappeared.

The proposed conclusion of the RILEM TC 159-ETC reported by Bentur and Alexander (2000) was that in concretes and mortars the effect of the ITZ quality was moderate and was overestimated in earlier investigations. Both the composite strength and transport properties depend on several of the abovementioned parameters that influence simultaneously the ITZ and bulk matrix. Particularly in high strength systems with silica fume and other microfillers, the ITZ may be strong or even nearly eliminated as a particular region and the control of the composite behaviour by ITZ is estimated as below 20–30%.

The composition and structure of the interface may be purposefully modified to improve the composite's mechanical properties, mainly by an increase of chemical affinity and mechanical interlocking of both phases. Examples of pre-treatment procedures applied to aggregate grains to improve the strength of the interfacial zone are described by Wu Xuenquan *et al.* (1988) and Wu Keru and Zhou Jianhua (1988), among others, and the following methods to increase bonding may be mentioned:

- coating of grains with high strength cement paste before mixing;
- coating with mixture of water glass and $CaCl_2$;
- increasing density of cement paste by silica fume;
- increasing the specific area of Portland cement.

The pre-treatment may be necessary for lightweight aggregates.

Higher bond strength obtained by the improvement of the aggregate-matrix interface also has an unfavourable effect: the brittleness of the composite material is increased because in such a case the fracture energy is diminished (Mindess 1988). In the design of a material it is necessary to consider all consequences of any modification and to select material properties that are appropriate for the purpose. In fact, not only compressive and tensile strength, but also brittleness is related to the quality of the aggregate-matrix interface. Other properties that should be taken into account are permeability and durability, resistance to abrasion, etc. Particular problems arise when the aggregate from recycled concrete is used. Properties of the ITZ depend considerably upon the aggregate surface, but no significant results related to that problem have been reported.

The mechanical role of the interface is visible in the overall behaviour of the cement-based composites. Both components – aggregate grains and hardened cement paste – behave separately as linear elastic and brittle bodies in a relatively large domain of applied stresses. However, these two components form a material with a non-linear and inelastic stress-strain relation almost at the beginning of the stress-strain curve. This is caused by the microcracks, which appear and develop under tension, shearing and bending mostly in the system of interfaces.

7.3 Fibre-cement-paste interface

The fibre-cement-paste interface has been studied by many authors, and for various kinds of fibres. Initially, attempts were made to use numerous results obtained for steel bars and cement mortar in reinforced concrete elements. However, it appeared that because of the different scales and roles of steel bars and thin fibres with respect to other elements of the material structure, such as sand grains, pores, etc., only few similarities existed. This helps to explain the nature and properties of the interface in fibre-reinforced composites.

For steel fibres the chemical bond plays a relatively small role and the interfacial layer is mostly influenced by:

• the natural roughness of the fibre surface;
• the shape of the fibre, special indentations and deformations;
• modification of the cement paste in the vicinity of the fibre surface, e.g. increased *w/c* ratio, higher porosity due to restraints in packing, etc.

The transition zone around steel fibre (as the aggregate-cement paste interface) is composed of a few different layers: duplex film, the CH layer, the porous layer of CSH and ettringite (Bentur 1988). This is shown schematically in Figure 7.9. The porous layer is characterized by lower micro-hardness and strength (Figure 7.10) and that is the weakest zone in which cracks between fibre and matrix are propagating. However, the composition of the ITZ in the case of cement paste depends considerably on the quality of mixing: in a very intensively mixed paste the ITZ, as a special and weak layer, nearly disappears. In general, the density of the ITZ and the value of bonding are higher when the matrix is composed only with sand; the average bond is less sensitive to the processing of the mix, but is more influenced by the sand content due to sand-fibre interlocking (Igarashi *et al.* 1996).

An example of the interface is shown in Figure 7.11. The upper part (black) is a steel fibre; the lower part is cement mortar; and in between there is a strongly heterogeneous transition zone. Schematic representation of the interface is shown in Figure 7.12.

The application of several methods to strengthen the interface around a 10 mm steel bar was reported by Chen Zhi Yuan and Wang Nian Zhi (1989). Improvement was obtained by an admixture of silica fume, by decreasing

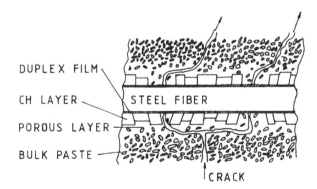

Figure 7.9 Scheme of steel fibre-cement paste interface with a crack propagating in transversal direction, after Bentur *et al.* (1985).

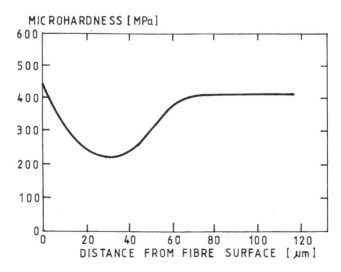

Figure 7.10 Microhardness of the cement paste measured from the steel fibre surface, after Wei *et al.* (1986).

the *w/c* ratio and by covering the surface of the bar with dry cement powder using a special technique. The results were analyzed by micro-hardness test, pore size distribution measurement, SEM observations and a pull-out test. It was concluded that:

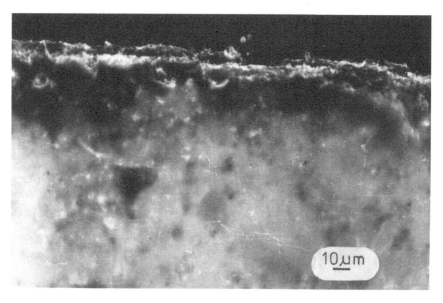

Figure 7.11 Example of steel fibre-cement mortar interface, image from optical microscope, from tests by Dr J. Potrzebowski in 1988.

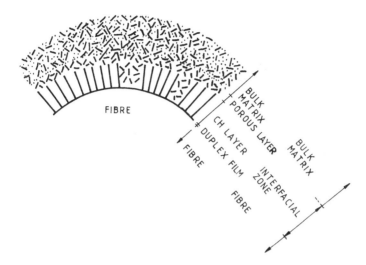

Figure 7.12 Scheme of the interfacial microstructure around a steel fibres, after Bentur (1988).

- the improved interfacial layer was denser and harder than the bulk cement paste and a minimum level of micro-hardness across the interface thickness was not observed;
- the interface was composed of smaller crystals than in non-treated specimens;
- cement grains remained with a possibility of their later hydration, therefore further increase of strength and density may be expected in the future;
- nearly 30% increase of pull-out force was observed with respect to non-treated specimens.

The fibre surface has a considerable influence on the composition of the transition zone. For highly corrosion resistant glass fibres with specially coated surface, like CemFIL2, this zone is very porous. This is in contrast to the strong interface formed around steel and asbestos fibres. In various composites with different kinds of fibres and matrices, the transition zone is formed as a result of chemical affinity, quality of the fibre surface and the penetration of the cement paste into the bundles of fibres. Furthermore, the ITZ may be different above and below a fibre due to bleeding and water lenses below fibres. Higher porosity of the ITZ around steel galvanized reinforcements than around ordinary steel rebars was observed by Belaïd *et al.* (2001).

The properties of the transition zone vary with time due to hydration of cement grains, shrinkage and eventual corrosion processes.

The effects of the weak zone around fibres are the mechanisms of crack arrest observed in many tests and described, for example, by Cook and

Gordon (1964). In these mechanisms, the interface fails ahead of the crack tip due to the tensile stress concentration, which is shown schematically in Figure 7.13a. The following sequence of events is then supposed, as shown in Figure 7.13b:

1   a crack approaches a weak interface;
2   interface fails ahead of the main crack;
3   crack stops in a T-shape or may be diverted, after Bentur (1988).

The measurements of fibre-matrix displacements in the interface are shortly described in Section 8.5.

    In contrast to the situation in unreinforced Portland cement systems, the quality of the ITZ may have a significant effect on the strength and permeability of FRC. This influence is particularly great in the composites reinforced with microfibres and with fibres in bundles. For fibres made of glass or polypropylene the main problems relate to the spaces between single fibres in a bundle. These spaces are only partly filled by hydration products and the transfer of stresses for internal fibres in a bundle is ensured only by point contacts. The chemical bond is lower than for asbestos fibres and that is another reason that empty spaces appear around single fibres. In that case, because of voids between single filaments of a bundle, the determination of the ITZ is complicated: inner and outer filaments are in different situations, e.g. glass fibres (Bentur and Alexander 2000). Moreover, the quality of the ITZ and the value of the fibre-matrix bond may decide whether, and to what proportion, the fibres will fracture or pull out at the critical load and fail.

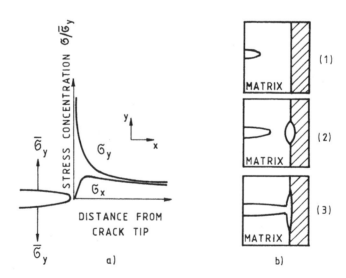

Figure 7.13 The Cook-Gordon arrest mechanisms of a crack, after Bentur (1988).

7.4 Interface between old and new composites

The strength of the interface between old and new materials is important in the repair of concrete structures. This concerns mostly structures of bridges, roads, dams, etc., which are exposed to the climatic actions of rain, frost, carbonation and various kinds of chemical agents. The problems related to repair are of increasing interest because every year, in many countries, large amounts of public money and effort are spent on repairing existing structures (cf. Section 14.5h). The quality of repair largely depends upon the strength and durability of the interface and there are several reasons for that relationship.

The interface should ensure appropriate transmission of forces between the old and new material. These forces are caused by external loads when new layers are expected to participate at load bearing capacity and stiffness of the old structure. In structures subjected to traffic or to sea waves, the external loads are of a dynamic character. Certain forces in the interface are induced, even without loading, by differential shrinkage and thermal deformations.

There is an unavoidable flow of moisture and heat across the interface, where, in new material, the processes of hydration and hardening develop, and later in service various climatic actions may have a different influence on the external repair and deeper layers of old concrete.

The stresses between old and new materials vary with time and considerably depend upon the differences in structure and composition. Even if both materials are specially matched in order to have similar properties, the difference appearing with age may prove to be the reason for additional stresses at the interface.

High impermeability should be ensured along the interface, if exposed on external actions, otherwise diffusion of moisture may endanger its durability.

The reasons described above lead to a conclusion that the interface may be the weakest element in the repair work, and as such, a source of future destruction. That is why there are many practical methods available to reinforce the interface and to ensure its high quality. Among the various known measures, special interface layers of the Portland cement paste, neat or reinforced with thin grids of steel or textile wires may be mentioned. Careful preparation of the surface of old concrete is always needed by cleaning and moistening. Intermediary layers may be prepared with increased Portland cement content, admixtures of polymeric materials or silica fume, or composed with epoxy resins.

Brandt *et al.* (1992) tested concrete slabs cast a few years ago with two kinds of concrete with basalt and limestone aggregate, with a maximum aggregate grain size of 20 mm and compressive strength approximately f_{28} = 25 MPa. On these slabs were cast layers of steel-fibre reinforced concrete (SFRC) after the old surfaces have been carefully cleaned and washed and their roughness has been artificially increased. The new material was reinforced with 1.5% of steel fibres of HAREX type (0.5 × 0.5 mm, ℓ = 16 mm). After seven days curing in 100% RH the slabs were stored for 18 months in laboratory conditions. The specimens for testing were sawn out with diamond saws and splitting tests were executed under static loading as shown in Figure 7.14. Line A–B of maximum tensile stress was situated differently in the specimens: in old concrete, in SFRC and just in the interface. The mean values of obtained results are given in Table 7.1, compared with splitting strength of concretes and SFRC separately. The conclusion was that even with careful execution of all the operations the ITZ remained the weakest link of the system, probably because no microfiller (fly ash, SF) had been added to it.

Carles-Gibergues *et al.* (1993) analysed the ITZ formed on small cylinders (20 mm in diameter) cored from old concrete on which a layer of new cement paste was cast. The results have shown that the properties of the ITZ were more dependent on the quality of cement of new paste than on the wetting of the old concrete surface. The difference in sulfate types between these cements was responsible for the different structure of the ITZ. No tests on the quality of bonding were performed in that investigation.

Li *et al.* (2001) did tests by splitting the joints made between old and new concretes. The ITZ was formed with different kinds of materials and after scanning electron microscopy (SEM) and energy dispersive spectroscopy (EDS) analyses the porosity and existence of large crystals were discovered. The bond strength of these kinds of ITZ may be ranged from the lowest to the highest: polymer modified binder, pure cement paste, expansive binder and mortar with fly ash. In all cases the joint was stronger than the old concrete. Later, in 2003, Li proposed the suggestion that for repair works of concrete structures the joints should be prepared with mortars with fly ash because of excellent strength at later age.

The tests reported by Júlio *et al.* (2004) on slant shear and pull-off specimens

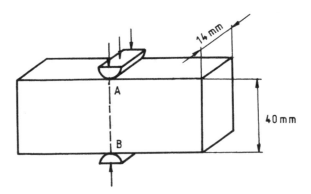

Figure 7.14 Specimen composed of old concrete and new SFRC subjected to splitting test along plane of the interface, after Brandt *et al.* (1992).

Table 7.1 Tensile strength (MPa) from splitting test (Brandt *et al.* 1992)

Concrete I with basalt aggregate (MPa)	Concrete II with limestone aggregate (MPa)	SFRC (MPa)	Interface	
			Concrete I/SFRC (MPa)	Concrete II/ SFRC (MPa)
2.69	2.85	4.05	2.10	1.78

have shown that sand-blasting was the best method for preparation of the substrate surface. The tests continued by Júlio *et al.* (2006) and performed on slant specimens built with old and new parts have proved the influence of the strength of the 'repair' concrete: the interface was much stronger when concrete of compressive strength of 100 MPa was used instead of 30 MPa concrete.

It may be concluded that even in laboratory conditions where specimens were thoroughly cleaned and wetted before casting the new concrete, the interface may be the weakest part in tension. Further tests are needed on specimens with different kinds of interface and exposure to other types of loading. It should be admitted that the influence of the scale of specimens examined is of some importance, particularly as it concerns stresses induced by differential shrinkage and thermal deformations.

For the repair of concrete slabs, the application of thin layers of SFRC was tested by Granju (1996). It has been proved theoretically and confirmed by experiment that the quality and durability of the system depends upon the crack resistance of the new layer much more than on bonding to old concrete: the cracking of the new layer preceded debonding and failure of repair. The strength of the repairing was considerably enhanced by the use of efficient reinforcement with dispersed steel fibres (SFRC) and on the basis of these results a method of repairing concrete pavements with thin SFRC layers was proposed. A similar repair technique was tested also by Bonaldo *et al.* (2005) with the ITZ improved by three different bond agents available on the market. They used a pull-off technique and concluded that if all preparatory operations were correctly performed, then the failure surface was generally situated in the substrate of old concrete, just below the ITZ. The technique of repair with thin SFRC layers is becoming the most effective for road pavements and bridge decks.

References

Akçaoglu, T., Tokyay, M., Çelik, T. (2004) 'Effect of coarse aggregate size and matrix quality on ITZ and failure behavior of concrete under uniaxial compression', *Cement and Concrete Composites*, 26(6): 633–8.

Aquino, M. J., Li, Z., Shah, S. P. (1995) 'Mechanical properties of the aggregate and cement interface', *Advanced Cement Based Materials*, 2: 211–23.

Asbridge, A. H., Page, C. L., Page, M. M. (2002) 'Effects of metakaolin, water/binder ratio and interfacial transition zones on the microhardness of cement mortars', *Cement and Concrete Research*, 32(9): 1365–9.

Belaïd, F., Arliguie, G., François, R. (2001) 'Porous structure of the ITZ around galvanized and ordinary steel reinforcements', *Cement and Concrete Research*, 31(11): 1561–6.

Bentur, A. (1988) 'Interface in fibre reinforced cements', in: Proc. Symp. *Bonding in Cementitious Composites*, Boston, 2–4 December 1987, Materials Research Society, Pittsburgh, Pennsylvania, pp. 133–44.

Bentur, A., Diamond, S., Mindess, S. (1985) 'Cracking processes in steel fibre reinforced cement paste', *Cement and Concrete* Research, 15(2): 331–42.

Bentur, A., Alexander, M. G., eds (2000) 'A review of the work of the RILEM TC15-ETC: Engineering of the interfacial transition zone in cementitious composites', *Materials and Structures*, 33: 82–7.

Bentz, D. P., Garboczi, E. J., Stutzman, P. E. (1993) 'Computer modelling of the interfacial zone in concrete', in: *Proc. Int. Conf. RILEM*, Toulouse, October 1992, E&FN Spon, pp. 107–16.

Bonaldo, E., Barros, J. A. O., Lourenço, P. B. (2005) 'Bond characterization between concrete substrate and repairing SFRC using pull-off testing', *International Journal of Adhesion & Adhesives*, 25: 463–74.

Brandt, A. M., Burakiewicz, A., Potrzebowski, J., Skawiński, M. (1992) 'Application of concrete composites to repair of overlays, decks and structures of road bridges' (in Polish), unpublished Report, Warsaw: IFTR.

Carles-Gibergues, A., Saucier, F., Grandet, J., Pigeon, M. (1993) 'New-to-old concrete bonding: influence of sulfates type of new concrete on interface microstructure', *Cement and Concrete Research*, 23: 431–41.

Chen Zhi Yuan, Wang Nian Zhi (1989) 'Strengthening the interfacial zone between steel fibres and cement paste', in: Proc. Int. Symp. *Brittle Matrix Composites 2* in Cedzyna, 20–22 September 1988, A. M. Brandt and I. H. Marshall eds, London: Elsevier Applied Science, pp. 342–51.

Cook, J., Gordon, J. E. (1964) 'A mechanism for the control of crack propagation in all brittle systems', *Proc. Roy. Soc.*, 282A: 508–20.

Darwin, D. (1995) 'The interfacial transition zone: "direct' evidence on compressive response', in: *Proc. of Mat. Res. Soc. Symposium*, Boston 1994, vol. 370, pp. 419–27.

Diamond, S. (2001) 'Considerations in image analysis as applied to investigations of the ITZ in concrete', *Cement and Concrete Composites*, 23(2–3): 171–8.

——.(2003) 'Percolation due to overlapping ITZs in laboratory mortars? A microstructural evaluation', *Cement and Concrete Research*, 33(7): 949–55.

Diamond, S., Huang, J. (1998) 'The interfacial transition zone: reality or myth?' in: Proc. 2nd Int. Conf., *The Interfacial Transition Zone in Cementitious Composites*, Haifa, March 1998, pp. 3–39.

——.(2001) 'The ITZ in concrete – a different view based on image analysis and SEM observations', *Cement and Concrete Composites*, 23(2–3): 179–88.

Elsharief, A., Cohen, M. D., Olek, J. (2003) 'Influence of aggregate size, water cement ratio and age on the microstructure of the interfacial transition zone', *Cement and Concrete Research*, 33(11): 1837–49.

Farran, J. (1956) 'Contribution minéralogique à l'étude de l'adhérence entre les constituants hydratés des ciments et les matériaux enrobés', *Revue des Matériaux de Construction et des Travaux Publics*, 490/491: 155–72; 492: 191–209.

Gao, J. M., Qian, C. X., Liu, H. F., Wang, B. Li, L. (2005) 'ITZ microstructure of concrete containing GGBS', *Cement and Concrete Research*, 35(7): 1299–1304.

Garboczi, E. J., Bentz, D. P. (1996) 'The effect of the Interfacial Transition Zone on concrete properties: the dilute limit', in: Proc. 4th Materials Engineering

Conf. *Materials for the New Millennium*, K. P. Chong ed., Washington, 10–14 November, American Society of Civil Engineers, pp. 1228–37.

Goldman, A., Bentur, A. (1993) 'Effects of pozzolanic and non-reactive microfillers on the transition zone in high strength concretes', in: Proc. Int. Conf. RILEM *Interfaces in Cementitious Composites*, Toulouse, October 1992, E&FN Spon, pp. 53–61.

Grandet, J., Ollivier, J. P. (1980a) 'Orientation des hydrates au contact des granulats', in: *Proc. of the 7th Int. Congr. Chem. Cem.*, vol. III, Paris, pp. VII63–8.

——.(1980b) 'Nouvelle méthode d'étude des interfaces ciment-granulats', in: *Proc. of the 7th Int. Congr. Chem. Cem.*, vol. III, Paris, pp. VII85–9.

Granju, J.-L. (1996) 'Thin bonded overlays', *Advanced Cement Based Materials*, Elsevier Science pp. 4, 21–7.

Hadley, D. H. (1972) 'The nature of the paste-aggregate interface', PhD Thesis, Purdue University.

Igarashi, S., Bentur, A., Mindess, S. (1996) 'The effect of processing on the bond and interfaces in Steel Fiber Reinforced Cement Composites', *Cement and Concrete Composites*, 18(5): 313–22.

Júlio, E. N. B. S., Branco, F. A. B., Silva, V. D. (2004) 'Concrete-to-concrete bond strength. Influence of the roughness of the substrate surface', *Construction and Building Materials*, 18: 675–81.

Júlio, E. N. B. S., Branco, F. A. B., Silva, V. D., Lourenço, J. F. (2006) 'Influence of added concrete compressive strength on adhesion to an existing concrete substrate', *Building and Environment*, 41: 1934–9.

Katz, A., Bentur, A., Alexander, M., Arligui, G., eds (1998) Proc. 2nd Int. Conf., *The Interfacial Transition Zone in Cementitious Composites*, Haifa, March 1998.

Kobayashi, K., Hattori, A., Miyagawa, T. (1998) 'Characters of interfacial transition zone in cement paste with admixtures', in: Proc. 2nd Int. Conf., *The Interfacial Transition Zone in Cementitious Composites*, Haifa, March 1998, pp. 311–18.

Kucharska, L. (1998) 'W/c ratio as an indication of the influence of ITZ on the mechanical properties of ordinary concretes and HPC' (in Polish), in: Proc. 2nd Conf. *Material Problems in Civil Engineering*. Mogilany, 17–19 June 1998. Cracow University of Technology, pp. 241–50.

Li, G. (2003) 'A new way to increase the long-term bond strength of new-to-old concrete by the use of fly ash', *Cement and Concrete Research*, 33: 799–806.

Li, G., Xie, H., Xiong, G. (2001) 'Transition zone studies of new-to-old concrete with different binders', *Cement and Concrete Research*, 23: 381–7.

Lusche, M. (1974) 'The fracture mechanisms of ordinary and lightweight concrete under uniaxial compression', in: Proc. Int. Conf. *Mechanical Properties and Structure of Composite Materials*, Jabłonna, 18–23 November 1974, A.M. Brandt ed., Ossolineum, pp. 423–40.

Lyubimova, T. Yu., Pinus, E. R. (1962) 'Crystallization and structuration in the contact zone between aggregate and cement in concrete' (in Russian), *Colloidnyi Zhurnal*, 24(5): 578–87.

Maso, J. C. (1980) 'Interfaces in cementitious composites', in: *Proc. 7th Int. Congress on the Chemistry of Cement*, Paris, vol. 1, pp. VII 1/3–1/15.

Maso, J. C., ed. (1993) *Interfaces in Cementitious Composites*, Proc. Int. Conf. RILEM, Toulouse, October 1992, E&FN Spon.

——.(1996) Interfacial Transition Zone in Concrete, State of the Art Report, RILEM TC 108, ICC, E&FN Spon.

Mehta, P. K. (1986) *Concrete: structure, properties, and materials*, 2nd ed., New Jersey: Prentice Hall.

Mehta, P. K., Monteiro, P. J. M. (1988) 'Effect of aggregate, cement, and mineral admixtures on the microstructure of the transition zone', in Proc. Symp. *Bonding in Cementitious Composites*, Boston, 2–4 December 1987, Pittsburgh, Pennsylvania: Materials Research Society, pp. 65–75.

Mindess, S. (1988) 'Bonding in cementitious composites: How important is it?' in: *Bonding in Cementitious Composites*, Proc. Symp. in Boston, 2–4 December 1987, Pittsburgh, Pennsylvania: Materials Research Society, vol. 114, pp. 3–10.

——.(1996) 'Mechanical properties of the Interfacial Transition Zone: a review', ACI SP 156 *Interface Fracture and Bond*, Detroit, pp. 1–10.

Mindess, S., Rieder, K. A. (1998) 'Interfacial fracture between concrete and rock under impact loading', in: Proc. 2nd Int. Conf., *The Interfacial Transition Zone in Cementitious Composites*, Haifa, March 1998, pp. 301–8.

Mindess, S., Shah, S. P. eds (1988) *Bonding in Cementitious Composites*, in: Proc. Symp., Boston, 2–4 December 1987, Pittsburgh, Pennsylvania: Materials Research Society, vol. 114.

Odler, I., Zurz, A. (1988) 'Structure and bond strength of cement-aggregate interface', in: Proc. Symp. *Bonding in Cementitious Composites*, Boston, 2–4 December 1987, Pittsburgh, Pennsylvania: Materials Research Society, pp. 21–7.

Roy, D. M., Jiang, W. (1995) 'Influences of interfacial properties on high-performance concrete composites', in: *Proc. of Mat.Res.Soc. Symposium*, Boston 1994, vol. 370, pp. 309–18.

Saito, M., Kawamura, M. (1986) 'Resistance of the cement–aggregate interfacial zone to the propagation of cracks', *Cement and Concrete Research*, 16: 653–61.

Scrivener, K. L., Crumbie, A. K., Laugesen P. (2004) 'The Interfacial Transition Zone (ITZ) between cement paste and aggregate in concrete' *Interface Science*, 12(4): 411–21.

Scrivener, K. L., Crumble, A. K., Pratt, P. L. (1988), 'A study of the interfacial region between cement paste and aggregate in concrete', in: Proc. Symp. *Bonding in Cementitious Composites*, Boston, 2–4 December 1987, Pittsburgh, Pennsylvania: Materials Research Society, pp. 87–8.

Struble, L. (1988) 'Microstructure and fracture at the cement paste–aggregate interface', in: Proc. Symp. *Bonding in Cementitious Composites*, Boston, 2–4 December 1987, Pittsburgh, Pennsylvania: Materials Research Society, pp. 11–20.

Wei, S., Mandel, J.A., and Said, S. (1986), 'Study of the interface strength in steel fibre reinforced cement based composites', Journal of American Concrete Inst., 83, pp. 597–605.

Wu Keru, Zhou Jianhua (1988) 'The influence of the matrix-aggregate bond on the strength and brittleness of concrete', in: Proc. Symp. *Bonding in Cementitious Composites*, Boston, 2–4 December 1987, Pittsburgh, Pennsylvania: Materials Research Society, pp. 29–34.

Wu Xueqan, Li Dongxu, Wu Xiun, Tang Minshu (1988) 'Modification of the interfacial zone between aggregate and cement paste', in: Proc. Symp. *Bonding in Cementitious Composites*, Boston, 2–4 December 1987, Pittsburgh, Pennsylvania: Materials Research Society, pp. 35–40.

Yang, C. C. (1998) 'Effect of the transition zone on the elastic moduli of mortar', *Cement and Concrete Research*, 28: 727–36.

Zimbelmann, R. (1985) 'A contribution to the problem of cement-aggregate bond', *Cement and Concrete Research*, 15: 801–8.

8 Strength and deformability under short-term static load

8.1 Models for unreinforced matrices

Analytical models of cement-based composite materials were initially based on the theory of elasticity with additional simplifying assumptions. The following notions were applied:

- Hooke's law is not analytically related to the ultimate values of stress and strain which mean the rupture of a continuous body;
- mean stress and mean strain, in which homogeneity of the material was assumed;
- simplified formulae for stress calculations, based on so-called technical theory of strength of materials;
- principles of superposition applied to composite materials.

That system is neither coherent nor experimentally verified, but it is simple enough for large applications and safe enough when combined with an imposed system of safety coefficients. These coefficients help to keep structures and elements made of concrete-like composites far away from ultimate states. Therefore, the inconsistencies of the system are only taken under consideration when they become unacceptable for various reasons; for example, stress concentrations, structure of ITZ, etc.

A compressive concrete strength may be examined as an example of such inconsistencies. According to the rules that have been established since the beginning of the twentieth century, the compressive strength was tested on specimens of various shapes and dimensions, always subjected to axial compression and calculated assuming that stress was distributed uniformly over the cross-section and that its value corresponded to the failure of the specimen. The main discrepancies between such procedures and reality are:

1 All cement-based composites are heterogeneous and the notion of a macroscopic stress has only very limited sense. Clear images of stress and strain variations in concrete were shown for the first time by Dantu (1958) and are presented in Figures 8.1 and 8.2. The fracture is initiated by the local stress concentrations where stress values are much higher than the mean value.

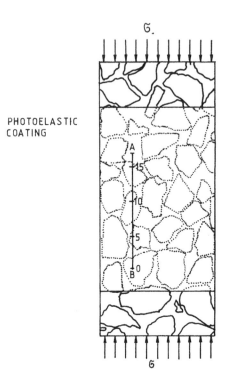

Figure 8.1 Concrete specimen subjected to axial compression $\sigma = 17$ MPa with one face covered with a photoelastic coating, after Dantu (1958).

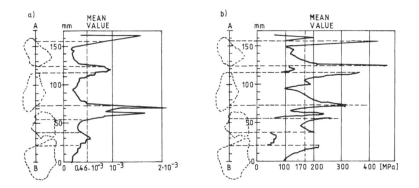

Figure 8.2 Diagrams obtained after measurement with a photoelastic coating on a lateral face of a compressed concrete element, after Dantu (1958):
   (a) Strain along AB line
   (b) Stress along AB line.

2   Nobody ever has seen a concrete specimen loaded by an axial compression where failure was caused by compression in its cross-section, because the distribution of other internal forces always produced different forms of failure, depending on conditions at the ends and other circumstances.

Another simple example is the direct tension test in which the load is again divided by the cross-sectional area, even though it is well known that a failure at tension is initiated at the weakest point, e.g. an initial crack or other defect.

The practical calculations of deformations and displacements of specimens and elements made of concrete-like composites are based on simple assumptions of Young's modulus and Poisson ratio, corrected by experimental observations. In such an approach, the strain-stress behaviour of concrete-like composites, as determined after the testing of specimens and elements, is generally considered as non-linear and non-elastic, as shown in Figure 8.3 for a case of bending or axial compression. Therefore, several different values of Young's modulus are distinguished, namely:

E_i – initial modulus is determined after the slope of the tangent at the origin of the strain-stress curve;

E_t – tangent modulus at any other point of the curve, which allows for non-elastic strain component;

E_c – secant modulus is determined at a conventionally accepted load level;

E_d – dynamic modulus, determined after measurement of the fundamental frequency of longitudinal vibrations, it is assumed that $E_d = E_i$.

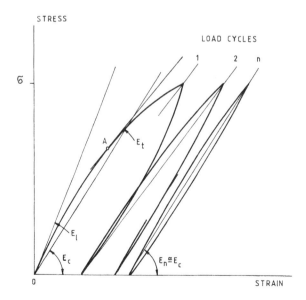

Figure 8.3 Example of a strain-stress curve for a cement based composite.

For any calculation of deformations or displacements the secant modulus E_c is usually considered. For plain concretes and mortars its value may be approximately determined after the material compressive strength E_c (f_c) and respective formulae are given in standards and recommendations.

For practical calculations it is also assumed that within well-defined limits the assumptions of linearity and elasticity may be accepted for cement-based materials. In that case the secant modulus E_c represents the approximate strain-stress behaviour of the material. Values of Young's modulus after the first, second, and n-th cycle vary and in Figure 8.3 their gradual increase is shown, which proves hardening of the material and decreasing creep in successive loading cycles. In such a case the behaviour at the n-th cycle may be considered as quasi-elastic and only slightly non-linear. It is the case of concrete structures normally used in practice, in which no accumulation of any damage should be observed. The opposite process occurs when the level of load exceeds a certain limit and cracks appear. Then, the decrease of the Young's modulus over successive cycles is a measure of the progressing process of fracture.

The Poisson ratio ν between transversal and longitudinal strain is usually assumed for concretes between 0.15 and 0.20, but for certain cases may fall to 0.11. Extensive review and test results on the values of Poisson ratio ν of HPC (high performance concrete) have been published by Persson (1999), and lower values than for OC (ordinary concrete) have been proposed. It has been found that $\nu = 0.13$ for concretes with quartzite aggregate and $\nu = 0.16$ for granite aggregate. These values are only slightly dependent on the conditions of curing and maturity of high performance concrete.

Both coefficients E and ν are approximations for much more complicated material behaviour. The relation between stress and strain varies with the load, as shown in Figure 8.3. Poisson ratio depends upon the quality of the material, but it also varies with position and direction in a considered element and with the level of loading. This more complex image of the strain fields has been examined by many authors and has been demonstrated by precise measurements executed inside concrete elements. A special measuring device with 9-gauges was used for this purpose (Figure 8.4), which measured all six strain components, and the possibility of evaluation of the scatter was ensured, thanks to the three additional gauges. Interested readers are referred to papers by Brandt (1971, 1973) and Babut and Brandt (1977).

The application of fracture mechanics to concretes, which was proposed in the early 1960s, was an important attempt to avoid the contradictions of homogeneity and continuity of cement-based composites, by Kaplan (1961, 1968). At the beginning, only linear elastic fracture mechanics (LEFM) was considered. The principal relations and formulae were taken from papers and studies concerning metals, and their application to concrete-like composites was attempted. It appeared obvious that the most direct and natural representation for the behaviour of brittle matrix composites should be based on the examination of the crack opening and propagation processes.

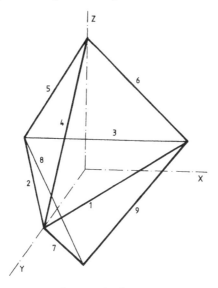

Figure 8.4 Spatial arrangement of gauges in the nine-gauge measuring device, after
Brandt (1973).

Further development of the application of fracture mechanics to concrete-like
materials is briefly described in Section 10.1.

The initial considerations concerned tensile and compressive strength and
these were developed later into more complicated strength criteria. Their main
objective was to analytically determine how the materials fracture in various
loading situations and using different strain or stress tensor components.
Because of the general conditions imposed on such criteria they should be
expressed by tensor invariants, and satisfy conditions of symmetry, etc.

In the simplest criteria, the material fracture was directly related to certain
limits of the principal values of stress or strain tensors. These criteria are
connected to the great names of Galileo, Leibniz, Navier, Lamé, Clapeyron,
Rankine and Clebsch for stress components and de Saint-Venant and Poncelet
for strain components, and they have rather historical importance. The
abovementioned notion of strength based on ultimate tensile or compressive
stress may be considered as the simplest form of the general criteria which,
by their simplicity and apparent relation to practical application in structural
design, are still used.

According to more complicated criteria, for example, Huber-von Mises-
Hencky, the energy of shearing strain is treated as representing the behaviour
of the material and is expressed as combinations of the first invariants of the
stress tensor. There are many other criteria that may be considered as more
or less similar to those mentioned here. For a more detailed description of
the strength criteria the reader is referred to books by Timoshenko (1953)
and Fung (1965), among others.

Only a few strength criteria were formulated with their application to brittle materials, such as natural rocks or concrete-like composites. They should take into account much lower tensile strength than the compressive strength, lack of effective plasticity which is replaced by microcracking, etc. Among these criteria the proposals made by Mohr (1906), and Caquot (1964) should be mentioned.

The application of classical strength criteria to brittle matrix composites gives rather unsatisfactory results (Voigt 1883), and at present it is reduced to two groups of consideration.

In the first group, the stress and strain distribution is calculated in the same manner as for homogeneous bodies and often only one stress or strain component is considered. Such calculations may only give valid results where homogenization is justified and the approximate values are acceptable for design or verification.

The second group of approaches, which derive from the single strength criteria, consists of various modifications of the law of mixtures. It is based on so-called parallel model (cf. Chapter 2) in which the Young's modulus of a two-phase (or more) composite is determined by the formula:

$$E_c = E_1 V_1 + E_2 V_2,$$

where the volumes of both phases V_1 and V_2 satisfy the condition:

$$V_1 + V_2 = 1$$

There are many proposals in the form of semi-empirical formulae based on the general concept of the law of mixtures and it is useful to review a few of them more in detail.

As applied to two-phase material – for example, cement based matrix and aggregate grains – the law of mixtures represents upper bound. That formula and other models are discussed in Section 2.5.

8.2 Strength and destruction of brittle matrices

The destruction of a brittle matrix consists of initiation and propagation of microcracks caused by local tensions. With increased load or imposed deformation, the dispersed microcracks are transformed into a system of macrocracks. Then, one or more major cracks divide the element into separate parts and the continuity of the element disappears in conjunction with a rapid decrease in bearing capacity.

The progress of the fracture process requires an energy input, at least until a certain stage; for example, in the form of an external increase in load. At different states of loading the destruction processes are influenced considerably by their duration, and also by the external constraints. The simplified description that follows is directly intended for elements subjected to axial tension, but it may be extended, with minor modifications, to other kinds of loading; for example, to the bending of concrete and fibre reinforced concrete elements.

In cement-based matrices, the internal microcracks exist from the very beginning and without any external load being applied (cf. Section 9.2). The cracks are concentrated in the interface layers between aggregate grains and cement paste, but intrinsic cracks also appear in the bulk cement mortar. If a load is applied, then its distribution is uneven because of the high heterogeneity of the material. Certain regions are subjected to much higher local stress and strain than others. Stress concentrations are the reason for the crack development even under relatively low average stress and low values of total load. That initial slow crack growth explains two phenomena, which are observed from almost the beginning of loading: limited acoustic emission (AE) and a small deviation from linearity of the stress-strain or load-displacement diagrams. The tests, in which various phenomena of fracture progress were monitored during the initial loading stages, were published several years ago by L'Hermite (1955).

Under a load corresponding to about 30% of its maximum value the AE events are already numerous and the deviation from linearity starts to be significant. The stress level of 30% is mentioned here as an approximate value, which varies in large limits, depending on material properties and the type of testing machine.

It is difficult to determine the maximum matrix strain ε_{mu} before cracks are open because it may vary considerably. In a more developed material structure, composed of inclusions, pores, fibres, etc., the value of ε_{mu} is higher than in homogeneous ones like cement paste. As to the heterogeneity of the matrix, not only are the differences between hard inclusions, weaker paste and voids considered, but also the fact that certain macroscopic composite regions have considerably different mechanical properties than others.

The influence of the biaxial or triaxial state of loading is observed in contrast to the axial tension that is theoretically applied. The influence of the rate of loading and weak or stiff testing machine is also decisive. The determination of the maximum tensile strain is subjected to different factors, such as the sensitivity of measuring devices and the magnifying power of optical instruments used for crack detection. Therefore, the definition of matrix failure is not precise enough to exclude all ambiguity. In general, it is assumed that Portland cement paste offers a maximum tensile strain of about $100\text{--}200.10^{-6}$ but smaller values of the order of 60.10^{-6} are also observed.

When the load is further increased the cracks gradually propagate and interconnect. On the load-deformation or stress-strain diagram (Figure 8.5) deviation from a straight line is due to microcracking that may be observed under relatively low stress.

As the load is increased to about 70–80% of the maximum, there is an abrupt acceleration in cracking and all other phenomena related to progressing

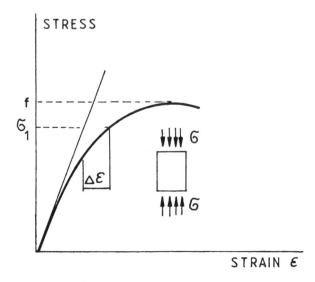

Figure 8.5 Schematic stress-strain diagram for a compressed concrete element. By δε apparent strain is denoted under stress σ₁ due to microcracks.

fracture. That level of stress is sometimes called the 'discontinuity point,' because it corresponds to qualitative modification of several processes:

1   rapid progress of the AE counts due to multiple cracks, which open and propagate;
2   inflexion of the transversal strain curve, caused by cracks which are measured as an apparent transversal deformation;
3   decrease of the ultrasonic pulse velocity, because ultrasonic waves slow down when they cross the cracks;
4   increase of the relative volume of material, in which voids start to be an important part of the apparent volume.

All these phenomena are explained by the crack formation and modification of material structure. When the load approaches its maximum value, the cracks cross the element, and this finally loses its bearing capacity in a more or less rapid way.

The above-mentioned phenomena are presented schematically in Figure 8.6.

The onset of the destruction of the material may be observed on the diagrams, but its definition is somewhat conventional, depending upon the assumed criteria. The shapes of particular curves in Figure 8.6 are related to properties of the components, and to the material's structure, etc.

Non-controlled crack propagation beyond the discontinuity point is considered as the material destruction. The slow-crack propagation before that level shows to what extent the behaviour of the real material differs from

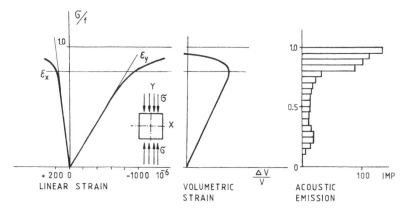

Figure 8.6 Schematic diagrams representing the behaviour of a brittle material under uniaxial compression.

that of the perfectly brittle one. The external energy furnished by the load is dispersed on the gradual creation of new surfaces of the cracks, which are stopped by inclusions of different kinds. The destruction is more abrupt in cement paste than in concrete (Figure 8.7), where hard aggregate grains are circled by propagating cracks at additional energy dispersion.

The form of the stress-strain or load-displacement curves not only depends upon the material properties, but also on the way the load is applied. There are two criteria for classification of testing machines: load or displacement control and stiffness corresponding to the capability of energy accumulation.

In 'soft' testing machines (load controlled) the energy accumulated in the loading system causes rapid fracture after cracking begins because load adjustment is impossible. The corresponding curve is modified by that effect and does not in fact reflect the material's properties. The behaviour of a soft

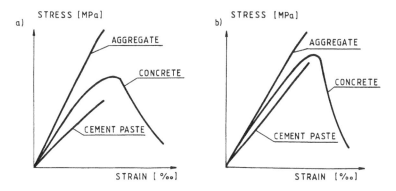

Figure 8.7 Load-deflection curves for materials more or less brittle.

machine may be compared to the action of a suspended weight directly load-ing a specimen in a tension test, and in fact such testing may furnish only information on the ascending part of the load-deformation (or stress-strain) curve. This is the case when testing concrete specimens in *in situ* laboratories where the concrete strength is determined.

Machines used in research laboratories are mostly displacement con-trolled. They are of sufficient stiffness and the actual post-cracking behaviour of brittle specimens may be represented thanks to an appropriate decrease of load, which follows gradual reduction of stiffness of the specimen and the deformation or deflection is shown more-or-less correctly. In early tests of that kind, special constructions were made to increase the inadequate stiffness of ordinary machines. Balavadze (1959) was probably the first to use such special stiffening in the form of unbonded steel bars, which were subjected to axial tension together with a nonreinforced concrete core (Figure 8.8a). He obtained a maximum strain of about $1,500.10^{-6}$ testing lightweight concrete specimens.

Steel frames were used by Evans and Marathe (1968) which deformed together with a specimen to control its deformations (Figure 8.8b). In such a frame, an explosive failure of the specimen such as would occur in an ordi-nary system where the specimen is placed under direct loading is impossible (Figure 8.8c). Balavadze explained the high extensibility of his specimens by the more uniform distribution of internal stresses in the specimen when fail-ure cannot be caused by rapid crack propagation from the weakest region.

In the non-reinforced brittle matrix specimen subjected to tensile stresses in bending or direct tension testing, the destruction is initiated at the tip of an intrinsic crack, which propagates across the specimen without obstruction. In the case of a single fracture, only a small amount of energy is required and it is reflected by the small area below the stress-strain curve as compared

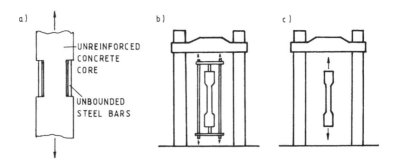

Figure 8.8 Schemes of types of loading equipment:
    (a) nonreinforced concrete core stiffened by steel bars and subjected to tension, after Balavadze (1959)
    (b) testing with special stiffening frame, after Evans and Marathe (1968)
    (c) ordinary testing of a specimen under tension.

with that of multiple cracking. To increase the energy requirement for crack propagation, reinforcement is necessary in the form of polymer impregnation, short or continuous fibres, traditional steel bars, etc.

The phenomenon of multiple-cracking confers on such a material's structure that any crack caused by local stress $\sigma > \sigma_{mu}$ cannot propagate because its tips are blocked by regions of much higher ultimate strength due to any kind of reinforcement for which $\sigma_{fu} \gg \sigma_{mu}$. Increased loading and new external energy produce the opening of new cracks in neighbouring weak regions. In such a way, a system of multiple cracks of reduced width appears in place of one wide crack. Instead of rapid failure a longer process is observed, in the form of a descending branch of the stress-strain curve. Such material behaviour is discussed in Section 9.4 and the dynamic character of cracking is analysed by Mianowski (1986).

It may be concluded that the main factor in the destruction of brittle materials is the propagation and development of cracks. That process is related to a material's structure and particularly to the distribution of weak and strong regions. It is therefore appropriate that fracture mechanics is proposed as the main approach for the explanation and modelling of fracture processes in brittle matrix composites.

In design considerations and for practical prediction of compressive strength basically at the age of 28 days, it is assumed that dependence is mainly on one parameter: this is the *w/c* ratio. A few well-known traditional formulae describe that relationship in the following form:

$$\text{Féret (1892): } f_{28} = A\left[\frac{V_c}{V_c + V_w + V_a}\right]^2$$

here V_c, V_w and V_a are volumes of cement, water and air, respectively.

$$\text{Abrams (1918): } f_{28} = A_1 \exp(-B\,w/c);$$

here w and c are masses of water and cement, respectively.

$$\text{Bolomey (1926): } f_{28} = A_2\,(c/w - B_1),$$

here A, A_1, A_2, B and B_1 are empirical constants.

These formulae are used with appropriate calibration of constants which allow for local conditions, specified properties of components (aggregate), quality of curing, etc. On the basis of 28 days compressive strength, other mechanical parameters like tensile strength or Young's modulus may be calculated from standardized formulae; these values are usually established experimentally.

8.3 Fibres in uncracked matrices

Application of the law of mixtures to fibre-reinforced composites as shown in Section 2.5 is possible after several simplifications and assumptions have been made, which transform the real behaviour of a highly heterogeneous material into that of an elastic and homogeneous one. Such an approach gives only approximate information which may, however, be useful as a general indication on the stress and strain values and on their reciprocal relations. For the simplest case of uniaxial tension or compression these assumptions are:

- continuity of the matrix, which means that there are no cracks appearing, i.e. that stress and strain are below cracking values

$$\sigma_m < f_m,\ \varepsilon_m < \varepsilon_{mu};$$

- a perfect bond between matrix and fibres is maintained

$$\varepsilon_m = \varepsilon_f = \varepsilon;$$

- only longitudinal stress and strain components are considered in the matrix as well as in the fibres σ_m, σ_f, ε_m, ε_f, without transversal components; i.e. Poisson ratio is equal to zero.

An element composed of brittle matrix and aligned continuous fibres may be considered as a two-component model subjected to uniform loading, as shown in Figure 8.9.

For unit total volume $V = V_m + V_f = 1$ and area $A = A_m + A_f = 1$ the equations are simple:

$$\text{load } F = \sigma A = \sigma_m A_m + \sigma_f A_f = (E_m A_m + E_f A_f)\varepsilon;$$

$$\sigma = \frac{F}{A_m + A_f} = \sigma_m(1 - V_f) + \sigma_f V_f \tag{8.1}$$

Because equal strain of both phases is assumed, therefore

$$\frac{\sigma}{E} = \frac{\sigma_f}{E_f} = \frac{\sigma_m}{E_m}$$

$$\text{and } E = E_m(1 - V_f) + E_f V_f \tag{8.2}$$

From equations (8.1) and (8.2) it is clear that for low fibre content the influence of fibres on stress or Young's modulus of composite is negligible (cf. Section 2.5).

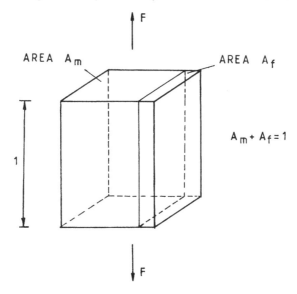

Figure 8.9 Two-phase composite material represented as a two-component model.

Consider the following numerical values: $V_f = 0.02$, $E_m = 40$ GPa, $E_f = 210$ GPa, which represent cement-based matrix reinforced with rather high content of steel continuous fibres. Then:

$$\sigma = \sigma_m (1 - V_f) + \sigma_m \frac{E_f}{E_m} V_f,$$

$$\frac{E}{E_m} = \frac{\sigma}{\sigma_m} = (1 - V_f) + \frac{E_f}{E_m} V_f = 1.085$$

In the above example the quantitative influence of fibre reinforcement on:

- average stress in a composite material corresponding to the cracking of the matrix, meaning to strain ε_{mu}, and
- Young's modulus of the composite material is equal to 8.5%.

When instead of continuous fibres, short fibres or wires are used, their efficiency should be accounted for by a coefficient $\eta_\ell \le 1.0$. Decreased efficiency is caused by the fact that the load is transferred from the matrix to the fibre by bond stress τ or friction on the fibre surface. Assuming that stress is uniformly distributed along the fibre length a concept of critical length ℓ_{cr} was introduced (Kelly 1973) as a boundary between shorter fibres which are pulled out of the matrix and longer ones which break under tensile stress. That critical length may be derived from the equilibrium between the anchorage force acting of half of the fibre length on one side of the crack and the tensile force:

$$\tfrac{1}{2} \ell_{cr} \pi d \tau = \tfrac{1}{4} \pi d^2 f_{tf}$$

$$\ell_{cr} = \tfrac{1}{2} d \, \frac{f_{tf}}{\tau} \tag{8.3}$$

where d – fibre diameter, τ – limit value of bond stress, f_{tf} – tensile strength of fibre.

The behaviour of the fibre in a tensioned element is presented in Figure 8.10 as a function of fibre length, assuming uniform distribution of the bond stress.

In these assumptions the efficiency coefficient was calculated as:

for $\ell \geq \ell_{cr}$ $\qquad\qquad \eta_\ell = 1 - \dfrac{\ell_{cr}}{2\ell}$ $\qquad\qquad$ (8.4)

for $\ell \leq \ell_{cr}$ $\qquad\qquad \eta_\ell = \dfrac{\ell}{2\ell_{cr}}$ $\qquad\qquad$ (8.5)

For brittle matrices the formulae (8.4) and (8.5) seem to be inadequate because the bond is not always perfect and cracking of the matrix modifies the phenomena completely. Several authors proposed different approaches; for example, Krenchel (1964), Riley (1968) and Allen (1971).

Laws (1971) indicated that in the pre-cracking stage the composite strain around fibres should be accounted for and the formula for the limit case when $\varepsilon_m = \varepsilon_{mu}$ was:

$$\eta_\ell = 1 - \frac{\ell_{cr}\varepsilon_{mu}}{2\ell\varepsilon_{fu}} \tag{8.6}$$

for long fibres with $\ell \geq \ell_{cr}$, where ε_{mu} and ε_{fu} are ultimate strain values for matrix and fibre, respectively.

For the post-cracking stage Laws proposed two formulae:

for long fibres with $\ell > 2\,\ell\,'_{cr}$ $$\eta_\ell = \frac{\ell_{cr}}{2\ell}\left(2 - \frac{\tau_d}{\tau_s}\right)$$ (8.7)

for short fibres with $\ell < 2\,\ell\,'_{cr}$ $$\eta_\ell = \frac{1}{2\ell_{cr}\left(2 - \tau_d/\tau_s\right)}$$ (8.8)

where $\ell\,'_{cr} = \frac{1}{2}\,\ell_{cr}\,(2 - \tau_d/\tau_s)$, τ_d is the static value of bond strength and τ_s is the friction during pull-out.

In the case of random distribution 2D or 3D, the fibres which are not parallel to the principal strain direction have considerably reduced efficiency. That reduction was determined experimentally and calculated using various geometrical and mechanical assumptions by several authors, without reaching coherent results. The values proposed by different authors for directional coefficient η_φ are shown in Table 8.1.

Certain authors proposed to multiply the coefficients to obtain overall efficiency coefficient $\eta = \eta_\varphi\,\eta_\ell$, for example, Chan and Patterson (1972), but others suggested a special combination of both, for example Laws (1971).

The considerations described above have only limited importance for the design and forecasting of the mechanical properties of brittle matrix fibre-reinforced materials prior to cracking because of the low fibre contents. Discussion on coefficients η_ℓ and η_φ may be considered as interesting refinements, which are difficult to verify quantitatively as to their influence on the composite's behaviour. Furthermore, a greater significance of technological factors is to be expected – the introduction of fibres always creates additional difficulties in compaction of the fresh mix and its porosity is usually higher than that of plain matrix.

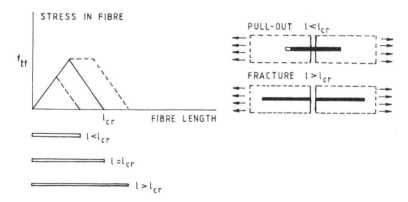

Figure 8.10 Behaviour of a fibre in an element subjected to tension for different fibre length.

Table 8.1 Directional efficiency coefficient η_0 for short fibres

Year	Authors	2D	3D
1952	Cox	0.330	0.167
1964	Romualdi and Mandel	–	0.4053
1964	Krenchel	0.375	0.2000
1971	Parimi and Rao	–	0.5–0.636
1972	Kar and Pal	0.444	0.333
1973	Aveston and Kelly	0.6366	0.500

It may also be observed that when the strength or stiffness of the composite have to be increased before its cracking, that task may be made easier by modification of w/c ratio than by the introduction of fibres.

For brittle matrix composites the influence of fibres on cracking stress and corresponding Young's modulus is negligible in the scope of the volumes of fibres that are usually applied. However, when the steel fibre volume fraction exceeds approximately 2%, then the tensile strength of the matrix is enhanced. Further increase of reinforcement, together with applications of appropriate technology, may modify the behaviour of the composite considerably, improving its deformability, Shah (1991). Slurry infiltrated fibre concretes (SIFCON) is one of the examples of such a material (cf. Section 13.5).

Investigations on the influence of fibres on the behaviour of advanced composite materials, where the fibre volume fraction is high and tends to be exploited mostly in an uncracked matrix, provided a system of reliable relations. In brittle matrix composites the situation is somewhat different: the fibre volume fraction is usually low and their influence in an uncracked state is small and covered by a large scatter, which is characteristic for these materials' properties. The fibres act on, and modify considerably, the brittle matrix behaviour only after the cracking.

Considerations concerning behaviour in an uncracked stage are presented here in limited extent only, and more information may be found in the book by Bentur and Mindess (2006). Fibre volume and behaviour of SFRC elements are verified after Eurocodes EN 14721:2000 and EN 14651:2005, respectively.

8.4 Bonding of fibres to brittle matrices

8.4.1 Nature of bonding and material design

The problem of the fibre pull-out was studied by many authors because the transfer of load from the matrix to the fibres is particularly important for the efficiency of the fibre reinforcement. A partial or approximate solution to this problem is necessary for any rational design of the fibre-reinforced composite material.

There are at least two objectives to studying the fibre-matrix bond in all aspects of this process. The first aim is to understand the nature of the bond and its relationship to the main material and technological parameters – quality of matrix, type of fibre surface and shape of a single fibre, chemical affinity of materials, influence of transversal compression due to matrix shrinkage and external load, etc. Similarly, the variation of the bond along the fibre length and its possible evolution with time are of interest. All these parameters may be considered as variables in a problem of material design or optimization and through their better understanding increased possibilities are open for purposeful material modifications. This is the justification for all the sophisticated tests and theoretical models proposed by many authors.

The second objective is to furnish necessary data for material design in the form of characteristics of the bearing capacity of one single fibre. Usually two values of the force in a fibre are sufficient: maximum value P_{max} corresponding to the end of the elastic behaviour and P_f characterizing the friction region of the force-displacement curve after debonding. These values of forces are often replaced for the purpose of calculation by the values of apparent stresses τ_{max} and τ_f, respectively, assuming a uniform distribution of stress along an embedded section of the fibre.

The fibre-matrix bond is also mentioned in other chapters; for instance, questions related to interface layers are examined in Section 7.3. The role of load transmission to fibres in cracking and fracture are studied in Chapters 9 and 10 respectively.

The fibre-matrix interaction is considered in more detail here in relation to short chopped steel-fibres. For fibres made with other materials the phenomena are different because they result from some other chemical affinity, shape of fibres and ratio of fibre and matrix Young's moduli; also the behaviour of fibres in bundles is different. Therefore these problems should be studied separately for each particular case, considering all the parameters involved.

8.4.2 Pull-out of steel fibres

In an elastic and uncracked matrix, the stress is transferred to short fibres by elastic bond stress. The fibres are considered as reinforcing inclusions. The situation of a single fibre in an imaginary cylinder cut out of the bulk matrix and subjected to tensile force in the direction parallel to the fibre was considered as shown in Figure 8.11.

Several models were all based more or less on the shear lag theory proposed by Cox (1952) who derived that the force P in the fibre at distance x from one end may be represented by the following equation:

$$P(x) = EA\varepsilon \left[1 - \frac{\cos h \, \beta \, (\ell/2 - x)}{\cos h \, \beta \, (\ell/2)} \right] \tag{8.9}$$

where $\beta = \sqrt{H/EA}$ and $H = \dfrac{2\pi G}{\log(d/D)}$,

E and A are Young's modulus and cross-sectional area of the fibre, respectively, ε is the uniform matrix strain, G is the shear modulus of the matrix. Corresponding diagrams of bond stress τ and tensile stress σ in the fibre are shown in Figure 8.11. Values of the tensile stress $\sigma(x)$ in the fibre are obtained directly from equation (8.9).

Equation (8.9) and diagrams in Figure 8.11 are simple and easy to apply to cement-based matrices, but are based on several assumptions which are not fully satisfied in brittle materials.

The brittle matrices are characterized by low tensile strength, the first cracks in the interface and fibre-matrix debonding occurs under a relatively small load. The system of microcracks is present from the beginning, which means before the application of any external load. Diagrams for τ_1 and σ_1 in Figure 8.11 concern elastic behaviour and complete fibre-matrix bond. The values of the elastic bond strength τ_u are relatively low and after debonding along a certain distance a, the diagram of the bond stress takes the form as shown for τ_2 and σ_2. The values of tensile stress σ in a fibre are also low due to weak bond stress and equal to only a small part of its strength. According to detailed calculations presented in Hannant (1978), for a bond strength equal approximately to 15 MPa, the stress in fibres does not exceed 200 MPa, which is a small part of the fibre strength. In most cases, the effective

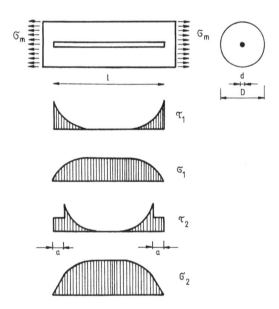

Figure 8.11 Behaviour of a steel fibre in elastic matrix subjected to tension.

bond stress and consequently the fibre tensile stress are significantly lower. Considerations given below concern the effective behaviour of steel fibres in cracked cement matrix and are aimed at conclusions derived for the design of these kind of materials.

The transfer of forces between the fibres and the matrix may be broken down for ease of understanding into a few mechanisms (Bartos 1981):

1   Elastic shear bond is a stress in the interface when its bond strength is not exceeded; it is supposed that the elastic shear bond is uniformly distributed along the fibre.
2   Frictional shear bond occurs when the bond strength is exceeded and certain relative displacement (slip) occurs.
3   Mechanical anchorage is caused by indentations or irregularities on the fibre surface, or is due to the hooks on the fibre ends.

These mechanisms are in fact complicated by the interference of the neighbouring fibres and other inclusions, by variation of the matrix strength as a function of the distance from the fibre surface, that is, by the quality of the interface (cf. Section 7.3) and by radial tensile or compressive stress, which may be exerted on the fibre.

The role of fibres in a cracked matrix is usually studied on various kinds of pull-out tests that simulate the situation of a fibre crossed by a crack as it is shown schematically in Figure 8.12 (also cf. Figure 2.5). After propagation across a non-reinforced matrix (a) a crack arrives to a fibre (b) and is stopped. Its further progress is conditioned by additional energy input; for example, by an increase of external load increasing tensile stress perpendicular to the plane of the crack. Because the crack is controlled by the fibre, its width at that point is possible only after partial debonding of the fibre and its elastic deformation (c). This phenomenon is a basis for the pull-out test in which a fibre is subjected to a force pulling-out from the matrix.

After debonding, the bond stress is gradually replaced by friction and there are different models proposed by several authors. One of the first was by Greszczuk (1969) who derived and solved an equation relating external load to the interface shear stress. Certain qualitative verification of this solution

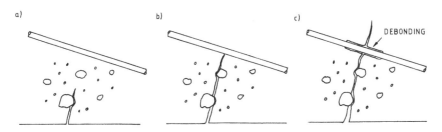

Figure 8.12 Successive stages of crack propagation across a steel fibre.

was made by a pull-out test of aluminium rods with variable embedment length from the epoxy resin matrix.

Pull-out tests were executed by numerous authors and excellent reviews of methods and results are published by Gray (1984) and Bentur and Mindess (2006). In Figure 8.13, two examples are given of the test arrangements for single fibres (a) and for groups of fibres (b) and corresponding force-displacement curves after the tests by Burakiewicz (1979). The curves for single fibres show the difference between the pull-out process for plain and indented fibres. The main difference focuses on the area under the curve; this means that to pull out an indented fibre a considerably larger amount of energy is required than for a plain one. The influence of the neighbouring fibres in a group is shown by the curves in Figure 8.13b for different volume fraction of the fibres: for V_f equal to 0.36%, 0.80% and 1.2%. The direction of the fibres expressed by the angle $\alpha = 30°$ with respect to the direction of the pull-out force also modifies the force-displacement curve.

In pull-out tests several modifications may be applied to account for real conditions:

1 influence of other neighbouring fibres by testing simultaneously groups of fibres;
2 fibres situated in such a way that an angle other than 90° appears between the fibre and the crack;
3 load which is applied in a dynamic way, producing fatigue, etc.

In pull-out tests, not only the fibre properties may be determined, but also those of the matrix, and of the technology of casting, etc.

From a pull-out test two values are usually considered as characteristic:

• maximum force denoted P_{max} corresponding to the end of linear displacement and beginning of the debonding process;
• force P_f corresponding approximately to the friction during the pull-out process itself.

Both these values, together with characteristic value for elastic displacement v_e (Figure 8.14) may be used for general characterization of the system fibres+matrix, and particularly for attempts at calculating the bearing capacity and fracture energy of fibre-reinforced elements (cf. Section 10.3). For that aim, in many studies the values of average stress were calculated, assuming uniform distribution of stress along the embedment length ℓ:

$$\tau_{av} = \frac{P}{\pi d\ell} \tag{8.10}$$

where d is the fibre diameter.

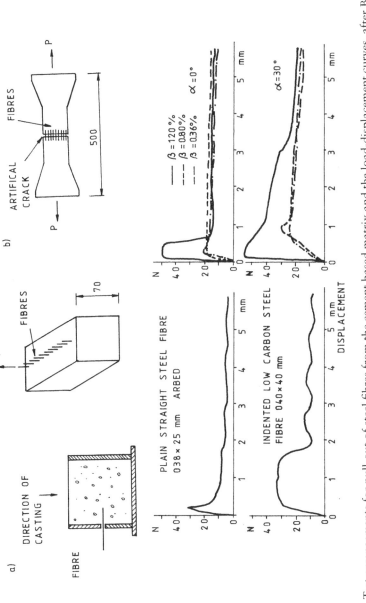

Figure 8.13 Test arrangements for pull-out of steel fibres from the cement based matrix and the load-displacement curves, after Burakiewicz (1979):
(*a*) single fibres in a specimen and examples of obtained load-displacement curves
(*b*) group of fibres subjected to simultaneous pull-out and examples of load-displacement curves.

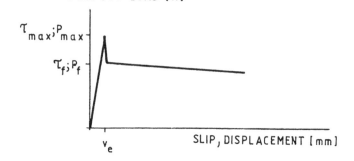

Figure 8.14 Typical diagram of pull-out load P versus displacement v, after Potrzebowski (1991).

Average interfacial shear stress τ_{av} has no physical meaning and is used only to compare the pull-out behaviour of fibres of different dimensions. The question of whether the stress is in fact uniformly distributed cannot be answered by a simple pull-out test only, because the measurements of force and displacement are executed outside the embedment zone.

The influence of the fibre-matrix bond in the case of deformed fibres was considered later by Sujivorakul and Naaman (2003) and modelled as an important parameter in the design of high performance composites, taking into account all three components that determine fibre-matrix cooperation: adhesion, friction and mechanical anchorage.

All these tests lead to the apparently justified conclusions that before P_{max} is reached the preceding linear part of the force-deflection curve corresponds to elastic shear bond, and that the debonding processes start just at the maximum pull-out force P_{max}. Such a simple image was completed by observations published already by Pinchin and Tabor (1975) who have shown that the debonding crack appears along the pulled fibre at a certain distance of a few micrometers from its surface. These observations in the fibre-matrix interface after the pull-out test corroborated well with the measurements of micro-hardness of the interface, which have shown that the weakest layer was situated at a distance from the fibre surface (cf. Chapter 7).

The problem how to observe the processes in the interface was studied by Potrzebowski (1985, 1991) and two series of combined tests have been successfully realized. In the first test, precise observations were made of the fibre-matrix interface simultaneously with the pulling out of the fibre from the matrix. For that purpose, special specimens were made in which the inter-face was partly exposed for external observation and measurements (Figure 8.15a). An example of the area observed by an optical microscope during test is shown in Figure 8.15b (cf. also Figure 7.11). The force-deflection curve on which points corresponding to observations are indicated (Figure 8.15c), together with obtained images, proved that this crack had already appeared during the linear part of force-displacement curve. By these observations it has been confirmed also that the crack was situated in the interface and not along the fibre surface.

In the next series of the tests described by Potrzebowski (1986, 1991), the relative displacements between fibre and matrix were measured in the interface by the method of laser interferometry. The results allowed the effective processes in the fibre-matrix interface to be represented in the following way: In Figure 8.15d three curves show the relative displacement between the points of the fibre and points situated on the matrix, on both sides. The measurements by the speckle method were precise enough to distinguish parts of a micrometer. It was observed (as it shown on the examples in Figure 8.15d) that already an initial part of the pull-out curve (Figure 8.15c, points 1, 2 and 3) corresponded to measurable displacements between fibre and matrix.

After these results, the pull-out behaviour may be described in the following manner: In the first stage, the elastic shear stress τ is transferred through the interface from the fibre to the matrix. This is reflected in the first part of the linear equation in Figure 8.14. In the next stage, the bond strength τ_u is reached at the beginning of the embedment zone and a crack starts to propagate in the interface along the fibre. The displacement between the fibre and the matrix, also called a slip, is observed and measured. However, the load is transferred across the crack by friction and the existence of a crack is not visible on the bond-slip or force-displacement diagram, which remains linear. Due to friction, the crack does not correspond to a decrease of the transferred force until its value P_{max} is reached. At this stage, according to measurements and observations, not only has the crack reached the end of the embedment zone, but a quite appreciable slip also occurs even at the end of that zone. The final stage corresponds to a gradual decrease of the force, accompanied with an increasing pull-out of the fibre. The rate of the force decrease is closely related to the quality of the fibre surface or to the efficiency of end anchorages.

The pull-out processes in brittle matrices were represented by different theoretical models. The linear model for the fibre-matrix bond was proposed by Nammur and Naaman (1989) in the form:

$$\tau = k\,s \tag{8.11}$$

where k is a constant which characterizes the fibre and matrix properties in this process, and s is the fibre displacement. This model was accepted and developed by several authors, but such a simple equation has not been confirmed by Potrzebowski (1991) where it has been proposed to consider the bond stress τ as a function of two variables

$$\tau = f\,[s(x), x] \tag{8.12}$$

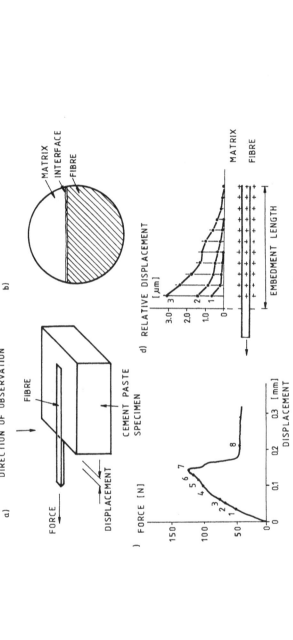

Figure 8.15 Load-slip curve combined with microscopic observations and measurements of the displacements in the interface, after Potrzebowski (1991):
(*a*) cement paste specimen with exposed fibre
(*b*) scheme of microscopic image of the interface
(*c*) force-displacement curve with points of observations and measurements
(*d*) curves of relative displacements of the fibre with respect to the matrix, corresponding to points 1, 2 and 3 in the force-displacement curve, after Potrzebowski (1991).

After the tests the model was proposed in the following form:

$$\tau(x) = \tau_{max} \tanh\left[A\left(\frac{s}{s_u}\right)^{\beta}\right] e^{(-Bx\ell_z)} \tag{8.13}$$

where τ_{max} is the maximum shear stress in the interface, of adherence or friction nature; s_u is the value of the slip s at the force equal to P_{max}; ℓ_z is the length of the embedment zone; A, B and β are coefficients which characterize the mechanical properties of fibres and matrix.

In these tests the coefficients were proposed as follows: $A = 3.5$, $B = 3.38$, $\beta = 0.92$ for $s_u = 40$ µm and $\ell = 21.3$ mm. The equation (8.13) may be represented by a 3D diagram shown in Figure 8.16. This image of the pull-out test is less simple than the previously accepted force-displacement curve, but it reflects a real process in which the equation $\tau = f(s)$ varies with the coordinate along the embedment zone.

8.4.3 Glass fibres

The bond between glass fibres and cement matrices depends mainly on chemical affinity and also on mechanical anchoring due to local surface

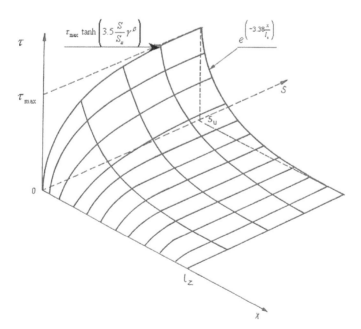

Figure 8.16 Three-dimensional diagram of relation $\tau = f(s,x)$ as in formula (8.13), after Potrzebowski (1991).

irregularities and entanglement of single filaments in a strand. Several tests have been and are being developed with the objective of measuring the interfacial properties in glass fibre reinforced composites (Bartos 1981). An extensive review of analytical studies of the fibre push-out test was proposed by Zhou *et al.* (1995) without considering any particular materials for fibres and matrix. There were basically two different theoretical models used for this problem. One is based on a criterion of shear bond stress, according to which debonding appears when the interfacial stress exceeds the shear bond strength. The second one is derived from the fracture mechanics concept of a crack that propagates along a fibre, according to the input of energy from external loading.

Chemical affinity depends on the quality of the glass. The ordinary E glass is subjected to accelerated corrosion due to high alkalinity of the Portland cement environment and the corrosion processes increase the bond as a result of gradual destruction of fibre surface. For alkali resistant zirconian glass fibres and for fibres provided with protective coating, special measures should be taken to ensure bonding. The more alkali resistant fibres are characterized by a weaker bond to the cement matrix (cf. Section 5.3).

Glass fibres are used mostly in bundles and the load transfer is complicated by voids and spaces between each single filament which are not filled up with hydration products (Bentur 1986a, 1986b). The quality of the interface layer with respect to the pull-out resistance is also dependent on the hydration conditions of the cement paste. The duration of cure in high humidity plays an important role in appropriate filling of the spaces around single filaments.

It has also been observed that the surface of the fibres influences the intensity of the hydration products' deposition. The difference between alkali resistant glass fibres and CemFIL 2 fibres provided with a special coating has been observed, which resulted in different cementation of single filaments in a strand.

Tests are usually executed on glass strands composed of multiple single filaments, embedded in a cement matrix and subjected by an intermediary of special grips to pull-out load. Such tests were reported among the others by Bartos (1982) and Oakley and Proctor (1985), and the following conclusions may be derived from these studies.

There is considerable variability in the quantitative results due to different conditions in the interface which are difficult to define precisely. The maximum ultimate elastic shear flow recorded was q_e = 62.9 N/mm, where q_e was understood as a product of unknown stress τ and ill-defined perimeter p: $q_e = \tau p$, and it corresponds to the pull-out force divided by the embedment length. After debonding, the maximum frictional shear flow recorded in these tests was q_f = 4.44 N/mm.

The length of embedment plays a considerable role in the kind of failure. For short lengths, pull-out occurred in most cases after complete debonding and longer strands are rather fractured.

The pull-out of a single glass filament is difficult because fibres mostly rupture rather than are pulled out from hardened cement matrix. Moreover, in the pull-out technique it is nearly impossible to determine the influence of other fibres in the bundle on the single fibre-matrix bond. The push-out (called also push-in) technique test involves axial loading of a single fibre using a punch until it completely debonds and slides out of the specimen. The push-out test is based on an indentation technique, which was originally developed by Marshall and Oliver (1990) for testing ceramic matrix composites. An effective use of such a test was reported for ceramic and metal matrix composites, Janczak *et al.* (1996) and Kalton (1998). The basic analysis of push-out data involves a determination of the maximum bond strength τ_m and the sliding friction stress τ_{fr}. The interfacial shear strength (bond strength) τ_m is defined by the stress corresponding to fibre debonding from the matrix. The sliding friction stress τ_{fr} is measured as the fibre moves out of the matrix. In a simple formulation the bond strength or the sliding stress are defined as follows:

τ_m and τ_{fr} are the bond strength and the friction stress, respectively; $\tau = P/(\pi D t)$;

P – is the axial load at the point of interest (debonding, sliding);

D – is the fibre diameter;

t – is the specimen thickness.

More detailed analysis of push-out data is possible considering the fibre expansion due to the effect of the Poisson ratio, contact length changes during progressive sliding of fibre out of the matrix or even residual stresses and thermal effects. The basic difference between pull-out and push-out techniques is in the influence of the Poisson ratio of the fibre whose influence of the stress distribution around the fibre has opposite signs (Zhu and Bartos 1997).

The main aim of several investigations was to determine the influence of the modification of matrices with various additions on the ageing of composites (Rajczyk *et al.* 1997, Marikunte *et al.* 1997, Brandt and Glinicki 2003). Extensive studies on the glass fibre reinforced cement composites were carried on by Glinicki (1998, 1999) and also the push-out technique was applied. In these tests, push-out tests were performed on thin slices about 440–600 µm, cut out of the specimens perpendicular to the direction of glass filaments. The specimens were made with a high addition of metakaolin in various proportions to Portland cement.

The schematic representation of the push-out technique is shown in Figure 8.17.

Specimens were mounted on a grooved stub that was then placed on the SEM stage. The displacement rate was about 0.2 µm/s, the maximum load applied was 70 N. A very thin (6.2 µm in diameter) diamond conical tip was used and any damage and splitting of the tested filaments were avoided. Single glass fibres in selected locations within the strand were loaded and pushed-out from the matrix. The selection of glass fibres was made on purpose to distinguish properties of the 'inner' fibres, surrounded by other fibres

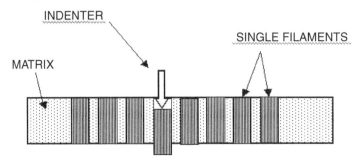

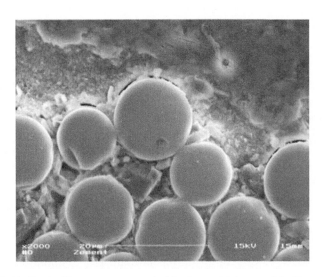

Figure 8.17 Schematic representation of the push-out technique. (Reproduced with permission from Glinicki M. A. and Brandt A. M., Quantification of glass fibre-cement interfacial properties by SEM-based push-out test; published by RILEM Publications S.a.r.l., 2007.)

and interstitial material, and the 'outer' fibres, located at the boundary of the strand (Figure 8.18).

The following conclusions have been drawn on the basis of those studies (Glinicki and Brandt 2007):

• For the matrices modified with metakaolin the increase of the maximum bond stresses τ_m was observed after accelerated ageing; approximately 50–140% for outer filaments and 40–120% for inner filaments.

Figure 8.18 Single filaments surrounded by interstitial material. (Reproduced with permission from Glinicki M. A. and Brandt A. M., Quantification of glass fibre-cement interfacial properties by SEM-based push-out test; published by RILEM Publications S.a.r.l., 2007.)

- The increase for the frictional stress τ_{fr} was also found, but reduced to approximately 40–100% and 25–60%, for outer and inner filaments, respectively.
- For increasing content of metakaolin a proportional decrease of the maximum bond stress τ_m and the frictional stress τ_{fr} after ageing was observed.
- The Marshall and Oliver (1990) model provided a very good description of push-out data.
- Results of identification of bond properties of glass fibres to modified matrices by push-out test may be applied in design.

In polymer modified cement matrices the bonding of fibres is considerably increased and consequently no pull-out was observed, but the fracture of brittle fibres was evident.

The influence of the angle of the pulled-out glass fibre strand with the crack simulated by the specimen edge was tested by Bartos (1982). Because of the brittleness of the glass strands their sensibility to an angle other than 90° was higher than observed for steel-fibres, and fracture occurred instead of plastic deformations, due to local bending. Characteristic load-displacement curves are shown in Figure 8.19. The two consecutive peaks are related to the breaking off of the loaded matrix edge and to the fracture of the strand, respectively.

8.4.4 Other fibres and hybrid reinforcement

An excellent bond to cement paste for asbestos fibres is of chemical origin, and its result is the linear and brittle behaviour of asbestos-cement products and their quasi-homogeneity, Allen (1971). A strong interface is formed around the fibres. Very small dimensions of these fibres already help in the efficient control of initial microcracking, provided that the bundles of fibres

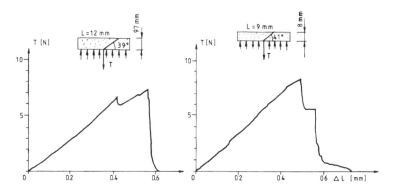

Figure 8.19 Characteristic load-displacement curves from the pull-out tests of glass fibre strands at an angle other than 90, after Bartos (1982).

are dispersed and good distribution in the bulk mix is ensured. Similarly, other kinds of thin dispersed fibres, for example, cellulose and carbon fibres, have this ability to control the crack nucleation and initial stages of crack propagation.

For several years carbon fibres have been considered to be the best reinforcement for cement matrices due to their durability in an alkaline environment, good stability at variable temperature and relatively high strength. Low modulus and low strength pitch-based carbon fibres are usually considered for these applications because they are cheaper than high modulus and high strength PAN (polyacrylonitrile) fibres (cf Section 5.7). The latter are used as reinforcement for polymer matrices in advanced high strength composite materials.

The problem of bonding is particularly important for carbon fibres, because their relatively high cost requires the maximum exploitation of their potential possibilities, but their hydrophobic nature and low affinity to cement paste makes the transfer of load between fibres and matrix rather unreliable and discontinuous. For analysis of this problem, Larson *et al.* (1990) used fibres 1.7 mm long and 18 ± 4 μm in diameter and the critical length was determined between 0.6 and 0.8 mm. The tests were arranged in such a way that the fibres of different lengths were pulled out and the critical length ℓ_{cr} was defined as corresponding to an equal percentage of fibres pulled out and broken. Therefore, bond strength may be calculated from the following equation:

$$\tau_{cr} = \frac{d\sigma_f}{4\ell_{cr}} \tag{8.14}$$

where d is the fibre diameter and σ_f is the tensile fibre strength. The bond strength calculated from this equation was between 2 MPa and 4 MPa approximately, which was confirmed by tests. Neither the variations of the w/c ratio with appropriate superplasticizer nor the addition of SF influenced the value of τ_{cr}. The increased bond was obtained with the styrene-butadiene latex admixture of 5–10% wt and by curing in hot water. However, it was observed that increased bonding and related improved bending strength did not correspond to increased flexural toughness, because the higher the bond strength, the more brittle the composite behaviour. A review of the results from testing of carbon fibre reinforced cements (CFRC) was published by Kucharska and Brandt (1997) and it was concluded that application of carbon fibres may be considered as an effective reinforcement of thin concrete plates. With the use of an appropriate cement matrix with microfillers and superplasticizers and adequate dispersion of fine carbon fibres, the control of microcracking is ensured as well as the durability of such products.

For organic vegetal fibres, due to surface irregularities the bond is based mainly on local mechanical anchorage. No reliable test results have been published on this subject.

In general, fibres with larger diameters and purposefully formed ends require higher energy at pull-out, but smaller fibres, because of their higher number for the same volume fraction, may exhibit better performance requiring smaller values of the bond strength. All these effects should be considered when fibre reinforcement is designed for a given application.

8.4.5 Influence of the fibre-matrix bond on composite behaviour

The relation of the fibre-matrix bond as determined in different pull-out tests, and the flexural or tensile behaviour of the composite elements is not completely established. In general, the better bond means higher first crack strength and ultimate strength as well. However, the relation between bond and composite strength is not linear. For various fibres and matrices it has been observed that doubled bond resulted only in a slight improvement of the ultimate strength (Gray and Johnston 1987). There are probably at least two reasons for this:

1   The fibre reinforcement efficiency is dependant not only upon the fibre-matrix bond quality, but on several other factors.
2   The bond strength established after the pull-out tests does not fully reflect the behaviour of the fibres in the matrix when the element is subjected to external load, particularly in the cracking state.

Extensive tests results were published by Markovic *et al.* (2003) on the role of pull-out force in the case of hybrid reinforcement. The considerable importance of the quality of the matrix was confirmed: density and strength of cement matrix increase pull-out force; that is, higher pull-out force and longer force-displacement curve correspond to lower *w/c* values and to application of microfillers that densify the matrix. This effect is enhanced by the presence of smaller fibres in the case of two or more kinds of fibres. In contrast, too high a density of the reinforcement may decrease the workability of the mix, which may have a negative influence on fibre-matrix bonding. Besides the bonding by adherence and friction, the role of end hooks was exhibited in these tests. As discovered earlier, with straight steel fibres a much smaller capacity was found compared with end-hooked fibres, where plastic deformation of the fibre require large amount of the energy at the hooks.

Tests of composite elements with specially prepared fibres to modify their bond in an artificial way are not considered in detail here. In an extreme case of fibres without any kind of adherence, their role is limited to inclusions transmitting only compressive stresses. It is enough to mention that approximately 25% improvement in the efficiency of steel fibres, in elements subjected to bending, has been obtained by chemical degreasing (Paillère and Serrano 1974).

8.5 Influence of fibre orientation on the cracking and fracture energy

8.5.1 Fibre orientation and fracture energy

Fibres are introduced in brittle matrices to control cracking but usually neither to increase the initial cracking stress (first crack load) nor to improve the ultimate composite strength. The first cannot be reached with a low volume of fibres, which is limited by technological reasons and the high cost of fibres. Moreover, it is easy to improve the matrix by other means; for example, by decreasing the w/c ratio. The increase of the ultimate strength of the composite is possible only by application of an efficient fibre-reinforcement of adequate volume and good fibre-matrix bond (cf. Section 13.4). However, that ultimate strength is achieved after considerable cracking, provided the cracks are sufficiently thin and correctly distributed to satisfy other requirements; for example, impermeability and durability.

The orientation of fibres is an important factor in the design of a composite material structure. Strong anisotropic effects may be created by fibres and there have been several attempts both to determine the influence of fibre orientation on mechanical properties and to optimize it. Much of the research was concentrated on advanced composites with ductile matrices and the objective function for design or optimization was the composite strength. These composites were extensively applied not just in aircraft and rockets, but also in the construction of cars; the review of these investigations is published by Ashby and Jones (2005).

The knowledge based on research in the field of advanced composites was partly used in brittle matrix composites, mainly for the stage of early cracking that appears because the ultimate matrix strain is much lower than that of the fibres: $\varepsilon_{mu} \ll \varepsilon_{fu}$.

For technological and economic reasons the volume fraction of fibres in cement-based composites is low – only with special technologies and for special applications the fibre volume fraction may be higher than about 3% for steel fibres or 5% for other kinds of fibres. It is then obvious that the fibres and their orientation play a completely different role in cement-based composites. Namely, the fibres control cracks opening and propagation, and substantially improve the material's toughness.

The problem of the orientation of fibres in cement matrices was considered first by Morton (1979) where it was shown that the work of a fracture may be considerably increased when fibres are not aligned with the direction of the principal tensile strain. These results have been based on previously published experimental works by Hing and Groves (1972), Harris *et al.* (1972) and Morton and Groves (1974). They have shown that the work of fracture, calculated as the amount of work of external load absorbed by the element, is the most important magnitude to be considered in the design of brittle matrix composites. This approach was developed in a proposal of formulae for energy calculation (Brandt 1982, 1984), and later in the solution of a simple optimization problem (Brandt 1985). An analytical solution was next compared with results of the experimental studies (Brandt 1986, 1987) in which control of the cracking process was taken into account in the design of the composite material. For that aim in these investigations the fracture energy as a function of the fibre orientation has been proposed as an objective function for optimization, formulated as a sum of a few components related to two main mechanisms:

* debonding of the fibres;
* pulling out the fibres from the matrix subjected to cracking.

The proposed equations are presented in Section 10.3.1.

The problem of the fibre orientation was also examined by Mashima *et al.* (1990), who tested tensioned specimens reinforced with unidirectional or bi-directional systems of polypropylene fibres. Because the polypropylene fibres behave differently in the cracked matrix, other fracture mechanisms were considered than those proposed for steel ones. Tests on glass fibres are mentioned in Section 8.4.3.

8.5.2 *Theoretical and experimental results for unnotched specimens*

The approach developed by Brandt (1985) was illustrated by examples of optimal solutions for an element subjected to tension and another to bending. As an objective function the fracture energy accumulated up to a specified limit state was selected. The results calculated were obtained after derivation of proposed simplified expressions with respect to the only variable – the angle θ of fibre system orientation. Later, the tests of specimens with various fibre orientation were executed and analyzed. All details of calculation and testing may be found in papers by Brandt (1986, 1991) and final results are summarized in Figures 8.20a and 8.20b. From these curves certain confirmation of theoretical results may be concluded, at least in the general shape of the curves and characteristic numerical values. However, there were no tests of elements with small values of angle $0 < \theta < 30°$ and the existence of an extremum for $\theta \neq 0°$ was neither confirmed nor excluded because of technical difficulties.

8.5.3 *Tests of notched specimens subjected to bending*

The specimens were prepared with the same mix proportions and fibre distribution as described in Brandt (1986). The reinforcement was made with mild steel round and straight wire 0.4 mm in diameter, continuous or chopped into 40 mm fibres. The results reported here concern only one volume fraction of continuous or short fibres $V = 0.67\%$. The fibre orientation of various series of specimens is shown in Figure 8.21, the angle θ being measured between aligned fibres and the longitudinal axis of the

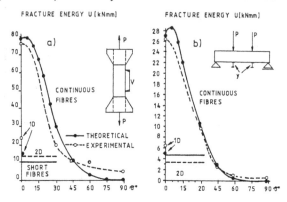

Figure 8.20 Variations of the fracture energy U with the angle β, V_f = 0.67%, results of calculations and tests, after Brandt (1991):
(*a*) specimens subjected to axial tension
(*b*) specimens subjected to bending.

specimen. The continuous fibres were aligned in several layers in the forms before concreting, each successive layer of fibres with angle + θ or − θ to avoid fracture of the plain matrix. The Portland cement mortar with 1.4 mm maximum aggregate grains was used as a matrix.

The specimens were sawn out of large slabs after hardening of the concrete and were stored in constant laboratory conditions of +18 ± 1 °C and 90 ± 2% RH. After two years curing, the specimens were notched and subjected to four-point bending in an Instron testing machine with the head displacements control. Load, central deflection and crack opening displacement (COD) were measured and recorded to calculate the characteristics of the cracking process.

The values characterizing the cracking of specimens are reported in Table 8.2. All data are the mean values obtained from testing between four and six identical specimens in a series.

The critical values of the stress concentration factor K_{Ic} were calculated from the classic formula of LEFM for the initiation of crack propagation at the notch tip and corresponding values of load P_{cr} were taken into account.

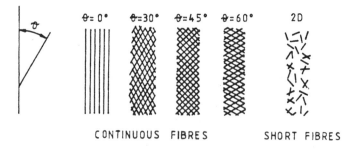

Figure 8.21 Fibre orientation systems in tested specimens, after Brandt (1991).

The curves in Figure 8.22 a show K_{Ic} and P_{cr} as functions of angle θ. For comparison, the values obtained for short fibres dispersed at random in horizontal planes (2D) are also given.

Values of the apparent stresses σ_{app} and σ'_{app} are calculated at the notch tip for two stages: crack opening and maximum load, respectively. These values have no physical meaning, as they are calculated assuming elastic behaviour and neglecting the cracks. Therefore, they are used only to compare the specimens between themselves, eliminating the influence of different dimensions. Corresponding curves are shown in Figure 8.22 b.

The fracture energy U' is calculated as the area under the curve load-deflection up to the maximum load. Specific fracture energy γ' is equal to U' divided by the area of fracture. That last magnitude was obtained in an approximate way by measuring the length of the crack, which propagated from the notch tip and was observed on both sides of each tested specimen. The area was calculated as a double product of the crack length and the beam width, without taking into account the roughness of the fracture surface. It is obvious that more exact values could be obtained considering the roughness and waviness of the fracture surface, which should be treated as a fractal object. However, it may be expected that the results obtained do have a certain significance, because the same matrix was used for all specimens and the roughness of all fracture surfaces was similar. In the same way U'' and γ'' were determined for the final state of fracture. The values of U' and γ' for different angles θ are shown in Figure 8.23.

In last columns of Table 8.2, analytical and experimental values of fracture energies U' and U'' are listed up to the maximum load and to the final fracture, respectively.

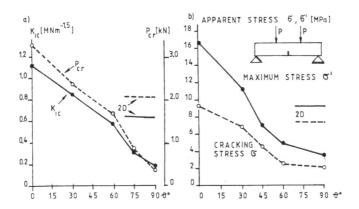

Figure 8.22 Variations of characteristic values with the angle $\beta > \aleph$
(a) variation of P – load corresponding to 1st crack opening and of K_{IC}
(b) variation of the apparent stresses σ and σ' corresponding to cracking initiation and to the minimum load, respectively, after Brandt (1991).

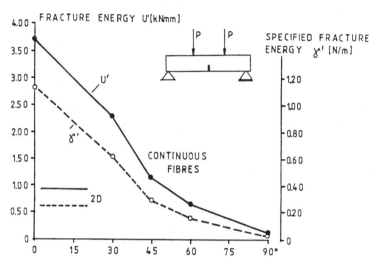

Figure 8.23 Variation of fracture energy *U*, and specific fracture energy *g* with the angle β, after Brandt (1991).

The importance of the angle θ between aligned fibres and the direction of principal tension has been confirmed for various measures of fibre efficiency. The strength and cracking characteristics depend directly on that angle. However, the shape of all obtained curves is essentially different from that shown for advanced composites, for example by Kelly and Davies (1965). The difference is due to such mechanisms as pull-out of fibres and passing of

Table 8.2 Calculated and experimental values of characteristic parameters for notched specimens

Experimental values							Calculated values	
Angle θ (°)	Force P_{cr} (kN)	Coefficient K_{Ic} (MNm$^{-1.5}$)	Stress σ_{app} (MPa)	Stress σ'_{app} (MPa)	Fracture U' (kNmm)	Fracture U" (kNmm)	Energy U' (kNmm)	Energy U" (kNmm)
Continuous fibres:								
0	3.26	1.165	9.2	16.6	3.73	5.82	2.73	4.02
30	2.36	0.850	6.8	11.2	2.32	3.10	2.62	3.03
45	1.72	0.591	4.6	7.0	1.16	1.76	1.92	2.70
60	0.85	0.327	2.6	3.9	0.66	1.19	1.01	1.39
90	0.42	0.198	2.1	3.6	0.08	0.10	0.00	0.00
Short fibres 2D:								
	1.66	0.815	7.3	9.3	0.88	2.82	0.49	2.69

Source: Brandt, A. M. (1986).

fibres across microcracks and cracks that develop quite differently in ductile and brittle materials.

The shape of the curves seems to indicate that maximum efficiency is related to angle $\theta \neq 0°$, probably between $10°$ and $15°$. This situation was not observed because there were no specimens available with such reinforcement, but such a supposition would be in agreement with theoretical considerations published in Brandt (1984, 1985).

In view of the influence of selected angle θ on the efficiency of the fibre reinforcement, it seems appropriate to consider this parameter in the design of the fibre-reinforced cements and concretes. The execution of a system of short steel fibres with arbitrary angle θ is possible using several methods: magnetic linearization of fibres in the fresh cement mix, casting of the mix through a kind of comb, shotcreting with appropriately designed parameters, etc.

In the case of any other kind of fibres or matrices it is also possible to determine adequate fracture mechanisms and to deduce analytical relations. Also the methods of execution should be adapted to properties of fibres and matrices.

The problems of fibre orientation examined here represent only a first step towards the optimization of the material structure of composites, which are also considered in Section 12.7.

8.6 Fibre-reinforced matrices as two-phase composites

The reinforcement of the brittle matrix by short dispersed fibres should be expressed in such a way to enable the designers to calculate structures and to ensure their safety and serviceability, considering the role of the fibres in a quantitative way.

The fibre-reinforced composites, namely asbestos cements and glass fibre-reinforced polymers, were analyzed as a two-phase material already by Krenchel (1964). He proposed two different theories to calculate Young's modulus and in the second the Poisson ratio is also given:

theory 1 $$E_c = E_m + V_f \left(\frac{3E_f}{8} - E_m \right)$$ (8.15)

theory 2 $$E_c = kE_m + \frac{3V_f E_f}{8} - \frac{\left(kvE + \frac{E_f V_f}{8} \right)^2}{kE_m + \frac{3E_f V_f}{8}}$$ (8.16)

$$\text{here} \quad v_c = \frac{kvE_m + \dfrac{E_f V_f}{8}}{kE_m + \dfrac{3E_f V_f}{8}} \tag{8.17}$$

where $k = \dfrac{1-V_f}{1-v^2}$ and Poisson ratio $v = v_m = v_f$.

Krenchel proposed also to determine the tensile strength of the fibre reinforced composite from a formula:

$$f_{tc} = f_{tf} \, \eta \, V_f + \sigma'_m \, (1 - V_f), \tag{8.18}$$

where f_{tf} is the tensile strength of fibres, η is the overall coefficient of efficiency and σ'_m is the matrix stress corresponding to the ultimate deformation of the fibres. In the materials tested by Krenchel, the ultimate strain of the fibres was always lower than that of the matrix. In other kinds of concrete-like composites reinforced with metallic, polymer or organic fibres, that relation is the opposite.

These formulae were partly verified experimentally. Namely, agreement within a few per cent has been found between calculated and observed values of Young's modulus and Poisson ratio. Also values of f_{tc} were only about 8% higher than the experimental ones. Other proposed formulae derived from the law of mixtures are briefly described below.

Kar and Pal (1972) proposed semi-empirical formulae for the ultimate tensile stress under bending (modulus of rupture) of fibre-reinforced concrete:

$$f_{fc} = f_{fm} \, 1.16 \, s_e^{-0.376} \text{ for } s_e \leq 1", \tag{8.19}$$

$$f_{fc} = f_{fm} \, (1.26 - 0.1 \, s_e) \text{ for } s_e > 1",$$

where s_e is effective fibre spacing expressed in inches, calculated from the following equation:

$$s_e = 9.545 \, \frac{d}{\sqrt{\dfrac{\eta v_f \ell}{d} \left(1 - \dfrac{\ell}{348d}\right)}} \tag{8.20}$$

here ℓ and d are length and diameter of a single fibre, η is the efficiency coefficient with values proposed by the authors for particular cases between 0.333 and 0.494. The formula (8.20) was derived assuming that the bond stress is distributed along the fibre length equal to $58d$. That assumption has not been justified. Both values η and s_e have no geometrical meaning and cannot be directly verified.

Another group of semi-empirical formulae has also been derived by Swamy and Mangat (1974a, 1974b). For the ultimate tensile stress and modulus of rupture of SFRC they proposed:

$$f_{tc} = f_{tm}(1-V_f) + 0.82\tau\frac{\ell}{d}V_f \text{ (tension)} \tag{8.21}$$

$$f_{tc} = \frac{\beta'}{\alpha'}f_{fm}(1-V_f) + 0.82\beta'\tau\frac{\ell}{d}V_f \text{ (bending)} \tag{8.22}$$

where $\alpha' = \dfrac{f_{fm}}{f_{tm}}$, $\beta' = \dfrac{f_{fc}}{f_{tc}}$,

τ is the mean ultimate bond strength, f_{tm} and f_{fm} are tensile and flexural matrix strength, respectively. The numerical coefficient 0.82 is equal to the double value of the efficiency coefficient derived by Romualdi and Mandel (Kelly and Davies 1965). The equation (8.22) may concern both stages of the composite material behaviour: the first crack opening and ultimate strength, and in its general form, the equation between composite strength f_c and matrix strength f_m is proposed as follows:

$$f_c = Af_m(1-V_f) + B\frac{\ell}{d}V_f \tag{8.23}$$

The numerical values of A and B were determined after several series of tests of elements made of mortar and concrete and reinforced with short steel fibres:

- for first crack opening A = 0.843; B = 2.93;
- for ultimate strength A = 0.97; B = 3.41.

The form of these formulae is obviously related to the law of mixtures, but other assumptions and coefficients are derived from tests of a given series of specimens. The general application of the formulae proposed would probably require an appropriate calibration of coefficients to examined elements for various shapes of specimens, their dimensions, material properties and other conditions.

Several other approaches to the behaviour of composite materials were based on the law of mixtures and for fibre-reinforced elements on such characteristics as ℓ/d, $\ell\,\tau/d_f$, $V_f\sqrt{(\ell/d)^3}$ and so on. In most cases, satisfactory agreement was obtained with experimental results, provided that the numerical coefficients were well selected.

It may be concluded that the application of the law of mixtures may have only local importance, that is, limited to accepted assumptions and introduced coefficients. There are two deficiencies of such an approach:

1   The fracture process is not considered and not included in proposed formulae, which in fact represent other fictitious processes.
2   The material's structure is considered in an approximate way.

The simplification of the material's structure in general may give unsatisfactory results for brittle matrix composites, because their internal structures are constructed just to increase the composite strength and fracture toughness. Also, the existence and influence of weak points where rupture is initiated are neglected. Consequently, a material's structure should be taken into account in design procedures and strength verifications.

From these various analytical approaches, confirmed by experimental data, several general observations have been proposed as to the influence of fibres on composite strength. On that basis, it was concluded by Swamy and Mangat (1974a) that the addition of 2% of steel fibres with $\ell/d = 100$ to a concrete mix with cement/sand ratio 0.4, *w/c* ratio 0.55 and 20% of coarse aggregate gives:

•   approximately twice the flexural strength;
•   an increase of compressive strength by 10–25%.

Another analytical approach to flexural strength is based on the stress distribution analysis in a cross-section and on the application of equilibrium conditions in an analogous way as it is in classical theories of reinforced concrete elements. That approach has the advantage of using the traditional methods and exploiting all available data concerning applied material components. Also test results and measurement data may be usefully introduced to improve the precision of the calculations. An example of such a comprehensive method for design and verification of fibre-reinforced elements for given limit states was also published by Babut and Brandt (1978). Its aim was to follow the basic procedure of a designer who should check bending moments in the main limit states: crack opening limit state and ultimate limit state (fracture). The procedure might be modified to suit various materials, limit states, beam dimensions, etc. Bending behaviour of the FRC is presented in Figure 10.18 showing various load-deflection curves for reinforcement with different efficiency.

Considerable progress in that approach, namely in formulating rules for structural design of elements, was achieved by the work of the RILEM Technical Committee 162-TDF 'Test and design methods for steel fibre reinforced concrete' led by L. Vandewalle (2000) and Vandewalle *et al.* (2003). The input from the fibres is based on the evaluation of the area under the respective portions of the load-deflection curves, related to the selected limit states, Section 10.2.4.

The final goal of how to design the fibre reinforcement and to take it into account by increasing the strength of reinforced cement matrix cannot be considered as solved, though the influence of the fibres is clearly visible in the tests. In the case of tension, the influence of the fibres is expressed by the value of so-called residual stress. In the design of concrete elements, in the majority of cases, the fibres are introduced by experience; their types and volumes are selected considering economic and technological arguments. The problem is discussed in more detail in Chapter 10.

References

Abrams, D.A. (1918) 'Design of Concrete Mixtures', Bulletin 1, *Structural Materials Research Laboratory*, Lewis Institute, Chicago.

Allen, H. G. (1971) 'Tensile properties of seven asbestos cements', *Composites*, 2, June: 98–103.

Ashby, M. F., Jones, D. R. H. (2005) *Engineering Materials*, parts 1 and 2, 3rd edition, Butterworth Heinemann.

Aveston, J., Kelly, A. (1973) 'Theory of multiple fracture of fibrous composites', *Journal of Materials Science*, 8(3): 352–62.

Babut, R., Brandt, A. M. (1977) 'Measurements of internal strains by nine-gauge devices', *Strain*, 13(1): 18–21.

——.(1978) 'The method of testing and analyzing of steel fibre reinforced concrete elements in flexure', in: *Proc. Int. Symp. RILEM Testing and Test Methods of Fibre Cement Composites*, R. N. Swamy ed., Sheffield: The Construction Press, pp. 479–86.

Balavadze, W. K. (1959) 'Influence of reinforcement on the properties of concrete subjected to tension' (in Russian), *Beton i Zhelezobeton*, 10: 462–65.

Bartos, P. (1981) 'Review paper: bond in fibre reinforced cements and concretes', *International Journal of Cement Composites & Lightweight Concrete*, 3(3): 159–77.

——.(1982) 'Bond in glass reinforced cements', in: *Proc. Int. Conf. Bond in Concrete*, Paisley, June 1982, P. Bartos ed., London: Applied Science Publishers, pp. 60–71

Bentur, A. (1986a) 'Microstructure and performance of glass fibre-cement composites', in: *Proc. Eng. Found. Conf. Research on the Manufacture and Use of Cements*, G. Frohnsdorff ed., Engineering Foundation, pp. 197–208.

——.(1986b) 'Mechanisms of potential embrittlement and strength loss of glass fibre reinforced cement composites', in: *Proc. Symp. Durability of Glass Fiber Reinforced Concrete*, S. Diamond ed., Chicago: Prestressed Concrete Institute, pp. 109–123.

Bentur, A., Mindess, S. (2006) *Fibre Reinforced Cementitious Composites*, 2nd ed. London: Taylor & Francis.

Bolomey, J. (1926) 'Determination of the compressive strength of mortar and concrete (in German). *Schweitzerische Bauzeitung,* Nos. 2 and 3, pp. 41–44 and 55–59.

Brandt, A. M. (1971) 'Les déformations du béton d'après la mesure de six composantes', *Cahiers de la Recherche,* 29, Eyrolles, Paris.

——.(1973) 'Nine-gauges device for strain measurements inside concrete', *Strain,* 9(3): 122–4.

——.(1985) 'On the optimal direction of short metal fibres in brittle matrix composites', *Journal of Materials Science,* 20: 3831–41.

——.(1982) 'On the calculation of fracture energy in SFRC elements subjected to bending', in: *Proc. Conf. Bond in Concrete,* P. Bartos ed., Paisley, 1982. London: Applied Science Publishers, pp. 73–81.

——.(1984) 'On the optimization of the fiber orientation in cement based composite materials', in: *Proc. Int. Symp. Fiber Reinforced Concrete,* Detroit, 1982, G.C. Hoff ed., American Concrete Institute, pp. 267–85.

——.(1986) 'Influence of the fibre orientation on the energy absorption at fracture of SFRC specimens', in: *Proc. of Int. Symp. Brittle Matrix Composites 1,* Jablonna, 1985, A. M. Brandt, I. H. Marshall eds, Elsevier Applied Science Publishers, pp. 403–20.

——.(1987) 'Influence of the fibre orientation on the mechanical properties of fibre reinforced cement (FRC) specimens', in: *Proc. 1st RILEM Congress* in Versailles, vol. 2, London: Chapman and Hall, pp. 651–8.

——.(1991) 'Influence of fibre orientation on the cracking and fracture energy in brittle matrix composites', in: *Proc. Euromech Coll. 269 Mechanical identification of composites,* A. Vautrin and H. Sol eds, London: Elsevier Applied Science, pp.327–34.

Brandt, A.M., Glinicki, M.A. (2003) 'Effects of pozzolanic additives on long-term flexural toughness of HPGFRC', in: *Proc. Int. Workshop on High Performance Fiber Reinforced Cement Composites HPFRCC-4,* A.E. Naaman and H.W. Reinhardt, eds, Ann Arbor, Michigan, pp. 399–408.

Burakiewicz, A. (1979) 'Testing of fibre bond strength in cement matrix', in: *Proc. RILEM Symp. Testing and Test Methods of Fibre Reinforced Cement Composites,* Sheffield, 1978, R. N. Swamy ed., The Construction Press, pp. 312–27.

Caquot, A. (1950) 'Commission d'études techniques de la Chambre Syndicale des Constructeurs en ciment armé de France', *Travaux,* October 1950, No. 192.

Chan, H. S., Patterson, W. A. (1972) 'The theoretical prediction of the cracking stress of glass fibre reinforced inorganic cement', *Journal of Materials Science,* 7: 856–60.

Cox, H. L. (1952) 'The elasticity and strength of paper and other fibrous materials', *British Journal of Applied Physics,* 3: 72–79.

Dantu, P. (1958) 'Étude des contraintes dans le milieux hétérogènes. Applications au béton', *Annales de l'ITBTP,* 11(121): 55–67.

Evans, R. H., Marathe, M. S. (1968) 'Microcracking and stress-strain curve for concrete in tension', *Materials and Structures RILEM,* 1(1): 61–4.

Fung, Y. C. (1965) *Foundations of Solid Mechanics,* Englewood Cliffs, New Jersey: Prentice-Hall.

Glinicki, M. A. (1999) 'Brittleness mechanisms and the durability of glass fibre reinforced cement composites' (in Polish), *IFTR Reports* 11.

——.(1998) 'Effects of diatomite on toughness of premix glass fibre reinforced cement composites', in: *Proc. 11th International Congress of GRCA,* Cambridge, 14–16 April, paper 4.

Glinicki, M. A., Brandt, A. M. (2007) 'Quantification of glass fibre-cement interfacial properties by SEM-based push-out test', in: *Proc. Int. Workshop on High Performance Fiber Reinforced Cement Composites HPFRCC-5,* A. E. Naaman and H. W. Reinhardt, eds, Mainz, 2007.

Gray, R. J. (1984) 'Analysis of the effect of embedded fibre length on fibre debonding and pull-out from an elastic matrix, Part 1: Review of theories, *Journal of Materials Science*, 19: 861–70; Part 2: Application to steel fibre – cementitious matrix composite system, *Journal of Materials Science*, 19: 1680–91.

Gray, R. J., Johnston, C. D. (1987) 'The influence of fibre-matrix interfacial bond strength on the mechanical properties of steel fibre reinforced mortars', *International Journal of Cement Composites and Lightweight Concrete*, 9(1): 43–55.

Greszczuk, L. B. (1969) 'Theoretical studies of the mechanics of the fibre-matrix interface in composites', in: *Interfaces in Composites, ASTM STP 452*, Philadelphia, pp. 42–58.

Hannant, D. J. (1978) *Fibre Cements and Fibre Concretes*, Chichester: J. Wiley & Sons.

Harris, B., Varlow, J., Ellis, C. D. (1972) 'The fracture behaviour of fibre reinforced concrete', *Cement and Concrete Research* 2: 447–61.

Hing, P., Groves, G. W. (1972) 'The strength and fracture toughness of polycrystalline magnesium oxide containing metallic particles and fibres', *Journal of Materials Science*, 7: 427–34.

Janczak, J., Stackpole, R., Bürki, G., Rohr, L. (1996) 'The use of push-out technique for determination of interfacial properties of metal-matrix composites', *17th Int. SAMPE Europe Conf. of the Society for the Advancement of Materials and Process Engineering*, Basel.

Kalton, A. F., Howard, S. J., Janczak-Rusch, J., Clyne, T. W. (1998) 'Measurement of interfacial fracture energy by single fibre push-out testing and its application to the titanium-silicon carbide system', *Acta Materialia*, 46(9): 3175–89.

Kaplan, M. F. (1961) 'Crack propagation and the fracture of concrete', *Journal of the American Concrete Institute*, 58: 591–610.

——.(1968) 'The application of fracture mechanics to concrete', in: Proc. Conf. *The Structure of Concrete and its Behaviour under Load*, A. E. Brooks and K. Newman eds, London, September 1965, Cement and Concrete Associates, pp. 169–75.

Kar, J. N., Pal, A. K. (1972) 'Strength of fiber-reinforced concrete', *Journal of Struct. Div., Proc. of ASCE*, ST5: 1053–68.

Kelly, A. (1973) *Strong Solids*, 2nd ed., Oxford: Clarendon Press.

Kelly, A., Davies, G.J. (1965) 'The principles of the fibre reinforcement in metals', *Metallurgical Review*, 10(37): 1–77.

Krenchel, H. (1964) *Fibre Reinforcement*, Copenhagen: Akademisk Forlag.

Kucharska, L., Brandt, A. M. (1997) 'Pitch-based carbon fibre reinforced cement composites', *Archives of Civil Engineering*, 43(2): 165–87.

Larson, B. K., Drzal, L. T., Soroushian, P. (1990) 'Carbon fibre cement adhesion in carbon fibre cement composites', *Composites*, 21(3): 205–15.

Laws, V. (1971) 'The efficiency of fibrous reinforcement of brittle matrices', *Journal of Physics D: Applied Physics*, 4: 1737–46.

L'Hermite, R. (1955) *Idées actuelles sur la technologie du béton*, Paris: La Documentation Technique du Bâtiment et des Travaux Publics.

Marikunte, S., Aldea, C., Shah, S. P. (1997) 'Matrix modification to improve the durability of glass fiber reinforced cement composites', in: *Brittle Matrix Composites 5*, A. M. Brandt, V. C. Li and I. H. Marshall eds, Cambridge and Warsaw: Woodhead Publishing–Bigraf, pp. 90–102.

Markovic, I., Walraven, J. C., van Mier, J. G. M. (2003) 'Experimental evaluation of fibre pullout from plain and fibre reinforced concrete', in: *Proc. Int. Workshop HPCFRCC 4 RILEM*, Ann Arbor, 15–18 June, pp. 419–36.

Marshall D. B., Oliver, W. C. (1990) 'An indentation method for measuring residual stresses in fiber reinforced ceramics', *Materials Science and Engineering*, A126: 93–103.

Mashima, M., Hannant, D. J., Keer, J. G. (1990) 'Tensile properties of polypropylene reinforced cement with different fiber orientation', *American Concrete Institute Materials Journal*, 87(2): 172–8.

Mianowski, K. M. (1986) 'Dynamic aspects in fracture mechanisms', in: *Proc. Int. Symp. Brittle Matrix Composites 1*, Jablonna, 1985, A. M. Brandt and I. H. Marshall eds, Elsevier Applied Science, pp. 81–91.

Mohr, O. (1906) 'Welche Umst, nde bedingen die Elastizit, tsgrenze und den Bruch eines Materials?' *Zeit. des VDI*, 44.

Morton, J. (1979) 'The work of fracture of random fibre reinforced cement', *Materials and Structures, RILEM*, 12(71): 393–6.

Morton, J., Groves, G. W. (1974) 'The cracking of composites consisting of discontinuous ductile fibres in a brittle matrix – effect of fibre orientation', *Journal of Materials Science*, 9: 1436–45.

Nammur, G. Jr., Naaman, E. (1989) 'Bond stress model for fiber reinforced concrete based on bond-slip relationship', *Journal of the American Concrete Institute*, 86(1): 45–57.

Oakley, D. R., Proctor, B. A. (1985) 'Tensile stress-strain behaviour of glass fibre reinforced cement composites', in: *Proc. of RILEM Symp. on Fibre Reinforced Cement and Concrete*, A. M. Neville ed., Construction Press, pp. 347–59.

Paillère, A. M., Serrano, J. J. (1974) 'Vers un nouveau béton armé', *Bull. Liason Labo.*, 72, June–July: 18–24.

Parimi, S. R., Rao, J. K. S. (1971) 'Effectiveness of random fibres in fibre reinforced concrete', in: Proc. Int. Conf. *Mechanical Behaviour of Materials*, Kyoto, 15–20 August, 5: 176–86.

Persson, B. (1999) 'Poisson's ratio of high-performance concrete', *Cement and Concrete Research*, 29: 1647–53.

Pinchin, D. J., Tabor, D. (1975) 'Mechanical properties of the steel/cement interface, some experimental results', in: *Proc. RILEM Symp. Fibre Reinforced Cement and Concrete*, A. M. Neville ed., The Construction Press, pp. 521–6.

Potrzebowski, J. (1985) 'Behaviour of the fibre/matrix interface in SFRC during loading, in: *Proc. Int. Symp. Brittle Matrix Composites 1*, Jablonna, A. M. Brandt and I. H. Marshall eds, Elsevier Applied Science, pp. 455–69.

——.(1986) 'Investigation of steel fibres debonding processes in cement paste', in: *Proc. Int. Conf. Bond in Concrete*, Paisley, June 1982, P. Bartos ed., London: Applied Science Publishers, pp. 51–9.

——.(1991) 'Processes of debonding and pull-out of steel fibres from cement matrix' (in Polish), *IFTR Reports*, 5, Warsaw.

Rajczyk, K., Giergiczny, E., Glinicki, M. A. (1997) 'The influence of pozzolanic materials on the durability of glass fibre reinforced cement composites', in: *Brittle Matrix Composites 5*, A. M. Brandt, V. C. Li and I. H. Marshall eds, Cambridge and Warsaw: Woodhead Publishing–Bigraf, pp. 103–12.

Riley, V. R. (1968) 'Fibre/fibre interaction', *Journal of Composite Materials*, 2: 436–46.

Shah, S. P. (1991) 'Do fibers improve the tensile strength of concrete?' in: *Proc. 1st Canadian Univ.-Ind. Workshop on Fibre Reinforced Concrete*, N. Banthia ed., Quebec: University of Laval, pp. 10–30.

Sujivorakul, C., Naaman, A. E. (2003) 'Modeling bond components of deformed steel fibers in FRC composites', in: *Proc. Int. Workshop HPCFRCC 4 RILEM*, Ann Arbor, 15–18 June, pp. 35–48.

Swamy, R. N., Mangat, P. S. (1974a) 'A theory for the flexural strength of steel fibre reinforced concrete', *Cement and Concrete Research*, 4: 313–25, 701–7.

——.(1974b) 'Influence of fibre-aggregate interaction on some properties of steel fibre reinforced concrete'. *Materials and Structures RILEM*, 7(41): 307–14.

Timoshenko, S. P. (1953) *History of Strength of Materials*, New York: McGraw-Hill Book Co.

Vandewalle, L. (2003) 'Test and design method for steel fibre reinforced concrete based on the σ-ε relation', in: *Some Aspects of Design and Application of High Performance Cement Based Materials*, A. M. Brandt ed., AMAS Warsaw, Lecture Notes 18, pp. 135–190.

Vandewalle, L. *et al.* (2000) 'RILEM 162-TC: Test and design methods for steel fibre reinforced concrete', *Materials and Structures*, 33: 3–5 and 75–81.

Voigt, W. (1883) *Ann. Physik*, 19: 44.

Voigt, W. (1915) *Ann. Physik*, 46: 657.

Zhou, L. M., Mai, Y. W., Ye, L. (1995) 'Analyses of fibre push-out test based on the fracture mechanics approach', *Composites Engineering*, 5(10–11): 1199–1219.

Zhu, W., Bartos, P. J. M. (1997) 'Assessment of interfacial microstructure and bond properties in aged GRC using a novel microindentation method', *Cement and Concrete Research*, 27(11): 1701–11.

Standards

EN 14651:2005 *Test method for Metallic Fibre Concrete – Measuring the flexural tensile strength (limit of proportionality (LOP), residual).*

EN 14721:2000 *Test method for Metallic Fibre Concrete – Measuring the fibre content in fresh and hardened concrete.*

9 Cracking in cement matrices and crack propagation

9.1 Classification of cracks

One of the main phenomena studied in the mechanics of brittle matrix composites is cracking. The cracks may be classified according to various criteria; for example:

- origin: cracks due to shrinkage and temperature variations in restrained elements or due to load producing local tensions;
- position in the material structure: cracks in matrix, aggregates, interface (bond cracks);
- width and importance: microcracks, macrocracks, major cracks that lead to final rupture;
- shape and pattern: single and multiple cracks, branching cracks, etc.;
- exploitation of element: admissible or inadmissible cracks with respect to the safety, serviceability or durability requirements;
- mode of rupture and of crack propagation: Modes I, II and III as well as mixed modes may be distinguished (cf. Figure 10.3).

A flaw or a crack is defined in the mechanics of solids as a plane 2D discontinuity in a solid body with its two dimensions much larger than the third one. Certain authors reserve the first term – flaw – for those discontinuities that are formed parallel to the smallest dimension without application of any tensile stress. In that convention the term 'crack' covers all other discontinuities just caused by tensile stress. In several manuals, this distinction is not used and both kinds of discontinuities in solids are called cracks; for example, Broek (1982). This practice is also followed here.

Cracks may be characterized by their width, length and pattern. The opening of a crack is always related to the creation of new free surface in the material, where free means that the surface is not loaded by any stress. An amount of energy is required to create a unit area of a new free surface of a crack. It is considered to be an important material characteristic and is called 'specific fracture energy,' which is difficult to directly measure, but may be calculated in an approximate way when the amount of released energy and crack dimensions are estimated and considered to be known.

The crack shape is usually irregular. The roughness or waviness of the crack surface is characteristic both for the material and for the origin of the crack. Because of this irregularity, the crack length and area of its surface may be considered as fractal objects. It means that they depend on the scale of observation or on the unit of measurement (cf. Sections 9.3 and 10.5.)

It is generally assumed that in brittle cement matrices the plastic phenomena, similar to those in metals, do not appear. In fact, when in specimens subjected to loading, certain apparently plastic deformations are observed; that is, an increase of deformation at a constant load, then this effect is caused by microcracks and therefore the term quasi-plasticity is used.

Brandtzaeg (1929) described the phenomenon of internal cracking in concrete after observing that the volume of a concrete specimen under compression decreases with the load increase up to so-called critical stress when the apparent increase of the volume occurs (cf. Figure 8.6). That increase is due to gradual development of microcracks. It was also observed that the value of the critical stress corresponds to the onset of intensive crack propagation.

The development of the fracture mechanics approach to metals lead to Kaplan's (1961) proposal to apply fracture mechanics to concretes, using also relations applied in other fields of materials science. The Griffith's approach to cracks in perfectly brittle and homogeneous materials was extensively followed in numerous studies of cracking processes in cementitious materials, (cf. Section 10.1.1). Then the question arose whether the cracks themselves, and which of them, were governed by the Griffith's approach. Later, however, it became clear that the situation in these materials is much more complex and requires a more diversified approach. The answer for the above question is therefore also more complex: the Griffith's approach is applicable to those materials, but with several restrictions and complementary assumptions. Various related questions are the subject of further developments of new theories and proposals (cf. Section 10.1).

The cracks in reinforced and non-prestressed concrete elements subjected to tension are unavoidable due to differences in Young's moduli and in values of ultimate strain in concrete and steel. It is the task of the designer to select appropriate remedies for the serviceability and safety reasons by adequate reinforcement or prestressing. If perfect impermeability is required, then the cracks are inadmissible or their width is strongly limited. In that case, also, the pore system in concrete should be completely blocked by special admixtures. Another method is to provide protective coatings or tight layers of other materials.

In most national and regional standards and regulations, the maximum crack width for different kinds of concrete structures is indicated (cf. ACI 224R-01).

In this chapter cracks are considered mostly in the context of the materials science and the analysis of the properties of concrete structures. The cracking of reinforced concrete structural elements, due to excessive loads or imposed deformations, are not examined here as this subject is abundantly treated in relevant manuals and recommendations for design of structures.

Studies of the crack systems may be directed at various aims:

- evaluation of the quality of a structural element from viewpoints of their safety, durability, impermeability, aspect, etc.;
- analysis of the influence of different external actions (load, thermal and humidity variations) on the tested concrete composition;
- influence of the concrete (or mortar) mixture composition on its resistance against cracking.

9.2 Cracks in the cement matrix

The cracking processes develop when a system of stresses of any origin is applied to a solid and its tensile strength is exceeded or, in other words, when the ultimate tensile strain is exceeded.

Theoretical tensile strength may be calculated from molecular cohesion in the atomic structure, but the effective strength of materials is, by a few orders, lower. The reasons, according to Griffith's theory, are local defects, cracks and flaws, with high stress concentrations at their tips. It is usually assumed that for ordinary Portland cement matrix the maximum tensile strain varies between 100 and 150.10^{-6} and may rarely attain 200.10^{-6}.

Also, an initial system of defects exists in all cement-based materials. These are cracks and microcracks, voids and pores, hard unhydrated cement grains, weaker and stronger regions, etc., which exist before the application of any load or imposed deformation and may be recognized with different methods. These defects cause concentrations of high stresses even under small external actions. The cracks propagate due to local exceeding of material tensile strength. Such systems of initial defects exist on a micro-, meso- and macro-level in all brittle materials.

The main causes of cracking in cement based composites are:

- volume changes during hydration of cement and hardening of matrices in presence of restraints, e.g. shrinkage, temperature variation, chemical influence on particular material components, swelling of reinforcing bars during their corrosion (these are intrinsic cracks);
- tensile stresses due to loads and deformations imposed during exploitation.

In the hardened cement gel, which is a framework of crystals of a different nature, the initial microcracks are formed between needle-shaped hydration products and also around sand grains. At the tips of these cracks only small amounts of energy are sufficient to enhance further separation of the material. Gradually, if the input of external energy is provided, larger macrocracks open and transform into major cracks, which lead to the complete separation of different regions from what was considered a solid body.

Different types of intrinsic cracks, which appear during processes of hydration at the early age of a cement-based matrix, may be distinguished.

Plastic shrinkage cracks are considered to be caused by the loss of water in a concrete mix because of evaporation to the air or suction by old, neighbouring concrete (cf. Section 11.4). This phenomenon is the origin of the formation of microcracks in cement paste (Kawamura 1978). Contraction during the hydration process develops in the CSH (calcium silicate hydrate) gel, but hard crystals $Ca(OH)_2$ are not subjected to volume changes related to wetting and drying. As the crystals are surrounded by a gel with low tensile strength, cracks appear, which may be detected both by microscopical observation and by increased permeability of the paste. It is estimated that crack width varies between 0.125 µm and 1.0 µm. Plastic shrinkage can be controlled by keeping the concrete surface wet since the very beginning. The direct cause is the evaporation and the cracks may be considerably reduced if the loss of water can be limited during at least first eight hours after casting. Cracks also develop in cement-based composites due to other kinds of shrinkage (cf. Section 11.4.2).

Thermal variations may cause cracks, because in these highly heterogeneous materials the thermal expansion of various components are slightly different and produce a system of tensile and compressive stresses due to local restraints. The overall deformations of the elements also produce stresses and possible cracks if the displacements are restrained by reinforcing bars, neighbouring members and layers of old concrete or other rigid material. The cement paste is restrained at a local scale by grains of sand and most of all by coarse aggregate grains. As a result, in any kind of construction, the material has no possibility of free deformation. Restraints may cause compressive and tensile stresses and because of low tensile strength of concrete-like materials, cracks appear – particularly at early age. It has to be admitted that microcracks in concrete are unavoidable, but the excessive cracks, due to shrinkage and thermal effects, are usually not acceptable and may be attributed to inadequate curing.

Restrained shrinkage cracking was tested by Malmberg and Skarendahl (1978), Dahl (1986) and Grzybowski and Shah (1990), among others, for fibre reinforced concrete. The shrinkage deformations were restrained by some kinds of stiff inner rings and the cracks were observed on outer rings made of tested material. The progress of cracks reflects the influence of material composition, structure and curing on the resistance against cracking.

The intrinsic cracks and flaws are dependent on material composition and ambient conditions – humidity, temperature and restraints – and are subjected to modifications in time together with these conditions. The opening of macrocracks and material fracture observed under load are closely connected to initial cracks and other defects.

The crack systems may be considered from various but interrelated viewpoints:

- as initiators and concentrators for cracks due to load in the service life of the element;
- as a reason of increasing permeability and lower durability;
- as phenomena inadmissible for structural serviceability for various reasons, e.g. external aspect, durability.

The structure of concrete composites is also macroscopically anisotropic in the direction of casting. The voids beneath aggregate grains are formed and partly filled with water. These voids create weak matrix-aggregate bonds in horizontal planes and are the origin of cracks when favourable stress fields are imposed due to external loading or other action (Figure 9.1)

The cracks in the matrix open and propagate under excessive local stress. This means that because of a very non-uniform distribution of stresses in the material structure (cf. Figures 8.1 and 8.2) the local stresses produce local cracks even if in the macro-scale the load is relatively small and is considered to be causing an average stress below limits of elastic behaviour. There are two characteristic points in the cracking process: crack initiation; that is, opening of the first crack in the matrix, and the beginning of unstable cracking. After the first crack opening, also called the discontinuity point, the stable crack propagation starts, provided that additional energy is supplied, for instance by an increase in load. When the unstable cracking stage is reached, the cracks progress rapidly without any input of new energy and lead to the failure.

In ordinary- and high-strength materials, the matrix has lower strength than the aggregate grains. Under load the cracks are crossing the matrix and

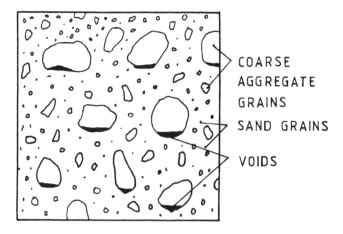

Figure 9.1 System of voids below aggregate grains in plain concrete.

contouring the grains. In lightweight composites the situation is opposite: the cracks propagate across the grains (Figure 9.2). It should be remembered that what is macroscopically called a matrix also contains grains and inclusions of lower level; that is, the mortar which, with the aggregate grains, forms the ordinary concrete structure is itself composed of cement paste and grains of sand. Also, non-hydrated grains of cement and particles of micro-fillers like fly ash and SF are surrounded by interfacial layers.

The cracks initiate and propagate in the form of decohesions in the aggregate-matrix interface and these regions deserve more particular attention. The physical and chemical composition, as well as the structure of the interface, is essentially different from these of the bulk matrix. Cracks are often arrested just at the interface layer a few micrometers before reaching an obstacle in the form of an aggregate grain or a fibre (cf. Chapter 7).

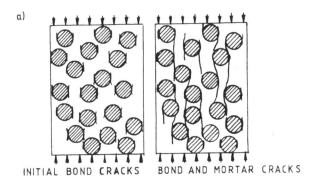

INITIAL BOND CRACKS   BOND AND MORTAR CRACKS

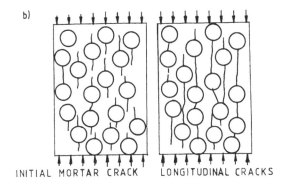

INITIAL MORTAR CRACK   LONGITUDINAL CRACKS

Figure 9.2 Cracking of concrete specimens with different ratio of strength of matrix to that of aggregate grains:
  (a) ordinary concrete
  (b) lightweight aggregate concrete.
  after Lusche (1974)

The increase in load corresponding to the first crack opening in elements subjected to bending or tension is an important objective in material design. It may be achieved by several methods: increase of the tensile strength of the matrix itself, improvement of bond strength to aggregate grains and reinforcement, transformation of major cracks to microcracks by adequate reinforcement, etc. If the first method improves resistance against crack opening, the others tend to increase the fracture toughness of the composite material. The problem in the design of concrete elements of how to consider that the fibre reinforcement may enhance the tensile strength of the matrix is discussed in Section 8.3.

It is possible to understand, at least partly, the origin of cracks after thorough examination of their pattern. Soroushian and Elzafraney (2004) observed that different damaging factors induced particular form and intensity of the crack pattern. With simple compression all dimensions of the cracks (length, width and area) increased, while fatigue enhanced the number of microcracks. Impact increased the crack width mostly, while cyclic freezing and thawing produced less tortuous microcracking.

9.3 Quantitative characterization of crack systems

The systems of cracks and microcracks in cement-based materials may be analyzed at different scales. An extensive discussion of scales of cracks is presented, for example, by Darwin *et al.* (1995) and by Ringot and Bascoul (2001). The methods of observation of cracks in specimens and elements are adapted to the selected scale and related to the purpose. At different scales, different systems of cracks and microcracks exist and may be observed and measured. For material design, testing and applications in building and civil engineering structures, but also in non-structural elements, the attention is concentrated at microcracks of width from approximately 10 μm to cracks a few millimetres wide. All features are strongly influenced by the sensitivity of observation and measurements. The lower bound of the crack width cannot be specified without ambiguity, because very thin cracks and thin crack tips disappear in pores and cavities of the cement paste structure.

The recognition and identification of cracks and microcracks with appropriate accuracy is necessary for their quantitative description. Thanks to recent technical tools and computer methods the acquisition of data and their processing is considerably developed in view to obtain reliable quantitative information. Certain methods used in the past are not described here, but modern processes in which computer image analysis plays a significant role are described. The review of various traditional methods may be found in the paper by Hornain *et al.* (1996).

Opening and propagation of cracks in cement pastes, mortars and concretes may be detected and followed by analyzing acoustic signals that accompany these events. The number and energy of the signals is recorded and analyzed. The acoustic emission events are produced by cracking and also the nature of the cracking can be deduced: debonding between two phases or fracture of one specific phase (cf. Section 6.2 and paper by Roy *et al.* 1991). Even the simplest counting of the acoustic events as a function of load, deformation or time gives important information about the processes developing in the tested element. An example of such analysis can be found in Figure 9.3 where two curves are shown from the bending test of a small beam 15 × 20 × 160 mm cut out from a plate cast with CFRC. These curves represent deflection of a beam and number of acoustic events as functions of the load. It is very clear how loading and unloading stages are accompanied by increase and decrease of the intensity of acoustic events. In analyzing these curves it is easy to identify the beginning of cracking, crack propagation and periods without any new cracks when the beam was unloaded. This analysis served to determine the influence of the reinforcement with different fibre volume fraction and the behaviour of the matrix (Kucharska and Brandt 1997). The relations between acoustic emission events and behaviour of specimens under loading are also analyzed in Section 9.2.

The first step for the direct and quantitative determination and identification of width, length and distribution of cracks in cement is the preparation of samples aimed at the exposure of cracks without interfering in their shape and dimensions. The specimen is carefully sawn out of cores or elements and then polished to obtain perfect surfaces. Later it is vacuum impregnated with

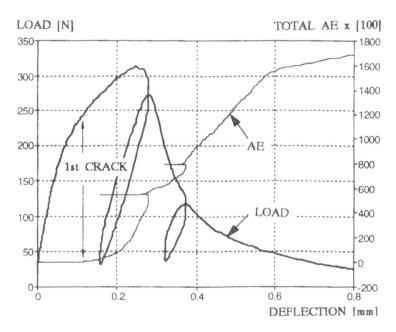

Figure 9.3 Load-deflection and acoustic emission (AE) – deflection curves for CFRC specimen under bending, V_f = 3%vol, after Kucharska and Brandt (1997).

epoxy resin containing fluorescent dye. After grinding and polishing again, a perfect surface with exposed cracks may be subjected to observation using scanning electron microscopy and optical microscopy. The obtained images are subjected to computer image analysis. This description is based on tests described in the papers by Glinicki and Litorowicz (2003) and Litorowicz (2006), and also by Ammouche *et al.* (2001).

A digital image taken on a concrete specimen in Figure 9.4 shows bulk cement paste, aggregates and cracks in a specially prepared specimen. The precision of the image acquisition determines the quality of results of all the procedure. To characterize the crack system, it is necessary to convert the colour image into a binary image through a series of operations called image segmentation, taking into account the influence of different colours and intensity of light. Using the image analysis system, the cracks appearing in a given field are extracted after a series of operations on the binary image. A series of filtering operations has to be performed to select cracks from among other objects, like aggregate grains, pores, etc. By applying appropriate shape criteria, cracks are automatically distinguished from these features. Finally, a skeletonized image is obtained by binary thinning in the form a set of thin lines ready for measurements of length, width and for orientation analysis.

A crack system as it is seen in an observed field may be characterized by several parameters:

- length of cracks L [mm] is the total length of all dendrites recognized in observed field;
- area of cracks A [mm²] total area of all cracks;

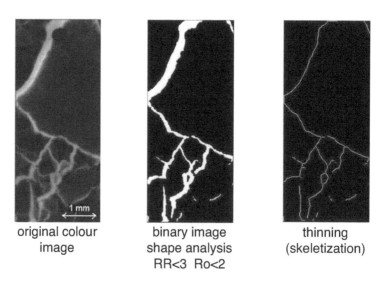

| original colour image | binary image shape analysis RR<3 Ro<2 | thinning (skeletization) |

Figure 9.4 Crack image processing, after Brandt and Kasperkiewicz (2003).

- average width of cracks W [mm] is the total crack area per total dendritic length;
- density of cracks L_A [mm/mm^2] ratio of the total dendritic length of all cracks to the total area of the observed field;
- areal fraction A_A [mm^2/ mm^2] ratio of the area of all cracks to the total area of the observed field;
- degree of orientation of the crack system that can be shown by means of 'rose of crack directions';
- distribution of crack width, shown as percentage object fraction in the given range of crack width.

After automatic calculations of all these parameters from the image of the crack system, its characterization allows the mechanical consequences to be determined, different materials' composition to be compared and the efficiency of fibre reinforcement to be estimated.

The crack width distribution and its average value can be defined and for that aim the cracks are first subjected to a sorting operation. Then the logical operation is applied between the image of sorted cracks and a system of parallel equidistant lines as shown in Figure 9.5. After the analysis of all intersections, the crack widths and lengths are calculated. The results provide the dendritic length of cracks and their total area, average width and density in the analyzed image, etc.

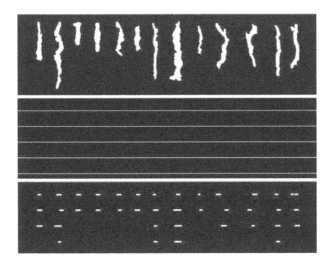

Figure 9.5 Logical operation of image of cracks sorted and aligned and image of parallel lines. Reprinted from *Cement and Concrete Research*, Vol 36, Litorowicz, A., 'Identification and quantification of cracks in concrete by optical fluorescent microscopy, p. 8., Copyright (2006), with permission from Elsevier.

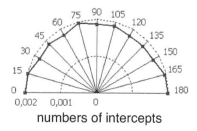

numbers of intercepts

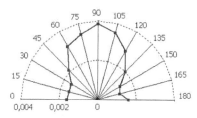

numbers of intercepts

Figure 9.6 Two different roses of directions of cracks. Reprinted from *Cement and Concrete Research*, Vol 36, Litorowicz, A., 'Identification and quantification of cracks in concrete by optical fluorescent microscopy, p. 8., Copyright (2006), with permission from Elsevier.

The degree of crack orientation is determined using the stereological method in which a system of oriented secants is superimposed on the image of dendritic cracks and number of intersections is counted. The results may be presented as a rose of intercepts; in Figure 9.6 two such roses are shown: one indicates quite uniform distribution of crack directions and in the other the cracks are oriented mostly in one direction.

The crack systems are analyzed on plane images and using stereological laws some conclusions on the 3D distribution of cracks may be formulated; for example, for determination of their connectivity, impermeability of the concrete elements, etc. Three dimensional images of internal cracks have been obtained using x-ray microtomography (Landis *et al.* 1996, 2007), and the aim of this approach is to get a better understanding of such a complex phenomenon as cracking in concrete. Because of various limitations, for example only small volumes are studied, this method represents only the first attempts to provide useful insight into spatial damage evolution of concrete and did not find larger application.

Ahead of the crack tip, a region of loosened material is observed and determination where the crack actually ends is rather conventional (Figure 9.7).

The analysis of larger cracks in concrete beams, columns or floors may be executed without microscopes and the basic features of the crack system may be determined visually. Quite often the measurement of such crack width

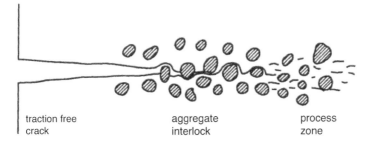

Figure 9.7 Scheme of a crack in concrete and the process zone ahead of the crack tip.

is a bit ill-defined because of its irregular shape; it may vary in large limits depending on where it is measured. That is the reason why the incremental measurements, made at arbitrarily selected points but at sequential stages of loading or at given time intervals, objectively describe the crack width development. Best practice is to trace a system of more-or-less regular lines on the surface where cracking is observed and to record when, and under what conditions, cracks are crossing the lines. Then the crack width at these intersections can be measured when it grows under load, variation of temperature, as an effect of ageing or other actions. An example of a system of lines for crack recording on an element subjected to axial tension is shown in Figure 9.8.

The appearance of the first crack is more or less clearly reflected on various curves, which characterize the behaviour of specimens and elements under external load: load-deflection, stress-strain, etc. The first crack and its propagation and development are observed after the changes of slope of the respective curves, because simultaneously the element stiffness is decreased. This change of slope is not always sufficiently clear in the form of a 'knee' to indicate the load or time of the crack opening. In an element with a highly

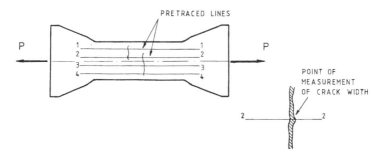

Figure 9.8 Measurement points of crack width at intersection with pre-traced system of lines.

efficient system of reinforcement and subjected to bending, a very smooth load-deflection curve may be obtained as a result of well-distributed micro-cracks, but without an offset indicating a certain threshold in the material behaviour.

Other methods of crack detection are based on physical phenomena closely related to cracking: increase of apparent volume and Poisson ratio, abrupt increase in the intensity of acoustic events, decreased velocity of ultrasounds, etc.

The location of the crack may also be detected using photoelastic or brittle coatings on external faces of tested specimens. Special techniques have been developed to visualize the cracks on the observed surfaces by increasing their contrast. This is obtained by filling up the cracks with special dyes or by spraying the surface with fluorescent fluids. Brittle coatings belong to the same group of experimental techniques that are nowadays less frequently applied.

The surface of a crack depends on the properties of the material and on loading characteristics. The surface may be more or less rough and developed. It appears that the surface of the crack in cement-based materials has fractal nature, which indicates that the effective determination of its area is related to the scale of magnification. General remarks about fractals and fractal dimension may be found in a book by Mandelbrot (1983). The methods on how to characterize the cracks and the fracture surfaces using the notion of fractal dimension is briefly described in Section 10.5.

9.4 Single and multiple cracking

Cracks open or propagate under local stresses, which furnish the amount of energy necessary for rupture of material bonds and for the creation of new surfaces. The replacement of a single crack by multiple cracks and microcracks is beneficial for a number of reasons. The advantages are: increase of strength and stiffness, increase of fracture toughness, improvement of durability through increased impermeability, better outward aspect, etc.

In composite materials, crack propagation is blocked by various kinds of inclusions: aggregate grains, pores, voids and fibres. On the crack path, the tensile strength of the matrix itself and its bond to inclusions must be overcome. The strength of the matrix against cracking is relatively low, and the abovementioned non-homogenities help to control cracks by increasing their length. A few examples of obstacles to crack propagation are shown in Figure 9.9. All these obstacles stop the cracks by arresting their propagating tips. These phenomena considerably increase the energy required for crack development and failure of the element, and are used in all methods of crack control.

When a crack meets one of the abovementioned obstacles it is stopped and a new input of energy by increase of load is required for its further propagation. Then, the crack may pass across the obstacle or may contour it to follow another path along a weaker region or layer. A crack may then be divided into several finer cracks; that is, crack branching is induced at an obstacle.

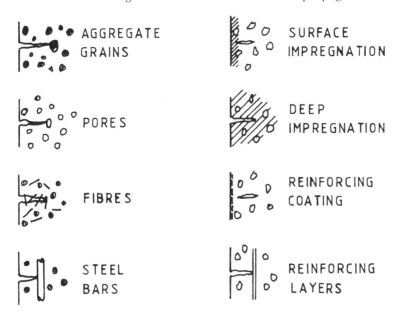

Figure 9.9 Example of obstacles against crack propagation.

Additional energy is also used for breaking the grain or for destroying the bond strength in the interface around it. All kinds of inhomogeneities, which produce crack deviation, branching or multiplication increase the total area of fracture surface which becomes several times the area of the cross-section of the element. The determination of effective fracture area is often impossible in an unambiguous way because it is a fractal object, depending on scale and unit of measurement.

The phenomenon of multiple cracking appears in brittle matrices reinforced by efficient systems of bars or fibres where the material's toughness is high. The opening of a new microcrack requires less energy than the propagation of an existing crack. This means, for example, that enough fibres with good bond may control the cracking process; this is schematically shown in Figure 9.10.

On the stress-strain curves the region corresponding to branching and multiplication of cracking is somewhat similar to the effects of plasticity of metallic materials – a smoothly curved line showing large deformation under constant or slowly increasing load. Multiple crack systems should be created, if large energy absorption and high toughness of the composite material is required. Therefore, in the design and composition of a material, multiple cracking is one of the main objectives. In Figure 9.11 an example is shown where deflections after the first crack opening (point A) are related to multiple cracking in the central part of the steel-fibre reinforced specimen subjected to bending. Discontinuous lines show the gradual decrease of the Young's

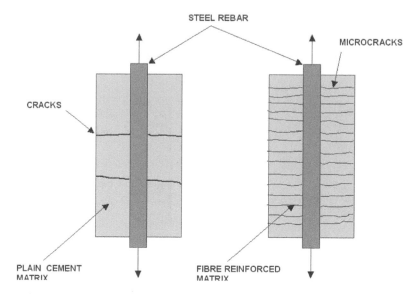

Figure 9.10 Schematic presentation of cracking in the elements subjected to tension.

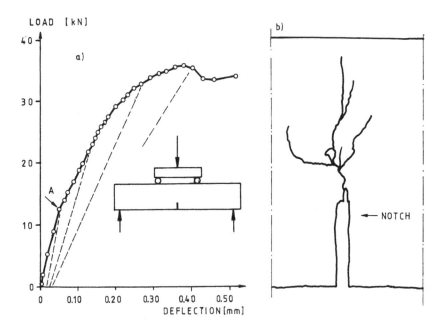

Figure 9.11 Multiple cracks:
(a) 'knee' at point A of the load-deflection curve
(b) multiple cracks on lateral surface of a steel fibre reinforced concrete beam.
after Brandt and Stroeven (1991)

modulus of the tested beam due to development of the cracks.

Crack control by fibres is also visible in those regions of the load-deflection curves where, after the unloading-reloading cycle, the elasticity of fibres bridging the cracks enables the recovery of the bearing capacity of the specimen before unloading.

The efficiency of fibres for crack control depends upon fibre dimensions, stiffness and distribution. An example of the diversified role of fibres has been revealed in the tests by Brandt and Glinicki (1992). Mortar specimens reinforced with various combinations of steel and carbon fibres were subjected to bending and acoustic emission (AE) was recorded. In Figure 9.12 diagrams are shown characterizing the behaviour of a specimen reinforced with 3.5% vol of Kureha pitch carbon fibres (3 mm length, 14.5 μm diameter, 720 MPa tensile strength and 32 GPa Young's modulus).

Similar diagrams shown in Figure 9.13 are obtained from a specimen reinforced with 7.5% vol. of steel Arbed fibres (0.25 × 25 mm) prepared with SIFCON technique (cf. Section 13.5).

On the left side diagrams in both figures, the load and AE counts (Nc) are shown as functions of time. On the right side diagrams (the relative scales are used) all values are divided by those that correspond to maximum load. In the specimen reinforced with carbon fibres the number of counts is quite small until about 85–90% of the maximum load was reached. In fact, the acoustic emission was only initiated at that stage. The acoustic emission in the specimen reinforced with a high amount of steel fibres (Figure 9.13) develops in a different way – the microcracks are recorded from virtually the beginning of loading, and at 30% of maximum load the total count is quite appreciable. For this specimen the ultimate strength is considerably increased by strong fibre reinforcement, but microcracks are not controlled at all. The carbon fibres in the tests shown in Figure 9.12 control microcracks and disperse

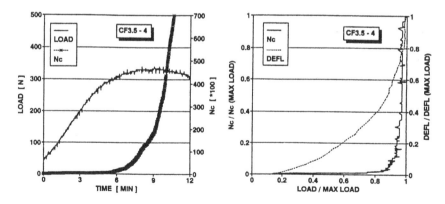

Figure 9.12 Comparison of curves for load and AE counts with relative deflections of a cement mortar specimen reinforced with 3.5% of carbon fibres, after Brandt and Glinicki (1992).

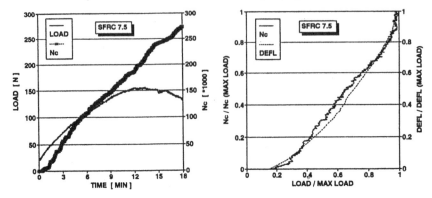

Figure 9.13 Comparison of curves for load and AE counts with relative deflections of a cement mortar specimen reinforced with 7.5% of Arbed fibres, after Brandt and Glinicki (1992).

them in such a way that they do not produce perceptible acoustic emission; that is, above the imposed discrimination level. However, the existence of microcracks is proved by highly non-linear behaviour of the specimen. The important role and efficiency of fine carbon fibres in controlling the microcracks is evident. Also steel fibres considerably decrease the maximum crack width, with positive consequences for the quality of the reinforced elements, as is shown in Figure 9.14.

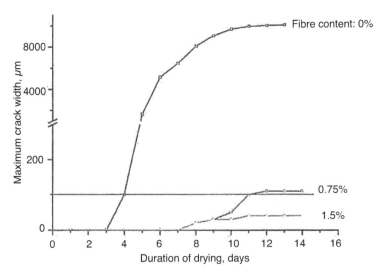

Figure 9.14 Influence of fibre reinforcement on the maximum crack width, after Wittmann (2006).

The analytical approach to the problem of multiple cracking in fibre-reinforced brittle matrix composites was examined by many authors and a few papers by Aveston, Kelly and their co-workers are particularly important (e.g. Aveston *et al.* 1971). That approach, called 'the ACK theory' is one of the well-known theories of fibrous composites. It was later described in several books, (e.g. Hannant 1978; Bentur and Mindess 2006) and that is why it is only briefly mentioned in Section 10.1.2.

9.5 Modelling of cracking

Cracks, their opening and propagation in the materials' structure may be modelled and reproduced by computer experiments. Such simulations are based on different assumptions as to the behaviour of matrix and inclusions. One of the first attempts in that direction was published by Zaitsev and Wittmann (1981), which considered 2D model structures with randomly distributed pores and inclusions in a cement matrix. A mainly linear behaviour is assumed, with some non-linear components. Cracks are induced by shrinkage and their propagation under external loads is modelled according to Modes I and II. Cracks also run across the matrix between hard aggregate grains and along interfaces between grains and matrix. In lightweight aggregate concrete models, the cracks run across the grains as well.

Computer simulations enabled the different relations between the strength of the matrix and the aggregate grains, influence of increasing load, quality of the matrix/aggregate bond, etc. to be investigated in a systematic way. Examples of crack patterns are shown in Figures 9.15 and 9.16.

Computer analysis of crack propagation through finite element grids was developed by several authors. Cracks are represented by discontinuities of the finite element mesh, and smeared crack models were also applied. Cement-based matrices were considered as linear elastic bodies up to the point where cracks open and later their behaviour becomes highly non-linear. Various methods are applied to represent non-linear and heterogeneous materials and to simulate their behaviour under load (cf. Petersson 1981). In discrete models, cracks are represented as discontinuities in the finite element mesh. This is also where smeared crack models are introduced.

The computer experiments with modelled materials with inclusions, cracks, etc., enable researchers to check different situations against various sets of parameters. However, the results obtained are limited to a large measure by their authors' assumptions reflecting the effective properties of material and the shapes of structures. No new physical elements may be expected as the computer simulations cannot exhibit more information than was initially introduced by their authors and for this reason they cannot replace experiments. The facility of calculation, various examples based on different sets of input data, their ease of modification and the possibility of finding the best model for experiments, represent the main advantages of the computer modelling.

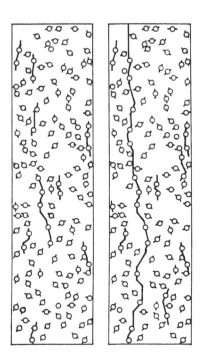

Figure 9.15 Crack patterns in hardened cement paste with randomly distributed initial pores and intrinsic cracks: initial load level and advanced load level, after Zaitsev and Wittmann (1981).

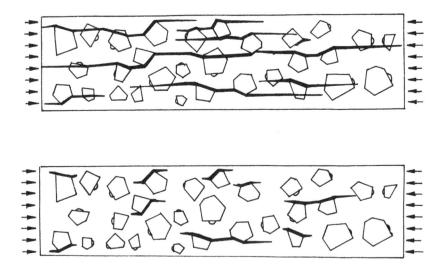

Figure 9.16 Crack patterns in high strength concrete under two levels of load, cracks run around or across the aggregate grains, after Zaitsev and Wittmann (1981).

9.6 Consequences of cracking

Cracks diminish the material strength and stiffness and may seriously influ-ence various material properties during ageing and exploitation. In the first place, the durability of structural and non-structural elements may be seri-ously endangered because the systems of interconnected capillary pores and cracks enable the transportation of gases and liquids through the concrete structure. The following processes may be mentioned here:

- permeation of gases, water, and water solutions;
- rise of water level by capillary suction;
- diffusion of water vapour.

Several aggressive substances may enter the material structure as a result of a combination of these processes. A measure of concrete durability was proposed by Hilsdorf (1989) in the form of an air permeability characteristic, which could complete standard strength in material characterization.

The decrease of strength in the presence of cracks may be described inde-pendently of their origin in the frame of the fracture mechanics on the basis of the Griffith's approach, or by using damage analysis. The decrease of stiffness is exhibited, for instance, by a decrease of slope of the strain-stress curve.

The danger that cracks may adversely affect the durability of the material is proportional to their width, strongly related to their connectivity, and depends on the environmental conditions. Also, the location of the cracks and function of the element are of importance. Table 9.1 lists approximate values of the crack width, which are considered as maximum permissible according to various national and regional standards or other regulations. These values are indicated as examples only and should be considered together with such aspects as the quality of execution and control, dimensions and importance of examined elements, their role as structural or non-structural ones, the minimum cover depth required for steel bars, etc.

The cracks in plain or reinforced elements, made of cement-based com-posites and subjected to tension, are unavoidable due to the difference in Young's moduli of hardened cement and steel. However, it is the task of the designer to select appropriate structural dimensions and material properties to ensure required impermeability by the material itself, or by additional coatings.

Some kind of healing of the cracks is observed in particular conditions (cf. Section 9.7), but in principle cracking is an irrecoverable process, par-ticularly in an advanced stage of crack propagation. After the discontinuity point, that is, after the first variation of the slope of the strain-stress (or load-deflection) curve, which is considered an indication of the first crack opening, the hypotheses of material continuity are no longer valid; thus the notion of stress is applicable with serious restrictions. The question as to whether these hypotheses are applicable at all should be also formulated.

Table 9.1 Maximum permissible crack width in concrete structures

Kind of structure and environment conditions	Crack width (mm)
Indoor structures dry air, impermeable coating provided	0.4–0.5
Outdoor structures medium humidity, no corrosive agents	0.3–0.4
Outdoor structures high humidity	0.2–0.3
High humidity corrosive agents (de-icing, sea water)	0.1–0.15
Reservoirs for fluids	0.10

9.7 Healing of cracks

The phenomenon of self regeneration of cracks in cement-based matrices, or so-called autogeneous healing, has been known for a long time. D. A. Abrams first published remarks on self-healing of cracks in reinforced concrete bridge elements exposed to external climatic influences over three years. He confirmed that under much higher loads new cracks opened but those previously healed did not re-open, Abrams (1913).

The healing of cracks is related to prolonged hydration of cement grains when cracked elements are maintained in sufficient humidity. In such conditions new crystallization products partly recover the cracks. It has been observed that the healing is inversely proportional to the crack width; that is, rather narrow cracks may exhibit healing. In most cases, these cracks were produced by shrinkage or thermal actions.

It has been established that probably three chemical reactions are initiated in cracks: formation of (i) calcium hydroxide; (ii) calcium silicate; and (iii) possibly calcium carbonate in the presence of carbon dioxide (Neville 2002). The composition of material within healed cracks constitutes an object of some discussion between researchers, but without any doubt it is a product of restarted hydration. Conditions for self-healing are:

- cracks should not be wide, the finer the better;
- water is necessary for chemical reactions, and high w/c ratio is helpful;
- unhydrated cement is also indispensable, so in young concrete healing is better;
- both surfaces of a crack should not be displaced;
- the age of concrete plays a role (younger concrete heals more effectively).

In all cases, it helps when cracks are filled or partly filled by some debris and dust.

In experimental conditions, healing of cracks 0.6 mm or even 0.8 mm were reported, but in the majority of tests, effective healing was observed when the crack width did not exceed 0.3 mm. When cracks are larger, then the process may last longer, provided that the favourable conditions are maintained. After healing, the recovery of initial tensile strength, Young's modulus and impermeability is observed. The recovery is not complete, and in some tests lower values of these characteristics were obtained. These observations were confirmed by Zamorowski (1985). The tests were made on young specimens subjected to tension up to the point of cracking and later cured in water under a slight pressure. Examples of the results achieved are presented in Figure 9.17 and it may be observed how strongly the age of concrete at cracking influenced the percentage of the strength recovery.

Gray (1984) has shown that the healing of the interfacial bond between cement mortar and steel fibres was greater than observed for cracks in plain mortar and concrete. The tests have been carried after 90 days of cure in water. In lightweight concrete, healing effects in both the steel-paste interface and in the cracks in cement matrix were observed by Mor *et al.* (1989).

A complete restoration of the fracture toughness and even an increase of strength was observed by Kasperkiewicz and Stroeven (1991) in plain concrete beams subjected to cracking under pure bending and later cured in fog room conditions over a two-year period. The crack width at the notch was initially equal to 0.3 mm or more. Similar results were also obtained by Hannant and Edgington (1975) for steel-fibre reinforced elements stored in natural environmental conditions and only the regain of the bearing capacity was measured. Concrete specimens exposed to freeze/thaw cycles and cracked were tested by Jacobsen and Sellevold (1996) and a loss of compressive strength up to 29% was obtained. After a two- to three-month cure in water, a recovery of strength of only 4–5% was obtained. However the values of Young's modulus were nearly completely regained; the results

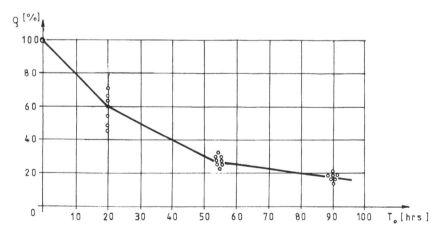

Figure 9.17 Strength restoration as a function of age of concrete at cracking, after Zamorowski (1985).

of strength tests and resonance frequency measurements were somewhat contradictory.

The self-healing process may be studied using acoustic emission measurements, but in most investigations only the recovery of tensile and compressive strength or permeability tests has been checked. Reinhardt and Jooss (2003) concluded that a higher temperature of 80°C favoured healing; in their tests crack width did not exceeded 0.20 mm and in the cracks of 0.10 the flow rate was decreased considerably after approximately 50 days. The proposed theoretical formulae were confirmed by the test results.

An extensive experimental program was reported by Granger *et al.* (2006), who tested ultra-high performance concrete (UHPC) beams subjected to bending up to cracking. Specimens were made with concrete with a *w/c* ratio equal to 0.2 and significant effects of healing were observed in bending using load-deflection curves and acoustic emission signals after 20 weeks of cure in water. The tests showed a high degree of recovery in strength and stiffness. It was evaluated that approximately 50% of cement grains took part in the second hydration processes. Similar tests repeated for ordinary concrete beams with a *w/c* ratio between 0.35 and 0.48 did not confirm the effects of healing.

The question of whether crack healing may be considered a phenomenon of appreciable importance for the durability of concrete elements may be answered positively, but all quantitative data published are obtained in laboratory conditions. Moreover, the results found in different laboratories do not supply quite a coherent image as to the scale of healing that may be expected in a given situation. That is certainly the reason that the healing of crack is considered as a process that is favourable for durability of concrete structures, but it is not accounted for in the design of concrete elements.

References

Abrams, D.A. (1913)'Tests of a 40 ft Reinforced Concrete Highway Bridge', *ASTM Proceedings*, 13.

Ammouche, A., Riss, J., Breysse, D., Marchand, J. (2001) 'Image analysis for the automated study of microcracks in concrete', *Cement and Concrete Composites*, 23(2): 267–78.

Aveston, J., Cooper, G. A., Kelly, A. (1971) 'Single and multiple fracture,' in: *Proc. Conf. 'The Properties of Fibre Composites,'* National Physical Laboratory. Guildford, U: IPC Science and Technology Press, paper 2: 15–24.

Bentur, A., Mindess, S. (2006) *Fibre Reinforced Cementitious Composites*, 2nd edition. Taylor & Francis.

Brandt, A. M., Stroeven, P. (1991) 'Fracture energy in notched steel fibre reinforced concrete beams,' in: *Proc. Int. Symp. Brittle Matrix Composite 3*, A. M. Brandt and I. H. Marshall eds, London and New York; Applied Science Publishers: pp. 72–82.

Brandt, A. M., Glinicki, M. A. (1992) 'Flexural behaviour of concrete elements reinforced with carbon fibres,' in: *Proc. RILEM/ACI Workshop 'High Performance Fiber Reinforced Cement Composites,'* H. W. Reinhardt and A. E. Naaman eds, Mainz, 1991. London; Chapman and Hall/Spon: pp. 288–99.

Brandt, A. M. and Kasperkiewicz, J. eds (2003) *Diagnosis of Concretes and High Performance Concretes by Structural Analysis* (in Polish), IFTR: Warsaw.

Brandtzaeg, A. (1929) 'Failure of a material composed of non-isotropic elements,' *Det. Kgl. Norske Idenskabers Selskabs Skrifter*, No. 2, Trondjem.

Broek, D. (1982) *Elementary Engineering Fracture Mechanics*, 3rd edition. The Netherlands: Martinus Nijhoff,

Dahl, P. A. (1986) 'Influence of fibre reinforcement on plastic shrinkage and cracking,' in: *Proc. Int. Symp. Brittle Matrix Composites 1*, Jablonna 1985, A. M. Brandt and I. H. Marshall eds, London and New York; Elsevier Applied Science: pp.435–41.

Darwin, D., Abu-Zeid, M. N., Ketcham, K. W. (1995) 'Automated crack identification for cement paste.' *Cement and Concrete Research*, 25(3): 605–16.

Glinicki, M. A., Litorowicz, A. (2003) 'Application of UV image analysis for evaluation of thermal cracking in concrete,' in: *Proc. Int. Symp. Brittle Matrix Composites 7*, A. M. Brandt, V.C. Li and I. H. Marshall eds, Warsaw, October 13–15: 101–9.

Granger, S., Loukili, A., Pijaudier-Cabot, G., Behloul, M. (2006) 'Self healing of cracks in concrete: from model material to usual concretes,' in: *Proc. 2nd Int. Symp. on Advances in Concrete through Science and Engineering*, Sept. Quebec City, QC; RILEM.

Gray, R. J. (1984) 'Autogenous healing of fibre/matrix interfacial bond in fibre-reinforced mortar.' *Cement and Concrete Research*, 14: 315–17.

Grzybowski, M., Shah, S. P. (1990) 'Shrinkage cracking of fiber reinforced concrete.' *ACI Materials J.*, 87(2): 138–48.

Hannant, D. J. (1978) *Fibre Cements and Fibre Concretes*, Chichester, UK; John Wiley & Sons.

Hannant, D. J., Edgington, J. (1975) 'Durability of steel fibre concrete,' in: *Proc. RILEM Symp. Fibre Reinforced Cement and Concrete*, Lancaster; Construction Press, pp. 159–69.

Hilsdorf, H. K. (1989) 'Durability of concrete – a measurable quantity'? *Proc. of IABSE Symp. Durability of Structures*, Lisbon, 6–8 September, 1: pp. 111–23.

Hornain, H., Marchand, J., Ammouche, A., Commène, J.-P., Moranville, M. (1996) 'Microscopic observation of cracks in concrete; a new sample preparation technique using dye impregnation.' *Cement and Concrete Research*, 26(4): 573–83.

Jacobsen, S., Sellevold, E. J. (1996) 'Self healing of high strength concrete after deterioration by freeze/thaw.' *Cement and Concrete Research*, 26(1): 55–62.

Kaplan, M.F. (1961) 'Crack propagation and the fracture of concrete,' *Journal of the American Concrete Institute*, 58(5): 591–610.

Kasperkiewicz, J., Stroeven, P. (1991) 'Observations on crack healing in concrete,' in: *Proc. Int. Symp. 'Brittle Matrix Composite 3'*, A. M. Brandt and I. H. Marshall eds, London and New York; Applied Science Publishers: pp. 164–73.

Kawamura, M. (1978) 'Internal stresses and microcrack formation caused by drying in hardened cement pastes.' *Journal of the American Ceramic Society*, 61(7–8): 281–3.

Kucharska, L., Brandt, A.M. (1997) 'Pitch-based carbon fibre reinforced cement composites: a review.' *Arch. of Civ. Eng.*, 43(2): 165–87.

Landis, E.N., Nagy, E.N., Keane, D.T., Shah, S.P. (1996) 'Observations of internal crack growth in mortar using x-ray microtomography,' in: *Proc. 4th ASCE Conf. Materials for the New Millennium*, Washington, DC, November 10–14, K. P. Chong ed., 1330–6.

Landis, E.N., Zhangm T., Nagy, E.N., Nagy, G., Franklin, W.R. (2007) 'Cracking, damage and fracture in four dimensions.' *Materials and Structures*, 40: 357–364.

Litorowicz, A. (2006) 'Identification and quantification of cracks in concrete by optical fluorescent microscopy.' *Cement and Concrete Research*, 36: 1508–15.

Lusche, M. (1974) 'The fracture mechanism of ordinary and lightweight concrete under uniaxial compression,' in: *Proc. Conf. 'Mechanical Properties and Structure of Composite Materials'*, A.M.Brandt ed., Jablonna 18–23 November 1974. Ossolineum, 423–40.

Malmberg, B., Skarendahl, Å. (1978) 'Method of studying the cracking of fibre concrete under restrained shrinkage,' in: *Proc. RILEM Symp. Testing and Test Methods of Fibre Cement Composites*, R. N. Swamy ed., The Construction Press: pp. 173–9.

Mandelbrot, B.B. (1983) *The Fractal Geometry of Nature*, New York; W.H. Freeman.

Mor, A., Monteiro, P.J.M., Hetsre, W.T. (1989) 'Observations of healing of cracks in high-strength lightweight concrete,' *Cement, Concrete and Aggregates*, CCAGDP, 12(2): 121–5.

Neville, A.M. (2002) 'Autogeneous healing – a concrete miracle'? *Concrete International*, 24(11): 76–82.

Petersson, P.E. (1981) *Crack growth and formation of fracture zones in plain concrete and similar materials*, Lund Inst. of Techn., Div. of Build. Mat., Rep. TVBM-1006.

Reinhardt, H.W., Jooss, M. (2003) 'Permeability and self-healing of cracked concrete as a function of temperature and crack width,' *Cement and Concrete Research*, 33: 981–5.

Ringot, E., Bascoul, A. (2001) 'About the analysis of microcracking in concrete,' *Cement and Concrete Composites*, 23(2): 261–6.

Roy, C., Allard, J., Maslouhi, A., Piasta, Z. (1991) 'Pattern recognition characterization of microfailures in composites via analytical quantitative acoustic emission,' in: *Proc. Int. Coll. 'Durability of Polymer Based Composite Systems for Structural Applications'*, A.H. Cardon and G. Verchery eds, London; Elsevier Applied Science, 312–24.

Soroushian, P., Elzafraney, M. (2004) 'Damage effects on concrete performance and microstructure.' *Cement and Concrete Composites*, 26: 853–9.

Wittmann, F.H. (2006) 'Specific aspects of durability of strain hardening cement-based composites,' in: *Restoration of Buildings and Monuments, 12(2) Bauinstandsetzen und Baudenkmalpflege*, Freiburg, Germany: pp. 109–18.

Zaitsev, Y.V., Wittmann, F.H. (1981) 'Simulation of crack propagation and failure of concrete,' *Materials and Structures*, 14(83): 357–65.

Zamorowski, W. (1985) 'The phenomenon of self-regeneration of concrete,' *The Int. Journ. of Cem. Comp. and Lightweight Concrete*, 7(2): 199–201.

Standards

ACI 224R-01 *Control of Cracking in Concrete Structures*.

10 Fracture and failure in the structures of the material

10.1 Application of fracture mechanics to cement matrices

10.1.1 Principles of linear elastic fracture mechanics (LEFM)

Fracture mechanics was first applied by A. A. Griffith (1921) as an approach to the analysis and evaluation of the material's behaviour. For the basic principles of fracture mechanics and its present development, the reader is referred to one of a number of available books and manuals: Anderson (2005). It is sufficient here to recall a few of the most important notions necessary for considerations of the brittle matrix composites.

Griffith's theory is based on two assumptions. The first concerns the considerable difference between observed and calculated tensile strength of materials. Effective strength is lower by at least one order of magnitude than the theoretical one calculated on the basis of the interatomic bonds. That difference is attributed to stress concentrations at microcracks and defects existing in every solid body before the application of any external load. In materials that exhibit plastic deformations, these stress concentrations disappear without appreciable reduction of material strength. In contrast, in brittle materials, a stress concentration initiates microcracks and their propagation, thus leading to fracture.

The second assumption is related to the condition of an energetic equilibrium at the crack tip when a crack may start to propagate. The relation between the strain energy release rate $\partial U/\partial c$ considered as the crack extension force and the rate of energy $\partial S/\partial c$ necessary to create new crack surface decides whether the crack is stable or is rapidly propagating. The strain energy is denoted by U and specific surface energy by S. The initial crack length is equal to $2c$. When:

$$\frac{\partial U}{\partial c} < \frac{\partial S}{\partial c} \qquad (10.1)$$

the crack driving force is too low and the initial crack remains stable. If the load is increased over the equilibrium state,

$$\frac{\partial U}{\partial c} \geq \frac{\partial S}{\partial c} \qquad (10.2)$$

then the crack propagates. The strain energy is transformed into the surface energy. This means that the crack propagation up to the critical state is defined by equation (10.1) and is conditioned by an energy input exceeding the energy required for the creation of new surfaces.

In an ideally elastic and brittle material, further crack propagation does not require additional energy and rapid crack propagation follows, which is the final fracture.

On that basis the theoretical value for tensile strength was derived by Griffith in the following form:

$$\sigma_f = \left(\frac{2E\gamma_f}{\pi c}\right)^{1/2} \qquad (10.3)$$

for a plate of infinite dimensions subjected to tensile stress σ and having an elliptical crack (Figure 10.1a) where γ_f is the fracture surface energy, considered as a kind of material property. However, the tests on real materials did not confirm the values of σ_f obtained from equation (10.3) because of stress concentrations at the tips of the microcracks. The value of the stress at the crack tip may be expressed by the equation

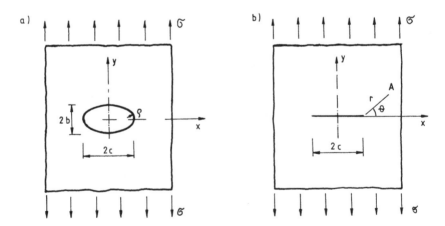

Figure 10.1 Schemes of an infinite plate subjected to tension considered by Griffith (1921):
(a) with an elliptical hole
(b) with a penny-shape crack.

$$\sigma_{con} = 2\,\sigma\left(\frac{c}{\rho}\right)^{1/2},\qquad\qquad(10.4)$$

where ρ is the curvature radius at the crack tip. When the crack becomes sharper, meaning with $\rho \to 0$, the effect of concentration considered as ratio σ_{con}/σ is increasing.

The concept of stress concentration at the crack tip is represented in Figure 10.2.

Theoretical stress concentration for a very sharp crack (the penny-shape crack, Figure 10.1b) in an elastic material leads to an infinite stress value (Figure 10.2a) which cannot be realized. If a local plastic zone appears around the crack tip, the value of the stress corresponds to the yield stress σ_y, which characterizes the material's behaviour (Figure 10.2b). In the concrete-like materials around crack tips, zones of microcracking appear and the corresponding stress distribution is supposed as shown in Figure 10.2c.

In real materials, much energy is used not only to create new crack sur-faces, but is also dispersed in zones of plasticized or microcracked material. Therefore, outside the region where the threshold value of σ_y is reached (Figure 10.2c) rapid crack propagation without an energy input does not occur at all, or is preceded by slow crack propagation, which requires new energy to overcome obstacles in material structure and in material inelasticity.

The propagation of an initial crack may be decomposed into three inde-pendent modes, which are most often studied separately before being considered as a mixed mode. These modes are shown schematically in Figure 10.3, but here only Mode I is described in more detail for two reasons: it is more suitable to analyze crack propagation at the macro-level, and in most cases other modes are not so important as Mode I, unless the load is applied in a special way, producing just Mode II or Mode III. Modes II

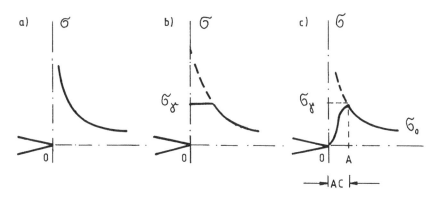

Figure 10.2 Stress concentrations at the crack tip

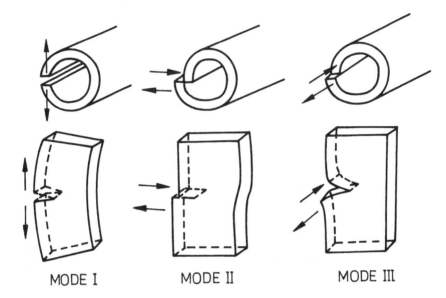

MODE I   MODE II   MODE III

Figure 10.3 Three modes of propagation of a crack.

and III are neglected with respect to Mode I, though it is obvious that in real heterogeneous materials all three modes are mixed together in a more-or-less unknown way (cf. Section 10.4).

The fracture mechanics equations derived by Griffith after his tests on glass specimens directly concern the brittle behaviour of materials and are certainly better justified for hardened cement paste than for any other cement-based composite. The general application of the fracture mechanics is therefore associated with the additional assumptions that plastic or quasi-plastic effects are negligible, or with appropriate modifications of the linear formulae in LEFM. In that context the linear and non-linear fracture mechanics approach should be distinguished.

In LEFM, the material is considered as homogeneous, isotropic and following the principles of linear elasticity. The derived formulae do not correctly reflect the behaviour of real materials which, to a large extent, present non-homogeneities, anisotropy and local plasticity.

In LEFM Mode I, two main notions are specified.

K_I [MNm$^{-3/2}$] is the stress intensity factor, which describes the intensity of the elastic stress field in the neighbourhood of the crack tip. It is expressed by a simple equation for an infinite element loaded by stress σ:

$$K_I = \sigma \, (\pi c)^{1/2} \tag{10.5}$$

and for elements with finite dimensions

$$K_I = \sigma \, (\pi c)^{1/2} \, Y(a/c) \tag{10.6}$$

where $Y(a/c)$ is a function expressing the influence of the shape of element, crack configuration and kind of loading. For several sets of such conditions the function $Y(a/c)$ was calculated and published, for example, by Brown and Srawley (1966) and by other authors, in the form of polynomials with respect to (a/c), where a is the characteristic dimension of the element and c is half length of the crack.

G_I [kNm⁻¹] is the elastic stress energy release rate, meaning that it describes the rate at which the energy for crack propagation is supplied. G_I as equal to $\partial U/\partial c$ is also called the crack extension force and is related to elastic properties of the material and element and to the applied load system.

For the simplest case of an infinite plate under tensile stress σ the strain energy release rate is given by equation

$$G_I = \pi c \frac{\sigma^2}{E} \qquad (10.7)$$

In such a way there are two approaches to apply fracture mechanics: the use of a stress intensity concept or an energy one. Both are equivalent and give identical results in the case of an ideal brittle material and they are related as follows:

$$\text{for plane stress } K_I^2 = EG_I \qquad (10.8)$$

$$\text{for plane strain } K_I^2 = EG_I \frac{1}{1-v^2}$$

where v is the Poisson ratio. Because its value may be assumed as approximately equal to 0.2 for various concretes, then the difference between the plane stress and plane strain values may be neglected in most of the practical problems in which precision of other parameters is often lower.

The notion of plane stress and plain strain are related to the geometrical stress distribution around the crack tip. For derivation of these equations the reader is again referred to the fracture mechanics manuals. It may be useful to mention here that plane stress configuration corresponds to a relatively thin plate loaded in its plane and plane strain, to a thick element where strain components perpendicular to its plane may be neglected.

G_I and K_I characterize the situation around the crack tip at the stable state described by equation (10.1) in which further increase of loading does not produce crack instability. When, due to the load increase, the threshold (10.2) is reached and the sign of the inequality is changed, then the onset of crack propagation is observed which corresponds to critical values G_{Ic} and K_{Ic}. These values form a kind of fracture criterion, completed by equation

$$G = 2 \, \gamma_f \tag{10.9}$$

derived from equations (10.3) and (10.7).

The values of G_{Ic} and K_{Ic} may be calculated from equations (10.5) or (10.6) and (10.7) when critical values of stress σ_{cr} corresponding to observed onset of crack propagation are introduced.

Kaplan (1961) attempted to apply the fracture mechanics approach to cement-based materials. His works were continued by many others, but already Kesler *et al.* (1971) correctly indicated that LEFM cannot be directly applied to concrete elements. A very comprehensive review of these works and the problems related to the introduction of fracture mechanics to these materials was published by Mindess (1983), and by other authors.

For heterogeneous materials with internal structure composed of aggregate grains or fibres, the fracture surface energy is considerably increased and crack propagation is more or less controlled in such a way that slow crack growth is observed and it is difficult to select an appropriate value for critical stress in the above equations. The fracture process zone created at the crack tip is not negligible in comparison to the dimensions of the specimen or element tested. For materials with multiple crack systems, thorough studies are carried out on one single crack and the phenomenon of multiple cracking cannot be directly considered within the frame of LEFM.

It has been shown by several authors; for example, Bažant (1980), that taking into account normal concrete inhomogeneities due to aggregate grains, the application of LEFM requires large specimens, for example beams or cubes with characteristic dimensions larger than 1.0 meter, because the fracture toughness increases with the decrease of specimen size; however, these should be kept sensibly larger than the plastic zone around the crack tip. Tests become expensive and a series composed of large specimens necessarily introduce other sources of errors that may considerably disturb the results.

In the situation described above, it is necessary either to verify that the non-elastic effects are small and negligible or try to consider them by appropriate modification of the formulae. For practical applications, it is useful to standardize the specimens and to compare only those with similar dimensions. It is also possible to use other parameters based mostly on experimental data in which the assumptions of an ideal elastic behaviour are not needed. All these ways of using the fracture mechanics for cement-based composites have been developed by several authors (cf. Section 10.2).

The use of the stress intensity factor K for concrete-like materials is preferable (Swamy 1979), because:

- K_{Ic} is a linear function of the load;
- K_{Ic} describes the stress field around the crack tip without attempting to represent the entire specimen or element;
- values of K_{Ic} due to different states of loading may be summed up.

The application of G_{Ic} and K_{Ic} as material constants to cement-based composites proved to be useful; however, there are several questions and limitations, which should be considered. In most cases the calculation of these parameters is applied for comparative purposes where certain shortcomings are less important.

Both these parameters have been derived for an ideal material and for particular conditions for crack dimensions and load application. Neither of these conditions is exactly satisfied, and calculated values may be considered as approximate ones only. Even if the geometrical conditions of element and loading are taken into account by appropriate determination of function $Y(a/c)$ as in equation (10.6), the behaviour of every material around the crack tip shows more-or-less important non-elastic effects in the form of local plastic deformations or microcracking.

Interested readers may find the basic description of LEFM in the collection of papers edited by Sanford (1997).

10.1.2 Application of fracture mechanics to elasto-plastic materials

The development of fracture mechanics aimed at consideration of non-elastic and non-linear effects is of particular importance for applications in composite materials in which the internal structure is actually designed to control the cracking and to increase the effective fracture surface energy. A few non-linear fracture approaches are described below.

The J-integral has been defined by Rice (1968) initially for non-linear elastic materials as a linear integral on a closed contour Γ_j:

$$J_\Gamma = \int_{\Gamma i} \left(W dy - T \frac{\partial u}{\partial x} \right) ds \qquad (10.10)$$

where x,y are rectilinear coordinates with y-axis perpendicular to the plane of the crack propagation, W is the strain energy density, T is the vector of traction which is perpendicular to Γ_j, u is the displacement vector and ds is a section of the curve Γ_j.

The Rice integral (10.10) is independent on the path, meaning that $J(\Gamma_1) = J(\Gamma_2)$ (Figure 10.4).

The J-integral is proposed as a fracture criterion for materials characterized by nonlinear stress-strain relations and its critical value denoted J_c corresponds to rapid crack propagation.

In linear elastic materials

$$J_{Ic} = G_{Ic}. \qquad (10.11)$$

Several methods were published for calculation of the J-integral and for determination of its critical value; for example, Mindess *et al.* (1977), Brandt

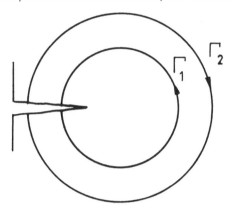

Figure 10.4 Integrating path for calculation of the *J* integral.

(1980). The tests carried out by various authors in different conditions furnished rather coherent results for J_{Ic} and it has been shown, that J_{Ic} is much more sensitive to the effects of fibre reinforcement than K_{Ic} for example. However, the dispersion of test results and dependence on arbitrarily estimated parameters do not allow them to be used for practical applications.

Another method for characterizing the behaviour of composite elements under load is based on the concept of crack opening displacement (COD). It consists of the measurement of displacements at an initial notch subjected to Mode I of cracking. The characterization of cracking behaviour of materials by COD is possible in LEFM as well as beyond the linear relation for slow crack propagation. The tests are undertaken in three- or four-point bending in displacement-controlled testing machines. However, the COD method does not allow the effective stress concentration to be calculated and may only be used as a valuable comparative measure.

In general, the experimental approach to the problems of crack propagation in non-linear and non-elastic materials is based on the observation and recording of behaviour of specimens with initial notches, and COD measurements were developed by several authors. Later, crack tip opening displacement (CTOD) was determined as a function of crack mouth opening displacement (CMOD) (Velasco et al. 1980), which was directly measured and recorded, for example, according to EN ISO 12737:1999 and ASTM E 399-06 (2006), (cf. Figure 10.5).

Brandt *et al.* (1989) applied double gauges for CMOD measurement to establish the angle and centre point of rotation during bending. The shape of the curves obtained is shown in Figure 10.6. The critical value of CTOD designed as CTOD$_c$ was established in different ways, as corresponding to the maximum load, to a local maximum of load, CTOD curve or to a rapid decrease of load, and the displacement is measured at the crack tip. This gives the possibility of using CTOD testing for various composite materials

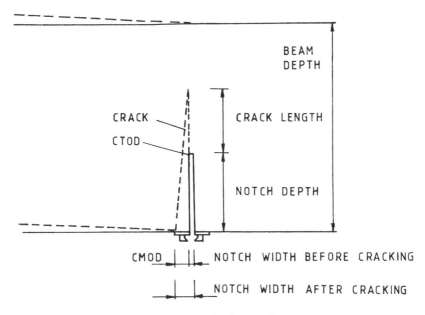

Figure 10.5 CTOD and CMOD in a notch after cracking.

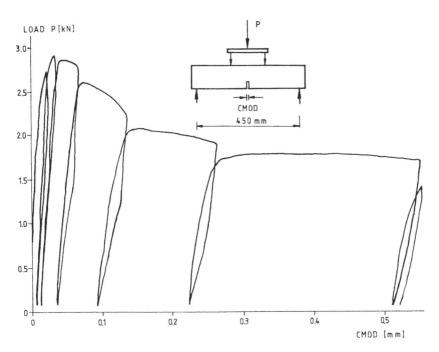

Figure 10.6 COD as a function of load for fibre-reinforced concrete specimens subjected to bending, after Brandt *et al.* (1989).

and for obtaining useful comparisons between the materials with different composition or internal structure.

Theoretical formulations for COD measurements are related to the model of a crack proposed by Dugdale (1960).

The postulate that critical values of $CTOD_c$ are characteristic for material behaviour and are independent from specimen dimensions and type of loading, is only partly satisfied and may give only comparative indications on the material properties. However, the validity of $CTOD_c$ testing is extended beyond the elastic behaviour of materials, and that is its main value.

Fracture in the process zone before the crack tip in elements under direct tension, or in tensile zones of elements subjected to bending, was represented by different models. In these models, the transfer of stress across the process zone is related to displacement in the crack by a so-called 'tension-softening diagram.' The form of this diagram represents a constitution law, which is characteristic for the considered material.

Considerable progress in the application of fracture mechanics parameters to concrete-like materials was achieved by Hillerborg *et al.* (1976) who proposed a notion of a fictitious crack (the so-called 'cohesive crack model'). For stress variation, as shown in Figure 10.2c, it is assumed that between point O where effective crack ends and point A where value σ_y is reached the material is partly microcracked. This effect is assimilated by an additional crack length ∂c. The partial destruction of the material ahead of the crack tip is confirmed experimentally by several tests and observations on cement matrix specimens. Discontinuous microcracks are clearly distinguishable in hydrated gel, which consists of entangled crystalline needles and amorphous material. In this intermediary region between points O and A the tensile stress is only partly transferred and its value varies between zero and σ_y. Behind point A the stresses decrease gradually to its mean value $\sigma = \sigma_0$.

Therefore, the situation at the crack tip may be described by two diagrams $\sigma(\varepsilon)$ and $\sigma(w)$, where w is the crack opening (Figure 10.7). The curve $\sigma(\varepsilon)$ is valid for the zone where $\varepsilon < \varepsilon_y$ and for $\varepsilon > \varepsilon_y$ the notion of strain has no more sense because of the discontinuities of the material and stress σ transferred across the fictitious crack is related to the crack opening w. At point O the critical value w_{cr} is reached and stress cannot be transferred; at that point effective crack tip already exists.

The experimental methods for the identification of crack behaviour in concrete materials were discussed by Mindess (1991) and later by Li and Maalej (1996), particularly with respect to the control of crack propagation by aggregate grains and the creation of a so-called 'cohesive zone.'

The experimental and theoretical studies are completed by the application of the finite element method (FEM) for numerical analysis of fracture processes. By that approach the region surrounding the crack tip can be approximately represented by an equivalent system of discrete elements. A simple example shown in Figure 10.8 is partly based on work published by Petersson (1981). The density of the mesh for finite elements, their shape (e.g.

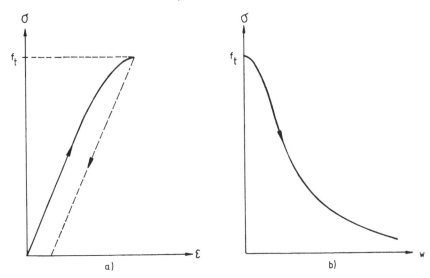

Figure 10.7 Curves σ(ε) and σ(w) for the fictitious crack model proposed by Hillerborg (1978).

triangles, rectangles, etc.) and forces at the nodes can be selected according to the required accuracy of the calculations, possibilities of available computers and other circumstances. It is also possible to correctly represent, though in an approximate way, the material property as expressed by curves σ(ε) and σ(w).

Various methods of determination of the tension-softening diagrams, which express the behaviour of different materials at fracture, were proposed; for example, the so-called 'crack band' approach was described by Bažant and Planas (1998), and a general review of the state of knowledge by Bažant (2002). The application of the notion of R-curves was again approached by Veselý and Keršner (2006) without conclusive results.

In the abovementioned studies on the fracture processes, there are several somewhat controversial assumptions and hypotheses. For the rather complicated structure of the composite materials on the micro-level, their homogeneity and isotropy is assumed on the macro-level. On the basis of simplifying assumptions, the structural and mathematical models are constructed to enable analytical and numerical treatment of cracking processes and to formulate problems which may be solved without much difficulty and expense in order to:

- obtain solutions expressed in stress and strain even if they are significantly different from the real material's behaviour;
- predict material behaviour for different sets of parameters, like material properties, external actions, etc., with appropriate coefficients and correcting factors;

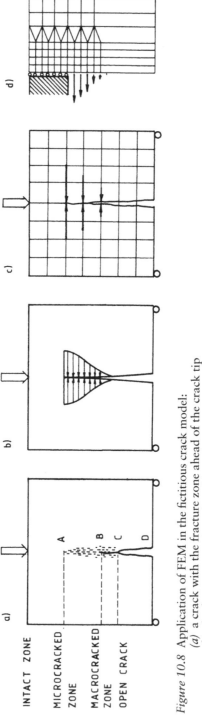

Figure 10.8 Application of FEM in the fictitious crack model:
(a) a crack with the fracture zone ahead of the crack tip
(b) the fracture zone is replaced by a fictitious crack and appropriate stresses
(c) system of finite elements and nodal forces which replace the stresses transferred across the crack, after Petersson (1981).

- establish the influence of various selections of components, technologies, and environmental conditions on the crack propagation.

When, instead of one single crack, the phenomenon of multiple cracking appears, which is the case in advanced composites with internal structure (fibres or polymers), the fracture mechanics models are less effective for the prediction of behaviour of the materials. Conditions for multiple cracking appear only when the energy required for unstable propagation of the first crack is greater than the energy required to form the next crack somewhere in the neighbourhood. That is the role of inclusions and fibres in the brittle matrices.

10.1.3 Role of dispersed reinforcement after the cracking of the matrix

The influence of fibres on brittle matrix behaviour may be reduced to two points: the control of crack propagation and the increase of the ultimate strain. Schematic image of a crack crossing a system of parallel fibres is shown in Figure 2.5 where different situations are presented of fibre cracking and debonding; these situations are represented by simplified modelling, corresponding to various properties of fibres and matrix.

In the 1960s and early 1970s, there were two theories that explained the behaviour of a brittle matrix reinforced with dispersed fibres. One of these theories was proposed in papers by Romualdi and Batson (1963) and by Romualdi and Mandel (1964), and was called RBM (Romualdi, Batson, Mandel). Another theory was published by Aveston *et al.* (1971) and Aveston *et al.* (1974) and was called 'the ACK (Aveston, Cooper, Kelly) theory.' Notwithstanding apparent differences, both approaches are rather close to each other and lead to similar conclusions. The main difference between these two theories is that in RBM the main role of dispersed fibres is to control crack propagation by providing the forces that keep the material together ahead of the crack tip. In the ACK theory the fibres limit the maximum opening displacement of the crack.

In the RBM approach the propagation of a single crack is considered in a matrix reinforced with a system of parallel fibres. The material of the matrix follows linear elasticity, and crack propagation is initiated from an existing intrinsic flaw when stress in the direction perpendicular to that flaw attains the tensile strength σ. If the flaw has the form of a penny-shaped crack of length $2c$ as it is considered in LEFM (cf. Section 10.1.1), then the critical value of the stress intensity factor is also attained:

$$K_\sigma = 2\frac{\sqrt{c}}{\pi}\sigma \qquad (10.12)$$

The system of aligned fibres arranged according to a square array is shown schematically in Figure 10.9. Before the application of load, the matrix and

A - A

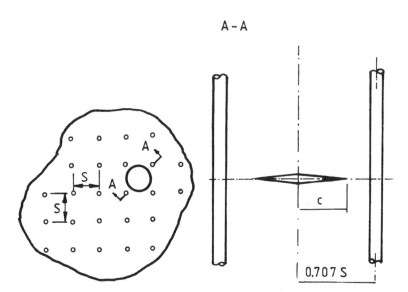

Figure 10.9 Assumed action of fibres in crack propagation control according to RBM
theory, after Romualdi and Batson (1963):
(*a*) cross-section with a penny-shape crack and square array of fibres
(*b*) two parallel fibres and a crack.

fibre strains are equal, but at crack opening the assumed inextensible fibres
present certain resistance against matrix displacements due to a perfect fibre-
matrix bond. Therefore, the fibres provide a stress p, which acts on the matrix
in an opposite direction to its virtual cracking. Finally, the expression for the
stress intensity factor is:

$$K_T = 2\frac{\sqrt{c}}{\pi}(\sigma - p)$$
(10.13)

The value of stress p is unknown but may be determined taking into account
the fibre spacing and the force imposed by a single fibre. The calculation was
proposed by Romualdi and Batson (1963) from a system of linear equations,
which expressed the equilibrium between displacements of non-reinforced
matrix in the cracked state and displacements due to the action of concentrated
forces from fibres. The calculated examples for various spacings derived
from the fibres volume fraction V_f (cf. Section 8.4) indicated the increase
of modulus of rupture caused by the fibre reinforcement. The comparison
of calculated curve and experimental data is shown in Figure 10.10 and a
general confirmation may be observed. Further development of the RBM
theory in which the critical value G_c of elastic energy release rate was related

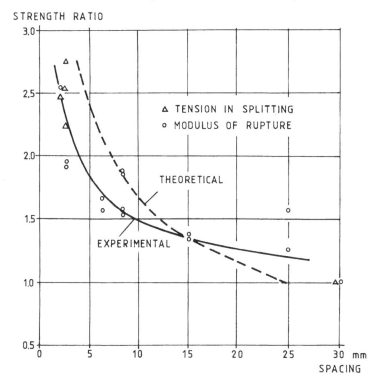

STRENGTH RATIO

Figure 10.10 Calculated and experimental results of relation between fibre spacings and matrix reinforcement for modulus of rupture and tensile strength in splitting, after Romualdi and Mandel (1964).

to parameters of fibre reinforcement is presented by Romualdi (1968). These considerations are not developed here in detail because its general utility for material design and strength verification is limited and the agreement with experimental results is inadequate.

The ACK theory in its general form covers elastic and non-elastic effects in fibres and matrix. Therefore the entire ascending branch of the stress-strain curve may be described. For that purpose, two cases are considered:

1   The fibres may be debonded from the matrix and the pull-out of fibres appears.
2   The fibres are bonded to the matrix and additional stresses are transferred from the fibres to the matrix.

In a composite element subjected to tension and reinforced with aligned fibres the stress-strain relation is initially linear. Its slope is given by elastic modulus determined by the law of mixtures (cf. Sections 2.5 and 8.3): $E_c = E_f V_f + E_m V_m = E_f V_f (1 + a)$.

The cracking strain in the matrix may be expressed by the formula derived from known properties and volumes of the matrix (E_m, V_m), of the fibres (E_f, V_f), the matrix surface work of fracture γ_m and fibre-matrix bond strength τ:

$$\varepsilon_{mc} = \left[A \frac{24\gamma_m \tau E_f V_f^2}{E_m^3 V_m^2 d} \right]^{1/3} \tag{10.14}$$

here constant $A = 1$ to 3 and expresses unknown values characterizing the fibre debonding processes and creation of new surfaces. The form of equation (10.14) was proposed by Kasperkiewicz (1983) on the basis of the ACK theory.

The matrix cracks when the critical strain for an unreinforced matrix ε_{um} or that for composite material ε_{uc} is reached, whichever is greater.

In an example of steel fibre reinforced cement-based matrix with $E_f = 210$ GPa, $E_m = E_c = 35$ GPa, $V_f = 0.03$, $\tau = 4$ MPa, $d = 0.4$, $\gamma_m = 10$ N/m the result is, $\varepsilon_{uc} = (224 \text{ to } 323)10^{-6}$, and this value should be compared with ε_{um} which is usually equal to about $(150 \text{ to } 200)10^{-6}$. This example shows that in a very densely reinforced matrix with $V_f = 0.03$, which is difficult to realize in normal technology, a slight increase of critical strain of a composite material with respect to plain matrix may be expected.

Similar results may be obtained when the critical volume fraction of fibres V_{fcr} is considered. It may be easily defined for aligned long or even continuous fibres in an element subjected to tension parallel to the fibre direction. This reinforcement may carry on the load after cracking of the matrix. Therefore, the value of V_{fcr} is derived from the simple equivalence of forces carried on by a composite element of cross-section area A_c and by the fibres: $A_c E_c \varepsilon_{mu} = A_c V_{fcr} \sigma_{fu}$, here $V_{fcr} = E_c \varepsilon_{mu}/\sigma_{fu}$, and the product $E_c \varepsilon_{mu}$ represents the cracking tensile stress in the matrix and σ_{fu} is the ultimate tensile stress in the fibres.

For example, for a high quality matrix with $E_c = 40$ GPa and $\varepsilon_{mu} = 100.10^{-6}$ and fibres made of mild steel with $\sigma_{fu} = 200$ MPa, the critical volume fraction is equal to 0.02, assuming that the fibres are fully loaded. In ordinary steel fibre reinforced concrete elements, usually short and randomly dispersed fibres are used. However, because of the low efficiency of such fibres, even if made with high strength steel (which is the case for most present applications), in normal technical conditions the achievement of V_{fcr} is nearly impossible. If other methods of concreting are applied, like SIFCON, or when fibres with increased bond and linearized are used, then the efficiency of reinforcement is significantly improved.

These conclusions may be generalized for elements subjected to bending, provided that the difference in loading states is taken into account and are confirmed by tests of steel fibre-reinforced cement elements.

The crack spacing may be calculated from another formula derived from the ACK theory giving the distance x of the load transfer from matrix to fibres

$$x = \frac{E_m\left(1-V_f\right)\varepsilon_{mu}d}{V_f\,4\tau} \tag{10.15}$$

if for steel fibre reinforced concrete $V_f = 0.02$, $E_m\,\varepsilon_{mu} = 5$ MPa, $d = 0.4$ mm, and $\tau = 2.0$ MPa, then $x = 12.25$ mm.

The actual crack spacing is between x and $2x$ and its most probable value may be estimated as equal to $(1.364 \pm 0.002)x$.

The ACK theory also covers other orientations of fibres and in Aveston and Kelly (1973) two formulae are given for 2D and 3D fibre distribution: $x'_{2D} = \pi\,x/2$, $x''_{3D} = 2x$.

After cracking, the composite Young's modulus is dependent on the fibre reinforcement:

$$E_c = E_f\,V_f.$$

The experimental verification of ACK theory is presented by Aveston *et al.* (1974) and the approximation obtained may be considered as quite satisfactory (Figure 10.11). However, it should be mentioned that specimens produced for these verifications were made with different materials than those used in normal practice.

Results of other verification have been published by Laws (1974) following tests on cement paste specimens reinforced with continuous and aligned glass fibres of various volume fractions (Figure 10.12). The tensile strain at fracture of the plain paste was not given; nevertheless, the influence of fibres is clearly visible and in general it confirms the ACK theory predictions, if

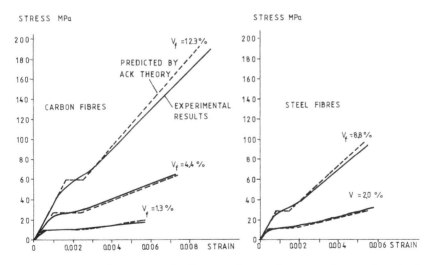

Figure 10.11 Tensile stress-strain diagrams for continuous carbon and steel wire reinforced cements, after Aveston *et al.* (1974).

an increase of strength with increase of reinforcement may be estimated as sufficient verification.

Both presented theories lead to similar conclusions concerning the cracking strain of reinforced matrices; however, both exaggerate the influence of fibres. After the crack opening, the ACK theory may be used to calculate the probable crack spacing and the composite's Young's modulus. Furthermore, in that theory, a few of the parameters used are taken from the testing of relevant composite elements and therefore it is to some extent more appropriate for verification and calibration than other proposed theories for fibre-reinforced composites.

The theories called RBM and ACK have two important limitations concerning their application to cement-based composites used in structural elements. The descending branch of the stress-strain curve is not covered, which means that the behaviour of composite materials after cracking is not described. Considering that the main purpose of fibres is to control cracking, and that their influence on that part of cement composites behaviour is the most meaningful, that limitation is important. Furthermore, both theories better describe the stress-strain or load-displacement diagrams the higher the fibre volume fraction. Or, in practice, because of many reasons described elsewhere, low volume fractions of fibre reinforcement are used.

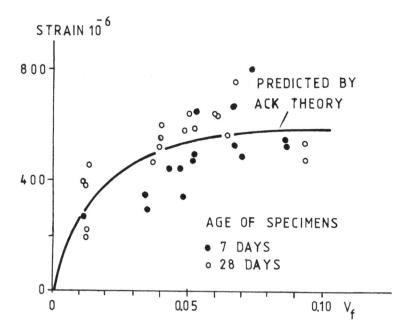

Figure 10.12 Matrix failure strain of glass fibre reinforced cement paste, after Laws (1974).

The formulae presented above are rarely applied and the input of dispersed fibres to the crack resistance of the composite elements is estimated using indirect notions of equivalent strength and other experimentally based approaches (cf. Section 12.5).

There is another notion derived for fibre-reinforced matrix and related to the post-cracking behaviour. It is the critical value of fibre volume fraction V_f which corresponds to the increase of the bearing capacity. For insufficient reinforcement, when $V_f < V_{fcrit}$, the load decreases after cracking of the matrix. The role of fibres is limited to the crack control, but its efficiency is not sufficient for any load increase beyond the level of the first crack in the matrix. The expression for V_{fcrit} was derived by Hannant (1978) on the basis of several simplifying assumptions for composites reinforced with short dispersed fibres. The numerical examples show that the values calculated in that way are much lower than the results obtained from tests. The notion of V_{fcrit} may be generalized for all kinds of reinforcement and used to estimate their efficiency.

It appears from comparison of various theoretical and experimental results that detailed calculations of fibre-reinforced concrete behaviour do not allow reliable data to be obtained without taking into account all influences and factors, such as the precisely determined efficiency coefficient, scale effects, etc. (cf. Bentur and Mindess 2006).

Toughness FRC as a result of steel fibre crack control was studied by Brandt (1996) who has shown that there was no single measure to characterize the ability of a composite material to absorb energy. All methods proposed by different authors have their shortcomings and the only way is to diligently use a few of them and carefully analyze the results.

An attempt to model crack propagation in FRC using the Mode I approach was presented by Zhang and Li (2004); they obtained good agreement with experimental results. The basic conclusion from this research is that the bridging ability of aggregate grains and fibres constitute an essential way to improve the strength of composite elements under bending.

10.2 Cement matrix composites under various states of loading

10.2.1 Role of components

In this section a short description is given of how the elements and specimens made of composite materials behave under compression, tension, bending, shear and combined states of stress. Because a large number of different test results are available in published papers and books, it is not intended to summarize all of them here, but rather to show characteristic behaviour under static short-term loading, and to explain the role of reinforcement, leaving the reader to follow particular aspects from the recent test reports.

The composite's behaviour is the main aspect described here; that is, how the heterogeneity of the materials and their purposefully designed structure influence their behaviour under external loads and other actions. It is not possible, on the other hand, to present the overall complexity of a material's behaviour and to pay full attention to the influence of the size of elements, rate of loading, stiffness of the loading machine and to many other important factors and conditions.

Inclusions of various types decrease the brittleness of the cement-based matrices. The aggregate grains add strength and hardness and these effects may be approximately calculated from the law of mixtures for elastic region of stress-strain curves. The role of these inclusions is slightly different when first cracks appear – they are stopped at aggregate grains and the failure is less brittle, because additional energy input is necessary to break or to contour the grains.

The influence of fibres depends on their efficiency; this means that the fibres that are long enough, or provided with additional anchorages and applied in sufficient volume, may slightly modify the composite's behaviour up to the point when the first crack opens. The increase of strength and of the Young's modulus is small, and related by the law of mixtures to the volume of fibres, which rarely and only with special technology may be larger than 3% per volume. In contrast, the behaviour after cracking is considerably modified. With efficient fibres the load bearing capacity may be substantially increased after the first crack has occurred and the fracture toughness as expressed by the area under the stress-strain or load-displacement curve is several times larger than that of the plain matrix.

The role of fibres depends on their volume, aspect ratio, strength and bond to the matrix. It has been shown in many tests how these fibre characteristics influence the composite's behaviour. In recent years, particular attention has been paid to the distribution of fibres – very small and well dispersed fibres may control the microcracks in the matrix form the very beginning of their opening and particularly high deformability of the composite may be obtained (cf. Section 14.5).

The polymers added to the cement matrices influence the overall behaviour of the composite in various ways. By improving the matrix-aggregate bond the strength is increased, which is accompanied by an increase in brittleness. Larger amounts of polymers, if they exhibit plastic behaviour, may, on the contrary, enhance the deformability of the composite material.

Considerable possibilities for designing the material according to well-specified requirements are opened up when a wide variety of modifications of the material structure and composition are considered and adequately applied.

It has been observed by Kim *et al.* (2004) that the fracture characteristics vary with the age of concrete; this observation concerns COD and CTOD that decreased, and also K_{Icr} and fracture energy G_F increased with concrete age tested up to 28 days.

More detailed information is given in subsequent chapters, according to different states of stress or loading.

10.2.2 Fracture under compression

The influence of fibres on the stress and strain characteristics of compressed elements was examined by several authors and main common conclusions were proposed. Certain particularities observed may be attributed to different kinds of fibres, mix proportions and testing techniques applied.

The increase of strength due to dispersed steel fibres is small and uncertain. This is observed by many researchers and may be explained by two controversial effects: steel fibres are reinforcing, but their introduction into the fresh mix modifies considerably its workability and usually increases the porosity. As a result, a certain decrease in strength is often observed. A 7% increase in strength was obtained by Hughes and Fattuhi (1977) when Duoform fibres were used, but other less efficient fibres gave no appreciable results. Polypropylene fibres decreased the matrix strength because of the lower Young's modulus. It may be concluded that inclusion of fibres is not the best way to obtain high compressive strength. Even if certain increases may be expected, the mix proportion modifications are more efficient by far. Also, the Young's modulus for the ascending branch of the stress-strain curve does not vary with the introduction of fibres.

The main aim in applying the fibres is to modify the descending branch of the stress-strain curve and to obtain quasi-ductility of the composite material. Moreover, fracture toughness is significantly increased because fibres control the opening and propagation of cracks. These effects are shown in Figure 10.13 from tests done in 1977; it is obvious that more energy is dissipated due to the pull-out of the fibres and total energy absorption is considerably larger.

In a large test program realized by Tanigawa *et al.* (1980), the limited but positive influence of steel fibres on strength and post-cracking behaviour of specimens under compression has been proved. Using a high rigidity compression testing machine a clear image has been obtained of descending branches of stress-strain curves. In Figure 10.14 three groups of curves are shown for various aspect ratios (height/diameter) of tested prisms: 1, 2 and 3, and for various volume fractions of fibres: 0%, 0.75% and 1.5%. In all cases of the aspect ratio of prisms the influence of fibres is visible and unambiguous; however the relative differences due to fibre reinforcement are rather small.

In tests executed in 1990 and subsequent years on HPFRCC specimens, it has been shown by several authors that by increasing the fibre content and by improving the fibre-matrix bond, after 28 days high compressive strength over 200 MPa may be obtained, cf. Behloul (2007), with maximum aggregate grain limited to 4 mm; the optimization approach is very suitable for the design of the mix composition.

The application of any type of fibre in structural elements subjected to compression is infrequent, because the limited advantages do not counterbalance increased cost and more complicated technology. However, by the addition of fibres to conventional reinforced concrete columns, a moderate

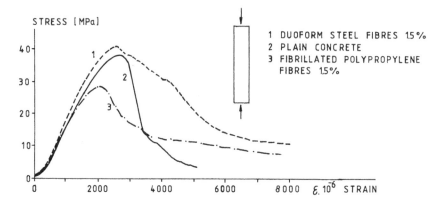

Figure 10.13 Compressive stress-strain curves for plain and fibre reinforced concrete
prisms 102×102×508 mm, V_f = 1.5 %, after Hughes and Fattuhi
(1977).

gain in strength and large increase of fracture energy may be obtained, which
is interesting for structures in seismic zones or in regions subjected to similar
events due, for example, to mining exploitation.

The behaviour of polymer concretes under compression depends on the
role and nature of the polymer component in the composite material, on the
kind of polymer and on its amount (cf. Section 3.2.5). Therefore it is possible

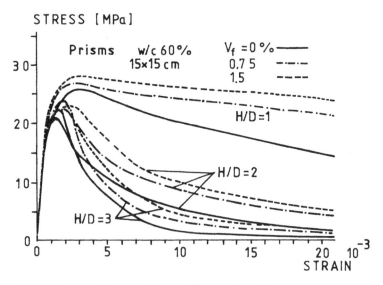

Figure 10.14 Stress-strain curves for steel fibre reinforced prisms with various volume
fraction of fibres and various height to width ratios, after Tanigawa
et al. (1980).

to design a composite material of more or less high strength and deformability. The examples of stress-strain curves of specimens subjected to compression are shown in Figure 10.15 for different material compositions from high strength and low deformability to less strong and more deformable ones. The specimens of type PIC (polymer impregnated concrete) were produced by impregnation of hardened Portland cement concrete specimens A and B with combinations of monomers: methyl methacrylate (MMA), n-butyl acrylate (BA) and trimethylolpropane trimethacrylate (TMPTMA), using mass loading from 4.3% to 5.2 %.

Other types of polymer concretes, that is, PC (polymer concrete) and PCC (polymer cement concrete), also give various possibilities for adjusting mechanical properties to the required ones.

10.2.3 Fracture under direct tension

The behaviour of plain and fibre-reinforced concrete elements under tension is considerably affected by the testing technique and it is difficult to eliminate all secondary effects. The specimen configuration and local stress concentrations, the type of gripping system and testing machine stiffness are among the most important factors. To avoid all secondary effects or to reduce their importance and to obtain reliable results, the tests should be

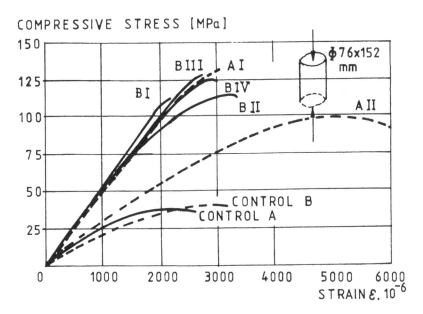

Figure 10.15 Compressive stress-strain curves for concrete impregnated with various compositions of MMA, BA and TMPTMA, after Dahl-Jorgensen *et al.* (1974).

well-instrumented and carefully executed. Even with all precautions the test results may be influenced by certain material eccentricities due, for instance, to the heterogeneity of the fibres' distribution. Local stress concentrations and accidental eccentricities may be decreased to some extent by special grips and the shape of specimens.

The strain energy accumulated in grips and testing machine is released at the first crack and the specimen may be broken without any control. In that case it is impossible to obtain descending branch of the stress-strain curve and even the maximum load is unknown, because after the first crack the element is rapidly ruptured. To avoid such accidents energy accumulation should be minimized and its release controlled by the application of testing machines with a closed-loop system and by sufficient stiffness of the grips and the machine itself.

In appropriately executed tests under direct tension the influence of different reinforcing may be clearly observed. The fibres improve strength and deformability provided that they are efficient enough. The influence of fibres is well-presented in the curves obtained by Kasperkiewicz (1979) and shown in Figure 10.16. It may be observed that the hooked fibres are more efficient than the straight ones. Next, the influence of the orientation of fibres is clear: only fibres aligned with the load direction increase both strength and deformability. The distribution of fibres considerably influences the resulting strength and a large scatter may be expected. The first crack appears sensibly at the same stress as in a plain matrix and this result confirms the law of mixtures up to that stress level. When the fibres are dispersed at random (2D), they slightly increase the deformability, but the first crack appears at a lower stress level, probably due to increased porosity. The fracture energy calculated as proportional to the area under the curve is related closely to the fibre efficiency.

Recommendations for testing fibre reinforced specimens with a circular cross-section are published in RILEM (2001). The specimen is notched and both its ends are glued to the metallic blocks, which are clamped to the testing machine. The machine itself should ensure the average measurement of the displacement over the notch as the feedback signal and provide a stable post-cracking response. In experimental practice, aluminium blocks are used because of lower Young's modulus for aluminium than for steel, therefore the stiffness of the blocks has less influence on the result (Rossi 1998). Also Lenke and Gerstle (2001) used cylindrical specimens.

The problem of secondary flexure in axial tensile tests is considered in view of the tension-softening process. The procedures aimed at elimination of that parasite effect were studied by Akita *et al.* (2001) and Novák *et al.* (2006) who have shown possibilities of nonlinear fracture mechanics simulation in analysis of the experimental data, including post-peak descending branch of the load-displacement curves, taking into consideration material imperfections and heterogeneities.

Examples of the behaviour of PIC elements subjected to tension by splitting

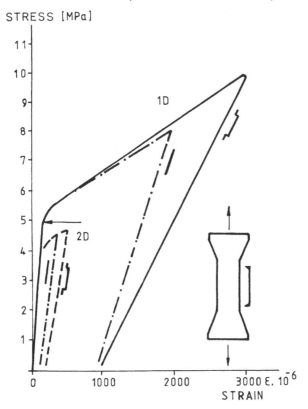

Figure 10.16 Average curves sigma-epsilon determined from direct tension tests, after Kasperkiewicz (1979).

are shown in Figure 10.17 for different compositions of impregnating agent.

The tensile strength may also be established after the splitting test, described by Bentur and Mindess (2006), among others (also cf. Chapter 7.4).

10.2.4 Fracture under bending

It has been shown on elements reinforced with fibres and subjected to bending how the role of the fibres is decisive for crack control.

The diagrams, as in Figure 10.18, are characteristic for a test when specimens are loaded by one or two concentrated and symmetrical forces. These two cases are: four-point or third-point with two symmetrical loads and three-point or centre-point with one force in the middle of the span. The first system is considered as better reflecting the bending conditions in the central part of the element, where pure bending without shearing is created. The second system is simpler to execute, but in fact the maximum effort is

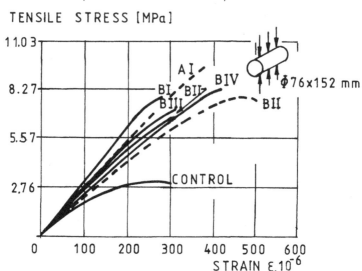

TENSILE STRESS [MPa]

Figure 10.17 Tensile stress-strain curves for concrete impregnated with various compositions of MMA, BA and TMPTMA (cf. Table 10.1), after Dahl-Jorgensen *et al.* (1974).

concentrated directly below the load and that small portion of the material is not necessarily representative for the tested element.

The curves in Figure 10.18 show examples of different efficiency of the fibre reinforcement. There are two characteristic points of each curve designed by bend over point (BOP) or limit of proportionality (LOP) and modulus of rupture (MOR). BOP is the departure of the load-deformation (or load-deflection) curve from linearity and point MOR is the maximum load after crack appearance. The first linear section of the load-deflection curve presents the brittle behaviour of a non-reinforced matrix. The influence of fibre reinforcement on the angle between that section of the curve and the deflection axis is negligible. This means that Young's modulus depends only very slightly on the efficiency of the reinforcement. Then, either the maximum load is reached at the BOP and after cracking the load is rapidly decreased (strain softening), or after BOP the load is increased up to MOR (strain hardening). The ascending part of the diagram after point A, that is, after the first cracks, shows increasing bearing capacity of the element due to efficient fibre reinforcement.

The fibres, depending on their efficiency, control further behaviour after BOP or MOR. Post-cracking hardening corresponds to so-called 'multiple cracking.' This effect is possible thanks to the reinforcement with high efficiency obtained by a combination of three factors: (i) a relatively high volume fraction of fibres; (ii) a good bond to cement matrix; and (iii) eventually an appropriate orientation of fibres with respect to the direction of principal tensile stress. The phenomenon of multiple cracking increases considerably

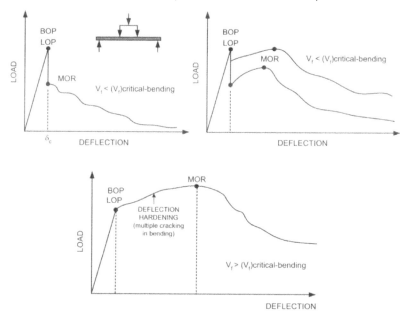

Figure 10.18 Typical load-deflection response curves of fibre reinforced cement composites. (Reproduced with permission from Naaman A. E., Strain hardening and deflection hardening fiber reinforced cement composites; published by RILEM Publications S.a.r.l., 2003.)

the area under the curves and, consequently, the total fracture energy (cf. Section 10.3).

Point BOP is often considered as corresponding to the first crack opening, but that question should be treated with all due restrictions concerning the determination of the cracks in cement matrices and its position is sometimes difficult to establish without ambiguity (cf. Section 10.3.2). It is considered as a limit for the region of linear stress-strain relation. After BOP, cracks appear and develop. Their opening and propagation is controlled by fibres, which break or are pulled out of the matrix, depending on which of these processes prevails.

The load at the MOR is used to calculate the stress due to bending of the cross-section assuming Hooke's law. MOR in fact expresses a fictitious value of the stress, but it is a convenient measure to compare flexural behaviour of elements of different dimensions.

Both characteristics depend on the type of fibres, their aspect ratio and tensile strength and on the fibre-matrix bond. All these factors contribute to the efficiency of the fibres, which determines the shape of the curves beyond point MOR. For steel fibres, in most cases the fibres do not break but are pulled out of the matrix.

The end point of the curve is sometimes difficult to establish, particularly when long fibres are pulled out in the fracture process and the load decreases slowly with very large deflections. In such circumstances the end point may occur in considerably modified conditions of loading and support.

The behaviour of FRC elements under bending may also be characterized by strain measurements. The curves shown in Figure 10.19 give the strain measured at the bottom and top faces of a steel-fibre-reinforced element for various volume fractions and for fibre orientation 1D and 2D. After a short linear zone, the slope of curves changes what corresponds to the onset of microcracking. Next, the gauges measured the apparent strain of the matrix with microcracks, which developed under increasing load. Various curves correspond to the different efficiency of the fibre-reinforcement.

The variety of behaviours of cement-based elements under bending shows the large possibilities of material design. According to requirements, different material properties may be obtained by varying properties of the matrix, quality, amount and distribution of the fibres, etc.

An approach to the problem of how to apply LEFM parameters to concretes, which neither behave as linear elastic bodies nor are they isotropes, and to specimens of relatively small dimensions is proposed in the *Draft Recommendations* published by the Technical Committee 89-FMT of RILEM (1990a) and (1990b). In the first of them the method of determination of

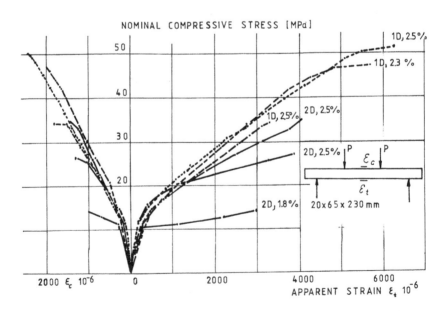

Figure 10.19 Nominal stress-apparent strain curves after measurements at the bottom and top faces of SFRC elements subjected to bending. The curves correspond to 1D and 2D fibre distributions and to various fibre volume content, after Babut (1983).

two fracture parameters K_{Ic} and $CTOD_c$ is proposed after test results of a three-point notched beam. In the second *Draft Recommendation,* a method is proposed to determine fracture energy and size of the process zone in concretes, using the same assumptions and executing tests on small beams of three different sizes. Both methods were tested in different laboratories and are considered as valid for characterization of concretes as to their resistance against crack propagation and fracture. That was an important step forward in the introduction of the methods of fracture mechanics to the design of brittle materials.

When the above fracture characteristics are known, then the so-called R-curves may be used in that problem. If in equations (10.1) and (10.2) the rate of strain energy is denoted by G (cf. also eq. 10.7), and the rate of change of energy necessary for crack propagation is denoted by R, then the state of the stabilization of the crack ends when the following condition is reached:

$$G = G_c = R.$$

For highly brittle materials like glass or cement paste R characterizes the material and any crack propagation from the initial notch or flaw leads to a rapid failure of the element. The situation is different in materials with an internal structure, which stops and may control the crack propagation, that is, in materials like mortar, concrete and fibre-reinforced concrete. The crack is propagating slowly until reaching the second condition:

$$d\frac{\partial G}{\partial c} = \frac{\partial R}{\partial c}$$

and that condition corresponds to the critical crack length c_{cr}. In these materials two values G_c and c_{cr} characterize the state of fracture. The value c_{cr} depends also upon the geometrical conditions, dimensions of the element and arrangement of the load, etc.

Figure 10.20 shows how the stress-strain diagram corresponds to the cracking in an element reinforced with fibres. Along section 0B of the curve only single cracks appear between the fibres. At point B a crack is crossing the element and that situation is reflected on the curve as a knee. At points C and D further cracking (multiple cracking) is controlled by the fibres and the external load may be increasing until the final fracture when the crack length is equal to its critical value c_{cr}. The consecutive stages of cracking have been confirmed by microscopic observations of thin sections of tested elements. As described elsewhere (Figures 10.13 and 10.14), according to the efficiency of the reinforcement, the stress-strain curves may have a different form.

In Figure 10.21 the R-curves correspond to more and less brittle materials. It has been shown among the others by Ouyang *et al.* (1990) and Shah (1991) that for quasi-brittle materials as concrete composites, the R-curves may be

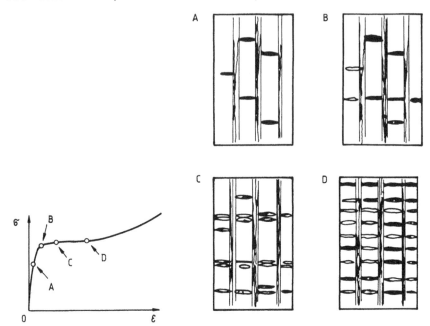

Figure 10.20 Stages of cracks in cement-based fiber reinforced composites, after
        Shah (1991).

determined knowing K_{Ic}, E and $CTOD_c$. When the R-curve is known, then
the fracture behaviour for any geometry can be described. In such a way using
R-curve approach, the crack control realised by the fibres in brittle matrices
may be simulated (see also Shah *et al.* 1995).

10.3 Fracture energy and fracture toughness

10.3.1 Work of fracture

Work of fracture is the measure of the energy necessary to break an element
made with a given material and is closely related to the kind of fracture: brittle
or ductile. Brittle fracture is characteristic for amorphous materials like glass,
and ductile fracture is for crystalline materials like metals. Depending on
conditions, the materials may behave in different ways, for example, metals
are brittle at low temperature.

Cement-based composites are characteristically brittle and their fracture
occurs usually along boundaries between inclusions and matrix, but cracks
can also pass across both mortar and bulk matrix.

Ductile fracture is accompanied by large deformation. In metals, there
are deformations along slip planes and in specimens under test, which are
subjected to tensile load, and can be observed as necking and horizontal
sections of the stress-strain curves. It is also called 'plastic behaviour.'

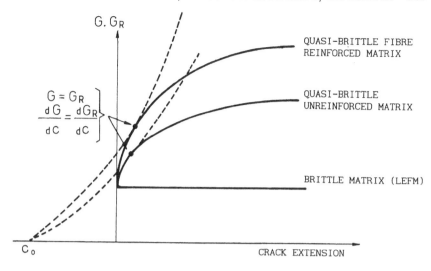

$$G = G_R$$
$$\frac{dG}{dC} = \frac{dG_R}{dC}$$

G.G_R

QUASI-BRITTLE FIBRE
REINFORCED MATRIX

QUASI-BRITTLE
UNREINFORCED MATRIX

BRITTLE MATRIX (LEFM)

C_0

CRACK EXTENSION

Figure 10.21 R-curves for different materials, after Shah (1991).

Brittle fracture appears in the planes perpendicular to the principal tension and is associated with small ultimate strain. Consequently, brittle fracture requires much less energy than ductile fracture; that is, the work of fracture is much smaller.

The work of fracture regarded as the energy required to break the material is a magnitude characteristic for every material, its quality, etc. However, as a material characteristic it is difficult to determine experimentally without ambiguity, because it strongly depends not only on the way of testing, but also on the presence and distribution of microcracks and other local inclusions and defects at different scales. As there is no single and universally accepted definition of brittle and non-brittle behaviour, the following description proposed by Hannant *et al.* (1983) may be used: brittle solids break without large plastic flow so that the total work of fracture measured in a controlled notch bend test type is less than 0.1 kN/m^2 and the strain at failure is less than 1%.

Real structural material may be classified somewhere in between ideal brittle and ductile materials. Examples of fracture energies for a few solids are shown in Table 10.1. In general, it may be assumed that the fracture energy of the cement-based composites is basically proportional to their strength, and more precisely, to the flexural tensile strength. A similar relation is also observed in the progress of hydration of cement with age, because obviously this process influences both strength and fracture energy. This general relation is modified to a certain measure by the characteristics of the material structure: distribution of dispersed inclusions, fibres, pores and other defects.

Table 10.1 Fracture energy of some solids

Material	Fracture energy kNm/m^2
Steel	300–500
Aluminium alloys	10–50
Vulcanized rubber	10.0
Polymethyl methacrylate	0.6
Resins (epoxy, polyester)	0.1
Graphite	0.05
Alumina	0.03
Firebrick	0.03
Glass	0.005

Source: Parratt, N.J., *Fibre Reinforced Materials Technology*, published by Von Nostrand Reinhold Co., 1972 [10.67].

Work of fracture is considerably increased when a brittle matrix is reinforced by a system of inclusions in the form of grains or fibres. An internal structure created purposefully transforms a brittle behaviour into a quasiductile one, characterized by large deformations and high fracture toughness. This transformation into a composite material is described here on many occasions.

The work of fracture is also related to bond strength in fibre-reinforced composites. A certain bond is necessary to ensure the composite's behaviour and thus to obtain an appreciable increase in the work of fracture and strength by the action of fibres. However, with further increase of the bond strength, there is a considerable decrease of work of fracture and the composite's behaviour becomes brittle; this is shown schematically in Figure 10.22.

An attempt to calculate the work of fracture in FRC elements at cracking was published by Brandt (1982, 1985) for elements under axial tension, and later extended to bending (Brandt 1986). The formulae were derived for calculation of the work in the fracture process in which fibres are pulled out of the matrix across a crack. For that purpose the following assumptions were accepted:

- The fracture process may be split into a few independent mechanisms.
- These mechanisms may be described by relatively simple formulae from which the particular components of total work of fracture may be calculated.
- The energy due to creation of new surface in non-reinforced matrix and to elastic strain is considered small with respect to the other energy components and may be neglected. Among other simplifications, the multiple cracking was not taken into account. Moreover, the angle of the aligned fibre system with the direction of principal tensions was treated as the only variable.

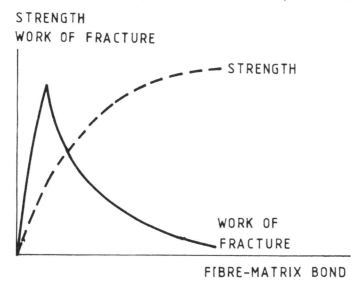

Figure 10.22 Schematic curves showing strength and work of fracture variation with the fibre-matrix bond strength, after Mindess (1988).

The total work of fracture was presented as a sum of five components W_i, $(i = 1, 2,...5)$:

W_1 debonding of the fibres from the matrix;

W_2 pulling-out of fibres against friction at fibre/matrix interface;

W_3 plastic deformation of the fibres;

W_4 yielding of the matrix under compression in the regions at the exits of fibres;

W_5 additional friction at the interface due to local compression.

Consequently, it is also assumed that other possible effects induce only negligible amounts of the fracture energy.

The formulae for calculation of particular components W_i were derived in above-mentioned papers for different situations: short or continuous fibres, random fibre distribution 1D, 2D or 3D, axial tension or bending. These formulae were also applied in optimization problems (cf. Chapter 12). A simple case of 1D fibres in an element subjected to tension with a crack of given limit width is described more in detail below according to Brandt (1985).

An element of steel fibre-reinforced concrete is considered as shown in Figure 10.23. The reinforcement is composed of a single system of parallel short fibres and θ is its angle with respect to the direction of the tensile loading. It is assumed for simplification, that in a neighbouring layer the respective angle is $-\theta$. In the case of symmetric reinforcement a single form of rupture is to be considered: a crack perpendicular to the direction of principal tension. The calculation of fracture energy is executed for a single crack of

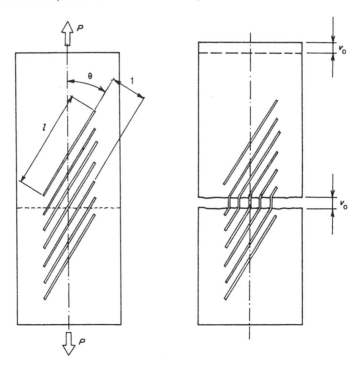

Figure 10.23 Fibre-reinforced element under axial tension before and after the crack
opening, after Brandt (1985).

width equal to v_0. The crack width is arbitrarily selected and may be related
to a certain structural or functional requirement.

The following formulae are proposed for all five energy components
(Figures 10.24, 10.25, 10.26 and 10.27):

$$W_1 = N_0 \cos^2\theta \frac{\ell\pi D}{8} \tau_{\max} v_e \tag{10.16}$$

$$W_2 = N_0 \cos^2\theta D\pi\tau \left[\frac{\ell}{4}(v_0 - v_e) - \frac{1}{2}(v_0^2 - v_e^2) \right] \tag{10.17}$$

$$W_3 = N_0 \cos\theta \frac{\pi D^2}{4} \theta v_0 \tau_f \tag{10.18}$$

$$W_4 = N_0 \left(\alpha \frac{f_f}{f_m} \right)^2 \eta f_m D^2 \left(\cos^2 \theta - \vartheta \frac{\cos \theta^2}{\sin \theta} \right) \tag{10.19}$$

$$W_5 = N_0 \pi \ell \tau D \phi v_0 \sin \frac{\theta}{2} \cos \theta \tag{10.20}$$

Here the symbols denote: ℓ and D– length and diameter of a fibre, α and η – coefficients, τ_f – shear yield stress of fibres, f_f and f_m – tensile strength of fibres and compressive strength of the matrix, ϕ – friction coefficient between fibre and matrix,

$$N_0 = \frac{4\beta}{\pi D^2} \quad \text{number of fibres,}$$

β – volume fraction of fibres (cf. Section 6.6.2 and equation 6.10).
The energy calculated from above formulae as a sum:

$$W = \sum_{i=1}^{i=5} W_i \tag{10.21}$$

was verified experimentally and results were published in Brandt (1991) and for different steel fibres of different shape also by Brandt and Stroeven (1991).

The proposed formulae are sufficiently precise to enable the analysis of different cases of steel fibre reinforced elements, however, their simplified form and assumed hypothesis leave scope for further development and improvement; for example, more sophisticated mechanisms may be used, without modifying the principle of the design method, based on the conviction that the best solution corresponds to the maximum area under the load-deflection or stress-strain curve. That maximum reflects the most effective crack control by the fibres.

10.3.2 Quantitative description of fracture toughness

The most important material property of brittle matrix composites is their ability to absorb energy during the loading process before final fracture. This property may be expressed quantitatively by that energy calculated as the amount of work of external forces, and it is called fracture energy. Fracture energy is defined in various ways related to the definition of the fracture, which sometimes is ambiguous and requires conventional assumptions. In brittle materials the fracture is immediate, after the opening of the first

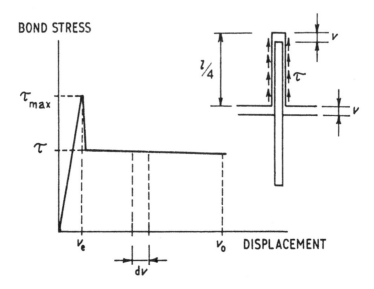

Figure 10.24 Bond stress as function of displacement in a pull-out test, after Brandt (1985).

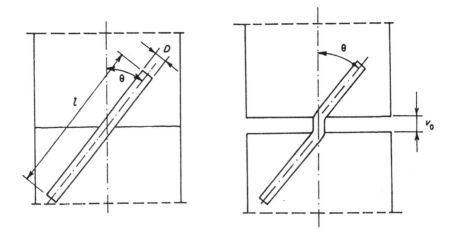

Figure 10.25 Plastic deformation of a fibre passing across a crack at an angle θ, after Brandt (1985).

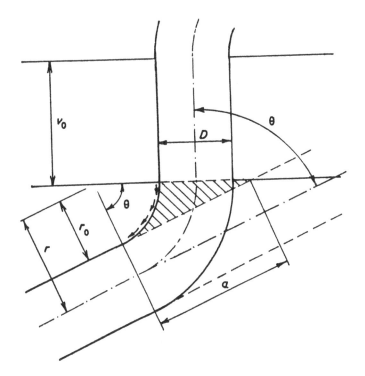

Figure 10.26 Yielding of the matrix due to local compression at the fibre bend, after Brandt (1985).

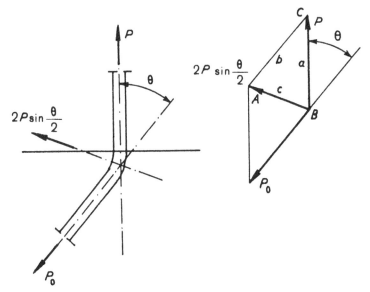

Figure 10.27 Additional force increasing friction at the fibre bend, after Brandt (1985).

crack and the corresponding load is close to the maximum one. In composite materials with internal structure the fracture may also be defined in the same way, but then it corresponds to the final stage of a long process of cracking. It is therefore often necessary to define the fracture as a given critical deflection, deformation or crack width. In such cases the notion of fracture energy also has rather conventional meaning as it concerns a certain degree of the rupture of the element.

The ability to absorb fracture energy is called fracture toughness and this term has various meanings, without a universally accepted definition. According to a review published by Kasperkiewicz and Skarendahl (1990), three main groups of meanings of the term 'fracture toughness' may be distinguished.

The absolute description was proposed by the Japan Concrete Institute, RILEM and ASTM. In this approach energy absorption is measured on prescribed specimens and also loaded in a well-defined way, and the energy calculated is considered to be proportional to the area under the load-deflection curve. In most recommendations, specimens subjected to bending under four-point loading are proposed; therefore the resulting magnitude is called 'flexural toughness' or just 'flexural energy absorption.'

The second group of proposals, made among the others also by ASTM in systematically renewed recommendations, is based on relative values; that is, without specifying the dimensions of tested specimens the ratio of the total area under the curve to that limited by the first crack opening is considered. Such a ratio is called the fracture index with some additional specifications.

In the third group of toughness descriptions various authors proposed to consider particular specimen dimensions and certain parts of the stress-strain or load-deflection curve. The term 'fracture toughness' is also used by certain authors for the critical value of the stress intensity factor K_{Ic}. For further details of various fracture toughness descriptions the reader is referred to the abovementioned book by Bentur and Mindess (2006). Only a few proposals are presented below that seem to have both physical meaning and practical importance and gained more universal acceptance than others.

Probably the idea to use a dimensionless magnitude called the fracture index as a possible material characteristic was first published by Henager (1978). The test beam standardized by ASTM of cross-section 102×102 mm and of 305 mm span was used. It was loaded in its centre up to the total deflection of 1.9 mm. The toughness index (TI) was calculated as the ratio of area A+B to area A as it is shown in Figure 10.28. For plain concrete or mortar the value of TI is 1.0 or a little more, and for strongly reinforced fibre composite it may reach values of 30 or even 46.

A similar method unrelated to any particular beam dimensions, but presented in the form of flexural toughness indices was proposed by ASTM and is discussed by Johnston and Gray (1986). That concept is shown in Figure 10.29 as a load-deflection curve obtained for a third-point loading

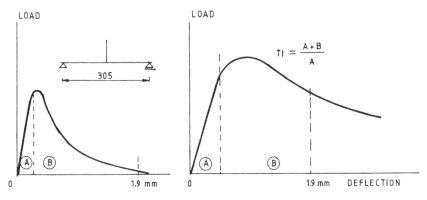

Figure 10.28 Schematic load-deflection curves for element under bending in calculation of the flexural toughness, after Henager (1978).

system. The deflection δ corresponding to the first crack is measured from point 0 and related to linear elastic behaviour. The indices are obtained by dividing the area under the load-deflection curve by the area under that curve up to the first crack. The calculated values of flexural toughness indices are independent of the dimensions and rigidity of the tested element (cf. also ASTM 2006). Index I_5 is determined at a deflection 3δ, I_{10} is determined at 5.5 δ and I_{30} at 15.5 δ. These values for an ideal elastic-plastic material are equal to 5, 10 and 30, respectively. For an ideal elastic-brittle material all indices are equal to 1.0 because in that case the total deflection is limited to δ.

The proposed ASTM method has been accepted also by ACI Committee 544.4 Report (1988). The bending test is recommended for establishing the toughness of the material, understood as the capacity of the energy absorption under static load.

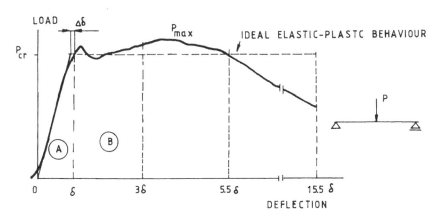

Figure 10.29 Schematic curve for determination of the flexural toughness indexes according to ASTM C 1018 (1997).

The system of indices reflects the toughening effect of the fibres and their values allow us to quantitatively compare the energy absorbed by the element simultaneously with both ideal limit cases mentioned above. The indices are used to evaluate the influence of all possible modifications introduced in the design of material or in the technology that is applied.

The suitability of fracture toughness to represent the quality of the cement-based composite materials is proved by its multiple applications both as absolute values of the fracture energy and in the form of dimensionless indices. However, it should be stressed that a fracture toughness index cannot be considered as a definite and universal characteristic of material toughness in the sense of its resistance to crack propagation and its post-cracking behaviour. One of the weak sides of fracture toughness determination is the difficulty in identifying the first crack occurrence. In strongly reinforced SFRC elements, for example, this fact is often impossible to recognize without some ambiguity. Observation of the specimen surface where cracks are expected is difficult and its results depend considerably upon the methods of crack recording. A characteristic offset of the curve and rapid change of its slope are also difficult to specify. As a result, the position of this point on the load-deflection curve is determined in an arbitrary way within relatively large limits. Consequently, area A (Figure 10.29) may be easily determined with large errors, resulting in a considerable final error of the toughness index.

Another disadvantage of toughness index determination is that the same values may be obtained for different load-deflection curves; for example, a short ascending branch or a long descending branch after 3δ may give the same result. Furthermore, early cracking and low value of area A will give higher values of the index than similar curves with only slightly later cracking.

These features of flexural toughness determination are related to the following two characteristics of fracture behaviour of brittle composites:

1   The first crack load and the susceptibility of the matrix to crack depend on its properties and not on the eventual reinforcement and its efficiency.
2   The post-cracking behaviour and the area under the descending branch are related mostly to the reinforcement.

Examples of load-deflection curves in Figure 10.30 show the possible deficiencies of this type. Curve *a* describes a small strain hardening while curve *a'* represents a rapid decrease of load after cracking and both are characterized by the same toughness index I_5. Similarly, for curves *b* and *b'* index I_{10} is the same and only index I_5 shows small difference, while the material's behaviour is completely different. Also, for curves *c* and *c'* only I_5 and I_{10} are different while the values of I_{20} are the same. These examples prove that the determination of the toughness index is only a useful method of comparison for the effective material's behaviour with that of an idealized linear elastic-plastic material. However, it cannot be considered as a perfect material characteristic.

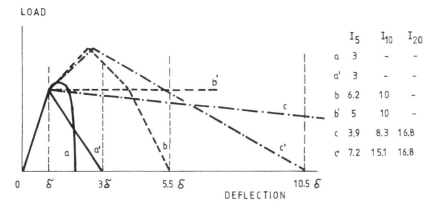

Figure 10.30 Examples of load-deflection curves with corresponding values of flexural toughness indices.

Further research aimed at the characterization of the ability of a material to absorb fracture energy, and of post-cracking behaviour, has been continued by many authors. Examples of calculation of work of fracture and toughness indices of tested elements are shown in Figures 10.31 and 10.32 where relevant curves are presented as functions of volume fractions of carbon fibres (Brandt and Glinicki 1992). The data obtained from tests of cement mortar elements, subjected to bending and following magnitudes, are shown:

1   WFt: total area under the load-deflection curve calculated up to the maximum load and δ_{max};
2   WF2.5: total area under the load-deflection curve calculated up to the deflection equal to 2.5 mm;
3   TI(3): toughness index calculated as ratio of respective areas up to deflection 3 δ_{max} and δ_{max};
4   I_5 and I_{10}: toughness indices calculated according to ASTM C 1018-97 (1997) as mentioned above.

It may be concluded from these tests that all applied measures of the efficiency of reinforcement gave similar results, which was the consistent increase of fracture toughness with an increase of fibre volume fraction up to 1%. Further increase of fibre volume induced a slight decrease of I_5 and TI(3) and negligible increase of I_{10}. In contrast, work of fracture increased up to the maximum fibre volume fraction. It appears from these tests that all applied measures may be of some use in particular case as they furnish slightly different information about the behaviour of tested material (Brandt and Glinicki 1999).

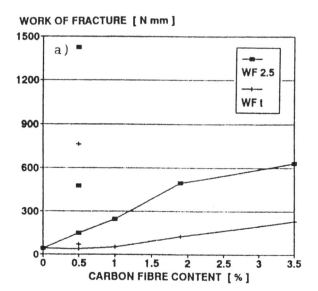

Figure 10.31 Values of the works of fracture WF2.5 and WFt plotted as functions of the volume fraction of fibre reinforcement, after Brandt and Glinicki (1992).

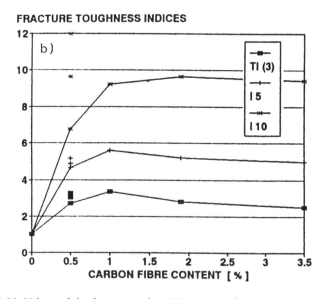

Figure 10.32 Values of the fracture indices TI(3), I5 and I10 plotted as functions of the volume fraction of fibre reinforcement, after Brandt and Glinicki (1992).

10.3.3 Influence of testing conditions

The energy absorption and overall behaviour of the composite elements subjected to external load is, to a large extent, related to the conditions of loading and of testing in general. The following should be mentioned among several factors, which influence specimens' strength and the fracture processes:

- rate of loading;
- size of specimens and elements;
- stiffness of the loading system.

Normal tests are executed with such slow loading that the effects of loading speed are neglected. One one hand, the results obtained in such tests are not relevant either for high rates of loading or for impacts and dynamic actions. On the other hand, the speed of loading must be limited if the introduction of rheological effects and fatigue are to be avoided. There are no test results related to the influence of the rate of loading on fracture toughness. It has only been established that with the increase of the loading rate, the strength of certain kinds of brittle matrix composites is improved. The various rates of loading applied in laboratory experiments are intended to simulate, and more-or-less closely represent, the natural situation in different kinds of structures (cf. Section 11.2).

The size of element has a certain influence on strength and on fracture processes. If the material characteristics are looked for, then the results of testing should be independent of the element size. It is therefore advisable to always test a series of elements of different dimensions and to establish within what limits the size influence may be considered as sufficiently small and negligible. Moreover, it is not only the size of the tested element itself that is of importance, but also the ratio of the element's stiffness to that of the testing system. In general, it has been shown that when the dimensions of tested elements are increased, then the resulting brittleness is also increased. Several authors proposed to determine the fracture toughness parameters for concretes on very large elements. That viewpoint is based on the observed relations of these parameters and dimensions of elements. However, the testing of very large elements is difficult, expensive and loaded by other sources of errors; for example, heterogeneity of material in a specimen and unavoidable differences between the specimens belonging intentionally to the same series, different curing conditions, etc.

The stiffness of the loading equipment is another factor that influences the toughness of tested elements. In a stiff testing machine, that is, with sufficient stiffness compared to the tested specimen, the elastic energy accumulated in the machine is not released after the first crack of the specimen. The test may be continued and a descending branch of the load-deflection (or stress-strain) curve may be established. In contrast, when the same element is tested on a less rigid machine, then after the first crack all the elastic energy accumulated is rapidly released and brittle fracture occurs at the maximum load (cf. Section 8.2).

The descending part of the characteristic curves may be established even better in special testing machines in which the position of the loading head is automatically adjusted in such a way that the rate of change of a certain magnitude, for example deflection or deformation, is kept constant. The results obtained in such a way cannot be directly compared with those obtained from ordinary equipment more commonly used for material control in production units.

10.3.4 Material brittleness

For obvious reasons, all structures should be designed to avoid brittle fracture of any kind; however, under particular conditions, such fracture may occur in practically all materials. For metals or plastics, if brittle fracture is conditioned by low temperature, fatigue, rate of loading, etc., it is a natural and common way of fracture for several kinds of matrices used in building and civil engineering materials. Matrices based on various cements and all kinds of ceramic materials are considered as brittle. Brittleness is the principal disadvantage, which should be controlled in all structural and even non-structural applications of these materials.

The brittleness of the matrices as inverse to the toughness is not defined in a universally accepted way. One possible definition was proposed by Knott (1973) as follows:

> The essence of fast fracture is that it is a failure mechanism which involves the unstable propagation of a crack in a structure. In other words, once the crack has started to move, the loading system is such that it produces accelerating growth ... We shall interpret a brittle fracture as one in which the onset of unstable crack propagation is produced by an applied stress less than the general yield stress of the uncracked ligament remaining when instability occurs.
>
> (Knott 1973)

A slightly different description of material brittleness was proposed by Hannant *et al.* (1983), (cf. Section 10.3.1). According to other proposals, the strain of 0.04 or 0.05 at failure may be considered as a limit of the brittle behaviour. Normally, the maximum strain of cement-based matrices does not reach 200.10^{-6}; that is, a few orders of magnitude lower. Therefore, it is universally considered that cement-based materials are brittle, although for concretes and mortars the brittleness is reduced by their internal structure of grain inclusions.

A quantitative description of the brittleness was proposed by Petersson (1981) and Hillerborg (1985) in the form of the characteristic length ℓ_{ch} called also ductility:

$$\ell_{ch} = G_f \, E / f_t^2 \; [\text{mm}]$$

where G_f is the fracture energy and f_t is the tensile strength. The brittleness is then defined as

$$1/ \, \ell_{ch} = f_t^2 / E G_f$$

The application of ℓ_{ch} for characterization of different materials is justified only when their σ-ε curves are of similar shape. The higher values of ℓ_{ch} mean lower brittleness and higher toughness. The average values of ℓ_{ch} for cement based materials are:

- Portland cement paste 5–15 mm;
- mortar 100–200 mm;
- ordinary plain concrete 200–400 mm.

The brittleness of cement-based composites increases with their compressive strength. It may be observed that in recent years the strength of conventional concrete has increased, thanks to improved technology and material composition. Also the extended use of micro-fillers like silica fume is the main reason that strength has increased considerably. However, the increase of E and G_f is quite moderate, and consequently ℓ_{ch} tends to lower values. In the design of high performance concretes their brittleness is of special concern (cf. Section 13.4.5).

Other measures for brittleness have been proposed by Wu Keru and Zhou Jianhua (1987), such as the brittleness index *BI* determined as shown in Figure 10.33. It is defined as the ratio of elastic strain energy to irreversible strain energy, corresponding to the peak point of the σ-ε curve obtained in a compression test; for ideal elastic and brittle material $BI = 0$; for ideal elastic and ductile materials $BI = 8$.

The brittleness depends not only on material properties and external conditions, but also on the size of the element (Bache 1989); therefore, this aspect should be considered in both material and structural design. Large elements exhibit more brittle behaviour than small ones when all other conditions are the same. Also a few indications are proposed on how to accommodate the increased brittleness of modern cement-based composites to structural elements:

- the geometry should be simple, without sharp corners and potential notches;
- smaller elements are less exposed to brittle fracture;
- eventual damage should concern parts of structure only; initial cracks and local stress concentrations should be prevented; for calculation of cracking and failure the theory of elasticity or LEFM is appropriate, (however, give comparative results only);
- carrying capacity may be determined by the weakest link statistics.

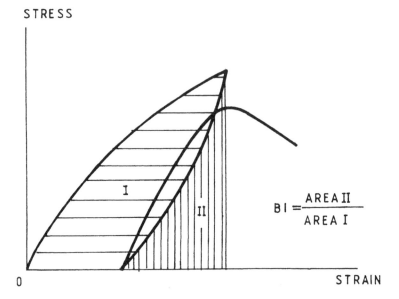

Figure 10.33 Complete σ-ε curve and brittleness index for a concrete specimen, after Wu Keru and Zhou Jianhua (1987).

The application of dispersed reinforcement and other measures of crack control are considered as an effective way to counterbalance the natural ductility of cement-based materials. This may lead to a high volume of reinforcement, for example SIFCON, where a high strength and very brittle matrix is transformed into a crack resistant and quasi-elastic material (cf. Section 13.5).

10.4 Modes of fracture

10.4.1 Mixed-mode fracture

There are three main types of failure by crack propagation as shown in Figure 10.3: I (opening), II (sliding), and III (tearing). In real elements made with composite materials and subjected to external load, all modes appear in different combinations and it is not completely clear how to deal with that situation. The models proposed at present seem to still be dependent on the scale: if in macro scale the tests executed according to these three modes are possible; the obtained results characterize only the effects in the same scale with mean values of some correcting parameters. On lower scales probably all modes appear simultaneously or at least Modes I and II act conjointly. The mixed-mode fracture parameters may be calculated for linear elastic materials within the frames of LEFM, but their direct application to the cement-based composites is questioned. The considerations below are partly based on the paper by Brandt and Prokopski (1990).

In a mixed-mode loading, when not only axial loading but also shearing is applied, it is assumed that fracture occurs when a total energy release rate G is larger than an energy consumption rate, R, and the fracture condition is given by $G = R$. The total energy release rate G for a linear elastic material is the following sum:

$$G = G_I \underset{+}{} G_{II} \underset{+}{} G_{III} = \frac{\left(1 - v^2\right)}{E}\left(K_I^2 + K_{II}^2 + \frac{K_{III}^2}{\left(1 - v\right)}\right) \tag{10.22}$$

For I – II mixed-mode loading, which may be adopted for the plane strain state, it is assumed that $K_{III} = 0$. This situation occurs in most cases when the dimensions of a loaded element admit plane strain state. Therefore

$$G = \frac{\left(1 - v^2\right)}{E}\left(K_I^2 + K_{II}^2\right) \tag{10.23}$$

here E – Young's modulus and v – Poisson ratio.

Several authors do not accept the existence of Mode II and III failure in concrete composites, because such materials fail easily in Mode I at local stress concentrations, and then cracks tend to follow the path of principal tensile strain (Arrea and Ingraffea 1982). In such case the above formula might be simplified to the form:

$$G = \frac{\left(1 - v^2\right)}{E}K_I^2 \tag{10.24}$$

as in nearly all published studies, it means that one of the following assumptions is accepted:

1   the material under consideration is characterized by $K_{II} = 0$; or
2   the Mode II of fracture does not exist at all in the examined fracture process.

Assumption 1) cannot be proved for any structural material, independently of whether its behaviour is considered as linear and elastic or non-linear and plastic, the fluids and gels being here excluded from the considerations.

Assumption 2) may be adopted for ideally homogeneous materials subjected to tensile loading. Its application to concrete-like materials seems inappropriate because of the high level of inhomogeneity. Even in the case of an external loading applied in a way to produce ideally Mode I (opening), the local effects between grains and voids produce all three modes of fracture. The assumption that Mode III may be neglected in a plane-strain state seems to be more admissible.

Comprehensive reviews of mixed-mode fracture of two-dimensional models in LEFM were published by Carpinteri (1987) and Taha and Swartz (1989). They classified the models into two categories: fracture mechanics oriented and failure theories oriented. In the first group the stress field at the crack tip is expressed using K_I and K_{II} and total stress intensity factor K^2_c as shown below in equation (10.25). The second approach is based on two critical values, which characterize the material by equation (10.26) or on determination of the effective stress intensity factor, which is obtained from the principal stress and not from the tensile stress component. The second category comprises classical failure theories based on maximum tangential stress or strain, minimum strain energy density, Mohr-Coulomb stress theory, etc. Other authors proposed to determine Mode II failure according to the maximum energy release rate rather than to the principal stress (Bažant and Pfeiffer 1986).

10.4.2 *Quantitative determination of mixed mode fracture*

Various experimental studies have shown that the cracks start from mortar-aggregate grain interface, even if the external load is applied as an ideal axial tension. Regarding these crack patterns and their model representation, it is difficult to admit that the crack propagation is limited to Mode I (e.g. Zaitsev *et al.* 1986).

Two examples of possible Mode II fracture in local situations around initial cracks or regions with different rigidity are shown in Figure 10.34 in the form of an initial crack or flaw (Figure 10.34a) in the matrix or a bond crack along the surface of an aggregate grain (Figure 10.34b). The nature of the external load – tension or compression – does not influence these possibilities. The deformations along lines a-a and b-b are different in such a way that the in-plane shearing appears at the tips of initial notches and the Mode II fracture begins.

A series of tests aimed at examination of Mode II failure in cement-based composites were executed and published, (e.g. Jenq and Shah 1988; John and Shah 1989; Shah 1989) using a simple arrangement for testing elements under bending to create various combinations of Modes I and II, shown in Figure 10.35.

A series of simply supported beams are provided with notches of different depth and location. The mode of cracking depends on the depth a of the notch and on its location defined by factor γ. The value of $\gamma = 0$ corresponds to pure Mode I and the other values of γ induce either a mixed mode at the notch tip or again pure Mode I when tension failure occurs at the midspan. Which of these possibilities is actually realized depends on both values γ and a. The results obtained were compared with calculations by finite element method assuming LEFM solutions for Modes I and II. The tests were executed under static and impact loading and the test and calculation results are shown in Figure 10.35. As the notch was moved away from the centre

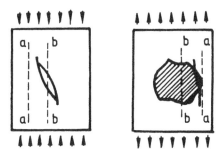

Figure 10.34 Local non-homogeneities as source of mode II fracture, Brandt and Prokopski (1990).

of the span, the maximum load increased. When $\gamma = \gamma_{cr}$, the beam failed at the central unnotched cross-section according to the pure Mode I. Similar behaviour was observed under impact loading, but for different value of γ_{cr}. The tests furnished interesting results, however strongly related to shape dimensions of specimens.

There is rather limited information about Mode III failure and only some experimental data are given by Bažant and Prat (1988). Also, in this subject, practical recommendations for material design and verification are not available.

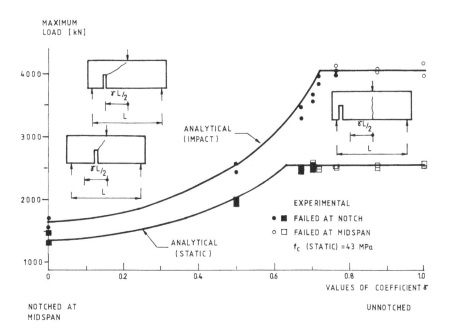

Figure 10.35 Influence of location of notch on maximum load and on mode of failure, after John and Shah (1989).

Interesting results for the mixed mode fracture were observed by Debicki (1988) from the tests of transversally orthotropic fibre-reinforced mortar cylinders 80 × 160 mm, subjected to axial compression. The compressive force *P* was applied at different angles θ with respect to the casting direction 1-1 (Figure 10.36). Two different criteria have been used for different regions of angle θ and excellent confirmation has been obtained by experimental results. For angle θ between 0° and 45° and between 75° and 90°, the splitting of the cylinder under compression was decisive. For angle θ between 45° and 75°, the shearing mode appeared because of weaker resistance against tangential stresses along layers perpendicular to the casting direction.

In strongly heterogeneous and anisotropic materials like concretes, the problem should be considered in a different but more complicated way. It is not a single crack subjected simultaneously to tension and in-plane shearing that produces a mixed mode of fracture, but a spectrum of loading conditions. In that spectrum, Modes I and II are only limiting cases; all the others are different combinations of both of them.

A model may be therefore proposed in which critical values of K_{Ic} and

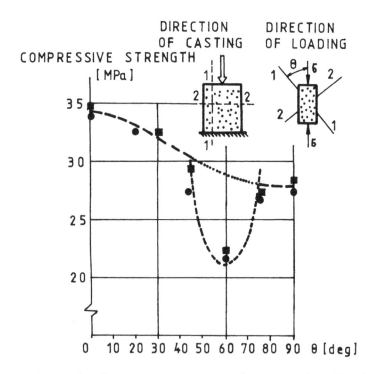

Figure 10.36 Results of compression tests: strength versus angle θ. Experimental results compared with calculation according to two strength criteria, after Debicki (1988).

K_{IIc} both determine a critical value of an overall factor K_c being a measure for the material crack resistance. It was written by Sih and Chen (1973) as a function describing the critical state:

$$f(K_{Ic}, K_{IIc}) = f(K_c), \tag{10.25}$$

without defining any particular form of that function.

Combined loading, which produced the mixed mode of cracking, was considered by Broek (1983) and two examples of fracture criteria were proposed:

$$K^2_{Ic} + K^2_{IIc} = K^2_c, \tag{10.26}$$

here it is assumed that $K_{Ic} = K_{IIc}$, and

$$\left(\frac{K_I}{K_{Ic}}\right)^2 + \left(\frac{K_{II}}{K_{IIc}}\right)^2 = 1, \text{ where } K_{Ic} \neq K_{IIc}. \tag{10.27}$$

For the second of these criteria, it was assumed that the crack propagated in a self-similar manner and remained in the plane of the original crack. In experiments it is usually observed that the crack extends along an angle with respect to its original direction.

One of the possible ways to find an acceptable form of the function (10.25) is to execute a series of experiments in which materials of known K_{Ic} and K_{IIc} will be combined. Different proportions of both materials and carefully arranged macro-regions in a tested element may help to determine the practically reliable function $f(K_{Ic}, K_{IIc})$. The high degree of indeterminacy of the system which represents a concrete specimen may cause such a problem and is quite difficult to deal with.

Tests in which the values of K_{IIc} were measured are not very numerous and the few available data collected in Brandt and Prokopski (1990) concern concretes with different composition and internal structure. These data from different authors are not very dispersed, which may be explained by the similar way of testing adopted. Values ranged from 0.42 (MNm$^{-3/2}$) for soil cement to 5.15 (MNm$^{-3/2}$) for concrete with granite aggregate. In most cases the higher values of K_{IIc} correspond to the materials for which higher strength and better aggregate-matrix bond was also observed, or may be supposed if the appropriate data are not given. This is expected, and reflects to some extent the material's resistance against shearing crack propagation. To verify these suppositions, a series of tests was executed by Brandt and Prokopski (1990) and more complex data were recorded.

The specimens for determining separately the fracture properties for both Modes I and II are shown in Figures 10.37 and 10.38 together with examples of load-deflection curves. The specimens were cast with ordinary Portland cement, fine aggregate and river gravel with a maximum grain size of 10 mm.

The critical values of stress intensity factors in Modes I and II were determined according to ANSI/ASTM (1978) formulae:

$$K_{Ic} = \frac{P_{cr}}{B\sqrt{W}} Y\left(\frac{a}{W}\right), \tag{10.28}$$

$$K_{IIc} = 5.11 \frac{P_{cr}}{2Bb} \sqrt{\pi a},$$

where P_{cr} is the value of the critical load P which initiated the crack propagation, W is specimen depth, b is ligament depth, a is notch depth and B is specimen width. That value of P_{cr} is identified on the curves (Figures 10.37 and 10.38) as a slight inflection for Mode II curves or as maximum value for Mode I curves that may be considered as the first crack opening. In certain cases such estimation of P_{cr} on the basis of a load-deflection curve is not precise and therefore calculated values of K_{Ic} and K_{IIc} may be ambiguous.

The above formulae are used here following Watkins (1983) and other authors, although they were initially intended for application for metallic specimens. Therefore the numerical results obtained should be considered as only approximate. The question of the applicability of LEFM to relatively small specimens is discussed elsewhere.

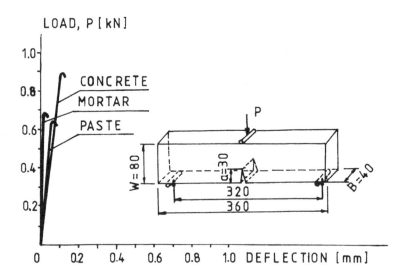

Figure 10.37 Examples of test results for K_{Ic} and dimensions of a notched beam subjected to bending, after Brandt and Prokopski (1990).

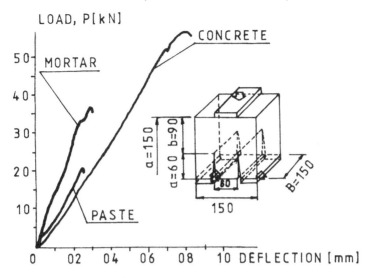

Figure 10.38 Examples of test results for K_{IIc} and dimensions of a punch-through shear specimen, after Brandt and Prokopski (1990).

The values calculated from formulae (10.28) are presented in Table 10.2. The dispersion of obtained results is particularly high for K_{IIc} and is probably related to the high influence of local crack resistance of the specimens. The highest values of the variation coefficient were observed for cement paste in both types of specimens. This result may be attributed to the relative homogeneity of that material's structure and its brittleness. The cracks were not blocked by any inclusion and critical load was primarily related to random local defects at the crack path.

On the load-displacement plots for concrete and mortar specimens, a slight quasi-plastic effect in the form of inflections appeared (Figure 10.38). This phenomenon may be related to rearrangement of the aggregate grains, after which an additional energy supply was required for further crack propagation. It was not observed on the curves for paste specimens.

The transgranular path of the cracks and their changes of direction when passing around the grains also required additional amounts of energy to be supplied by the external load. This was probably the reason why higher values of fracture loads were necessary for concrete and mortar specimens than for paste ones.

It may be concluded that the results obtained confirm the data published previously by other authors. The values of K_{Ic} and K_{IIc} for the same materials may be used in establishing a criterion for mixed-mode fracture in concretes. Soroushian *et al.* (1998) investigated the problems of mixed-mode fracture in concrete elements with steel and polypropylene fibres. Modes I (opening) and II (sliding) were considered and the conclusions supported application

Table 10.2 Test results of paste, mortar and concrete specimens

Parameters	Types of material		
	paste	*mortar*	*concrete*
Compressive strength (MPa)	56.9	30.5	35.3
coefficient of variation (%)	2.2	5.3	3.4
number of specimens	3	3	3
Critical value of stress intensity			
factor K_{Ic} (MNm$^{-3/2}$)	0.472	0.557	0.935
coefficient of variation (%)	13.2	7.5	4.3
number of specimens	9	9	9
Critical value of stress intensity			
factor K_{IIc} (MNm$^{-3/2}$)	1.480	2.430	3.648
coefficient of variation (%)	23.0	9.5	8.5
number of specimens	7	7	7

Source Brandt, A.M. and Prokopski, G. (1990).

of fibres even in a small volume fraction (as 0.3% in these tests) to control crack propagation in both modes.

All three modes and their combinations were experimentally analysed on specially prepared and loaded specimens (Song *et al.* 2004). The results were analysed within LEFM and were aimed at application in large structures; for example, arch concrete dams. In the last decade the problems related to the mixed-mode fracture in concrete-like materials attracted less attention in research, which may be explained by lack of direct application in structural design for practical purposes.

10.5 Relationship between fracture toughness and fractal dimension in concretes

The irregular shape of cracks and fracture surfaces of cement-based composites is at the origin of the question as to whether they may be considered as fractal objects. The application of fractal analysis should be aimed at a better understanding of the fracture mechanics of these materials, by possible quantification of the fracture surface roughness, length and surface of cracks and all other irregular elements that may characterize the materials and their behaviour.

The development of the fractal characterization of irregular lines and surfaces was initiated by Mandelbrot, who introduced in two books in 1977 and 1983 the fractal concept to many fields of science.

Man-made objects have in most cases linear or curvilinear contours and surfaces, corresponding to Euclidean geometry in which points, lines, surfaces and volumes have topological dimensions represented by integers 0,

1, 2 and 3, respectively. However, natural objects like coastlines, clouds and rock surfaces are irregular and composed of 'mountains' and 'valleys.' It is easy to observe that at higher magnification further families of 'mountains' and 'valleys' appear. The effective length of such an irregular line and the effective area of an irregular surface seem to be related to the magnification, and therefore cannot be determined in an objective way. At the discontinuities these curves and surfaces are non-differentiable.

The word 'fractal' has been proposed by Mandelbrot (1977) together with fractal dimension D, which characterizes the irregularity of the fractal objects. The lines have non-integer values of D between 1 and 2. The irregularity of surfaces is expressed by their fractal dimension, varying between 2 and 3. The more irregular is the object, the higher is its fractal dimension. The fractal dimension exceeds the Euclidean dimension.

The fractal dimension is defined by the equation corresponding to a segment of a line from which fractal line is generated:

$$D = \frac{\ln N}{\ln 1/r} \quad \text{or } N = (1/r)^D$$

here N is the number of subparts for which the initial segment of unit length is divided at each step and $1/r$ is the scaling factor, r being the length of each subpart. The generation of two simple fractal lines is shown in Figure 10.39.

The generation of a fractal line may be repeated indefinitely and the total length of the line is expressed as function of r and D

$$L = L_0\, r^{-(D-1)}, \tag{10.29}$$

here L_0 is the length of initial straight segment. The total length increases indefinitely with decreasing r, e.g. for $D = 1.5$ it is:

r	1	0.5	0.1	0.01	0.0001	0.0001
L	1	1.41	3.16	10.00	31.62	100.00

More general relation for lines and surfaces has the following form:

$$L = L_0\, E^{-(D_f - D)} \tag{10.30}$$

here L and L_0 are the total and initial length and area, respectively, E is scale of measurement, D_f and D are fractal and topological dimensions.

The natural lines and surfaces with apparent irregularities are not always the fractal objects with given fractal dimensions. The necessary requirement is self-similarity; that is, within certain limits of magnification a larger region of the object should appear exactly or approximately similar to a smaller

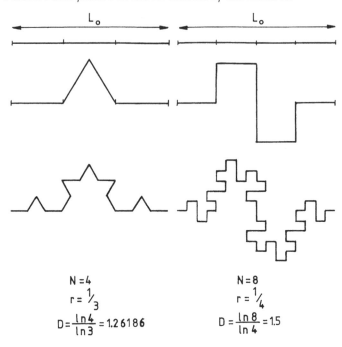

Figure 10.39 Generation of two examples of simple fractal lines.

region observed with appropriate magnification. In natural objects, the self-similarity is not extended over all ranges of magnification; below and above certain levels the object has no fractal character. The reasons are different; for example, the structure of the object may be significantly different at different levels.

The fractal dimension as a quantitative measure of the irregularity of the objects may be applied in various fields of materials science. It has been shown that the self-similarity over two orders of magnitude occurs for fracture surfaces of tempered steel and alumina. A linear relation between the fractal dimension and fracture toughness for these materials was also proposed.

There are many other papers in which correlations between mechanical properties of solids and fractal dimensions of fractured surfaces were investigated within different orders of magnitude of scale. In the paper by Winslow (1985) fracture surfaces of cement paste were examined and their fractal character was shown, indicating its limits. Also a relation between the *w/c* ratio and the roughness of surfaces of fractured specimens was observed. Saouma *et al.* (1990) studied concrete specimens with different maximum aggregate grain size from 20 to 75 mm, representing concretes used for construction of dams. They also observed fractal character of examined concretes without significant differences in fractal dimension for different aggregates. This last conclusion was attributed to the identical origin of all aggregates used in tested specimens.

The roughness of a fracture surface is studied along a system of profiles. Their orientation is of great importance. Measurements are made with a specially designed profilometer, which was displaced over the fracture surface along two families of orthogonal lines. The other technique is based on execution of a polymer replica of the fracture surface. The replica is then sawn into slices by parallel sections. The contours obtained are called fracture profiles and are subjected to close examination and analysis – the length of each fracture profile is measured with varying step *r* and the results are plotted in the logarithmic system of coordinates using equation (10.29) in the form

$$\log L = \log L + (1 - D) \log r. \tag{10.31}$$

If the result obtained is approximately a straight line, it means that the fractal dimension equal to its slope is constant over certain levels of magnification. This is called the vertical section method. Another approach called slit-island method is explained among others by Pande *et al.* (1987) and was used for analysis of fracture surfaces of metallic specimens.

To obtain statistically reliable results from the analysis of a fracture surface by the vertical section method, a large number of sections should be examined. The parallel sections executed, for example, along rectangular coordinates may give information about possible orthotropic properties of the fracture surface. In most cases random sections oriented at different angles should be studied.

The characterization of the fracture surface after the analysis of fracture profiles is based on an assumption that having determined the profile roughness parameter

$$R_L = \frac{L}{L_0}$$

it is possible to define the fracture surface roughness parameter

$$R_S = \frac{S}{S_0}$$

here L_0 and S_0 are apparent projected length and area respectively, and projection was on the mean or average topographic direction or plane, (Figure 10.40). Therefore, an equation of the following type is needed

$$R_S = \overline{R_L \, \Psi} \tag{10.31}$$

here $R_L \Psi$ is an expected or average value of the product and Ψ is the profile structure factor, which expresses the position and orientation of each elementary segment of the analyzed fracture profile.

The equation (10.31) is general and is not based on any assumption concerning the nature of the examined surface. The values of R_L and Ψ are independent and should be obtained in a number of vertical sectioning planes oriented differently to ensure the statistical representation of the product, where an efficient sampling procedure is proposed for estimation of fracture surface roughness from the measurements. The profile roughness parameter R_L should not be correlated to fracture toughness and any correlations observed may be misleading. The approximate formula proposed by Underwood and Banerji (1983) has the form:

$$R_S = 4/\pi \ (R_L - 1) + 1 \tag{10.32}$$

and provides, according to these authors, the best fit to experimental data. When comparing equation (10.31) and equation (10.32) the following remarks should be made:

- R_S is an important quantity in respect to its possible relation to the fracture properties of the materials, but it is inaccessible by simple experimental measurements;
- R_L is easy to be determined experimentally, but its relation to the surface roughness as given by equation (10.31) is not so simple;
- equation (10.32) has been proposed and discussed for metallic specimens and its application to cement-based materials has not been yet verified.

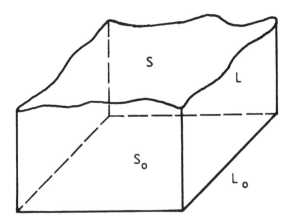

Figure 10.40 Symbols for irregular lines and surfaces with respect to their projections.

It should be also observed that R_L does not characterize completely and without ambiguity even a profile – different profiles may have identical values of R_L (e.g. Figure 10.41). Also, two fracture surfaces with the same value of R_S are not necessarily similar.

The calculation of the fractal dimension D as well as of the fracture surface roughness parameter R_S is a valuable attempt to quantify the irregularities of natural fracture surfaces. Its application in the fracture mechanics of cement-based composites is promising. However, the quantitative conclusions from such calculations have only relative importance and should be limited to the materials of the same kind. For example, the roughness of the fracture surfaces of a series of mortar specimens with variable *w/c* ratio, age or grading curves may reflect correctly their fracture toughness. In contrast, any comparison between fractal dimensions of cement pastes and concretes may give misleading results because of different nature of these materials.

The question as to whether there is a general and reliable relationship between the fractal dimension of a surface and the fracture toughness of the material is considered by a few authors and it has been shown that a crack surface profile can be effectively described in terms of fractal geometry (e.g. Lange *et al.* 1993).

The results for concrete elements fractured under combined compression and internal pressure are compared in Table 10.3 with these obtained for other materials. The results show a relation between fractal dimensions of the fractured surfaces and the critical values of the stress intensity factors K_{Ic} determined experimentally. The observed relation may be expressed as follows: in a group of similar materials those with higher values of K_{Ic} have

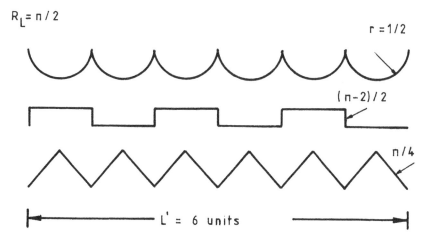

Figure 10.41 Three different profiles having identical values of the profile roughness parameter $R_L = \varpi/2$, after Banerji (1988).

a higher fractal dimension for their fracture surfaces. It may be considered as an indication that fractal dimension of materials reflects their fracture toughness. The results in Table 10.3 are incomplete and do not allow for more detailed discussion. As mentioned above, it is incorrect to compare mechanical data and fractal dimension for different materials; for example, no conclusion may be formulated from the fact that alumina and Ocala chert have similar values of D and different values of K_{Ic}. Duxbury (1990) indicated that the correlation between the fracture toughness and fractal dimension may be either positive or negative and some additional specifications as to the compared materials are necessary before formulation of conclusions of useful character. In concrete specimens with an important pore system, the negative correlation might be expected and positive correlation for a system of dispersed hard grains. The second possibility is typical for ordinary structural concretes.

The tests of concrete specimens subjected to Mode II fracture were aimed at a further investigation of relations between fractal dimension and roughness of the fracture surface after Mode II crack propagation and are presented below based on paper by Brandt and Prokopski (1993).

The specimens were cast with concretes made with three kinds of coarse aggregate: crushed basalt, river gravel and crushed limestone. Other specimens were prepared with cement mortar and paste. The specimens were subjected to shearing as shown in Figure 10.38.

The values of K_{IIc} were calculated according to the formula proposed by Watkins (1983):

$$K_{IIc} = \frac{5.11 P_{cr}}{2Bb} (\pi a)^{1/2} \tag{10.33}$$

here P_{cr} is the value of the critical load P which initiated the crack propagation, b is the ligament depth, a is the notch depth and B is the specimen width (Figure 10.39).

The fractured surfaces were used to prepare replicas with acrylic resin. The replicas were sawn into slices as shown in Figure 10.42. Then, the profile lines of each replica were subjected to the computer image analyzer Magiscan (Joyce-Loebl) and the length of each profile line was measured with varying steps equal to 0.45, 0.30, 0.15, 0.075, 0.05 and 0.0375 mm. The image of a profile line was transferred on a monitor and covered with a system of orthogonal lines 512×512 pixels. The modification of scale of the profile lines caused a modification of the steps. To measure the length of the profile lines the erosion function has been applied of unit width equal to one pixel.

Mean values from four measurements of profile lines were used to calculate the fractal dimension from formula (10.29). Specimens of dolomite and gravel concretes were selected for fractal analysis from specimens with different values of K_{IIc}.

Table 10.3 Mechanical properties and fractal dimensions *D* for different materials

Authors / Material test	Material specification	D	K_{Ic} (MNm$^{-3/2}$)	γ (J/m^2)	Temperature (°C)
Underwood and Banerji (1986)		1.085	–	–	200
		1.091	–	–	300
		1.090	–	–	400
Steel	AISI 4030	1.072	–	–	500
bending		1.084	–	–	600
		1.079	–		700
Mecholsky *et al.* (1983)					
Alumina		1.15	2.5	–	–
bending	UCC, Lucalox	1.31	4.0	20	–
Anstis *et al.* (1981)					
Alumina		1.21	2.9	11	
tension	AD 90, AD 999	1.31	3.9	19	
Neilson (1981)					
Alumino silicate	GA Tech				
bending		1.18	2.2	–	
Hellmann (1986)	WESGO				
Alumina	(A1500) GEND	1.20	3.6	23	
tension		1.23	3.9	27	
Mecholsky (1982)	Zinc-silicate 1	1.05	1.6	22	
	Zinc-silicate 2	1.09	1.8	22	
	Zinc-silicate 3	1.11	2.2	27	
Glass-ceramics tension	Lithium borosilicate	1.18	2.7	40.5	
Mecholsky and Mackin (1988)		1.32	1.55	–	20
		1.26	1.46	–	300
Ocala chert		1.24	1.25	–	400
bending		1.15	1.05	–	500
Saouma *et al.* (1990)					
Concrete		1.07			
		1.12			
compression					

Source: Brandt, A.M. and Prokopski, G. (1993).

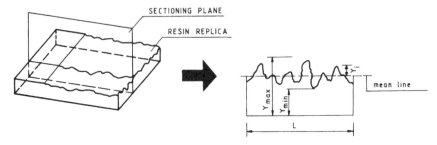

Figure 10.42 Preparation of replicas for analysis of profile lines, after Brandt and
Prokopski (1993).

The results of tests are presented in Figure 10.43 and in Table 10.4. The
following conclusions may be proposed after the tests.

The fracture surfaces are fractal objects within the scope of analysed scales
and are characterised by fractal dimensions. The coefficents of the expo-
nential correlation for lines in Figure 10.43 are close to 0.99.

It has been here assumed that the profile lines analysis allows the fracture
surface roughness to be deduced. It is therefore expected that approximate
relationship (10.32) may be used as characteristic for the comparison of
surfaces of different materials, belonging however to the same group.

The relations between values of D and K_{Ic} confirm the previously mentioned
conclusion. For Mode II fracture in cement-based composite materials it has
been observed that higher fracture toughness is accompanied by higher values
of fractal dimension of the fracture surfaces.

The differences in fractal dimensions for concrete, mortar and paste specimens may be explained by the influence of the smallest grains of sand and
cement which have completely different roles in these composite materials.

The above results are at least partly confirmed by Roh and Xi (2001) who
tested concrete specimens with different aggregate subjected to compressive
and splitting loads with different loading rates. Fractal dimensions of
fractured surfaces were determined. Interesting conclusions are:

- with aggregate grain size increasing, strength and fractal dimension also
  increased;
- with increasing loading rate the fractal dimension increased.

The obtained test results and their analysis support the hypothesis that the
fracture surfaces of concrete-like composites are fractal objects. However, the
fractal dimension of fracture surfaces is not sufficiently sensitive with respect
to strength, toughness and fracture energy of cement-based composites to
allow some practical applications. It is not yet possible to use the data related
to the fractal dimension in design of cement-based materials, for example,
to design a material for required fractal dimension or to estimate strength or
other mechanical property on the basis of fractal dimension only.

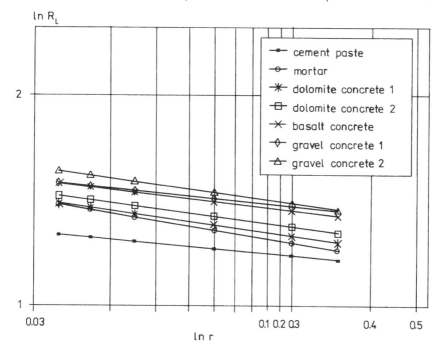

Figure 10.43 Fractal plots for profiles of specimens with different materials, after Brandt and Prokopski (1993).

After the test results discussed above and data presented in Tables 10.3 and 10.4 it appears that fractal dimension and other parameters characterizing the fracture of concrete-like materials do not reflect with enough precision and without ambiguity their mechanical properties. This is the reason why, in the last few years, less research attention has been paid to the application of fracture mechanics to cement-based materials.

Table 10.4 Fractal dimensions and fracture toughness of cement-based composites

Material	D	K_{IIc} (MNm$^{-3/2}$)
Cement paste	1.033	1.60
Cement mortar	1.060	3.37
Dolomite concrete 1	1.050	3.90
Dolomite concrete 2	1.054	4.36
Gravel concrete 1	1.038	2.74
Gravel concrete 2	1.051	3.40
Basalt concrete	1.043	5.16

Source: Brandt, A.M. and Prokopski, G. (1993).

References

Akita, H., Koide, H., Sohn, D., Tomon, M. (2001) 'A testing procedure for assessing the uniaxial tension of concrete,' in: *ACI SP-201 Fracture Mechanics for Concrete Materials: Testing and Applications*, C. Vipulanandan, W.H. Gerstle, eds Farmington Hills, MI: pp. 75–91.

Anderson, T. L. (2005) *Fracture Mechanics: Fundamentals and Applications*, 3rd edition. London; Taylor and Francis.

Anstis, G. R., Chantikul, P., Lawn, B. R., Marshall, D. B. (1981) 'A critical evaluation of indentation techniques for measuring fracture toughness: I. Direct crack measurements,' *J.Amer.Ceram.Soc.* 64(9): 533–8.

Arrea, M., Ingraffea, A. R. (1982) *Mixed-Mode Crack Propagation in Mortar and Concrete*. Ithaca, NY: Department of Structural Engineering, Cornell University, Rep: pp. 81–13.

Aveston, J., Cooper, G. A., Kelly, A. (1971) 'Single and multiple fracture,' *Proc. Nat. Phys. Lab. Conf. 'The Properties of Fibre Composites,'* Guildford, UK; IPC Science and Technology Press Ltd.: pp. 15–24.

Aveston, J., Kelly, A. (1973) 'Theory of multiple fracture of fibrous composites,' *Journal of Materials Science*, 8(No. 3): 352–62.

Aveston, J., Mercer, R. A., Sillwood, J. M. (1974) 'Fibre reinforced cements – scientific foundations for specifications,' in: *Proc. Nat. Phys. Lab. Conf. Composites – Standards, Testing and Design*, Teddington, Middlesex, UK; IPC Science and Technology Press Ltd.: pp. 93–102.

Babut, R. (1983) 'Bearing capacity and deformability of SFRC elements subjected to bending' (in Polish), in: *Proc.Conf. Mechanics of concrete-like composites*, J. Kasperkiewicz ed., Jablonna 3–8 December 1979: 71–146.

Bache, H. H. (1989) 'Fracture mechanisms in integrated design of new ultra-strong materials and structures,' in: *Fracture Mechanics of Concrete Structures. From Theory to Application. Report of TC90 of RILEM*, L. Elfgren ed., London; Chapman and Hall: pp. 382–98.

Banerji, K. (1988) 'Quantitative fractography: a modern perspective,' *Metallurgical Transactions* A, 19A: 961–71.

Bažant, Z. P. (1980) 'Material behaviour under various types of loading,' in: *Proc. Workshop High Strength Concrete*, 1979, S.P. Shah ed., University. of Illinois at Chicago; Circle: pp. 79–92.

——.(2002) 'Concrete fracture models: testing and practice,' *Eng. Fract. Mech.*, 69(2): 165–205.

Bažant, Z. P., Planas, J. (1998) 'Fracture and size effects in concrete and other quasibrittle structures,' Boca Raton, FL, CRC Press.

Bažant, Z. P., Pfeiffer, P. A. (1986) 'Shear fracture tests of concrete,' *Materials and Structures RILEM*, 19(110): 111–21.

Bažant, Z. P., Prat, P. C. (1988) 'Measurement of Mode III fracture energy of concrete,' *Nuclear Energy Eng. and Design*, 106: 1–8.

Behloul, M. (2007) 'HPFRCC field applications: Ductal recent experience', in: *Proc. Int. Workshop on High Performance Fiber Reinforced Cement Composites HPFRCC-5*, H.W. Reinhardt and A.E. Naaman eds, Mainz, Germany, pp. 213–222.

Bentur, A., Mindess, S. (2006) *Fibre Reinforced Cementitious Composites*, 2nd edition London; Taylor & Francis.

Brandt, A. M. (1980) 'Crack propagation energy in steel fibre reinforced concrete,' *Int. Journal of Cement Composites*, 2(No. 3): 35–42.

——.(1982) 'On the calculation of fracture energy in SFRC elements subjected to bending,' in: *Proc. Int. Conf. Bond in Concrete*, Paisley, June 1982, P. Bartos ed., London; Applied Science Publishers: pp.73–81.

——.(1985) 'On the optimal direction of short metal fibres in brittle matrix composites,' *Journal of Materials Science*, 20: 3831–41.

——.(1986) 'Influence of the fibre orientation on the energy absorption at fracture of SFRC specimens,' in: *Proc. Int. Symp. Brittle Matrix Composites 1*, A. M. Brandt and I. H. Marshall, eds, London; Elsevier Applied Science Publishers: pp. 403–20.

——.(1991) 'Influence of fibre orientation on the cracking and fracture energy in brittle matrix composites,' in: *Proc. Euromech. Coll. 269 Mechanical Identification of Composites*, A. Vautrin and H. Sol, eds, London and New York; Elsevier Applied Science: pp. 327–34.

——.(1996) 'Toughness of fibre reinforced cement based materials,' *Arch. of Civ. Eng.*, 42(4): 471–93.

Brandt, A. M., Prokopski, G. (1990) 'Critical values of stress intensity factor in Mode II fracture of cementitious composites,' *Journal of Materials Science*, 25: 3605–10.

Brandt, A. M., Stroeven, P. (1991) 'Fracture energy in notched steel fibre reinforced concrete beams,' in: *Proc. Int. Symp. Brittle Matrix Composites 3*, A. M. Brandt and I. H.Marshall, eds, London and New York; Elsevier Applied Science Publishers: pp. 72–82.

Brandt, A. M., Glinicki, M. A. (1992) 'Flexural behaviour of concrete elements reinforced with carbon fibres,' in: *Proc. Int. RILEM/ACI Workshop High Performance Fiber Reinforced Cement Composites*, H. W. Reinhardt and A. E. Naaman, eds, Mainz 1991, Chapman and Hall/Spon: pp. 288–99.

Brandt, A. M., Prokopski, G. (1993) 'On the fractal dimension of fracture surfaces of concrete elements,' *Journal of Materials Science*, 28: 4762–6.

Brandt, A. M., Glinicki, M. A. (1999) 'Investigation of the flexural toughness of fibre reinforced composites (FRC),' *Arch. of Civ. Eng.*, 45(3): 399–426.

Brandt, A. M., Stroeven, P., Dalhuisen, D., Donker, L. (1989) 'Fracture mechanics tests of fibre reinforced concrete beams in pure bending,' *Report 25.1-89-7/C4*, Faculty of Civil Eng., Delft University of Technology.

Broek, D. (1983) *Elementary Engineering Fracture Mechanics*, Martinus Nijhoff Publishers: The Hague.

Brown, W. F., Srawley, J. E., eds (1966) 'Plane strain crack toughness of high strength metallic materials.' *ASTM Special Technical Publication*: No. 410.

Carpinteri, A. (1987) 'Interaction between tensile strength failure and mixed mode crack propagation in concrete,' *Rep. Subcomm. C., RILEM TC89-FMT*, Italy; Dept.of Struct. Eng., Politecnico di Torino.

Dahl-Jorgensen, E., Chen, W. F., Manson, J. A., Vanderhoff, J. W., Liu, Y. N. (1974) 'Polymer-impregnated concrete: Laboratory studies,' *ASCE/EIC/RTAC Joint Transportation Engineering Meeting*, Montreal, July 14–19.

Debicki, G. (1988) 'Contribution á l'étude du rôle de fibres dispersées anisotropiquement dans le mortier de ciment sur les lois de comportément, les critères de résistance et la fissuration du matériau,' *Thèse de docteur d'Etat*. Lyon; Institut National des Sciences Appliquées.

Dugdale, D. S. (1960) 'Yielding of steel sheets containing slits,' *Journal of the Mechanics and Physics of Solids*, 8(2): 100–8.

Duxbury, P. M. (1990) 'Breakdown of diluted and hierarchical systems', in: *Statistical Models for the Fracture of Disordered Media*, H. J. Herrmann and S. Roux, eds, North Holland; Elsevier Science Publisher B.V.: pp. 189–228.

Griffith, A. A. (1921) 'The phenomenon of rupture and flow in solids,' *Phil. Trans. Roy. Society*, A221:163-98.

Hannant, D. J. (1978) *Fibre cements and fibre concretes*, London; J. Wiley and Sons.

Hannant, D. J., Hughes, D. C., Kelly, A. (1983) 'Toughening of cement and other brittle solids with fibres,' *Phil. Trans. Roy. Soc.*, London; A310: pp. 175–90.

Hellmann, J. K. (1986) 'Alumina processing and properties characterization workshop,' *SAND-86-1224*, Albuquerque, NM; Sandia National Laboratory.

Henager, C. H. (1978)'A toughness index of fibre concrete,' in: *Proc. Int. RILEM Symp Testing and Test Methods of Fibre Cement Composites*, R. N. Swamy ed., Sheffield; Construction Press Ltd: pp. 79–86.

Hillerborg, A. (1978) 'A model for fracture analysis.' *Report TVBM-4005*, Lund; Div. of Build. Mat., Univ. of Lund.

——.(1985) 'Determination and significance of the fracture toughness of steel fibre concrete,' in: *Proc. of US-Sweden Joint Seminar Steel Fiber Concrete*, S.P. Shah and A. Skarendahl eds, Stockholm; CBI: pp. 257–71.

Hillerborg, A., Modéer, M., Petersson, P. E. (1976) 'Analysis of crack formation and crack growth in concrete by means of fracture mechanics and finite elements,' *Cement and Concrete Research*, 6: 773–8.

Hughes, B. P., Fattuhi, N. I. (1977) Stress-strain curves for fibre reinforced concrete in compression. Cement and Concrete Research, 7(2): 173–84.

Jenq, Y. S., Shah, S. P. (1988) 'Mixed mode fracture of concrete,' *Int. J. of Fracture*, 38: 123–42.

John, R., Shah, S. P. (1989) 'Mixed mode fracture of concrete subjected to impact loading,' *Journal of Structural Engineering, ASCE,* January: pp. 585–602.

Johnston, C. D., Gray, R. J. (1986) 'Flexural toughness and first crack strength of fibre reinforced concrete, using ASTM Standard C1018,' in: *Proc. RILEM Symp. Developments in Fibre Reinforced Cement and Concrete*, R. N. Swamy, R. L.Wagstaffee, D. R,Oakley eds, Sheffield: paper 5.1.

Kaplan, M. F. (1961) 'Crack propagation and the fracture of concrete,' *Journal of the American Concrete Institute*, 58: 591–610.

Kasperkiewicz, J. (1979) 'Ultimate strength and strain of steel fibre reinforced concrete under tension,' (in Polish), *Mech. Teoret. i Stos.*, 17(1): pp. 19–34.

——.(1983) 'Internal structure and fracture processes in brittle matrix composites' (in Polish), *ITFR Reports*, Warsaw: No. 39.

Kasperkiewicz, J., Skarendahl, A. (1990) 'Toughness estimation in FRC composites,' *CBI Report* 4.

Kesler, C. E., Naus, D. J., Lott, J. L. (1971) 'Fracture mechanics – its applicability to concrete,' in: *Proc. Int. Conf. on the Mechanical Behaviour of Materials*, Kyoto. The Soc. of Mat. Sc., vol IV: pp. 113–24.

Kim, J. K., Lee, Y., Yi, S. T. (2004) 'Fracture characteristics of concrete at early ages,' *Cement and Concrete Research*, 34: 507–19.

Knott, J. F. (1973) *Fundamentals of Fracture Mechanics*, Butterworths, London.

Lange, D. A., Jennings, H. M., Shah, S. P. (1993) 'Relationship between fracture surface roughness and fracture behavior of cement paste and mortar,' *J. of Am. Ceramic Soc.*, 76(3): 589–97.

Laws, V. (1974) 'On increase of tensile failure strain of Portland cement reinforced with glass fibre bundles,' *Disc. Proc. Nat. Phys. Lab. Conf. Composites – Standards, Testing and Design*, Guildford, UK; IPC Science and Technology Press Ltd.: pp. 102–3.

Lenke, L. R., Gerstle, W. H. (2001) 'Tension test of stress versus crack opening displacement using cylindrical concrete speciemens,' in: *ACI SP-201 Fracture Mechanics for Concrete Materials: Testing and Applications*, C. Vipulanandan, W. H. Gerstle, eds Farmington Hills, MI: pp. 189–206.

Li, V. C., Maalej, M. (1996) 'Toughening in cement-based composites. Part 1,' *Cement and Concrete Composites;* 18: 223–37.

Mandelbrot, B. B. (1977) *Fractals: Forms, Chance and Dimension*, San Francisco, CA; W. H. Freeman.

——.(1983) *The Fractal Geometry of Nature*, San Francisco, CA; W. H. Freeman.

Mecholsky, J. J. (1982) 'Fracture mechanics analysis of glass ceramics,' in: *Advances* in *Ceramics, vol.4, Nucleation and crystallization in glasses*, J. H. Simmons, D. R. Uhlmann, G. H. Beall, eds Columbus, OH; Amer. Ceram. Soc.

Mecholsky, J. J., Mackin, T. J. (1988) 'Fractal analysis of fracture in Ocala chert,' *J. Mater. Sci. Lett.*, 7: 1145–7.

Mecholsky, J. J., Passoja, D. E., Feinberg-Ringel, K. S. (1983) 'Quantitative analysis of brittle fracture surfaces using fractal geometry,' *J. Amer. Ceram. Soc.*, 72(1): 60–5.

Mindess, S. (1983) 'The application of fracture mechanics to cement and concrete: a historical review,' in: *Fracture Mechanics of Concrete*, F. H. Wittmann, ed., Amsterdam; Elsevier: 1–30.

——.(1988) 'Bonding in cementitious materials: how important is it?' in: *Proc. Symp. Bonding in Cementitious Composites*, Boston 1987, S. Mindess and S. P. Shah, eds, Pittsburgh, PA; Mater. Res. Society: pp. 3–10.

——.(1991) 'The fracture process zone in concrete,' in: *Toughening Mechanisms in Quasi-Brittle Materials*. Dordrecht: Kluwer Academic: pp. 271–86.

Mindess, S., Lawrence, F. K., Kesler, C. E. (1977) 'The J-Integral as a fracture criterion for fiber reinforced concrete,' *Cement and Concrete Research*, 7: 731–42.

Naaman, A. E. (2003) 'Strain hardening and deflection hardening fiber reinforced cement composites,' in: *Proc. Int. RILEM Workshop High Performance Fiber Reinforced Cement Composites HPFRCC4*, H. W. Reinhardt and A. E. Naaman eds, Ann Arbor, MI: pp. 95–113.

Neilson, C. L. A. (1981) 'Investigation of high temperature increase of alumino-silicate refractories,' *M.Sc. Thesis*, Atlanta, GA; Georgia Institute of Technology.

Novák, D., Podroužek, J., Akita, H. (2006) 'Virtual 3D nonlinear simulation of uniaxial tension test of concrete,' in: *Proc. Int. Symp. Brittle Matrix Composites 8,'* A. M. Brandt, V. C. Li and I. H. Marshall, eds, Warsaw; Woodhead Publ. and ZT RSI: pp. 205–12.

Ouyang, C., Mobasher, B., Shah, S. P. (1990) 'An R-curve approach for fracture of quasi-brittle materials,' *Engineering Fracture Mechanics*, 37: 901–16.

Pande, C. S., Richards, L. E., Louat, N., Dempsey, B. D., Schwoeble, A. J. (1987) 'Fractal characterization of fractured surfaces,' *Acta Metall.*, 35(9): 1633–7.

Parratt, N. J. (1972) *Fibre Reinforced Materials Technology*, New York; Van Nostrand Reinhold Co.

Petersson, P. E. (1981) 'Crack growth and development of fracture zone in plain concrete and similar materials,' *Rep. TVEM-1001*, Lund: Lund Inst.of Techn., Div. of Build. Mat.

Rice, J. R. (1968) A path independent integral and the approximate analysis of strain concentration by notches and cracks. *Journal of Appl. Mech., Trans. ASME*, 35: 379–86.

Roh, Y. S., Xi, Y. (2001) 'The fracture surface roughness of concrete with different aggregate sizes and loading rates,' in: *ACI SP-201 Fracture Mechanics for Concrete Materials: Testing and Applications*, C. Vipulanandan and W. H Gerstle, eds Farmington Hills, MI; American Concrete Institute: pp. 35–54.

Romualdi, J. P. (1968) 'The static cracking stress and fatigue strength of concrete reinforced with short pieces of thin steel wire,' in: *Proc. Int. Conf. The Structure of Concrete*, London, September 1965, disc 204-6: Cement and Concrete Association: pp. 190–201.

Romualdi, J. P., Batson, G. B. (1963) 'Mechanics of crack arrest in concrete,' *Journal of Engineering Mech. Div., Proc. ASCE*, 89(No. EM3): pp. 147–68.

Romualdi, J. P., Mandel, J. A. (1964) 'Tensile strength of concrete affected by uniformly distributed and closely spaced short lengths of wire reinforcement,' *Journal of the American Concrete Institute*, 61: 657–71.

Rossi, P. (1998) *Les Bétons de Fibres Métalliques*, Paris: Presses de l'Ecole des P.et Ch.

Sanford, R. J. ed. (1997) *Selected Papers on Foundations of Linear Elastic Fracture Mechanics*, Adelphi, MD; University of Maryland.

Saouma, V. E., Barton, C. C., Gamaleldin, N. A. (1990) 'Fractal characterization of fracture surfaces in concrete,' *Engineering Fracture Mechanics*, 35(no. 1/2/3): pp. 47–53.

Shah, S. P. (1989) 'On the fundamental issues of mixed mode crack propagation in concrete,' in: *Fracture Mechanics of Concrete Structures: From Theory to Application. Report of TC90 of RILEM*, L. Elfgren, ed., London; Chapman and Hall: pp. 27–38.

——.(1991) 'Do fibers improve the tensile strength of concrete?' *Proc.of First Canadian University-Industry Workshop on Fiber Reinforced Concrete*, Quebec City, QC; Universite Laval: pp. 1–30.

Sih, G. C., Chen, E. P. (1973) 'Fracture analysis of unidirectional composites,' *J. of Comp. Mater.*, 7(2): 230–44.

Song, L., Huang, S. M., Yang, S. C. (2004) 'Experimental investigation on criterion of three-dimensional mixed-mode fracture for concrete,' *Cement and Concrete Research*, 14: 913–16.

Soroushian, P., Elyamany, H., Tlili, A., Ostowari, K. (1998) 'Mixed-mode fracture properties of concrete reinforced with low volume fractions of steel and polypropylene fibers,' *Cem. and Concr. Comp.*, 20: 67–78.

Swamy, R. N. (1979) 'Fracture mechanics applied to concrete,' in: *Developments in Concrete Technology – 1*, F.D. Lydon, ed., London; Applied Science Publishers: pp. 221–81.

Taha, N., Swartz, S. (1989) 'Crack propagation models for mixed-mode loading,' in: *Fracture of Concrete and Rock*, S. P. Shah, S. E. Swartz and B. Barr, eds, Cardiff, Elsevier Applied Science.

Tanigawa, Y., Hatanaka, S., Mori, H. (1980) 'Stress-strain behaviour of steel fiber reinforced concrete under compression,' *Trans. of the Japan Concr. Inst.*, 2: 187–94.

Underwood, E. E., Banerji, K. (1983) 'Statistical analysis of facet characteristics in a computer simulated fracture surface,' in: *Proc. 6th Int. Congr. for Stereology, Acta Stereol.*, 2(Suppl. 1): pp. 75–80.

——.(1986) 'Fractals in fractography,' *Mater. Sci. and Engng.*, 80(4): 1–14

Velasco, G., Visalvanich, K., Shah, S. P. (1980) 'Fracture behaviour and analysis of fiber reinforced concrete beams,' *Cement and Concrete Research*, 10: 41–51.

Vesely, V., Keršner, Z. (2006) 'R-curves from equivalent elastic crack approach: effect of structural growth on fracture behaviour,' in: *Proc. Int. Symp. Brittle Matrix Composites 8*, A. M. Brandt, V. C. Li and I. H. Marshall eds, Warsaw; Woodhead Publishing and ZT RSI: pp. 527–36.

Watkins, J. (1983) 'Fracture toughness test for soil-cement samples in Mode II,' *Int. J. of Fracture*, 23: R135-R138.

Winslow, D. N. (1985) 'The fractal nature of the surface of cement paste,' *Cement and Concrete Research*, 15(5): 817–24.

Wu Keru and Zhou Jianhua (1987) 'The influence of the matrix-aggregate bond in the strength and brittleness of concrete,' in: *Proc. Symp. Bonding in Cementitious Composites*, Boston, S. Mindess and S. P. Shah, eds, Mater. Pittsburgh, PA; Res. Society: pp. 29–34.

Zaitsev, Y. V., Ashrabov, A. A., Kazatski, M. B. (1986) 'Simulation of crack propagation in various concrete structures,' in: *Brittle Matrix Composites 1*, A. M. Brandt and I. H. Marshall, eds, London; Elsevier Applied Science Publishing: pp.549–57.

Zhang, J., Li, V. C. (2004) 'Simulation of crack propagation in fiber-reinforced concrete by fracture mechanics,' *Cement and Concrete Research*, 34: 333–9.

Standards

ACI 544.2R-89 (reapproved 1999) *Measurement of properties of fiber reinforced concrete.*

ACI 544.4R-88 (reapproved 1999) *Design considerations for steel fiber reinforced concrete.*

ANSI/ASTM E 399-78, (1978) *Standard test method in plane strain fracture toughness of metallic materials.*

ASTM C 1018 –97 (1997) *Test method for flexural toughness and first-crack strength of fiber-reinforced concrete (using beam with third-point loading).*

ASTM C 666/C666M–03 (2003) *Standard Test Method for Resistance of Concrete to Rapid Freezing and Thawing.*

ASTM C 672/C672M–03 (2003) *Standard Test Method for Scaling Resistance of Concrete Surfaces Exposed to Deicing Chemicals.*

ASTM E 399-06 (2006) *Standard test method for linear-elastic plane-strain fracture toughness K_{Ic} of metallic materials.*

EN ISO 12737:1999 *Metallic materials. Determination of plane-strain fracture toughness.*

RILEM Draft Recommendation, (1990a) 'Determination of fracture parameters (K_{Ic} and $CTOD_{IC}$) of plain concrete using three-point bend tests,' *TC89-FMT: Fracture Mechanics of Concrete – Test Methods, Materials and Structures*, 23: 457–60.

RILEM Draft Recommendation, (1990b) 'Size-effect method for determining fracture energy and process zone size of concrete,' *TC89-FMT: Fracture Mechanics of Concrete – Test Methods, Materials and Structures*, 23: 461–5.

RILEM Recommendation, (2001) 'Uni-axial test for steel fibre reinforced concrete,' *TC162-TDF: Test and Design Methods for Steel Fibre Reinforced Concrete, Materials and Structures*, 34: 3–6.

11 Behaviour of cement matrix composites in various service conditions

11.1 Loads and actions

The elements and structures made of cement-based composites are subjected to loads and actions of various origins and natures. Internal forces, that is axial and shearing stresses and bending moments, are due to dead weight, service loads of static or dynamic nature, and imposed deformations. Besides, external actions of chemical (corrosion) and physical (low and high temperature, moisture) origin can be foreseen. Various combinations of these actions, together with ageing processes, are considered in structural and material design and in execution.

Ordinary concrete elements cast in situ or in a pre-cast plant have to support their own weight at a relatively early age, together with the dead weight of other parts of the structure and of non-structural elements. In contrast, the service loads and different other actions due to environmental conditions are usually applied much later and then they are combined with the dead weight. During exploitation, the structures are loaded slowly and only a part of the peak stresses in structural elements is caused by rapid actions, for example, traffic on bridges or runways, which induce dynamic effects.

In normal conditions, it is expected that structures support loads and various actions over a long period of time, expressed in tens of years; for ordinary structures a period of 50 years is usually considered and for special structures longer periods are required. This implies long-term resistance of materials and durability of structures and their parts. On the other extreme end there are temporary structures and situations in which loads are imposed with very high rates, for example, due to accidents of various kinds.

The behaviour of cement-based materials is a result of several processes, simultaneous and interrelated, starting with the hydration of Portland cement developing over a long period of time with an appropriate increase of the hardened material strength. The permanent flow of heat, gases and fluids between the material and the environment is characteristic for all composites based on Portland cement. Initially, these processes are of high intensity, for example, production of heat during hydration. Later, the exchange of heat and moisture with the environment is slow and goes in both directions from and to the hardened cement paste with a trend towards equilibrium with the environment. These flows are the sources of variations of the material's volume, for example, shrinkage or swelling, creating local internal stresses due to restraints by hardened regions.

A system of microcracks appears during the initial period of hardening of the cement paste. It is later subjected to the action of external loads and imposed deformations, which vary over time. Local stress concentrations contribute considerably to crack propagation also under relatively low to average stresses. In certain conditions, the microcracks may exhibit healing (cf. Section 9.7). In most cases, however, they remain and may develop into major cracks leading to disintegration of elements.

Several processes are caused by the action of detrimental agents from the environment. Corrosion of reinforcement and cement matrix, leaching, excessive abrasion and cavitation, overloading and impacts; these are examples of degradation mechanisms acting simultaneously with normal wear from external loads, and are accelerated by existing crack systems.

All these processes develop in very heterogeneous and disordered materials as cement-based composites. Random distribution of all phases, heterogeneity of interfaces due to pore and crack systems and varying environmental conditions; all these create the reason for a complex material behaviour. The coefficient of variation of the value of ultimate strength under a static loading of the simplest configuration may reach 15%, depending on quality of casting and testing. This dispersion of properties should be taken into account in all material and structural design.

11.2 Behaviour under dynamic loads

11.2.1 Rates of loading

Various kinds of concrete structures may be subjected to impacts and dynamic loadings with different rates. Because all cement-based composites are sensitive to the rate of loading, this problem has been investigated over a long period of time. Impact loading may produce higher damage of structures than those of static loading. The influence of the rate of loading on the strength of cement-based materials was observed by Abrams (1917); later important test results were presented by Watstein (1953) and many other authors.

High rates of stress or strain encountered in the cement-based materials appear in various accidental situations, for example, earthquake events, impacts of vehicles and projectiles, explosions, particular kinds of exploitation in different industrial facilities or the action of waves on off-shore platforms and bridge piers. The influence of dynamic effects in concrete may be considered either in terms of rate of loading or strain rate. In the case of rate of loading, the rate is indicated in MPa/s and in the case of strain rate, in s^{-1}. The relation between these two groups of values is always approximate as the notion of the Young's modulus is not well defined. From the engineering viewpoint the loading rates are more meaningful because of their direct relation to situations of structures subjected to various kinds of external actions. In testing, however, the values of strain are measured directly and are related to observed variations of strength. The classifications are sometimes confusing, and based on conventional criteria. A coherent proposal of possible rates of stresses and strains in engineering structures is shown for different situations in Table 11.1 where both measures are used. Tensile stress and strain are more carefully analyzed because in brittle matrices these are the direct reasons for all kinds of destruction and fracture. Impacts occur either during normal exploitation or due to accidental events, and are considered in the design of structures having in mind the survival of people involved in the possible structural collapse. Two examples characterize different situations: an accident where a column of a viaduct is hit by a car; or fulfilment of the function by a structural element, such as a head of a pile driven by the impact of a hammer.

11.2.2 Influence of the rate of loading on impact resistance

As a brittle material concrete is stress-rate sensitive. The high rate of loading has an influence on several mechanical properties of concrete: compressive and tensile strength, Young's modulus, critical strain, Poisson ratio and the energy absorption capacity. The general conclusion is that the strength of a cement matrix increases with an increase in the rate of loading. The values of ultimate strain ε_u vary with the rate of loading according to the curve shown in Figure 11.1 from the tests by Wittmann (1984). The minimum is reached at rates corresponding approximately to the standard static tests. There are large variations of this effect for compressive and tensile strength, fracture toughness and Young's modulus. Also, the structure of the material, its composition and environmental factors like temperature and moisture influence the material's sensitivity to the rate of loading.

Table 11.1 Ranges of rates of stresses and strains in different situations of concrete structures

Sources of loading	Rates of tensile	
	stress $\dot{\sigma}$ [MPa/s]	strain $\dot{\varepsilon}$ [s^{-1}]
Collision of a ship	10^{-1}–10^0	10^{-5}
Collisions in rail and road traffic	10^0–10^1	10^{-4}
Gas explosion	10^0–10^1	10^{-4}
Aircraft collision	10^2–10^3	10^{-2}
Earthquake	10^2–10^5	10^{-2}–10^{-1}
Pile-driving	10^3–10^4	10^{-2}–10^{-0}

after Reinhardt (1982).

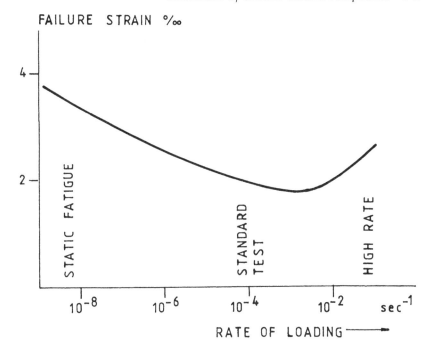

FAILURE STRAIN ‰

STATIC FATIGUE

STANDARD TEST

HIGH RATE

10^{-8} 10^{-6} 10^{-4} 10^{-2} sec^{-1}

RATE OF LOADING ──➤

Figure 11.1 Failure strain as function of rate of loading, after Wittmann (1984).

The explanations for the increase of strength with the rate of loading and the proposed models for prediction of the material behaviour are not fully defined. One of the reasons proposed is that the slow crack growth, which appears in cement-based composites already under about 50% of the short-term strength, has not enough time to develop and to decrease the strength; the system of internal cracking is less developed and there is no possibility for creep and stress relaxation.

On the basis of fracture mechanics and damage theory, Sukontasukkul *et al.* (2004) have proved how different stress-strain curves are for static loading and for impact loading (Figure 11.2). Curves show behaviour under static loading and under impact of a hammer from 250 mm and 500 mm heights to provide two different striking velocities of 2.25 m/s and 3.13 m/s, respectively. Concrete behaved more brittle under impact and all parameters: strength, toughness and modulus of elasticity increased with the rate of loading. Georgin and Reynouard (2003) suggested that concrete under high strain rate should be modelled considering viscoelastic and viscoplastic parameters. It was observed that the specimens subjected to high rate loadings exhibited flattened fracture surfaces because multiple micro-cracks had to pass across hard aggregate grains; these processes required more fracture energy (Yan and Lin 2006). Using the instrumented Hopkinson bar, Brara and Klepaczko (2007) have found that the tensile strength increased

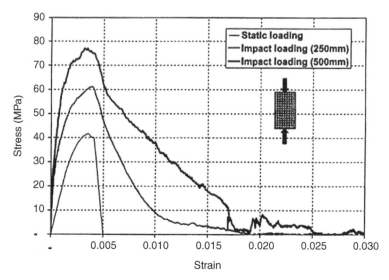

Figure 11.2 Stress-strain curves of plain concrete subjected to static and impact loading. Reprinted from *Cement and Concrete Research*, Vol 34, Sukontasukkul, P.; Nimityongskul, P.; Mindess, S., 'Effect of loading rate on damage of concrete', pp. 8., Copyright (2004), with permission from Elsevier.

from approximately 20–52 MPa for stress rates from approximately 900–4600 GPa/s when the wet specimens were tested.

Impact strength may be measured by the number of blows exerted on a tested specimen before it fails to exhibit rebound, that is, becomes a composition of separate parts rather than a solid body. Many tests were performed according to ACI 544.2R89 (1989), but as there are no universally agreed methods of testing, the published results have a relative and conventional value. A modified version of the ACI method has been proposed by Badr and Ashour (2005).

A few general conclusions have been established, which form the present state of knowledge. There are two groups of parameters which influence the impact strength:

1   internal structure of material (porosity, aggregate grading, dispersed reinforcement and its bonding to the matrix, etc.);
2   age of the cement matrix, conditions of curing and state of hydration; these parameters decide also on the material's static strength.

The concrete specimens cured in water exhibit usually higher impact strength than air-cured specimens. Also, smaller aggregate grading improves the impact strength with all other conditions maintained the same.

The fracture processes are different for different rates of loading. The differences are related to the progress of microcracks under loading, to the creep deformations in cement paste and to the relaxation of stress concentrations. All these phenomena develop with various rates – initially as slow and delayed processes, and later as very rapid ones. Consequently, the rate of loading is important; namely, for static loading, the slow crack growth develops and this gradually decreases the strength of the cement matrix. Cracks in the matrix propagate already under mean stress equal to 50–60% of the ultimate strength because of stress concentrations at the corners of hard aggregate grains or at the tips of microcracks. The local stress may reach very high values (cf. Section 8.1).

Simple phenomenological explanations based on work by Rüsch (1960) and other contemporary authors seem insufficient for sound quantitative predictions, and an explanation based on fracture mechanics was proposed by Wittmann (1984), who concluded that the parameters were the same for impact strength as for static strength. However, it seems that the state of initial cracking, due to multiple debonding of aggregate grains and to shrinkage in the cement matrix, are more detrimental for impact strength than for short-term static loading. It has been shown by Suaris and Shah (1983) that the work of fracture commonly used for characterization of the fracture toughness of specimens subjected to static or quasi-static loading is only weakly sensitive to the rate of loading. In contrast, it was proved that the fracture toughness expressed by K_c, namely,

$$K_c = \sqrt{\pi a},$$

increases with loading rate.

The rate theory and stochastic approach was proposed in the late 1970s and was intended as a basis for quantitative prediction (Mihashi and Izumi 1977). However, these models do not give a fully satisfactory agreement with experimental results.

The impact strength of cement-based composites is tested on relatively small specimens. The following methods applied are related to strain and stress rates:

1 in hydraulic testing machines with rates up to 10^3 MPa/s;
2 in Hopkinson bar up to 4.10^6 MPa/s;
3 in Charpy hammer with special equipment allowing for tensile loading, up to 8.10^2 MPa/s;
4 by a falling weight (drop tests) up to 10^4 MPa/s;
5 by rotating hammer with impact velocity up to 30 m/s.

The highest rates were obtained using rotating hammer by Radomski (1981); wave propagation events are somewhere beyond this scale.

The experimental results are collected in Figure 11.3, where the scale of the strain rate $\dot{\varepsilon}$ is added assuming a constant Young's modulus and of the

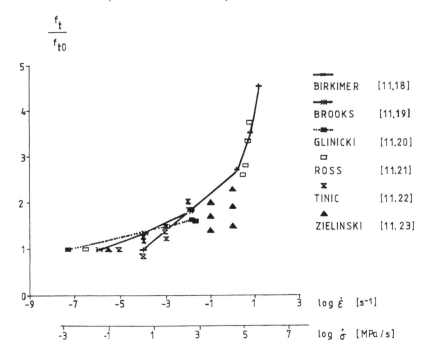

Figure 11.3 Ratio of concrete dynamic to static strength at axial tension as function of the rate of loading, after Glinicki (1991).

same value for all compared tests. All test results show consistent increase of strength with the rate of loading. The following approximate formulae were proposed by Zielinski (1982) for tensile strength as a function of the rate of stress $\dot\sigma$

for concretes $\ln f_t = 1.51 + 0.042 \ln \dot\sigma$;

for cement mortars $\ln f_t = 1.23 + 0.045 \ln \dot\sigma$.

The increase of material strength with higher loading rates might be taken into account in design for the assessment of the structural integrity and safety in accidental situations. It is also of primary importance for structures such as driven concrete piles, foundations for machinery with dynamic actions, etc. Another application of sound data on the strength increase in high loading rates may be for analyses of accidental destructions of structures where an assessment of the causes and of the sequence of events during the failure is looked for. In several structures, concrete is subjected to impact, but the element is confined and a kind of multiaxial loading is created. Plain and steel-fibre-reinforced specimens were tested in an instrumented drop-weight machine under compressive impact (Sukontasukkul *et al.* (2005). The ultimate strength and strain increased with increasing confinement and higher rate sensitivity was found than in the case of free unconfined specimens.

The present state of knowledge consists mostly of laboratory tests without adequate confirmation in natural scale by measurements from real structures. That is why the proposed models and formulae do not create a sufficient basis for the prediction of the material's behaviour, or for design of structures taking into account the increase of strength under high rate loading and the understanding of the effects remains largely qualitative.

11.2.3 *Influence of fibre reinforcement and other inclusions*

Available test results indicate that dispersed fibres of all kinds increase impact resistance of cement-based matrices and improve energy absorption capacity. This concerns the influence of the material's structure in general.

In the investigation carried on by Ramakrishna and Sundararajan (2005), only natural fibres (coir, sizal, juta and Indian hemp) were applied with different amounts from 0.5% to 2.5% weight of cement. These fibres increased the impact resistance by 3–18 times, respectively, compared to plain cement mortar specimens. The coir fibres reinforced slabs have shown the best performance.

Wang *et al.* (1996) tested the influence of polypropylene and steel-fibres on the fracture energy absorbed due to the impact of a hammer. It was concluded that there was a critical volume fraction of steel fibres somewhere between 0.5% and 1.0%, that is, below that volume the influence of both kinds of fibres was negligible, and above this volume the fibres became less efficient (Figure 11.4). Also steel fibres were used with high performance matrices by Yan *et al.* (1999) to test impact resistance according to the ACI 544.2R-89 method of a freely falling ball. Concrete beams with 1.5% volume of steel fibres exhibited impact indices increased by a factor of two with respect to high strength concrete without fibres. Fibres were found as mitigating the stress concentration at the crack tips and delayed the damage process under impact. Applying the same method, Brandt *et al.* (1996) verified the usefulness of steel-fibre reinforcement of concrete substrate for underground railway tracks. Under dynamic loadings the strength of fibre-reinforced cement-based materials is estimated as 3–10 times greater than that of plain matrix.

Using the instrumented impact machine and testing specimens under uniaxial tension, Banthia *et al.* (1996) have shown that the higher the strength of the matrix itself, the less effective the fibres are in improving the fracture energy absorption.

The influence of steel fibres on the behaviour of high performance concrete slabs subjected to high velocity impact of projectiles was tested for construction of shields in military applications. The reinforced concrete slabs exhibited smash failure, while the slabs reinforced with high volumes of fibres (7% and 10% volume, pre-placed in the moulds) remained only with minor radial cracks on both sides. The projectiles were either embedded in or rebound from the targets. Several experiments were performed on slabs and plates subjected to perforation by projectiles or falling weights by Luo *et al.*

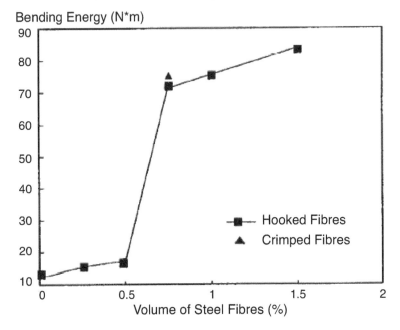

Figure 11.4 Fracture energy of concrete beams with steel fibres subjected to impact loading. Reprinted from *Cement and Concrete Research*, Vol 26, Wang, N.; Mindess, S.; Ko, K., 'Fibre reinforced concrete beams under loading', pp. 14., Copyright (1996), with permission from Elsevier.

(2000). Aramid fibre mesh was also used as reinforcement of thin mortar sheets and subjected to impact by projectiles (Kyung and Meyer 2007). The problem of various body-armour products is of considerable importance because of the increased number and intensity of natural disasters, and more importantly, terrorist attacks. The results have shown that brittle matrix composite materials reinforced with fibres and composed of several thin layers have considerable capacity of energy absorption. There are ready-to-use design methods for such structures and already a vast amount of knowledge has been accumulated in this field.

Impact behaviour of elements reinforced with PE (polyethylene) and AR (alkali resistant) glass fibre fabrics was investigated in view of its application in such structures as piles, pavements, industrial floors, etc., reaching an absorption of 20–50% of total input energy (Genconglu and Mobasher 2007).

The drop-weight impact machine was applied by Banthia *et al.* (1998) to determine the influence of steel and carbon fibres, which are also used as hybrid reinforcement. It appeared that steel macrofibres were better than carbon microfibres; however specimens with hybrid reinforcement (1% + 1%vol.) were the toughest. The tests were also performed in subnormal temperature –50°C, but no special difference was observed. Similar tests carried on by Ong *et al.* (1999) confirmed the superiority of steel hooked fibres over polyolefin and polyvinyl alcohol ones.

Glinicki (1991) tested several series of specimens made with materials characterized by the same tensile strength and having intentionally introduced differences in the structure:

1   cement mortar (MR);
2   model concrete made with aggregate of discontinuous grading composed of sand ≤ 2.0 mm and crushed grains 8–12.5 mm (CE);
3   cement mortar reinforced with Bekaert fibres V_f = 1.2% (BF);
4   cement mortar reinforced with Harex fibres V_f = 1.7% (HF);
5   porous cement mortar with artificial pores 0.4–4.0 mm V_f = 14.5% (PM).

The tests were realized on dog-bone specimens subjected to axial tensile loading in an Instron 1251 machine. The results in Figure 11.5 show a considerable increase of relative strength for different materials with increasing porosities; their densities are indicated on the right hand side in the figure. The conclusion is that for the same static tensile strength the increase of the impact resistance was inversely proportional to the material density. The explanation that may be suggested for the above relationship is based on the existing system of microcracks in the cement paste. In more porous

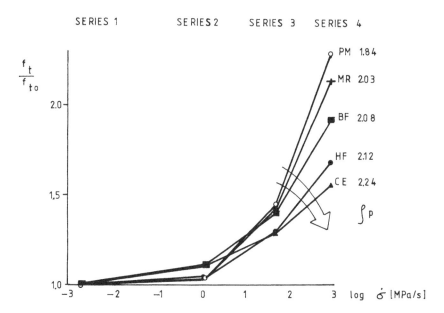

Figure 11.5 Relative increase of tensile strength as function of the loading rate å for materials with different apparent density, after Glinicki (1991).

materials there is less stress concentration at the crack tips and at higher loading rates the cracks have no possibility to propagate as at lower rates. That effect is less important for higher density materials.

The observed increase of impact strength was non-linear with respect to the loading rate; it was larger for higher rates. It may be supposed after these tests, as well as after other authors' investigations, that for the loading rates above 10^3 MPa/s the observed increase in strength will be even greater.

Tandon and Faber (1999) tested the specimens made with concrete, mortar and cement paste that were subjected to impact loading with various rates, and recorded the fracture processes. Both values of fracture toughness K_{max} and roughness number R_q were measured and the results are shown in Figures 11.6 and 11.7. The values of K_{max} decreased with the increase of the loading rate on the concrete specimens, so the fracture surface was becoming less rough and tortuous because cracks crossed the aggregate grains. This effect was not observed in mortar and cement paste specimens. Roughness number R_q was calculated from the formula:

$$R_q = \sqrt{\frac{h_1^2 + h_2^2 + ... + h_n^2}{n}}$$

where h_1, h_2,..., h_n are heights of the individual coordinate points on the roughness plane and n is the total number of points. Comparing the results it appears that the specimens tested at 0.25 µm/min were rougher than the one tested at 10 µm/min by more than a factor of two. For the fast loading concrete specimens surface roughness approached that of mortar specimens. A similar difference in roughness was evident between mortar and paste specimens. These results help to explain the process of fracture in cement-based material with different structure.

11.2.4 Seismic loading

Dispersed fibres significantly improve ductility over traditionally reinforced concrete structures when subjected to cycle loading due to a seismic event. Increased tensile strength, dispersion of cracks into microcracks and enhanced deformation capacity are the important features required for design and retrofit of buildings in regions of high seismic activity. Dispersed steel fibres increase ductility, energy absorption, shear resistance and stiffness, and so-called strain hardening behaviour is obtained through use of higher volume fractions of various types of steel fibres: hooked or twisted, and also PE and polyvinyl alcohol (PVA) fibres in 1.5–2.0 % volume fractions. Such use of fibres not only improve strength, but also provides confinement without increasing the transversal reinforcement volume; therefore substantial simplification in reinforcement is achieved (Parra-Montesinos 2006). The coupling beams and structural walls are the structural members that ensure the stability of

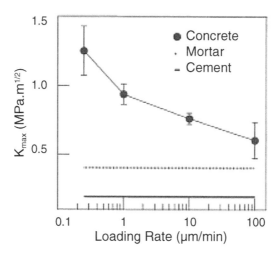

Figure 11.6 Values of fracture toughness K_{max} for concrete, mortar and cement paste versus loading rate. Reprinted from *Cement and Concrete Research*, Vol 29, Tandon, S., Faber, K.T., 'Effects of loading rate on the fracture of cementitious materials', pp. 10, Copyright (1999), with permission from Elsevier.

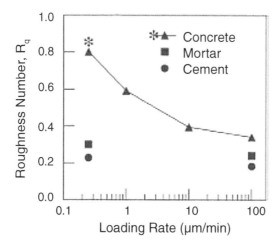

Figure 11.7 Roughness number R_q of fractures surfaces versus loading rate, after Tandon and Faber (1999) *Fig.12, page 404.*

a building in earthquake-induced loadings. The fibres may partly replace traditional reinforcement of a beam-column joint and strengthen it, and as a result it provides stronger joints, allows the hoop spacing to increase and to simplify construction. In fact, HPFRCC material behaves like confined concrete, adding enhanced shear resistance of the walls to increased plastic hinge rotation capacity. Such behaviour was experimentally verified.

Tests performed by Wight *et al* (2007) confirmed enhanced strength when a high volume of fibres was applied, and dispersion of cracks into microcracks were observed. Similar results related to punching shear resistance have been obtained by Cheng and Parra-Montesinos (2007) after tests of slab-column systems that were subjected to monotonic loading that simulated the effects of a seismic event.

Fibre-reinforced elements are also used to strengthen the concrete elements and joints in existing structures, for example, in Japan. The experimental results show the improvement in performance with FRC retrofit. Ultra-high-strength SFRC with 200 MPa resistance in compression has been tested (Kimura *et al.* 2007). The problem of retrofitting numerous structures using so-called 'seismic retrofit technology' is especially important in highly developed regions where buildings and infrastructure from past decades were not properly designed to resist seismic loadings and should be prepared to survive possible events of that kind. According to present knowledge, their strengthening is necessary. Retrofitting with FRC pre-cast panels also seems to be a useful technology. The material properties of HPFRCC make it particularly suited for that purpose.

Buyle-Bodin and Madhkhan (2002) tested concrete piles reinforced with steel fibres and subjected to alternative cyclic bending with and without axial load. It appeared that classic reinforcement was necessary to avoid brittle failure and was more efficient than fibres when there was no axial load on a pile. However, with axial load, steel fibres better dissipated energy. Similar results have been obtained by other researchers.

Application of HPFRCC in structures that should be earthquake-resistant is a relatively new but already confirmed option. It has been proved that ductile behaviour, increased resistance and damage tolerance can be achieved, thus reducing necessity of a special anti-seismic reinforcement by traditional bars or prestressing.

11.3 Fatigue

11.3.1 Factors and effects

Fatigue of a material is a progressive fracture, which consists of the accumulation of damage resulting from sustained, long-term loading of a constant, variable or cyclic nature. Local modifications that appear in the material structure under load may lead to a kind of stabilization, for example, in the form of permanent deformation and stress or of elastic deformation.

However, the word fatigue is used for such situations when there is no stabilization, but rather a gradually increasing damage. Fatigue under constant load is considered in more detail in Section 11.4, and here fatigue under cyclic load is described.

Two kinds of cyclic loading should be distinguished: low-cycle and high-cycle. In the former, a few cycles with high load values are applied, for example, at seismic events, while in the latter, a large number of cycles with lower load values occur, as in road and airport pavements.

In cement-based brittle materials, cracks spread out from a system of initial microcracks and multiply under relatively low loads (cf. Chapter 9). The opening and propagation of microcracks determine the material's behaviour under sustained and cyclic loads.

The fatigue strength of materials (called also endurance limit for flexural loading) means the maximum value of stress, which may be supported indefinitely and is indicated in relation to the static strength under single, short-term loading. The majority of concrete structures are built for static loads, and if subjected to cyclic loadings these are of rather limited amplitude and structures are not designed against fatigue. However, in multiple applications in industrial buildings and civil engineering structures, cyclic loading plays an important role. It is then usually assumed that the maximum load supported during 10^6–2.10^6 cycles may be considered as the fatigue strength of the material in relation to cyclic loading.

The strength of plain concrete under cyclic loads is always represented as decreasing with time and it may be stated that the fatigue strength does not exist for such materials. It means that after a sufficiently high number of cycles, failure occurs. That assumption is supported by the fact that permanent damage appears in heterogeneous brittle materials under quite low stress. To accommodate these apparently contradictory assumptions the application of the notion of probability is helpful. In that sense, the fatigue strength corresponds to a situation, and to a kind of loading, in which after approximately 2.10^6 cycles the probability of failure is high. However, for fibre-reinforced cement-based materials it is believed that fatigue strength does exist. This means that if a specimen withstands approximately 10^6–2.10^6 cycles, then it is highly probable that it will resist an infinite number of similar cycles.

The strength depends on the nature of the load (compression, tension or flexion, constant or cyclic, range between maximum and minimum stress in a cycle, etc.) and on the temperature and humidity. In the case of a cyclic load its frequency may also have some importance. For long-term application of the load of either nature, a certain reduction of strength should be foreseen.

The fatigue strength under sustained loads and cyclic loads is an important material characteristic. In all structures, the long-term stress should be maintained below the fatigue strength and only for infrequent and exceptional accidental situations may higher stress be admitted without serious risk of a failure. Also, in many structures, variable loads and their range between the maximum and minimum values determine their behaviour; for example, bridge structures subjected to traffic loads and marine structures exposed to waves.

The present state of knowledge of the fatigue strength of brittle matrix composites under various kinds of loads is rapidly developing, but not without difficulties. The main reason is the variety of possible parameters related to the material's structure and to the loading conditions: frequency, stress levels, etc. Moreover, the observations of the structures do not furnish all information necessary for such an assessment and the testing of elements under fatigue is particularly long and expensive. Basic phenomena in gradual degradation of the material structure are strongly related to the randomness of all involved factors and incomplete understanding is even stronger in fibre-reinforced composites. As a result, published test results and conclusions related to the fatigue of cement-based composites are frequently different.

11.3.2 Fatigue of plain and fibre-reinforced concretes

The fatigue processes in brittle materials are entirely different from those in metals and ductile matrices of high strength composites. In a highly brittle matrix, such as glass, ceramics or cement paste, the cracks appear at very low tensile strain and stable crack growth is nearly non-existent. Therefore, the first crack strength is very close or practically the same as fatigue strength. In cement-based composites, however, the cracks are partly controlled by such elements of their internal structure like aggregate grains and reinforcing fibres.

Microcracks of various dimensions exist in the cement matrix from the very beginning. They propagate and multiple cracks appear caused by an additional input of the external energy. Initially, the load is supported by the matrix and later by reinforcement, if it exists. Usually, the stress at local concentrations (tips of initial cracks, inclusions) exceeds the tensile matrix strength (cf. Section 8.1). That is the reason why the cracks propagate at each load cycle even if the mean stress values in a cross-section are relatively low. By the accumulation of damage, the strength and stiffness of the composite material decrease. The transfer of load from matrix to fibres is ensured by the bond and friction in the process called pull-out, in which gradual damage and fatigue is also observed (cf. Section 8.4). The fatigue consists of a decrease of strength and stiffness resulting from different phenomena during the load cycles.

The strain necessary for initiation of multiple cracking and crack propagation is lower for a cyclic load than for a static and constant load. The process of fatigue in fibre-reinforced cement composites is similar to the static increase of load in the sense that debonding and pull-out also occur gradually. The fibres used as reinforcement for cement matrices usually do not show fatigue at all because the stress in fibres is maintained at quite a low level as related to their strength, in steel or carbon fibres, for example.

For both plain concrete and fibre reinforced composites subjected to cycle of bending fatigue loads three stages of behaviour can be distinguished:

- stage I damage (micro-cracking) nucleation;
- stage II stable damage expansion due to slow crack propagation;
- stage III unstable damage development and rapid crack propagation.

Proportions between these stages depend on various parameters characterizing materials and loading conditions.

Different approaches have been proposed for predicting the fatigue strength taken from the so-called 'advanced composite materials,' from continuum mechanics characterization of damage, through application of fracture mechanics to fatigue reliability. The studies of fatigue behaviour of cement-based composites cannot be entirely based on these results and are rather limited to tests of specimens and to the formulation of practical conclusions for the prediction of the fatigue strength.

The damage to the material is also recognized under cyclic load as a set of permanent micro-structural modifications caused in a material by physical or chemical actions. These changes in the material's structure may have various forms, for example matrix cracks, fibre breaks, fibre/matrix or inclusion/matrix debondings and – in most cases – combinations of all three.

The micro-structural damage may be randomly distributed and oriented in the material or may have a preferred orientation and position. Cracks occur mostly along fibre/matrix and aggregate/matrix interfaces, at the tips of initial cracks and at stress concentrations around hard inclusions. In the first stage, non-interacting cracks develop in the brittle cement matrix. In the next stages, either a kind of stabilization occurs in which strength and stiffness do not decrease with consecutive load cycles, or on the contrary, cracks interact and join each other to form major cracks, leading to a rupture. The stabilization mentioned corresponds to such a slow rate of decrease of strength that the safety is ensured up to approximately the standard number $N = 2.10^6$ cycles.

The cycles of loading are defined either by the range of the cycle stress, which means by $\sigma_{max} - \sigma_{min}$ or by the ratio of the stress $\rho = \sigma_{max}/\sigma_{min}$ For brittle materials it has been observed that the sign of ρ is important, it means that the existence of compression and tension in one cycle adversely affects the fatigue strength. Also, the value of σ_{max} plays a decisive role.

According to approximate estimations based on experimental results represented on the Goodman diagram as shown in Figure 11.8 for N cycles, the fatigue strength under cyclic loading is inversely proportional to the amplitude between maximum and minimum stress related to the concrete strength: σ_{max}/f_c, σ_{min}/f_c. The diagram allows, for example, σ_{max} to be determined for given σ_{min} and for required number of cycles N which concrete can withstand. The curves for different values of N are non-linear, which indicates that for higher values of stress the admissible range is rapidly decreasing. In other similar diagrams some quantitative differences may be observed that confirm the approximate character of that kind of representation.

The fatigue strength as established after 10^6 cycles is equal to approximately 50% of the static short-term strength. Other representation of the progress of fatigue may have the form of a three-stage curve of deformation versus total life cycle of the element, in the case of the process leading to failure, similar to the curve shown in Figure 11.18 for creep.

The frequency of cycles has a smaller influence than the range of stress, but the fatigue strength of steel-fibre-reinforced concretes also depends upon the loading rate, which confirms the general observations made in that respect.

The considerable influence of the range and nature of loading on fatigue strength was observed already by Batson *et al.* (1972) in tests of beams reinforced with approximately 3% volume of steel fibres. The fatigue strength of 74% and 83% of the first crack static flexural strength at 2.10^6 cycles for complete reversal and non-reversal cycles of loading were obtained, respectively. The beams failed by pulling out of the fibres. The fatigue strength of only 50% of the static strength was determined at 10^6 cycles.

Physically, the fatigue accumulation in cement matrices is based on the gradual progress of cracks after each cycle of loading in which stress

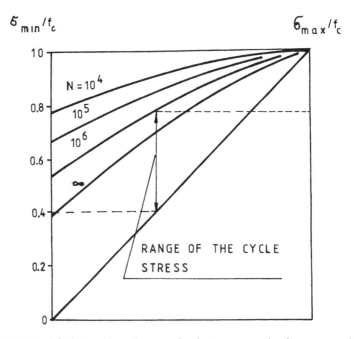

Figure 11.8 Modified Goodman diagram for fatigue strength of concrete subjected to compression with 2.10 cycles, after Neville (1997).

concentrations occur. Crack propagation has an immediate and obvious consequence: a decrease in strength and toughness, together with a decrease of stiffness and Young's modulus.

The fatigue strength depends considerably upon the material's structure and its sensitivity to the accumulation of damage in successive cycles. When a material's structure is provided with mechanisms that control the propagation of cracks, then the fatigue strength is considerably increased. The development of cracks under cyclic loading is also controlled by the kind of constraints that exist, for example, in road and air field pavements where biaxial combinations of compressive and tensile stresses appear under cyclic loading due to traffic. A simplified physically based model of crack growth is proposed by Subramaniam and Shah (2003). The anisotropic character of the microcracking in concrete and the accumulation of damage were considered in a 3D model describing the fatigue behaviour and was based on damage theory, presented by Alliche (2004).

It is generally assumed that the slow and stable crack growth at the interface of the aggregate grains is the most important process which determines the fatigue failure and the characteristic endurance limit of the composite. The debonding process at sand particles was identified as a decisive factor. A modelling approach to the prediction of fatigue behaviour was proposed by Zhang *et al.* (1999), based on the damage development in the interfaces fibre/matrix and aggregate grain/ matrix.

The fatigue of concretes made with lightweight aggregate is basically the same as for ordinary concretes. The introduction of 1% by cement mass of nano-particles TiO_2 and SiO_2 as additives to the concrete mix was tested by Li *et al.* (2007). Nano-particles were added to water together with a water-reducing agent and then poured to aggregate, and mixed separately. The flexural fatigue strength for the best combination of components was 4.75 and 2.67 times higher compared to plain concrete, for stress levels of 0.85 and 0.70, respectively. The technical and economic aspects of such additions should be considered before larger application.

The fatigue strength is increased with the fibre addition and the conclusions may be summarized as follows:

- in flexion the fatigue strength may reach 70–80 % of the static strength;
- the steel-fibres do not break but are pulled out as in the static tests.

However, the test results obtained by Cachim *et al.* (2002) have shown that fibres that are too long decrease the fatigue strength; this may result from an increase of porosity because of lower workability.

In the review paper Lee and Barr (2004) have shown that in the elements subjected to cyclic compression, the influence of fibre reinforcement on the fatigue life cannot be expected, while it is significant in the elements under flexural loading. This is shown in Figures 11.9 and 11.10 where diagrams of *S-N* type are given; here *S* is the dimensionless term – the percentage of

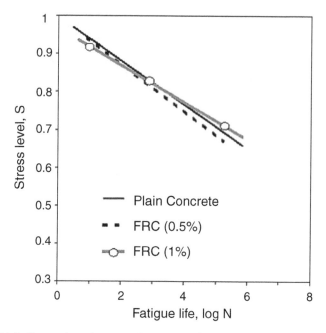

Figure 11.9 Comparison between *S-N* curves for plain and SFRC concretes (0.5% and 1%vol) under compression, Lee and Barr (2004) *Fig. 1d, page 302.*

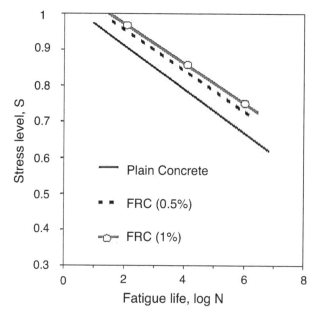

Figure 11.10 Comparison between *S-N* curves for plain and SFRC concretes (0.5% and 1%vol) under flexural loading, Lee and Barr (2004) *Fig. 2d, page 303.*

static strength obtained for the same elements and N is number of cycles in logarithmic scale.

Results obtained by Ramakrishnan and Lokvik (1992) concerned concrete specimens subjected to flexure and reinforced with four types of fibres: straight, corrugated and hooked steel-fibres, and polypropylene fibres up to 1% volume. The relations between number of cycles N and maximum fatigue stress f_{fmax} divided by the modulus of rupture f_r are shown in Figure 11.11 as estimated regression lines for different kinds of fibres. The proposed formula for prediction of the fatigue behaviour is the following:

$$f_{f\,max}/f_r = C_0 N^{C_1} \qquad\qquad (11.1)$$

where C_0 and C_1 are two coefficients to be determined experimentally. These and other tests indicate that the polypropylene fibres considerably improve the fatigue strength even if added in small volume of 0.5 % or even 0.25%, while these low modulus fibres do not appreciably improve the static short-term strength of the cement matrix. Polypropylene fibres have a beneficial influence on both the number of cycles to failure and the total amount of dissipated energy with respect to plain concrete, subjected to low-cycle

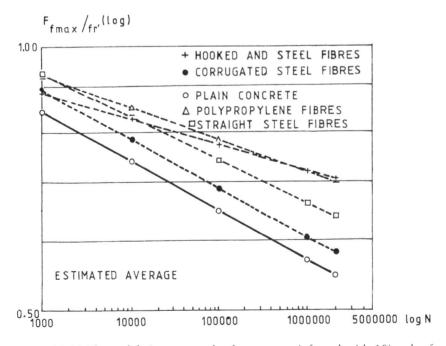

Figure 11.11 Flexural fatigue strength of concrete reinforced with 1% vol. of different fibres, after Ramakrishnan and Lokvik (1992).

fatigue; higher volume fraction has no positive influence, probably because of the less dense structure of the matrix.

Sąsiadek (1980) performed a series of tests in which the fatigue strength of plain concrete was compared to that of steel-fibre-reinforced concrete. The fatigue strength was stabilized after for 2.10^6 cycles and these results may be summarized in the form of relations between fatigue strength f_F and static strength f_{st};

- for specimens under flexion:
  plain concrete $f_F = 0.575\ f_{st}$
  fibre concrete $f_F = 0.714\ f_{st}$
- for specimens under compression:
  plain concrete $f_F = 0.54\ f_{st}$
  fibre concrete $f_F = 0.62\ f_{st}$

The fatigue strength of fibre concrete depends upon two conditions:

1   that the reinforcement is efficient, i.e. at least 1.3–1.5 % vol. is necessary, which may be considered as a rather high volume, considerably influencing the material costs;
2   that the fibres are distributed in a uniform way, because local regions without fibres considerably decrease the composite material strength.

The influence of the nature of aggregate was observed by Sąsiadek (1991) who tested specimens under bending at 2.10^6 cycles with the ratio of the stress cycles $\rho = 0.2$. It appeared that the concrete with limestone aggregate had fatigue strength equal to $0.49\ f_{st}$ and with granite aggregate to $0.63\ f_{st}$; such a difference is meaningful for special applications.

The orientation of fibres is certainly an important factor influencing the fatigue strength, but experimental confirmations were published only from the tests of advanced composites with ductile matrices. Similar tests for cement-based composites are not yet available because in most applications, fibres are randomly distributed.

The structures of cement-based composites to be exposed to cyclic loading are designed using rare experimental data of fatigue strength and large safety factors are necessary, as no reliable computation models have been developed, based on thorough studies of the processes in all their complexity. The safety factor should also allow for a particularly large scatter of fatigue properties as compared with that for plain concretes, which is believed to result from non-uniform distribution and orientation of fibres and of their variable bonding properties.

Beams made with high early-strength fibre-reinforced concrete were tested by Naaman and Hammoud (1998) and steel and polypropylene fibres were applied separately and as a hybrid reinforcement. Loading range was 10–90% and 10–80% of the ultimate static strength. In Figure 11.12, the

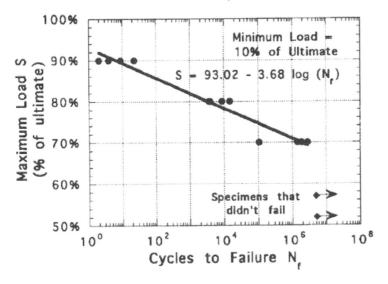

Figure 11.12 Maximum applied load in percent versus number of cycles to failure, after Naaman and Hammoud (1998) *Fig. 6, page 359.*

maximum load as a percentage of the ultimate load is shown against the logarithm of the number of cycles to failure and an approximate line is traced. The main result was proposed that the 2% addition of hooked steel fibres was most efficient and that a stress range of 65% can be considered as the endurance limit in design.

Very high fatigue strength of so-called engineered cementitious composites (ECC) was at the origin of the concept of layered beams. Concrete beams were repaired with thin top layers of ECC reinforced with PVA fibres and subjected to bending. The results allowed the authors to propose the application of that technique to increase the fatigue life of old ordinary concrete pavements (Zhang and Li 2002). Similar results have been obtained in the tests of composite elements with bottom layer between 25–50% of total depth made with ECC, composed with sand, up to 1000kg/m^3 of binder and 1.7% volume of PVA fibres. The cost of such composite was 4–5 times higher than ordinary concrete. Such beams tested under four-point flexural cyclic loading shown excellent fatigue strength. At a stress level of 90%, the increase of the fatigue lifetime was approximately two orders of magnitude for a 100 mm deep beam with a 25 mm ECC layer, and three orders of magnitude for a 50 mm layer (Leung *et al.* 2007). Nonlinearity of damage accumulation was observed already at 75% of stress limit, in contrast to Palmgren-Miner's rule (Grzybowski and Meyer 1993).

Damage behaviour of CFRC was studied by Wang *et al.* (2006) and conclusions similar to that for other cement-based materials were proposed.

Measuring the increase of electric resistance may monitor the progress of damage in CFRC.

Also, a very high strength material called Multi-Scale Cement Composite (MSCC) was tested for fatigue loading by Parant *et al.* (2007). Three kinds of steel fibres (up to 11% vol.) were used with a high dosage of Portland cement and SF. Different lengths of fibres from 5 mm to 25 mm ensured high tensile strength and control of micro-cracking. After preliminary tests of beams under flexure, the endurance limit was established as equal approximately 0.65 of static tensile strength.

Singh and Kaushik (2003) analyzed different formulae for determination of the fatigue stress limit for fibre-reinforced concrete:

$$S = a - b \log (N)$$

$$S = 1 - b(1 - f_{min}/f_{max}) \log N$$

$$S = C_1 (N)^{-c_2}$$

$$S = C_1 (N)^{-c_2 \left(1 - f_{min}/f_{max} \right)}$$

where a, b, C_1 and C_2 are experimental coefficients, N is number of cycles to failure. In the last equation, the influence of minimum and maximum fatigue stress is also considered. The fatigue life of steel-fibre-reinforced concrete is described by the two-parameter Weibull distribution and it can be used to calculate the equivalent fatigue lives corresponding to different survival probabilities that may be selected by a designer.

A theoretical model based on micromechanics of short fibres bridging cracks in the cement matrix has been proposed by Li and Matsumoto (1998). The model enables us to predict fatigue crack growth, which determines the fatigue life FRC. The Paris type law is proposed to describe the situation at the crack tip:

$$\frac{da}{dN} = C \left(\Delta K_{tip} \right)^n$$

where a – crack length, N – number of load cycles, C and n – Paris constants, δK_{tip} – crack tip stress intensity factor amplitude (Paris and Erdogan (1963).

The knowledge of the fatigue life of such a complex material as cementitious materials with fibres is still deficient and there are a lot of conflicting test results and quantitative conclusions. The mechanisms of degradation of the fibre/matrix frictional bonding that result from cyclic loading, taking into consideration various kinds of fibre reinforcement, are not fully understood. Intensive investigations are needed before the design methods are available for professional application. Nevertheless, the increased fatigue strength of fibre concretes is confirmed as the fibres provide better control of the development of microcracks.

11.4 Long-term behaviour and ageing

11.4.1 General remarks

The strength and Young's modulus of cement-based materials under sustained loads and actions depend on several factors and conditions, and may be considered as a result of the following processes:

- development of damage due to local stress concentrations under permanent and service loads;
- slow hydration of cement grains, which is believed to continue indefinitely if there are adequate conditions for the chemical reactions;
- shrinkage and creep of cement matrix depending on moisture and loading conditions, respectively;
- possible degradation due to detrimental action of external agents, e.g. carbonation, chemical attack, etc.

The rate of hardening of Portland cement paste decreases with age and it is considered that the final level is practically attained after 90 days. Later, hydration becomes very slow for two main reasons: the amount of available cement grains is decreasing, and the transfer of free water is reduced by the hydration products that fill the capillary pores (cf. Section 6.3). The remaining non-hydrated grains increase the composite strength of the matrix as hard inclusions.

All these processes are developing simultaneously and are interrelated, even based on simplifying assumptions, they are often tested and analyzed separately. The results may be considered in terms of strength, cracking and durability, but also in further economical analysis the outward aspect, safety and serviceability should be taken into account. In every case, the final results depend upon the quantitative importance of each particular process and are not only related to the properties of the material itself but also, in large measure, to structural constraints. Free deformations do not cause stresses, but they do not exist in real structures, where constraints occur at various levels and in different forms.

11.4.2 Shrinkage and creep

According to the generally accepted hypothesis, shrinkage and creep are independent and additive. It means that shrinkage is the time-dependent deformation in an unloaded concrete at constant temperature and creep is defined as the change of deformation since the application of load, corrected for shrinkage, and is related to the stress state and its intensity. Consequently, it is possible, by suitable arrangement of testing procedures, to specify each phenomenon separately and to measure relevant strain: shrinkage on specimens free from any stress, and creep on specimens without loss of humidity. Even if this assumption is not perfectly valid, it is nevertheless very useful.

(a) Shrinkage

Shrinkage of the cement-based matrix is the change in its volume due to moisture variation or chemical reactions and thermal effects during and after the hydration process. The cement paste is the source of shrinkage while other components are inert and only may control the deformations due to shrinkage.

Three basic phenomena have been described by Mindess *et al.* (2003) as being responsible for shrinkage in cement paste: capillary stress, disjoining pressure and changes in surface free energy. The origin of the volumetric variations is the appearance of tensile stresses in the menisci created in the fresh concrete as it is drying (plastic shrinkage) or in the hardened concrete due to self-desiccation (autogeneous shrinkage) and drying (drying shrinkage). Disjoining pressure is exerted by water in micropores; that pressure keeps neighbouring CSH layers apart. Without entering into a detailed analysis, the following kinds of shrinkage of cement paste may be distinguished:

- plastic shrinkage is observed in fresh cement paste and it increases with cement content and decreases with volume fraction of aggregate;
- autogenous shrinkage occurs when drying is due to hydration and water is used for chemical processes, but all loss of water to environment is excluded;
- drying shrinkage corresponds to the loss of water of the hardened cement matrix;
- carbonation shrinkage occurs in hardened concretes exposed on the influence of carbon dioxide from atmosphere.

Various kinds of shrinkage may occur simultaneously. Their separate analysis is possible only in special laboratory conditions or is based on simplified assumptions. The absolute volume of the hydrates is smaller than the sum of the absolute volumes of cement particles and water (Aïtcin 2003). Particularly, all kinds of high-strength and high-performance concretes with low values of w/b ratio are exposed to plastic shrinkage and should be cured in high humidity immediately after casting. Autogenous shrinkage is due to the loss of water used for hydration even if no drying to the environment occurs. It is more important in high performance concretes with low values of *w/c* ratio and use of SF, than in ordinary ones where it is practically negligible. The proposed explanation is based on the fact that the water is maintained in fine capillaries in which high capillary forces appear when it is used for hydration.

Self-desiccation is the process in which very fine porosity drains water from the larger capillary pores where water is not as strongly bonded.

Drying shrinkage is more important by far in ordinary concretes. It is believed that its origin lies in the increase of capillary forces caused by loss of water. The forces are supported by the hardened cement skeleton. Detailed analyses of the shrinkage in Portland cement paste were proposed by many authors and comprehensive reports were already presented by L'Hermite (1957). Various descriptions are formulated in slightly different ways, justified by a multitude of conditions, methods of testing and kinds of cements, but the main conclusions are the same. Due to developed network of capillary pores, the progress of drying shrinkage is relatively rapid from the external layers inwards.

Modelling shrinkage as a result of increased stress in capillaries was proposed by Freyssinet and it has been both developed and criticized. Several other concepts are discussed and the phenomenon of shrinkage is still far from being fully understood (Kovler and Zhutovsky 2006).

The carbonation shrinkage starts much later according to circumstances; that is, it develops according to the amount of CO_2 in the atmosphere, and is less important as it concerns only external layers of concrete elements. Carbonation shrinkage may be reduced by the addition of SF (Persson 1998).

Total drying shrinkage in high performance concretes is considerably smaller than in ordinary concretes (approximately 50%), but it develops more rapidly. This different behaviour should be considered in structural design to avoid the cracking of young concrete.

The strain, due to shrinkage, is partly irreversible as it is developed during the hardening process. Other parts may be recovered when the humidity of the environment is increased. It is generally assumed that the final value of shrinkage depends upon:

- total Portland cement content in the mixture composition and its properties;
- w/c ratio and amount of water loss during hydration and hardening;
- conditions of curing at the early age of cement hydration;
- environmental conditions in service life;
- internal structure of material in which inert inclusions (aggregate grains, fibres) may reduce the volume changes and control the deformations.

Final values of strain due to shrinkage may be predicted from various empirical formulae, based on different assumptions and simplifications, some of which were developed many years ago. The equations shown below should be considered as examples. Guyon (1958) proposed the following formula:

$$\varepsilon_{sh} = \frac{1300 V_w}{E V_s} \log \vartheta \qquad (11.2)$$

where V_w and V_s are volumes of water and solid part, respectively, in the concrete composition, ϑ is atmospheric relative humidity (RH) and E is Young's modulus.

Three different formulae were proposed by L'Hermite (1955, 1957), each related to other measurable quantities:

$$\varepsilon_{sh} = \frac{\varepsilon_{cp}}{\dfrac{2800}{V_c} - 1.3} \tag{11.3}$$

here ε_{cp} is shrinkage of neat cement paste and V_c is the cement content,

$$\varepsilon_{sh} = \varepsilon_{cp} \frac{V_c}{V_c\left[1 - \left(V_c + V_w + V_p + V_d\right)\right]a_{sh}} \tag{11.4}$$

here V_w, V_p and V_d are contents of water, pores and smallest particles, respectively, and a_{sh} is a coefficient depending on the quality of aggregate – its value is smaller the more compressible is the aggregate. For siliceous aggregate and continuous sieve curve a_{sh} is between 0.8 and 1.0.

The last equation was obtained from the preceding one for medium Portland cement content V_c between 300 and 400 kgs/m³:

$$\varepsilon_{sh} = b_{sh}\left(V_c + V_w + V_p + V_d\right)^{1.5} \tag{11.5}$$

here b_{sh} is another empirical coefficient and for exponent different authors proposed values from 1.32 up to 1.52.

The development of shrinkage in time is dependent on the duration of moisture curing, humidity of environment and on the dimensions of the particular element. In small elements, the rate of loss of water is quicker, yet in the large ones it may last years. The respective curves of deformation over time have the general form as it is shown in Figure 11.13. Swelling may be expected in fully saturated air with RH = 100% or when specimens are in water. The rate of shrinkage is the main factor, which determines the importance of possible cracking and other destructive processes in concrete structures, because it is developing simultaneously with hardening; early shrinkage cannot be supported by fresh concrete with low tensile strength.

Typical values of shrinkage of a neat cement paste as indicated by Neville (1997) are:

- 1300.10⁻⁶ after 100 days (env. 3 months),

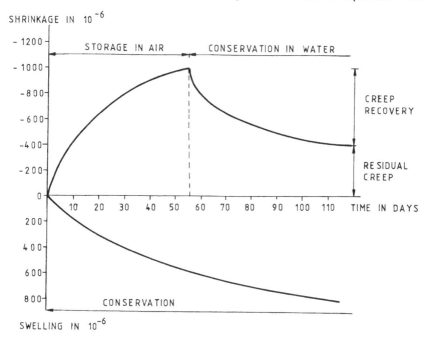

SHRINKAGE IN 10^{-6}

STORAGE IN AIR   CONSERVATION IN WATER

CREEP RECOVERY

RESIDUAL CREEP

TIME IN DAYS

CONSERVATION

SWELLING IN 10^{-6}

Figure 11.13 Characteristic development of shrinkage and swelling of concrete elements in average conditions.

- 2000.10^{-6} after 1000 days (env. 3 years),
- 2200.10^{-6} after 2000 days (env. 6 years).

Corresponding values for concrete are 10–20 times smaller because of the introduction of aggregate as an inert component and the stiffening inclusions. Also, curing in high humidity is a way to decrease the final values of shrinkage because the cement paste may reach a further stage of hydration. Immediate cure is necessary for all kinds of concretes with low w/c where early-age shrinkage is responsible for excessive cracking as concrete has not gained enough strength (Holt 2005).

These data are in agreement with old Guyon's (1958) general indication on percentage development of shrinkage: 33% after 10 days; 50% after 1 month; 90% after 1 year, and 100% after 2 or 3 years.

The influence of secondary components is more complicated: the addition of fly ash may decrease the autogeneous shrinkage, but increase the drying shrinkage. Partial replacement of Portland cement by fly ash decreases early strength, therefore cracking due to shrinkage may be enhanced. The addition of SF may increase autogenous shrinkage because the pore system is refined, while reducing drying shrinkage. The values of shrinkage strain may be mitigated by several measures (Kovler and Zhutovsky 2006):

- special brands of low-shrinkage cements with increased amount of gypsum, partial replacement of cement by GGBS (Jianyong and Yan 2001);
- expansive additions together with high belite cement (Collepardi *et al.* 2005);
- admixtures that reduce the influence of both mechanisms related to drying shrinkage – surface tension and capillary tension; these are so-called shrinkage reducing admixtures (SRA – a blend of propylene glycol derivatives) (Folliard 1997; Bentz and Jensen 2004; Rongbing and Jian 2005);
- addition of SF that decreases drying shrinkage but may increase autogenous shrinkage (Brooks and Megat Johari 2001);
- internal curing of concrete.

That last measure merits more detailed description. Internal desiccation of concretes with low values of w/b ratio may be reduced by application of pre-soaked lightweight aggregate (LWA) grains or super-absorbent polymers (SAP) as water reservoirs. Such particles are dispersed in bulk concrete and water is used gradually for cement hydration, thus acting in a somewhat 'intelligent' way.

The capillary forces of the cement paste are high enough to absorb water from LWA or SAP grains and to transport it to the drier cement paste regions, where the reaction with unhydrated cement may advance. The suction forces in the capillary pores are inversely proportional to the radius; the smaller the capillary pores in the cement paste the higher are the suction forces. In Figures 11.14 and 11.15, particular zones around aggregate grains are visible where moisture from grains is active in cement hydration. With advanced hydration of cement and increasing density of the structure the transport of water and vapour slows down and stops when the relative humidity in the LWA grain and in the hardened cement paste are in equilibrium (Weber and Reinhardt 1997; Jóźwiak-Niedźwiedzka 2005; Kovler 2008).

More efficient SAPs may have a water absorption capacity up to 5,000 times their own weight, and particles of such materials are dispersed in concrete during mixing. Various kinds of SAP are available on the market (Bentur and van Breugel 2003; Lura *et al.* 2006).

In recent years, extensive tests on shrinkage of high performance concretes were executed when these kinds of concretes were studied extensively (cf. Section 13.4). The research programs were closely related to the application in outstanding structures where delayed deformations were of great importance. In high performance concretes with $f_{c28} \geq 60$ MPa and very high performance concretes with $f_{c28} \geq 100$ MPa, the final values of shrinkage are smaller than in ordinary concretes. Also, cracking in the external layers is reduced due to the smaller amount of free water. However, a higher rate of shrinkage may occur, for example, for high performance concrete with SF as admixture and $f_{c28} = 80$ MPa already at approximately 70% of shrinkage after 10 days may

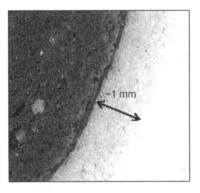

Figure 11.14 Coloured corona expanding around LWA aggregate saturated with weak ink solution and cast in the white Portland cement paste of water cement ratio 0.3; Lura *et al.* (2006).

be expected. This explains why extensive cure in high moisture is much more important than for ordinary concretes.

The reinforcement of a cement-based matrix with different kinds of fibres does not significantly modify shrinkage and various authors obtained different test results: a small decrease of 10–20% or no influence even with relatively high fibre volume. In contrast, fibres of high Young's modulus

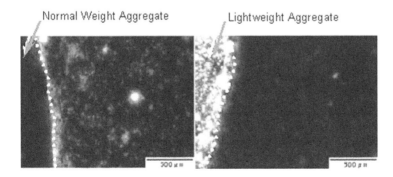

Figure 11.15 Fluorescence microscope images of internal transition zone between cement paste and normal weight aggregate (left) and saturated LWA (right). Lura *et al.* (2006).

greatly improve crack resistance of concrete and allow a decrease in the maximum crack width and cracks due to shrinkage.

The differences in conclusions based on experimental results may be explained by the complicated influence of fibres on the matrix: the increase of tensile strength combined with an increase of porosity due to modification of workability. Therefore, all comparisons are difficult in the sense that they concern slightly different materials, even if authors try to maintain the same composition and conditions.

The influence of polypropylene (PP) fibres on early shrinkage (plastic shrinkage) cracking was tested by many authors. A rather low fibre content, 1.0 kg/m³ or so, is considered sufficient to prevent early shrinkage cracking. Various kinds of synthetic fibres are used to control cracking (cf. Section 5.6).

The difference in deformations of neighbouring elements or layers due to different shrinkage induces tensile stress, which also may exceed local strength. This occurs, for example, between the external layer and internal core of every cast concrete element, because conditions of water loss are different if careful humidification is not ensured during curing. The phenomenon called 'differential shrinkage' also occurs when a layer of fresh cement-based mix is put on an old concrete element during repair.

Tensile stress caused by the shrinkage of cement-based matrix is related to existing constraints, which preclude free deformations. When the tensile stress exceeds concrete strength, cracking appears, and this is detrimental for strength, durability and aesthetics of concrete elements. There are a number of methods to help decrease this influence of shrinkage:

- reducing the amount of Portland cement in the mixture composition, e.g. by partial replacement of Portland cement with other binders, by application of special brands of low-shrinkage cements with appropriate amount of gypsum, etc.;
- adequate design of material structure and composition, with correctly selected aggregate and introduction of fibres;
- intensive cure in high humidity during long periods, immediately after casting and at least a few days later, or by internal curing;
- careful structural design allowing for free displacement of particular elements without unnecessary constraints.

Deformations and cracking due to shrinkage are tested on specimens in the form of small beams or plates, but most frequently the so-called 'ring tests' are carried out (Figure 11.16). Specimens are cured in constant conditions and deformations are measured at determined intervals. Data on the first crack appearance are used to characterize the material from the shrinkage viewpoint (Grzybowski and Shah 1990; Shah and Weiss 2006). Determination of shrinkage deformations of specimens in variable ambient conditions provides new information on the time-dependent behaviour of concrete structures (Vandewalle 2000).

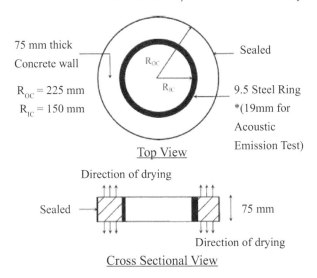

75 mm thick Concrete wall

$R_{OC} = 225$ mm
$R_{IC} = 150$ mm

R_{OC}
R_{IC}

Sealed

9.5 Steel Ring
*(19mm for Acoustic Emission Test)

Top View

Direction of drying

Sealed

75 mm

Direction of drying

Cross Sectional View

Figure 11.16 Geometry of the ring specimen. (Reproduced with permission from Shah H. R. and Weiss J., Quantifying shrinkage cracking in fiber reinforced concrete using ring test; published by RILEM Publications S.a.r.l., 2006.)

(b) Creep

The creep of the hardened cement matrices is explained by the gradual transfer of external load from the solid skeleton and the water in capillaries to the skeleton alone, due to the evacuation of water. By definition, creep of a material occurs when the deformations are increasing under constant stress, tension or compression. The strain due to creep and designated ε_{cr} are represented as additional to that which appears immediately after loading and which is called instantaneous strain ε_i. The opposite situation, when the stress is decreasing with no strain, is defined as relaxation and in fact it is another form of creep when constraints for displacements and deformations do exist. Extensive reviews on possible sources of creep and their respective relations are given by Mindess *et al.* (2003), among others.

Tests and observations carried on by several authors have led to the following general conclusions concerning the deformations due to creep in cement-based composites:

1   Creep depends directly upon the imposed stress but for stress above 50% of the material strength the relation is highly nonlinear.
2   Creep is influenced by external humidity and is higher in dry cured specimens. A part of creep is called 'drying creep' and is probably caused by the loss of water so that its separation of shrinkage is doubtful.
3   The quality and nature of concrete and its components also have an influence on creep.

4   Values of strain ε_{cr} are comparable to ε_i and under high stress may be equal to 3 ε_i.
5   As a consequence of the above, the age of concrete when the load is applied is of great importance, particularly for young concretes when it depends on their degree of hardening.

The phenomenon of creep has different consequences in structural design, for example:

- the creep of concrete is one of the sources of decrease of the prestressing force in post-tensioned elements;
- creep helps to decrease the importance of local stress concentrations and to distribute internal forces in the structures.

When the amount of aggregate is increased then a lower creep of mortar and concrete may be expected, but the nature of aggregate is also relevant; concretes made with sandstone aggregate exhibit higher creep than those with limestone. Creep is lower when various kinds of lightweight aggregate are used, which maintain humidity. For compositions with ordinary Portland cement, creep is higher than for high alumina cement and rapid hardening cement. Also, the application of air-entraining agents and plasticizers may increase creep.

The typical development of creep in concrete specimens is represented by curves in Figure 11.17 where different possibilities are indicated and delayed deformations are combined with instantaneous ones on an example of a

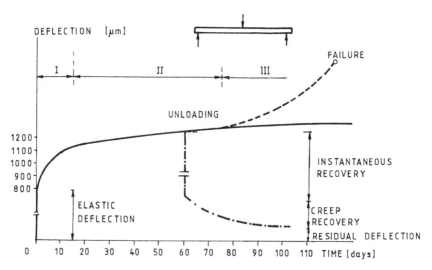

Figure 11.17 Creep of plain concrete beam in flexure as example of creep development in time. Consecutive stages and characteristic features are specified.

plain concrete beam subjected to flexure (Brandt 1965). After instantaneous deflection, stage I creep is observed in which deflection rose at a decreasing rate. Next, deflections are also slowly increasing in stage II of quasi-stabilization. Then two possibilities are demonstrated: either (i) an effective stabilization to a certain constant value of deflection is achieved; or (ii) in stage III a rapid increase of deflection occurs, leading to a failure. The unloading at a given age causes instantaneous recovery and creep recovery with certain residual deflection. More complex behaviour may be also hypothesized with modifications of environmental conditions and with various sequences of loading and unloading, but according to usually adopted assumptions the final results may be deduced from the principle of superposition of effects caused by different actions. In this research a step-wise curve was obtained at early age after loading (Figure 11.18), which can be interpreted as a result of crack and slippages in the material structure.

There are several published expressions for creep $\varepsilon_{cr}(t,T)$ as a function of time t since loading and age T of cement-based matrix at the loading. The creep $\varepsilon_{cr}(t,T)$ may also be related to the final value of creep $\varepsilon_{cr}(\infty,T)$, based on an assumption that the creep rate is in a certain way proportional to a creep value still to be developed. In that approach, the decreasing creep rate was taken into account. One of the expressions proposed by Glanville (1933)

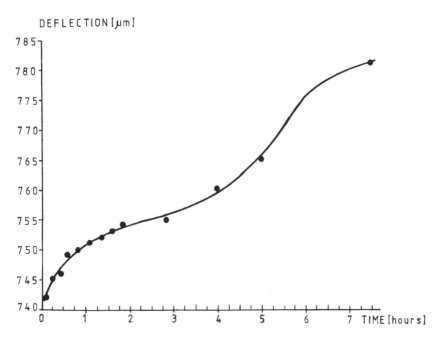

Figure 11.18 Step-wise curve reflecting the flexural creep immediately after loading, after Brandt (1965).

and Le Camus in 1947 maintains its validity even after many years and has the following form:

$$\varepsilon_{cr}(t,T) = \varepsilon_{cr}(\infty,T)\left[1 - e^{-(k_1 \log((T+t)/T)+k_2 t)}\right], \tag{11.6}$$

where after Glanville $k_1 = 0.62$, $k_2 = 0.30$ and after Le Camus $k_1 = 0.75$, $k_2 = 0.28$. Similar formulae were later published by many authors:

$$\varepsilon_{cr}(t,T) = \varepsilon_{cr}(\infty,T)\left[1 - e^{-A(t-T)}\right]^B, \tag{11.7}$$

here again A and B are constants to be determined after experimental data. The age of concrete is not considered here in a direct way.

A different form of function was indicated in French recommendations (BPEL and BAEL) for prestressed and ordinary reinforced concretes for creep deformation:

$$\varepsilon_{cr} = \varepsilon_i K_{cr}(T) f(t-T), \tag{11.8}$$

where K_{cr} is the creep coefficient understood as ratio of creep to instantaneous strain $\varepsilon_i = \sigma_i / E$, and

$$f(t-T) = \frac{(t-t_0)^{1/2}}{B + (t-t_0)^{1/2}} \tag{11.9}$$

A different expression was proposed by the ACI 209R-92 (2008) and a so-called creep coefficient was used:

$$\frac{\varepsilon_{cr}}{\varepsilon_i} = A\frac{(t-T)^D}{B + (t-T)^D} \tag{11.10}$$

where B and D are constants and A is the ultimate creep coefficient,

$$A = \frac{\varepsilon_{cr}(\infty,T)}{\varepsilon_i} \tag{11.11}$$

(see also ACI 209.1R-05)

The test results and models adopted for shrinkage and creep were verified again when new kinds of cement-based composites (high performance concrete (HPC) and others) were introduced. It appeared that in fibre-reinforced concretes and in high performance concretes both the final values and development in time were different. After Acker and de Larrard (1992) creep coefficient K_{cr} in equation (11.8) varies from 1.05 to 1.67 for high performance concrete specimens sealed to avoid drying and from 1.96 to 2.96 for non-protected specimens. In the case of ordinary concretes the standard value of K_{cr} is close to 1.0 and between 3.0 and 4.0, respectively. The coefficient B varied from 1.7 to 11 for high performance concretes and is equal to 10 for ordinary concretes.

Polymer cement concretes (PCC) usually exhibit higher creep, which may even increase in elevated temperature. This is not the case for polymer impregnated concretes (PIC) in which a hardened skeleton does not allow for increased creep, particularly when a low volume of polymer is used for impregnation.

Apparently, dispersed fibres do not appreciably modify the behaviour of cement-based elements subjected to creep. There have been few experimental studies of this problem and the results obtained are inconclusive. As mentioned above for shrinkage, modification of the internal structure of composite materials causes quite complex results and comparisons are difficult. In several reports, similar behaviour of plain and fibre-reinforced elements was observed under creep conditions, both qualitatively and quantitatively. However, extensive studies executed by Swamy *et al.* (1977) proved that fibre-reinforcement decreased creep considerably. Similar conclusions were formulated by Brandt and Hebda (1989) who tested elements under eccentric compression in long periods of time. The creep was small for reinforced specimens. The most important factor relating to the final creep values was the level of load with respect to the material's strength and that factor was also influenced by the volume of fibre reinforcement. That was the reason why the creep recovery measurements carried out by different authors did not furnish consistent results.

Creep tests performed on gypsum specimens reinforced with glass fibres by Bijen and van den Plas (1992) have shown similar behaviour to that of cement-based concretes.

A prediction of the final values of shrinkage and creep is needed to determine when a state of certain stabilization is achieved. However, perfect stabilization is possible only in artificial conditions created in a laboratory, because in the natural environment the hygrothermal conditions vary and a flow of heat and moisture to and from hardened cement paste is continued indefinitely. Predictions may also concern the development in time and the rate of both processes.

Various sets of data may be considered as bases for prediction, but all are not always available: mixture composition, hygrothermal conditions of cure and during service life, values characterizing shrinkage and creep as measured over a short period of time, etc. Furthermore, most of the data are subject to stochastic distribution and are random variables.

The reliable formulae and extensive test results for polymer and fibre-reinforced composites are scarce. Simple relations for the prediction for shrinkage and creep of ordinary concretes for application in structural design were given already in CEB-FIP Model Code (1990) with coefficients to allow for actual conditions. Later, so-called Model B3 was recommended by ACI (1999) and extensively discussed by Bažant and Baweja (1995). A slightly different approach was proposed in EN 1992 (2004) and detailed comments may be found in Vandewalle (2000).

Further studies of shrinkage and creep are necessary because of development of various kinds of high performance concrete with extensive use of superplasticizers, SF, and other secondary components. A great variety of admixtures are available in the market, which may have different effects when used together with various kinds of Portland cement and aggregate. All this creates new fields for investigation. In general, considerable reduction of final values of shrinkage and creep may be obtained in new kinds of concrete. For practical application in the design of concrete structures, detailed recommendations are available in EN Standards.

11.4.3 Strength under sustained load

The notions of sustained load and of strength under sustained load are of great importance for obvious reasons: in all structural applications the cement-based composites are subjected to long-term states of loading. As mentioned in Section 11.3, the phenomenon of fatigue occurs not only under cyclic loading, but a sustained load may also cause failure under an average stress below the short-term static strength of the material.

Rüsch (1960) has already shown the influence of the duration of loading on the strain-stress curve and strain development over time. In Figure 11.19 a few curves represent the results of axial compression imposed by constant strain rate on concrete specimens. For slow loading the influence of creep and propagation of microcracks is important, while for higher loading rates strength is apparently increased (cf. Section 11.2.1).

In Figure 11.20 curves for different values of ratio σ_c/f_c show development of strain in time for compressed specimens. It appears that for maximum load when $\sigma_c/f_c = 1$ failure occurred in about 20 minutes, for $\sigma_c/f_c = 0.8$ after a few days and for lower values of this ratio the increase of strain was stabilized and no failure was observed. The same test results are presented in Figure 11.21 in a slightly different way.

The long-term strength is a result of two processes: (i) the increase in strength of Portland cement paste due to the progressive hydration of cement, with the possible healing of cracks under compression; and (ii) the accumulation of local damage due to the slow growth of microcracks. Both processes develop differently in ordinary concrete and in high performance concrete. In fact, the relative increase of strength after 28 days is smaller in high performance concrete because of its higher rate since the beginning with

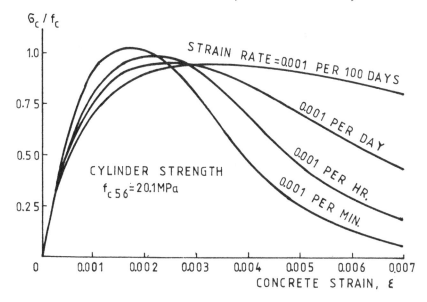

Figure 11.19 Strain-stress curves for various strain rates of axial compression of concrete specimens at an age of 56 days, after Rüsch (1960).

lower *w/c* ratio and the addition of SF and other admixtures. Moreover, the non-homogeneity of the stress-strain field and resulting redistribution of stress due to creep should be considered (Ožbolt and Reinhardt 2001). The process of the increase of strength with age is closely related to hygrothermal

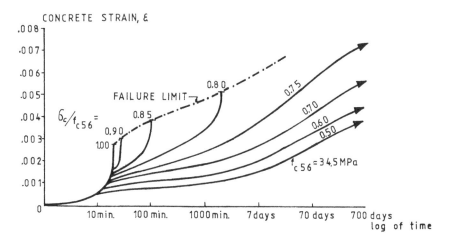

Figure 11.20 Development of strain in time under sustained compression of concrete specimens with different ratio of σ_c/f_c, after Rüsch (1960).

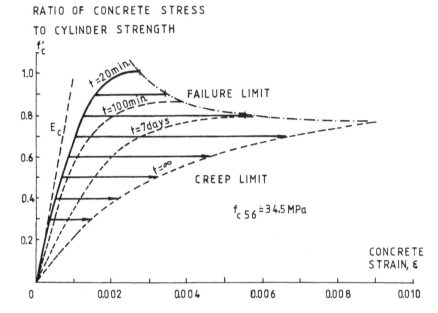

Figure 11.21 Influence of ratio σ_t/f_c and duration of load on concrete strain under compression, after Rüsch (1960).

conditions during hydration and hardening, which cannot be precisely predicted during design of structures.

It is generally believed that sustained strength ranged between 70% and 95% of the short-term strength. Similar limits for tensile strength may be suggested, but no reliable test results are available and in the case of important structures, the experimental verification is recommended in local conditions and with actual materials.

The application of dispersed fibres and polymers creates a more complex internal structure and certain control in the cracking process may increase relative long-term strength. There is, however, not enough experimental evidence to propose general indications. In many cases, special tests of representative specimens are needed.

11.5 Durability

11.5.1 Definitions and importance of durability

Durability may be defined as a property of a structure to fulfil its expected function during foreseen service life without the necessity of excessive repair works, or as 'the ability of a building and its parts to perform its required function over a period of time and under the influence of agents (that cause deterioration),' (BSI 1992). The definition can also be presented in several different ways, but the main element – the maintenance of safety and serviceability – is always included.

The period of time in which durability is ensured is more-or-less consciously related to the lifetime of the structure, which is not precisely defined for buildings and the majority of civil engineering structures, as is often the case, for example for structural elements of aircraft and ships. The durability of a structure is considered adequate if it fulfils its intended functions related to serviceability, strength and stability, without excessive and unforeseen maintenance throughout the intended lifetime for which it was designed. The durability is closely related to an important requirement imposed on our civilization – sustainable development. For some 20 years, the durability of structures is considered in a larger aspect: as a lifecycle design of structures, i.e. taking into account all stages of the service of a structure, including its construction and demolition.

It is perhaps more correct to discuss the long-term behaviour of the materials and keep the term durability for elements or structures, because the same material may be considered durable under one set of conditions, and not durable under other sets. Durability of materials is, however, a universally accepted term which is understood correctly, through analogy, to the abovementioned definitions of the durability of structures. That is why the term durability, even though not exactly correct, is also used here to describe a material's properties in long-term exploitation. In this chapter, durability is limited to the behaviour of materials and as far as possible any relation to structures is not developed. The durability of a structure depends also upon several non-technical conditions that are not considered here; the main lines of modern approach are presented by Brandt and Kucharska (2001).

Durability is a very important feature of the general performance of materials, and performance is the behaviour of a product related to its use. From both technical and economic reasons the durability, in the full meaning of its term, should be taken into account in design, selection and execution of brittle matrix composites as materials for building and civil engineering structures.

The durability is always examined in situations when certain destructive processes are to be foreseen as unavoidable due to environmental influences and as a result of exploitation. However, one particular aspect of all damage and destruction processes should be examined carefully for every practical problem: its rate. The rate of a destruction process decides whether the effects are or are not dangerous during the lifetime of the structure. Consequently, the appropriate measures to prevent appreciable negative results should be applied to ensure the safety and serviceability of the structure (Sarja 2000; Naus 2003). Analytical, experimental and simulative methods are used to solve the problem of how to predict the performance of a structure in given conditions. In those methods, the interaction of damage modes should be considered properly and incorporated in predictions.

All processes involved in the behaviour of materials and in their durability are random, and have uncertain causes and effects. Not only are the composition of a material and its initial properties subjected to random variations, but also the occurrence and intensity of exposure of the structure and of all external actions are random. There is only a limited amount of data from experience and tests concerning the ageing of materials, actions on structures, etc., and there is not enough data for a complete statistical analysis. Finally, the extent of simplifications in description and modelling of the various phenomena and of cause and effect relationships is unknown. These are the reasons why an analysis of durability of composite materials, taking into account their stochastic nature, are always only approximate, and all attempts in that direction are based on several simplifications. Durability may be considered in a deterministic way by the application of appropriate safety coefficients in practical calculations in order to allow for unfavourable values for certain random parameters.

Selected groups of problems related to durability and the long-term behaviour of concrete-like composites are examined below. Particular questions related to the alkali-aggregate reaction (AAR) in concrete and concerning the durability of glass fibres in the Portland cement matrix are described in Sections 4.2.2 and 5.3 respectively.

11.5.2 Durability as a quality of cement-based materials

The main conditions for high durability of a cement-based composite may be summarized as high impermeability and density of the matrix, chemical and thermal compatibility of all material components and the existence of an internal structure that can control microcracking (Mather 2004). Density and impermeability of the matrix may be achieved by appropriate material design and often by application of secondary and ternary binder blends with fly ash, SF, metakaolin, etc. (cf. Chapter 4). For the control of cracking, adequate material composition and appropriate curing are necessary. Moreover, excessive stresses and detrimental external actions should be avoided or reduced by adequate structural detailing.

Cement-based materials are subjected to complicated processes during ageing and exploitation, which influence their properties. Hydration and hardening of cement paste continue over long periods of time, but at a decreasing rate during the entire lifetime of a structure. In favourable conditions, it is believed that the strength of the cement matrix is continuously increasing. All other processes in materials during their ageing tend to be somewhat detrimental and are caused by various kinds of external actions of a mechanical, chemical, biological and hygrothermal nature related to exploitation and environmental conditions. All these actions have a negative influence on material durability, particularly if their intensities are increasing beyond certain limits. The following are some examples:

- corrosion of reinforcing steel and swelling of the corrosion products;
- freezing and thawing, heating and cooling, and soaking and drying

cycles, which cause excessive internal forces in a heterogeneous material structure and in constrained elements;

- processes that produce additional internal stresses, like shrinkage and swelling of matrix, chemical reactions with aggregates (AAR), etc.;
- penetration of CO_2, Cl ions, sulphates and various corrosive fluids and gases, which cause destructive processes in matrix or reinforcement, also UV radiation for certain materials;
- fatigue due to long-term or repetitive application of excessive loads and other actions;
- abrasion and wear due to normal exploitation;
- leaching, cavitation and erosion related to the kind of water and to the intensity of flow.

These effects appear separately or simultaneously and their results may be summarised with negative synergism in the form of cracks, spalling and scaling, discoloration, etc., which require repair. Advanced cracking and corrosion are dangerous for the safety of structures because of the decrease in their bearing capacity. All detrimental effects usually increase the rate of further destruction in the form of a feedback; for example, local cracks or spalling open the way to accelerated corrosion of reinforcement and cracking of the matrix. Here also, various cases that qualify as misuse are to be considered: overloading, local accidental impact, etc. Corrosion of reinforcing steel is considered as the main single cause of the deterioration and destruction of reinforced concrete structures. This is particularly frequent in the marine environment.

There is an appreciable difference between ordinary concretes with w/b ratio equal to 0.5 or higher, and high performance concretes with $w/b \leq$ 0.4. HPC are characterized by their very dense microstructure and if microcracking is controlled by dispersed fibres and adequate curing immediately after casting, then very low permeability is obtained, which is necessary for improved durability. In contrast, in ordinary concrete with $w/b \geq 0.6$ and executed often with improper placing technique and poor curing, there are open systems of pores and microcracks exposed to all external chemical and physical agents. It is possible to combine low- or medium strength concretes, with their improved durability, even in harsh environments (Aïtcin 2003). Various kinds of special concretes, such as strain hardening fibre reinforced cementitious composites, may be used for particularly exposed elements, together with steel reinforcement to improve the impermeability of the external layers against the ingress of chloride ions, and increase its durability (Ahmed and Mihashi 2007).

To represent the process of concrete structure performance as a function of time, a two-stage model was proposed by Mehta (1997, 1999), as shown in Figure 11.22. Stage 1 represents normal exploitation and at its end some kind of damage is initiated. During stage 2, damage is developed due various actions and finally the critical level D of damage is reached. For durability of a

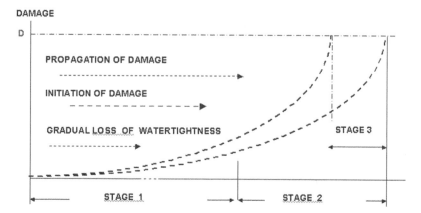

Figure 11.22 Model of two-stage damage of concrete: stage 1 – induction of damage, stage 2 – development of damage. Stage 3 is added to show that damage process reached D value, after Brandt and Kucharska (2001) and Mehta (1997).

structure, it is essential that stage 1 should be maintained as long as possible, because at initiation of stage 2 the structure is condemned to reach a final damage. Stage 3 is in fact already the ultimate limit state of a structure.

The durability of concrete and concrete structures is related not only to the solution of chemical and physical problems, but also to economic aspects. This subject is discussed in Section 14.6.

The maintenance of concrete structures comprises more or less continuous execution of small repairs and discrete major repairs, which often take a large part of the financial resources allowed by the owner. The following groups of maintenance operations may be defined and scheduled during the ageing of the structure and its decreasing performance:

- corrective maintenance (occasional repairs, cleaning and replacements);
- scheduled maintenance (periodic inspections and repairs);
- rehabilitation (restoration of initial performance);
- upgrading (improvement of initial performance).

The initial cost of a structure, the cost of exploitation and maintenance, together with cost of demolition and dismantling a structure, are all components of the total cost. How to properly split the total cost into particular allocations is an economic problem rather than a technical one. It is obvious that with a higher initial expenditure for rational improvements of materials and structure quality, cheaper maintenance may be expected. The cost and quality of concretes is only a part of the problem of structural durability. The repair of concrete structures with new cementitious materials is briefly considered in Chapter 14.

11.5.3 Durability with respect to carbonation

Carbonation is a process in which calcium oxide CaO and calcium hydroxide $CaO \cdot 2H_2O$ in hardened cement paste are converted to calcium carbonate by carbon dioxide CO_2 penetrating by diffusion from the atmosphere into the system of pores and microcracks:

$$CO_2 + Ca(OH)_2 = CaCo_3 + H_2O$$

In this reaction, the alkaline hydrates take parts that contain Ca, NaOH and KOH, producing additional micro-crystals of calcium carbonate. The carbonation products fill the matrix structure and reduce the porosity and the specific surface area. Finally, there are two effects of carbonation:

1   Density of cement paste increases as does its compressive strength and Young's modulus;
2   Passivity of inter-crystalline water decreases considerably, from $pH \geq 12.5$ sometimes to $pH \leq 9.0$.

If the first effect can be considered beneficial for the durability of the matrix itself, the second one increases the danger of accelerated corrosion of steel-reinforcement, which is by far the most important reason for deterioration of reinforced concrete structures exposed to open air. The steel is prevented from corrosion as long as the passive oxide film is maintained, but the carbonation of concrete destroys that film. The depth and density of the cover concrete over the steel reinforcement decides on the structure's durability.

The final effects of carbonation are strongly dependent on the quality of Portland cement and on the permeability of the hardened cement paste. Cements with a small amount of alkalis are less exposed to carbonation, and the same concerns cements blended with fly ash and blast-furnace slag. Carbonation occurs at the highest rates at relative humidity about 40–70%. Near 0% or 100% there is little or no carbonation.

Large capillary pores and microcracks increase the carbonation rate particularly in open air and humidity between 50% and 70% RH. Moreover, it has been observed that cyclic soaking and drying with exposure to an atmosphere with high CO_2 content produces quicker carbonation than the less variable conditions. It is therefore concluded that with a low w/c ratio, for example, below 0.25 as for high performance concrete, the carbonation is very slow and does not appreciably decrease the durability of reinforced concrete structures. The long cure of a fresh concrete in high humidity increases its density and only reduced carbonation may be expected. Specimens made with high performance concrete of compressive strength equal to 65 MPa or more and a w/c ratio ≤ 0.35 do not exhibit any measurable carbonation after accelerated tests that are equivalent to several years of natural exposure. When subjected to the same testing procedure, specimens made with the same components and workability, but having a higher w/c ratio > 0.5 and lower strength of an order of 40 MPa, have shown 13 mm and 23 mm of carbonation depth after moisture and dry curing, respectively (Lévy 1992).

The dependence of advance of carbonation and the compressive strength of concretes f_c is justified by the relation of both properties and the cement matrix density and strength. The tests were performed by Nischer (1984) on various concretes of different quality subjected to open air during a three-year period. The results are summarized in Figure 11.23 and confirm earlier established relations shown by Schiessl (1983) in the form of straight lines for various w/c ratios representing carbonation depth as a function of time (Figure 11.24). Carbonation is a process with a decreasing rate, because the density of external layers increases with carbonation products. There are several models proposed that describe its progress in time, for example, Steffens *et al.* (2002).

The testing of concrete specimens with respect to the level of carbonation is carried out by applying a phenolphthalein solution to a freshly fractured or sawn surface. Noncarbonated areas become red while carbonated areas remain grey. The rate of carbonation in ordinary concrete elements exposed to the atmosphere is schematically shown in Figure 11.25. It is assumed that the depth of cover d of steel-reinforcement should be bigger than the depth of carbonation d_c estimated after 100 years to ensure safety of reinforcement (CEB 1992). In normal conditions, half of that depth can be reached within 15 years. A simplified formula proposed for the carbonation rate is: $d_c = 10\,b\,\sqrt{t}$, here d_c is in mm and t in years; b is a numerical coefficient that characterizes the quality of concrete, for example, for a very good quality concrete $b = 0.15$. Carbonation is usually deeper on those sides of a structure that are exposed to sunshine because the process of conversion into calcium carbonate $CaCO_3$ by carbon dioxide CO_2 is quicker in frequently variable hydro-thermal conditions.

The predictions for d_c are based on accelerated tests and on experience gained after the examination of old structures. These predictions are expressed in the form of standardized requirements for depth of reinforcement cover in concrete structures. In EN 1992 (2004) nominal values of concrete cover are given in a prescriptive way, considering all circumstances, namely, environmental conditions and quality of steel, while some modifications are left for particular decisions of member countries.

Another aspect of the phenomenon of concrete carbonation is described by Pade *et al.* (2007) who have shown that CO_2 is reabsorbed during the life cycle of cement-based concretes and mortars. Consequently, the contribution of the cement and concrete industry to net CO_2 emissions may be partly counterbalanced. The effect of carbonation is enhanced by the way the concrete structures are crushed and handled after demolition.

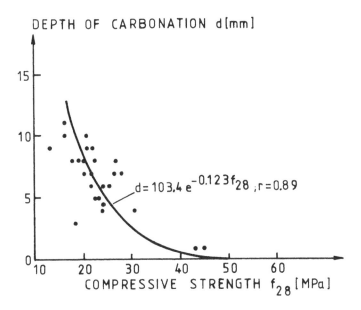

Figure 11.23 Influence of compressive strength f_c on carbonation depth after open air exposure during 3 years, after Nischer (1984).

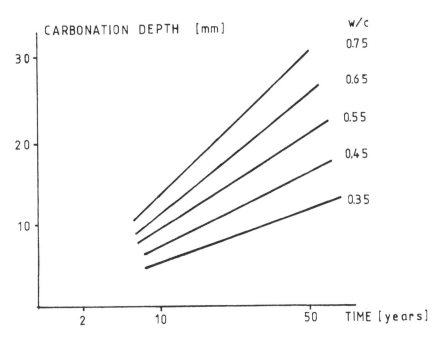

Figure 11.24 Carbonation rate as function of *w/c* ratio, after Schiessl (1983).

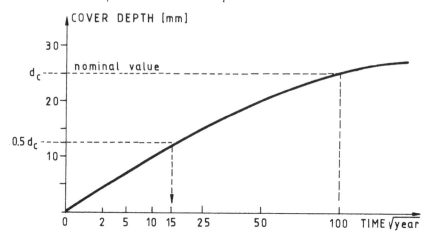

Figure 11.25 Rate of carbonation or chloride penetration depth in cement mortar, after CEB (1992).

11.5.4 Durability with respect to other external actions

The sensitivity of concrete structures to sulphate attack is strongly related to the exposure conditions. Structures in an environment of high sulphate content in the air or in water, for example sewage tunnels, are particularly vulnerable. After sulphate ions penetrate the pore system of cement paste, complex reactions start with C_3A leading principally to two kinds of processes: gypsum corrosion and sulphoaluminate corrosion (Mindess *et al.* 2003). The products of sulphate reactions with cement expand and can cause cracking and destruction. The permeability of the material's structure and the quality of cement decide upon the rate of these processes. Special Portland cements as well as high alumina cements may be used for elements exposed to sulphates (cf. Section 4.1.1).

Acid attack from air or water may produce the conversion of calcium in Portland cement paste into soluble salts. As a result, the binding capacity of hardened mortars and concretes is reduced.

Chloride ions Cl⁻ in concrete materials act as a catalyst for processes of corrosion of steel reinforcement. Their influence on cement paste itself is less important. A thin oxide passivation film around reinforcing bars provides electrochemical protection to the steel. When it is destructed by chloride ions the corrosion may be rapidly enhanced.

Chloride ions may be introduced to the fresh mix from mixing water, marine aggregate, or admixtures with chloride components. The chloride ions also penetrate to the hardened concrete materials from the air through pores, bleed-water channels, entrapped-air voids and cracks. The density of the material and its impermeability is decisive to improving the durability in the case of possible chloride ions attack under service conditions. The intensity of penetration of chlorides is considered as proportional to the value of *w/c* ratio. Density of concrete structures may be significantly increased with the application of various additional components like pozzolans; for example, ground granulated blast-furnace slag, or SF and metakaolin as additional binders. Kleen and Niepmann (2007), among others, reported positive results of tests and obtained considerable reductions of chloride penetration in ternary and quaternary blend binder concretes. It has been observed also that curing the concrete at +50°C increased the penetration of chloride ions from marine water if compared with concretes which were cured in environmental temperature.

High chloride content in the air, combined with soaking and drying cycles, create the dangerous conditions which may appear, for example, in industrial or marine environments in hot climates. Concrete structures in harbours and sea fronts at seaside localities are strongly exposed, but the wind direction may extend the danger zone far inland. When melamine-based superplasticizers are applied, then the risk of corrosion of steel resulting from chloride penetration is reduced, especially when sulphate-resistant and cement that is poor in C_3A is used. The rate of penetration of chloride ions and the depth of destroyed material are similar to or even greater than those that are characteristic for carbonation. Concrete exposed to the influence of the marine environment requires special precautions.

The electrical resistivity of concrete may indicate its susceptibility for corrosion of reinforcement as it characterizes the microstructure and is used to control the possibility of chloride penetration. The current is carried by ions dissolved in the pore liquid.

When concretes with the same moisture content are compared, resistivity is higher for the lower *w/c* ratio, for higher levels of hydration and for the use of additional components like fly ash, SF, etc. Resistivity increases with age and hardening of concrete because the pore system becomes less permeable; there is an inverse correlation between concrete resistivity and the chloride diffusion rate. This fact is exploited for the measurement of concrete resistivity in view of assessing its susceptibility to chloride penetration and, consequently, to corrosion of reinforcement for both its initiation and rate of progress. The resistivity of external concrete cover is the most important and is expressed by the formula:

$$(11.12)$$

$$\rho = \frac{RA}{L}, \; [\Omega m],$$

where R is measured resistance of a concrete element with a length L [m] and a cross-sectional area A [m²]. The resistivity of concrete varies in a wide range from 10 to 10^5 Ωm as a function of the moisture content and the material composition and structure.

The measurement is relatively simple and may be executed by many methods, in situ or in laboratory conditions, without any destruction of concrete (cf. Polder 2001). The most reliable measurement is based on the application of four electrodes equally spaced that are in contact with the concrete surface (Wenner or 4-point method) and alternating current (AC) is used. If the influence of the moisture is eliminated, then the results indicate values of the resistivity. The difficulties of such measurements are caused by the heterogeneity of concrete, the significant influence of moisture conditions and also by such details as quality of contacts, specimen shape, etc. For details readers are referred to specialized reports (cf. Chung 2003; Ferreira and Jalali 2006).

The use of dispersed fibre reinforcement is an efficient measure to control cracking and avoid premature corrosion of steel reinforcement. The durability of organic fibres as reinforcement may be endangered by the destructive action of alkaline pore solution or by biological agents. In unfavourable conditions, for example high alkalinity of cement, high environmental moisture and permeability of cement matrix, the result will be simply the disappearance of fibres and of their reinforcing effects. A certain number of failures of roofs made of cement sheets reinforced with vegetal fibres of local origin have already occurred in 1980s (Gram *et al.* 1984). Additional precautions may sometimes be difficult in local conditions, but they are necessary, and natural fibres cannot be used as a simple replacement for asbestos fibres, which have excellent durability in an alkaline environment (cf. Section 5.9).

Steel-fibres do not corrode in a cement-based matrix provided it is of adequate density and without cracks of excessive width. However, when fibres are partly exposed to external influences they corrode quickly and may spoil the appearance of a concrete surface with reddish dots. In the marine environment and other regions particularly exposed to corrosion, carbon fibres that do not corrode are particularly appropriate. The application of fibres and polymeric admixtures generally improve durability in the same measure as they increase impermeability and control cracking (cf. Chapter 5 and Chapter 9).

All kinds of chemical agents that attack the concrete-like materials are dangerous if allowed to penetrate into the material's structure and if there is enough moisture for chemical reactions. Impermeability of the matrix stops the penetration of chemicals and migration of water to such an extent that the reaction becomes slow with respect to the lifetime of the structure and the material has improved durability. As mentioned elsewhere, high performance materials with a low w/c ratio, low capillary porosity and high density are durable in conditions in which ordinary materials may exhibit sensitivity to destructive agents.

11.6 Behaviour in high and low temperature

11.6.1 Limits of temperature

The limits for normal exploitation of concrete-like composites in temperate climates are from +50°C to –30°C. The upper limit corresponds, for example, to a bridge deck or building cladding exposed to solar radiation in the summer time and the lower limit corresponds to an outdoor structure in winter. Both limits may be extended for polar and tropical regions by approximately 10–20°.

For special structures and accidental conditions, the above limits are extended considerably. For refractory elements a temperature of +300°C is admitted as normal. In the case of a fire in a building, the temperature of elements may exceed +1000°C. In the reservoirs for liquefied gases the temperature of a concrete wall may reach a level as low as –170°C, due to an accident. This is the range of temperature to which cement-based composites may be exposed occasionally, in the form of cyclic actions or in rare situations.

Besides the value of peak temperature, other conditions are also of importance: the rate of temperature variation (heating and cooling), the number of heating/cooling cycles, material humidity and heterogeneity of concrete components' behaviour in high temperature. Furthermore, there is no linear relationship between the temperature variation and its influence on a material's properties because there are several factors acting with different intensity and in opposite directions.

The behaviour of cement-based composites in a hot and dry climate is considered with respect to following processes:

1   major cracking due to excessive differences of temperature in neighbouring regions of large concrete elements and inefficient heat diffusion;
2   local cracking, spalling and fractures of the matrix due to internal stresses caused by expansion of material components and accelerated shrinkage.

Spalling is described as the detaching of layers or pieces of concrete from its surface when it is exposed to high and rapidly rising temperature. The vulnerability of cement-based composites to deterioration by excessive cracking due to thermal effects is increased. The influence of temperature should be taken into account in the curing of concretes and special measures are applied at early age cement hydration and hardening to avoid any damage. The cooling of aggregate and water, as well as special installations with cold water inside concrete structures, are sometimes applied. At early ages, elements are particularly exposed because of the high shrinkage rate and relatively low tensile resistance. Hot weather also accelerates processes of corrosion of steel-reinforcement in the presence of aggressive salt in the air. Basic modifications of the material mechanical properties due to elevated environmental temperature are to be considered. For information related to that part of concrete technology, the reader is referred to specialist publications, manuals and standards.

In cold seasons with temperature below +10°C, hydration of Portland cement may be slow or completely stopped. For special measures to ensure that concrete attains expected mechanical properties, the reader should again consult the abovementioned manuals and standards. The influence of low temperature on hardened concrete is discussed in Section 11.6.3 together with the scaling of concrete surface due to freezing and thawing cycles in the presence of de-icing agents.

11.6.2 Cement-based materials in elevated temperature

When analysing the influence of accidental high temperature on concrete structures, two situations should be considered separately: (i) behaviour of a structure during fire, that is, concrete strength and other mechanical properties; and separately, as another state (ii) the situation after the fire and the possible cooling processes. During a fire the structure may collapse because of a reduction in concrete strength. After the fire, the properties of concrete should be diagnosed in view of future exploitation and necessary repair, that is, the residual strength is of primary importance.

According to various sources (Hager and Pimienta 2004, ACI 216.1-97), there is a sequence of processes when cement-based materials are exposed to elevated temperature:

- evaporation of water over +100°C;
- destruction of cement gel due to dehydration at +180°C;
- dehydration and loss of the non-evaporable water at +250°C;
- first sizable degradation in compressive strength +200–300°C; at +300°C strength reduction is between 15% and 40%;
- decomposition of Portland cement clinker at +500°C;
- strength reduction by 55–70%;
- +550°C is critical as calcium hydroxide dehydration starts;
- transformation of quartzite at +570°C;
- decomposition of CSH at +700°C;
- decarbonisation of limestone aggregate over +800°C and decrease of strength to approximately 20%;
- melting of concrete components begins at +1150°C;
- complete destruction occurs at approximately +1200°C or +1300°C.

This schematic list does not contain all accompanying processes in cement paste and aggregates which occur simultaneously often as consequences of these above. Destruction of a material structure is accompanied by variation of colour, as shown schematically in Figure 11.26. The modifications of colour are used in forensic research to assess the extent of damage and to define scope of necessary repair.

At temperature over +100°C the cement paste exhibits a moderate expansion up to +150°C and then some contraction up to +600°C. At temperature over +600°C again expansion is observed. Examples of variation of the compressive strength of concrete as a function of temperature are shown in Figures 11.27 and 11.28. These are results of tests on specimens with lime-

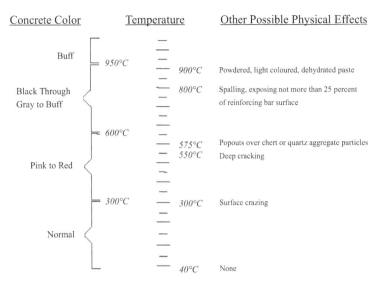

Concrete Color	Temperature	Other Possible Physical Effects

Buff

950°C

900°C   Powdered, light coloured, dehydrated paste

Black Through
Gray to Buff

800°C   Spalling, exposing not more than 25 percent
of reinforcing bar surface

600°C

575°C   Popouts over chert or quartz aggregate particles
550°C   Deep cracking

Pink to Red

300°C

300°C   Surface crazing

Normal

40°C   None

Figure 11.26 Visual evidence of temperature to which concrete has been heated. (Reproduced with permission from Georgali B. and Tsakirides P. E., Microstructure of fire-damaged concrete: a case study; published by Elsevier Limited, 2005.)

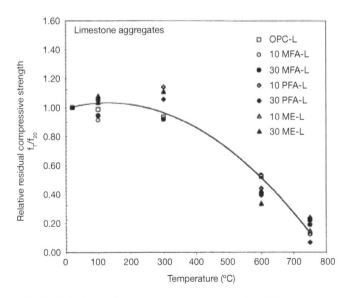

Figure 11.27 Relative residual compressive strength of limestone concretes, after Savva *et al.* (2005)

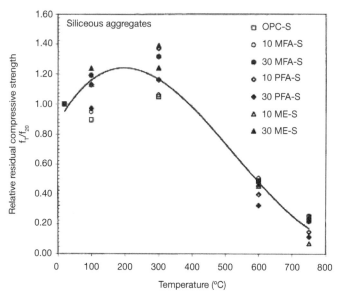

Figure 11.28 Relative residual compressive strength of siliceous concretes. after Savva *et al.* (2005)

stone and siliceous aggregates and with various kinds of fly ash as partial cement replacement. The Young's modulus decreases rapidly down to 5% of its initial value, and the Poisson ratio decreases between 0.15 and 0.23, down to 0.02 at temperature of +1100°C (Lau and Anson 2006).

Internal pressure in concrete is caused by the restrained expansion of various material's components that form heterogeneous cement-based composites. The value of linear coefficients α of thermal expansion of aggregate varies according to its mineralogical origin from 5 to 11×10^{-6} per 1°C and the approximate values are shown in Table 11.2. For example, the expansion of basalts at +100°C is approximately half of that of sandstones and quartzites. The values of the thermal expansion coefficient for plain cement paste vary between 10.8 and 21.6×10^{-6} according to Mitchell (1953). While these incompatibilities of expansions do not cause microcracks in the interface layers when the temperature is maintained below +100°C, significant destruction is produced in elevated temperature during fire. At over +400°C the differential expansions of aggregate grain decide upon the level of material destruction. Also, thermal deformation of steel reinforcement should be taken into account, and usually admitted constant value of respective coefficient α is an approximation, which is not acceptable in extreme temperature.

The value of α_c coefficient for a composite material results from values characterizing all components, taking into account their respective volume fractions. For example, the value of α_c coefficient for a concrete may be calculated from the equation proposed by Dougill (1968)

Table 11.2 Coefficients α of thermal expansion of rocks for temperature +20°C to +100°C

Types of rock	Values of α
Granites, rhyolites	$8 \pm 3 \times 10^{-6}$
Andesites, diorites	7 ± 2
Basalts, gabbros, diobases	5 ± 1
Sandstones	10 ± 2
Quartzites	11 ± 1
Limestones	8 ± 4
Marbles	7 ± 2
Slates	9 ± 1

after Skinner (1966).

$$(\alpha_c - \alpha_a) = (\alpha_p - \alpha_a)(1 - g)^n, \qquad (11.13)$$

where α_a, α_c and α_p are coefficients for aggregate, concrete and cement paste, respectively, g is the volume fraction of aggregate and n is a numerical coefficient, which is determined experimentally. Furthermore, the expansion coefficient for cement paste varies with its moisture content; namely it exhibits minimum for both extreme cases – very dry and completely saturated material. As a result, the thermal expansion of concrete composites and induced stresses may be greater than might otherwise be anticipated (Marshall 1990).

The resulting internal pressure may cause cracking if it exceeds the tensile strength of a cement matrix. As is described elsewhere, this tensile strength is relatively low, unless additional internal structures of fibres or polymer links are created. Therefore, at elevated temperature extensive cracking and dangerous spallings have to be foreseen if the internal pressure is not reduced by other measures. Improved mechanical properties of FRC due to steel fibre reinforcement are retained up to +1200°C. Also, at all levels of temperature, high performance concrete shows a higher residual strength than ordinary concretes. For example, significant damage in high performance concrete is usually observed above +600°C, but for ordinary concrete it is already evident above +400°C. In contrast, spalling, in its explosive form as detaching of fragments of the surface layers, is a serious risk in high performance or ultra high performance concrete where special densifying components are used. In these materials with high relative impermeability when subject to fire, water vapour cannot be readily dissipated and its pressure within the concrete induces stress that may overcome the concrete tensile strength.

The behaviour of concretes reinforced with steel-fibres was studied by Purkiss (1988), also by computation of fracture toughness parameters. Thermal expansion was measured on specimens reinforced up to 2.5% volume and subjected to up to +800°C. However, the influence of plain fibres was rather limited: up to +300°C thermal strain was reduced by fibres and that reduction was proportional to fibre volume. At a higher temperature, between +300°C and +800°C, thermal expansion was apparently bigger than for comparable plain concrete specimens. It is believed that the fibres may participate in the distribution of heat, but the interaction between fibres and matrix at high temperature is fairly complex and no explanation was furnished for observed relations.

The influence of brass-coated, crimped and plain steel-fibres on the compressive strength of cement composites at high temperature is presented in Figure 11.29. It appears that up to +400°C the fibres efficiently controlled cracking and reinforced the matrix against a decrease of strength. In contrast, their influence at higher temperature between +400°C and +800 °C was quite negligible and the strength of reinforced specimens was nearly equal to that of plain concrete ones. Similar results were obtained for specimens under flexure. This conclusion was only partly confirmed by the tests executed by Lau and Anson (2006) who observed some compressive strength benefit to concrete from the addition of steel fibres used in 1% volume. However, at temperature over 1100°C in both reinforced and non-reinforced concretes

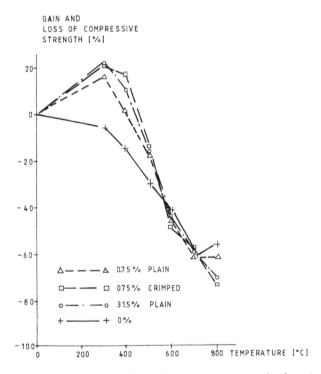

Figure 11.29 Variation of gains and loses of compressive strength of specimens tested in different temperature, after Purkiss (1984).

their strength was very low. In contrast, Poon *et al.* (2004) observed that steel fibres (1.5% vol.) were effective in reducing the degradation of compressive strength of the concrete after exposure to 800°C. The steel-fibre-reinforced concretes also had the highest toughness values after the high-temperature exposures.

After several years of testing, the general conclusion is that cement composites reinforced with steel fibres are more resistant than non-reinforced materials in moderately high and even very high temperature. The differences related to the type, shape and volume of the fibres are of lower significance. It is also believed that stainless steel fibres are better than fibres made of ordinary steel and are applied in concrete elements exposed to elevated temperature in normal exploitation, for example, in various refractory structures.

Reinforcement with glass fibres used up to 1.5% volume does not modify the behaviour of the cement matrix in high temperature (Purkiss 1985).

All kinds of polymeric fibres are rather sensitive to elevated temperature and cannot directly strengthen the matrix. For example, PP fibres melt at approximately +170°C and are absorbed by the matrix. Then empty channels are developed in the cement matrix that may facilitate an exit for water vapour. The channels are combined with existing pores and porous ITZ and were observed on SEM images. As a result, PP fibres prevent explosive spalling by significant reduction of internal pressure; their optimum dosage between 1.0 kgs/m^3 and 2.0 kgs/m^3 is suggested (after Noumowé [2005]; appr. 1.5 kg/m^3) and that amount may lead only to a small decrease in the composite strength and stiffness (Kalifa *et al.* 2001). High performance concrete specimens with 0.9 kgs/m^3 and 1.75 kgs/m^3 (0.1% and 0.2% vol.) of PP fibres were tested at temperature varying from +20°C to +600°C and the fibres increased the relative compressive strength with respect to the specimens without fibres; the influence was particularly significant at +250°C. Hager and Pimienta (2004) tested the specimens at elevated temperature and simultaneously compressed and observed that for higher volume of fibres the Young's modulus was slightly decreased.

Quantitative results obtained by different authors are not completely coherent. The differences between published conclusions are probably due to slightly different conditions of carrying the experiments and the different materials tested.

Extensive numerical studies were carried out by Fu *et al.* (2007) to model the thermal stresses and cracks in a cement-based matrix with inclusions under elevated temperature. A mechanical thermal-elastic damage 2D model with heat transfer was built to study these phenomena. It has been established that the temperature gradient is dependent on the heating rate and the thermal conductivity. The cracks open and propagate when the thermal stress reaches the tensile strength of the matrix. The crack pattern is closely related to the heating rate and thermal coefficients of the phases. The cracks play a double role: they increase the permeability of the matrix and allow the pore pressure to decrease, but the cracks also induce destruction of the concrete element, for example, producing explosive spalling. An example of the crack pattern obtained from numerical simulation is shown in Figure 11.30. Observations of spalling in various situations were published by Hertz (2003).

The influence of elevated temperature is always taken into consideration in design and detailing of reinforced concrete structures for buildings where a necessary period of time for evacuation of the inhabitants before possible collapse must be defined (EN 1992 (2004). At the present state of knowledge it is possible to predict the fire resistance of reinforced concrete elements with an accuracy that is adequate for practical purposes, including elements made with high performance concrete and fibres.

11.6.3 Durability in low temperature and under freeze-thaw cycles

When concrete structures are subjected to variations of temperature below zero and to the freezing of water solutions in the pore system, then exposure to destructive processes of the cement-based matrices is serious. An increase in the volume of water in the pores when it is transformed into ice is approximately 9% and it may produce important internal stresses, inducing cracks and spallings. This is particularly dangerous in the following situations:

1   Due to the large value of *w/c* ratio, a high volume of capillary pores in the matrix accumulate free water that is subjected to freezing.

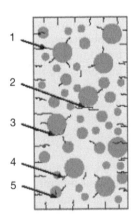

1-Temperature-induced shrinking crack;
2-Temperature gradient-induced horizontal crack;
3-Tangential crack; 4-Radial crack; 5-Inclusion crack

Figure 11.30 Crack patterns in cement-based composites at elevated temperature, after Fu *et al.* (2007) *Fig. 13, page 114.*

2   Outdoor structures like bridges, roads and jetties are directly exposed to environmental moisture due to rain, sprinkling by sea water or steam condensation. This concerns structural and also non-structural elements such as external claddings and decorations.

3   In large geographical zones, for example, Central Europe, North America and Japan, the temperature variations across 0°C are frequent in winter and also occur in spring and autumn, so that the number of dangerous freezing and thawing cycles is of the order of 100 or more in a year.

4   De-icing agents used on roads and runways have a detrimental influence on the cement-based matrix and initiate corrosion of steel-reinforcement.

5   In ordinary concretes, the tensile strength of the matrix is not sufficient to support local tensions due to freezing, particularly at early ages.

Destruction of concrete due to frost has two distinctly different forms: cracking of the bulk concrete and scaling on the concrete surface exposed to external actions. These two phenomena should be considered separately. The former is less complicated as to its mechanism and is better known to apply reducing measures. The latter, so-called 'frost salt attack,' is more complex and less understood. Perhaps recent experimental and theoretical results create a better basis for efficient methods to control the detrimental effects of freeze-thaw cycles on the outdoor concrete structures.

The durability of concrete elements subjected to freezing depends on the intensity of internal forces which appear when pore water is frozen and on the tensile strength of the matrix. The mixture composition of concrete considerably influences resistance against freezing. Density of the internal structure can be increased with appropriate application of various additives like SF, fly ash, ground granulated blast-furnace slag, metakaolin, etc., and with decrease of the w/b ratio. HSC and high performance concrete, which are carefully designed with so-called ternary or quaternary components, are considered by many researchers as resistant against freezing and thawing cycles without any additional precautions.

In ordinary concretes, the air-entrainment is considered as necessary for durability of outdoor structures in northern parts of Europe, America and Japan. Not only additional content of air is needed, but also correct distribution of diameters of the pores in a cement matrix is essential to control the intensity of internal stress due to freezing. The main parameter that characterizes the pore distribution is the so-called spacing factor $\overline{L}$ according to ASTM C457 (2006). It is equal to the average maximum distance of any point of the matrix to the nearest pore, that is, to half the distance between neighbouring pores. Larger pores characterised by a larger spacing factor are not helpful because tensile stress at freezing cannot be balanced by the tensile strength of the matrix and microcracks are unavoidable. The maximum value of $\overline{L}$, yet corresponding to the safe supporting of the tensile stress produced in the matrix by transformation of water into ice is called critical $\overline{L}_{cr}$ and should be maintained below approximately 200 µm for mortars and below

300 μm for concretes. Therefore, in ordinary concrete, special air-entraining agents are added (cf. Section 4.3.6) to produce an artificial system of spherical air bubbles uniformly distributed in the hardened matrix with spacing factor $\overline{L}$ equal to 200 μm as a maximum. The spacing factor is the most important parameter characterizing the resistance of cement-based materials to freeze-thaw cycles, but other parameters (e.g. α, A_{300}) also characterize the air-entrained pore system and their values should be maintained within prescribed limits; all are described in detail in Section 6.5.

The air-entrainment effects are related to the workability of the fresh mix: with higher slump the volume of air is increased, but for high slump (≥ 150–175 mm) the mix may be too fluid to retain the air bubbles before hardening. The proper amount and nature of an air-entraining admixture is needed to obtain the required effects and this is usually established by testing.

The total volume of air voids is not the most important parameter, but it gives certain indication as to the freeze-thaw resistance. It is measured in a fresh mix and in hardened material, but between these two measures large differences may occur during casting, vibration, etc. When the total air volume in the fresh mix is below 4% it indicates that there is not enough air and probably $\overline{L} > \overline{L}_{cr}$. When the air volume is greater than 4%, then with appropriate air bubbles distribution a satisfactory situation may be expected, that is, that $\overline{L} \leq \overline{L}_{cr}$ and other entrained air parameters may have satisfactory values also in hardened concrete; all this should be verified, for example, by testing cores.

It has already been suggested by Gagné *et al.* (1990) (among others) that for high performance concretes with a low *w/c* ratio approaching 0.25, the critical value $\overline{L}_{cr}$ is close to 750 μm and no air-entrainment is needed; such a conclusion is confirmed in the tests by Sahmaran and Li (2007). This is an important advantage of high performance concrete because the pores produced by air-entraining agents decrease strength (appr. 5.5 MPa in compressive strength per 1% of air), and the procedure of application of air-entraining admixtures is considered as difficult in the ready-mix concrete-producing plants, thereby increasing the final cost of the composite material. In Table 11.3 a set of criteria is presented relating the values of *w/c* ratio and required spacing factor $\overline{L}$ for durable concrete structures.

Entrained pores are quasi-closed and even in a completely saturated matrix are not entirely filled with water, but they are available for expansion when water is freezing. This is the reason why entrained pores considerably decrease all detrimental results of frost-thaw cycles.

The freezing of water in capillary pores depends on their dimensions. As is shown in Figure 11.31 the freezing point decreases with pore diameter and it has been found that a temperature of about –15°C is required for nearly 100 % of water to be frozen, but water in pores which are smaller than 20 nm freezes at temperature below –43 °C. The freezing process is complicated by the diffusion of water from smaller pores to larger ones and by simultaneous water evaporation and capillary suction – the latter being of considerably

Table 11.3 Proposed criteria for frost resistance as a function of high performance concrete *w/c* ratio

w/c ratio	Recommended average spacing factor	Permitted maximum spacing factor	Number of cycles of freezing and thawing
Greater than 0.40	230 μm	260 μm (for scaling resistance)	300 cycles ASTM C666 50 cycles ASTM C
Comprised between 0.35 and 0.40	350 μm	400 μm (for scaling resistance)	300 cycles ASTM C666 50 cycles ASTM C
Comprised between 0.30 and 0.35	450 μm	550 μm	500 cycles ASTM C666
Below 0.30	As long as no more data are available, the previous criteria could be used – except if ASTM C666 Procedure A test shows that this criteria could be relaxed for a particular concrete		

Source: Aïtcin (1998) Table 6, page 432.

higher rate. Moreover, in the natural condition of outdoor structures external temperature vary, for example the water freezing process during the night may be interrupted by higher temperature periods during the day. The rate of cooling and frequency of freeze-thaw cycles are of primary importance.

The danger of water freezing is related to pore saturation, which also varies in time in real conditions of outdoor structures. In completely saturated pores, freezing produces high tensile stress, while for the pores that are partly filled or empty this stress may reach negligible values (cf. Figure 11. 32).

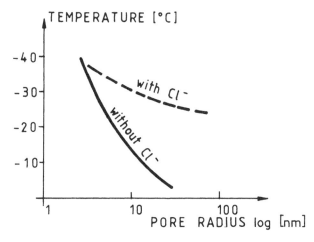

Figure 11.31 Freezing of water in capillary pores as function of their diameter, after CEB (1992).

Figure 11.32 Partially and completely water-saturated pores.

The scaling of a concrete surface appears when de-icing agents are used on road pavements to facilitate road traffic in severe climatic conditions. When frequent freezing in the nights occurs for long periods during a year, water freezes nearly in the same range of temperature in pores of all diameters in the top concrete layers. Therefore, the possibility of water distribution during these processes is strongly reduced and the damage of the cement matrix is more serious than without de-icing agents. Moreover, de-icing agents that are based on salts NaCl and CaCl increase the corrosion of steel reinforcement, and on airfield pavements are these products prohibited. Such a situation causes scaling, which is the damage of the surface of concrete elements in the form of small flakes of the cement matrix detached from the concrete surface. This process progresses and leads to serious destruction of concrete elements. Through such ruined external concrete layers other detrimental agents such as chloride ions have easy ingress down to the steel reinforcement.

The mechanism of concrete scaling under freeze-thaw cycles in the presence of the salt solution on the concrete surface is not yet fully understood, but basic factors controlling this phenomenon are known. The reason for scaling is also the increase in volume when water is transformed into ice. Internal stresses are created that are larger than the tensile strength of the cement matrix. Therefore, as in the case of cracking in the bulk concrete, scaling depends on the pore system and matrix strength. Extensive reviews of the problems related to scaling were published by Valenza and Scherer (2007). According to Çopuroğlu and Schlangen (2008) the scaling of a concrete surface is caused by cracking of the ice layer (Figure 11.33) and its volume increases with the thickness and stiffness of the ice layer (Figure 11.34). The pessimum effect for salt concentration is around 3%, because for solutions much below 3% neither ice nor concrete surfaces crack, and for higher concentrations above 3% the ice layer is too soft to induce concrete cracking. The influence of preliminary carbonation was also observed for ordinary and high performance concretes: for ordinary concrete the carbonation increases the resistance to scaling and for high performance concrete the influence is apparently opposite.

The rate of freezing also influences its effects on concrete: with a higher rate the internal damage may be more serious, while for lower rates and longer duration of temperature below –10°C the volume of scaling increases.

Internal curing applied to control shrinkage (cf. Section 11.4.2) is an active approach to reduce internal stresses in concrete due to freezing and

(a)   (b)

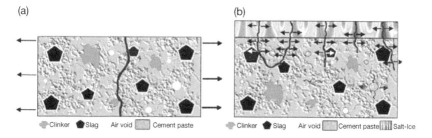

Figure 11.33 Schematic image of a cement paste under tensile stress (a) and of the cracks induced by the cracking of external layer of ice (b), after Çopurodlu and Schlangen (2008).

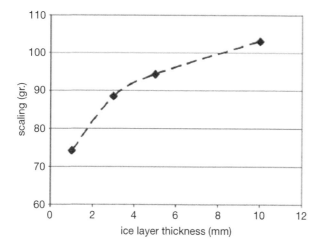

Figure 11.34 Cumulative mass scaling of the identical specimens with various depth of the ice layers after three freezing/thawing cycles, after Çopurodlu and Schlangen (2008).

thawing cycles. The application of pre-soaked lightweight aggregate (LWA) grains or SAP (super absorbent polymers) as water reservoirs considerably increases frost resistance of concretes exposed to outdoor weather conditions, because additional water is added in an 'intelligent' way where it is needed for better hydration of cement, which causes higher tensile strength of the cement matrix. Laboratory tests prove that internal curing was efficient against the detrimental results of multiple freezing cycles of both processes: surface deterioration in the form of scaling and internal cracking of concrete elements (Jóźwiak-Niedźwiedzka 2005).

The resistance of concrete-like composites to freezing is verified by standard freeze-thaw cycles in a form of accelerated tests. According to EN 12390 (2007) the material lost from a tested specimen should not exceed 1 kg/m^2 after 56 cycles.

Two recommendations, ASTM/C666M (2003) and ASTM/C672M (2003), are relevant for the testing of both destructive processes defined above, namely:

1 'Resistance to rapid freezing and thawing': cycles in water;
2 'Scaling resistance of concrete exposed to de-icing chemicals': resistance against surface spalling in exposure to freezing with de-icing agents.

Results of the work of TC 176-IDC 'Internal damage of concrete due to frost action' was published in RILEM (2004).

All these standard procedures of testing give valuable but only approximate indications as to the actual frost resistance, verified on many structures in severe climatic conditions. Their practical importance is also related to a possibility of testing different concretes to obtain comparative results and thus to select the best material for the purpose.

11.6.4 Cement-based materials in cryogenic temperature

Concrete has been used since the early 1950s as a structural material for liquefied natural gas (LNG) storage in specially built tanks of various constructions. The gases are cooled to cryogenic temperature, around $-165°C$ (appr. 108K) and liquefy, thus reducing their volume approximately 600 times as necessary for transportation and storage. In most cases, concrete is used for secondary containments and the concrete tank may have contact with the cryogenic liquids only if the primary tank and insulation layers fail. However, in certain constructions concrete is exposed on long-term contact with cryogenic temperature. The majority of the research related to this technology and to the concrete behaviour at cryogenic temperature, were carried on till the late 1970s and an extensive review of the results was published by Marshall (1982). Later, there was a considerable drop in construction of LNG concrete tanks and also in the research in this field. At present, this interest is resurging and the concrete properties at cryogenic temperature are again being considered, for example with the review paper by Krstulovic-Opara (2007) which covers the majority of investigations published until 1990.

Deep freezing of concrete causes multiple processes in its structure, namely the contraction of the solid skeleton and expansion of ice in the pores. The pores are not completely filled with water, which freezes at different temperature in different sized pores. The expansion of the ice is not linear with decreasing temperature. In real situations, water filtration and diffusion continue with temperature variations. Formation of ice may disrupt certain parts of the concrete structure, but larger voids filled with ice represent hard inclusions, which increase the strength of the composite material. In general, the properties of concrete result from the amount of freezeable water in the pore system, the distribution of pore dimensions and the rate of freezing.

In the investigation published by Burakiewicz and Marshall (1987) specimens of plain and fibre concretes were subjected to compression at different temperature down to –165°C. The consistent method of testing specimens at very low temperature was established and applied. The strength, strain at maximum stress and strain energy were recorded and analyzed (Figure 11.35). It has been established that the effect of fibres is appreciable only if at least 0.7% volume of fibres is added and a considerable decrease of all three mechanical parameters was observed at temperature approaching –165°C.

The investigations with fibre-reinforcement were executed by Rostasy and Sprenger (1984) in view of its application in the construction of outer containments for LNG. The specimens were tested at a temperature of –170°C (appr. 103K) and cycles between +20°C and –70°C (appr. 293K and 203K) were also imposed. The influence of various volume fractions of Wirex fibre was examined together with the application of Portland cement and blended cement with blast-furnace slag (up to 70%). No appreciable

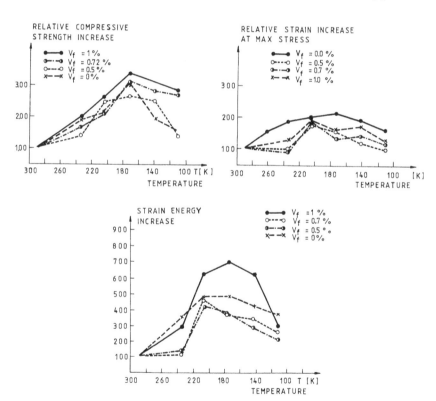

Figure 11.35 Relationships between test temperature and mechanical properties, after Burakiewicz and Marshall (1987):
(*a*) compressive strength
(*b*) strain at maximum stress
(*c*) strain energy.

influence of fibre-reinforcement on thermal strain was observed as shown in Figure 11.36; however, only one cooling cycle was applied. In the tests under compression at different temperature from +20 °C (appr. 293K) down to –170°C (appr. 103K) only a slight increase of strength and ultimate strain was observed for specimens reinforced with fibres. More important results of fibre-reinforcement were obtained in specimens subjected to several (up to eight) cycles of cooling prior to loading. As is presented in Figure 11.36, the specimens without fibres exhibited clear decrease of tensile strength due to cooling cycles. Fibres used at 1% and 2% volume reduced that decrease considerably. Also, in specimens subjected to axial compression, an appreciable influence of fibre-reinforcement was observed.

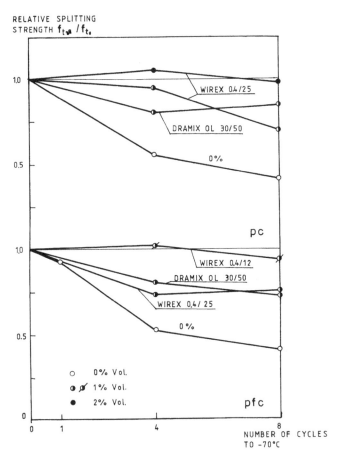

Figure 11.36 Relative splitting strength of specimens made of plain concrete and of SFRC after low temperature cycles, pc – Portland cement, pfc – blast-furnace slag cement, after Rostasy and Sprenger (1984).

Rostasy and Pusch (1987) performed extensive testing of specimens made of structural lightweight concrete subjected to cycles of temperature between +20°C and –85°C (appr. 293K and 188K). It has been found that lightweight concretes of f_{c28} between 50 MPa and 60 MPa behaved similarly to ordinary concretes; that is, their sensibility to cooling cycles was related to their degree of water saturation: the higher saturation, the more damage was observed in the material's structure. But in these kinds of materials, evaporable and freezeable water is stored in pores in the aggregate grains and the strength after cooling may be considerably diminished proportionally to the total moisture content.

In general, concrete strength is increased as the temperature is lowered to cryogenic values. The extent of improvement depends on the amount of water in the concrete pore system, that is, on *w/c* ratio, and on pore size distribution that determines the volume of freezeable water that is transformed into ice. Krstulovic-Opara (2007) proposed the following relationships for approximate calculations, based on the results of the investigations carried on in Japan:

$$\text{for } t > - 120 \text{ °C (appr. 153K)} \quad f_c(t,mc) = f_c + 12\,(\text{MPa}) \cdot \left[1 - \left(\frac{t+180°C}{180°C}\right)^2\right]$$

for t < – 120 °C (appr. 153K) $f_c(t,mc) = f_c + 10.7$ MPa $\cdot$ *mc*

where f_c is the compressive strength of concrete at +20°C, *t* is temperature in degrees Celsius and *mc* is the percent of the moisture content by weight. This means that below –120°C the increase of strength is proportional to the moisture in concrete (Miura 1989).

All these phenomena cause relatively complex behaviour in cement-based composites at cryogenic temperature. Nevertheless, the abovementioned and many other research results allow us to formulate certain conclusions as to the concrete's properties. However, it is not surprising that certain conclusions from experimental investigations are partly different; this is probably due to differences in material properties and testing techniques. It is generally believed that most of the mechanical properties are improved in low temperature and that this increase at cryogenic temperature is proportional to the moisture content. Approximate indications on the basis of different investigations and the consideration of the concrete humidity as an important agent are presented in Table 11.4.

Thermal cycling, including cryogenic temperature, is not considered here and it is possible that it may cause damage to a material's structure leading to an appreciable decrease of strength. However, there is not enough available information on that problem, and also for storage purposes cycling is usually not considered because such accidents with repeated contact with cryogenic temperature loading have never occurred to concrete elements.

Table 11.4 Typical properties of ordinary concrete at –100°C and below compared with respective values at +20°C (appr. indications)

Properties	Moist concrete	Dry concrete
Compressive strength	increase approximately by 10 MPa	increase approximately by 10 %
Tensile strength (splitting)	increase by 5 MPa	10% increase
Tensile strength (bending)	increases by 10 MPa	10% increase
Impact strength	doubled already at –30°C	
Young's modulus	50% increase	10% increase
Poisson ratio	0.20–0.22	0.18
Thermal conductivity	slight increase when ice is present	
Creep	halved at –30 °C	
Permeability	negligible	0.3 darcys (3×10^{12} m^2)

Source: after Krstulovic-Opara (2007).

11.7 Transportation of liquids and gases through hardened composites

In hardened cement-based composites the transportation of liquids and gases through pore and microcrack systems plays a very important role in many processes, such as hydration of Portland cement, pozzolane effects of microfillers, carbonation, corrosion of cement paste and reinforcement due to reaction with external agents, shrinkage and creep, etc. These processes are partly described in respective Sections 4.1, 4.3, 6.5 and 11.5. Only basic information is reiterated below concerning the flow of liquids and gases through concretes and mortars.

Permeation or pressure flow is due to external pressure, which causes flow of fluids or gases through systems of pores and microcracks. In the case of water permeability, it may be described by Darcy's law as volume $V[m^3]$ which passes in time $t[s]$ through area $A[m^2]$

$$V = K_w \frac{\Delta H}{\ell} At \qquad (11.14)$$

here: $\delta H/\ell$ is hydraulic gradient [m/m] and K_w is coefficient of permeability [m/s], which may be related to concrete strength f_{c28},

$$K_w = 10^{-(f_{c28}/9+8)},$$

or to the *w/c* ratio:

$K_w = 10^{(-18+10(w/c))}$ for $0.4 < w/c < 0.7$

For practical calculations, the permeability of concrete is considered as equal to that of hardened paste. In that assumption, the existence of microcracks and aggregate grains in concrete compensate each other for water flow. For higher concrete strength and low w/c ratio the coefficient K_w and permeability itself decrease considerably. A further decrease may be expected for high and very high performance concretes with low values of w/c ratio and with extensive use of microfillers.

In the case of gas permeability the formula (11.14) becomes:

$$V = K_g \frac{A}{\ell} \frac{(p_1 - p_2)}{\eta} \frac{p_m}{p} t \tag{11.15}$$

here: $(p_1 - p_2)$ – difference of pressures [N/m²], p_m and p – mean and local pressure, η – gas viscosity [Ns/m²]. For air and oxygen: $0.02 \times 10^{-16} < K < 2 \times 10^{-16}$ [m²].

Diffusion when the motion of particles caused by difference of concentration is represented by Fick's first law. The respective formula proposed by CEB-FIP (1990) is derived in the following form for the amount of transported mass Q as a function of difference of concentration $(c_1 - c_2)$

$$Q = D \frac{c_1 - c_2}{\ell} At \tag{11.16}$$

Here D may be calculated as a function of f_{c28} and gas properties. For oxygen and carbon dioxide values of D are between 0.5×10^{-8} and 6×10^{-8} [m²/s]. For chloride ions D is between 1×10^{-12} and 10×10^{-12} [m²/s] for Portland cement paste, and between 0.3×10^{-12} and 5×10^{-12} for pulverized blast-furnace slag (PBFL) cement.

Capillary flow is described by Washburn's law (Marshall 1990):

$$V = \frac{r\gamma_L}{4d\mu} At \cos \vartheta$$

here V is volume of fluid, r – capillary radius, γ_L surface tension, d – depth of penetration, μ – viscosity of the fluid and θ – contact angle. Both γ_L and μ depend on temperature, for example the viscosity μ at +30°C is about half of that at +5°C. Also, the purity of the fluid is important and pore water is less fluid than distilled water, but admixtures may have some influence.

The above relations cannot be applied directly in calculations of the volume of gases or fluids that may pass through the element of composite material under consideration even if it was the intention of the authors to provide practical formulae for designers (cf. CEB-FIP 1990). The formulae may only indicate the influence of the pore system, which enhance or reduce the transportation of a gas or fluid (Meng 1994), but the permeability depends on several parameters characterizing the transported medium, on pore and microcrack systems of material and on external conditions like concentration, pressure and temperature. For example, by decreasing the *w/c* ratio, considerable reduction of the coefficient of water permeability of Portland cement paste may obviously be expected, as shown in Figure 6.10. Other measures enhancing the impermeability of a composite material would be the better cure of fresh mix to reduce capillaries or the application of micro-fillers.

Reinforcement with fibres may have different results with regard to permeability. Obviously, the control of cracking is better and transport through a system of cracks and microcracks may be reduced. In contrast, the introduction of fibres to the fresh mix usually increases its porosity.

Gas permeability of concrete to air and other gases is important for structures like sewage tanks, various containers and pressure vessels in nuclear reactors. Also in ordinary reinforced concrete structures exposed to carbonation or chloride ions ingress, the permeability of concrete cover may decide on the danger of corrosion of steel reinforcement. There are many ways to improve impermeability of concretes, mentioned above, based on the application of various microfillers and on a limitation of w/b ratio. It may be added that air-entraining, applied usually to increase frost resistance, is beneficial in considerably reducing penetration of oxygen and other gases (Mahoutian *et al.* 2007).

Measurement techniques for concrete cover are presented by Torrent (1999) and the gas permeability of concretes ranges are indicated in Table 11.5 where results of both methods are indicated: k_0 means the coefficient of permeability to oxygen according to the so-called 'Cembureau method', and k_T is the coefficient of air permeability determined after the Torrent method; the depth of the cover is determined simultaneously. The measurements by the so-called 'Torrent permeability method' may be realised on site directly on the concrete elements without any kind of their destruction, or in a

Table 11.5 Classification of covercrete gas permeability

Quality grade of concrete	$k_0 \ [10^{-16} m^2]$	$k_T \ [10^{-16} m^2]$	Quality
1	< 0.1	< 0.01	Excellent
2	0.1–0.5	0.01–0.1	Very good
3	0.5–2.5	0.1–1.0	Medium
4	2.5–12.5	1–10	Poor
5	> 12.5	> 10	Very poor

Source: Torrent (1999).

laboratory on specimens and cores. The sample preparation and its drying process significantly influence the measured value of gas permeability and should be carefully executed. Repeatability of the measurements is quite satisfactory and confirmed by different laboratories. The verification in situ of the quality of the concrete cover is an important element of ensuring the durability of the infrastructure structures. The methods for measurement of the gas permeability of concrete are recommended by RILEM (1999).

Percolation of capillary pores in cement paste. A few remarks are needed as a reminder that all the processes considered develop in highly disordered media. High porosity does not necessarily mean high permeability, because only interconnected pores represent channels for the possible flow of gases and fluids. The relationships governing the flow are linear only within narrow limits and more often high non-linearity should be expected. For basic information on percolation processes in strongly disordered media, the reader is referred to Mandelbrot (1982), but the application of the notion of percolation to cement-based materials is still studied by many authors.

Pores may be considered as a system of objects which are not completely connected and a flow of a gas or a fluid is a process of percolation in which a threshold exists. This creates a clear difference between two situations: the system is permeable to water, or it is not. The degree of connectivity of the phases decides on the flow through the cement-based materials. At the beginning, when cement hydration is initiated, the pores create a fully connected system and only progressively the capillary pores become partly filled and connections are reduced. The threshold is expressed by a certain level of hydration when the pores are not connected and the permeability disappears; it is called the 'depercolation threshold.'

According to Powers *et al.* (1959) the capillary porosity in cement paste is characterized by a percolation transition (from connected to disconnected) at a volume fraction of about 20% porosity, but this opinion is not confirmed by the tests shown by Ye (2005), who proved that the same porosity may correspond to the permeability different by one order of magnitude, as shown in Figure 11.37 (curves for $w/c = 0.40$ and $w/c = 0.60$ indicate the same porosity equal appr. 30.9%).

The discussion over these processes in cement pastes continues; it was shown that the creation of a depercolation threshold depends on preparation of specimens, conditions of curing, etc. Tests carried out by Banthia and Mindess (1988) indicate that for cement pastes with w/c ratio over 0.30 percolation is not reduced completely even after one year curing, which means that the depercolation threshold does not exist in such materials; similar results have been obtained by other authors.

Experimental results are completed by modelling and numerical simulations, presented by Bentz (2006) and Ye (2005), among others.

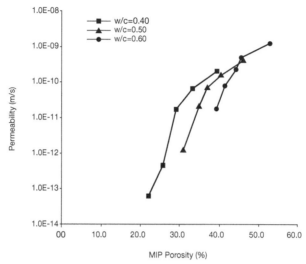

Figure 11.37 Relationship between permeability and total porosity measured by MIP, after Ye (2005).

11.8 Concrete under accelerated testing

The long-term behaviour of materials and their durability with respect to certain required performance levels for structures and buildings are forecast after experience, tests and observation in shorter periods of time. Results of observations in 50 or 100 years of exploitation are not available for modern composite materials. Observations of traditional materials such as brick, wood and stone are partly irrelevant because present environmental and exploitation conditions did not exist before and any forecast may not be reliable. Real-time testing to correctly understand the long-term behaviour is the best solution, but appreciable results of ageing can only be expected after many years. However, research programmes lasting 10 or more years are rare for organizational and economic reasons. That is why accelerated tests are useful and often necessary.

The accelerated tests of specimens or elements subjected to special conditions in which, after a short time, the results obtained are equivalent to natural long-term exposure and are applied with moderate success. However, it is generally believed that such methods are needed and should be improved as far as possible. Examples of the properties that should be evaluated after long-term exploitation are strength, deformability, stability of dimensions, as well as the results of various processes like carbonation, chloride ions transmission, and many others.

Accelerated tests are composed of different actions or cycles of actions, which should produce measurable and appreciable effects after a certain number of hours, days or months. Typically, they do not exceed a few weeks or months of steady action or a certain number of cycles.

According to standard recommendations in ASTM C684 (2003), three methods are applied to evaluate long-term properties of concretes:

1   In the warm water method, the concrete specimens are cured in water at 35 ± 3°C and tested at the age of 24 hours;
2   In the boiling water method, after a curing during 23 hours in moisture at 21 ± 6°C the specimens are put into boiling water for another 3.5 hours and tested at the age of 28.5 hours.
3   In the autogeneous method, the specimens are subjected to heat due to the hydration of cement in a completely insulated environment over a period of 48 hours.

Other actions may also be used in particular situations:

• freezing and thawing cycles in clear water or in water with a de-icing agent, which simulates day and night temperature variations in winter for bridges and road overlays;
• soak and dry cycles with sea water or other aggressive water solutions corresponding to real conditions of exploitation, while drying can simulate periods of exposure to sunshine;
• warm and cool cycles with higher temperature imitating natural exposure, e.g. of refractory materials;
• exposure to highly concentrated gases or liquids.

The sequence of various treatments in consecutive cycles should, in principle, simulate natural exposure whilst accelerating its effects in a standardized way. In most test procedures, elevated temperature is applied to accelerate chemical reactions. No single accelerated test can fully simulate the full range of ageing processes and therefore a series of tests of a different character is usually executed, after identification of the main ageing mechanisms, which may be considerably accelerated in that way.

It is difficult to determine a relationship between the results of accelerated tests and of the natural exposure of certain durations. However, many attempts are published in which a safe life cycle is estimated from the results of accelerated tests. For example, it has been proposed that exposure of concrete elements to pure CO_2 over a period of 36 days with a ratio of concentration 3000 times higher may give the same carbonation of concrete as 300 years of service in natural atmosphere (Lévy 1992). In the tests carried out by Sisomphon and Franke (2007) it has been derived on the basis of the second Fick's law that the carbonation in natural exposure with concentration CO_2 of 0.03% is approximately 10 times slower than in the accelerated tests at concentration CO_2 of 3%.

In the case of the alkali-aggregate reaction (AAR) that causes serious damage to structural elements in many countries it is particularly important to select non-reactive or mildly reactive aggregate that could safely be used for concrete. The generally accepted accelerated mortar bar test is based on subsequent storage in hot water (+80°C) and NaOH solution. In that way, the testing in real time, that is during a period of 6–12 months, may be replaced by a test of a few weeks' duration, and tests carried out by Shayan *et al.* (2001) confirmed this method.

Accelerated testing was applied for glass-fibre-reinforced cements to study the durability of such systems. It has been shown by Purnell and Beddows (2005) that different regimes should be selected for different matrices, for example a temperature over +65°C is not appropriate for matrices with metakaolin, and hot water in general should not be used for polymer-modified matrices because the results may considerably overestimate their durability.

Because the simulation of natural processes is far from perfect and all these forecasts are uncertain, it is essential to combine accelerated tests with real-time observations, if possible. Real-time test results even after a relatively short time may help considerably to calibrate a forecast based on accelerated tests.

Standardized accelerated tests are useful for carrying out comparative tests of different composite materials. It is easier to determine which material is better than to forecast its effective durability. For example, it is believed that freeze-thaw tests as imposed by ASTM (2003) or other standardization institutions could ensure acceptable durability of cement-based materials in natural conditions during their lifetime in the climate of Northern America and Central Europe, even if the validity of simulation of natural conditions is doubtful. The possibility of using accelerated tests for glass-fibre-reinforced composites are discussed in Chapter 4.

General recommendations for testing the durability of cement-based fibre-reinforced composites are not yet available. Basic requirements may be tested in the same way as for plain concretes; however, problems concerning compatibility of fibres and matrices and processes at the interface should be solved by special methods. The same concerns high and very high performance concretes and fibre-reinforced concretes. For these materials, it is particularly important to know, well in advance, their strength and other properties at a given age. On many occasions, structural elements are subjected to reduced loading at a very early stage, for example, after three or seven days. In that case, different formulae should be used in the accelerated methods than for ordinary concretes because the development in time of strength, Young's modulus and other properties is different. Such methods are not yet fully reliable, however several proposals have already been discussed.

References

Abrams, D. A. (1917) 'Effect of rate of application of load on the compressive strength of concrete,' *American Society for Testing of Materials, Proc. 17*, part II: pp. 364–77.

Acker, P., de Larrard, F. (1992) 'Fluage des bétons à hautes et à très hautes performances,' in: *Les Bétons à Hautes Performances*, Y. Malier ed., Paris; Presses de LCPC: pp.165–76.

Ahmed, S. F. U., Mihashi, H. (2007) 'A review on durability properties of strain hardening fibre reinforced cementitious composites (SHFRCC),' *Cement and Concrete Composites*, 29: 365–76.

Aïtcin, P. C. (1998) 'The influence of the spacing factor on the freeze-thaw durability of high performance concrete,' in: *Int. Symp. on High-Performance and Reactive Powder Concretes*, August, Canada; Université de Sherbrook: pp. 419–33.

——.(2003) 'The durability characteristics of high performance concrete: a review. *Cement and Concrete Composites*, 25: 409–20.

Alliche, A. (2004) 'Damage model for fatigue loading of concrete,' *International Journal of Fatigue*, 26: 915–21.

Badr, A., Ashour, A. F. (2005) 'Modified ACI drop-weight impact test for concrete,' *ACI Materials Journal*, 102(4): 249–55.

Banthia, N., Mindess, S., Trottier, J.-F. (1996) 'Impact resistance of steel fiber reinforced concrete,' *ACI Materials Journal*, 93(5): 472–9.

Banthia, N., Yan, C., Sakai, C. (1998) 'Impact resistance of fiber reinforced concrete at subnormal temperatures,' *Cement and Concrete Composites*, 20: 393–404.

Batson, G., Ball, C., Bailey, L., Landers, E., Hooks, J. (1972) 'Flexural fatigue strength of steel fiber reinforced concrete beams,' *Journal of the ACI*, 69(11): 673–7.

Bažant, Z. P., Baweja, S. (1995) 'Justification and refinements of Model B3 for concrete creep and shrinkage,' *Materials and Structures*, 28: 415–30 and 488–95.

Bentur, A., van Breugel, K.P. (2002) 'Internally cured concrete,' in: *Early Age Cracking in Cementitious Systems, Report 25, RILEM TC 181-EAS*: pp. 339–53.

Bentz, D. P. (2006) 'Influence of alkalis on porosity percolation in hydrating cement pastes,' *Cement and Concrete Composites*, 28(5): 427–31.

Bentz, D. P., Jensen, O. M. (2004) 'Mitigation strategies for autogenous shrinkage cracking,' *Cement and Concrete Composites*, 26: 677–85.

Bijen, J., van den Plas, C. (1992) 'Polymer modified glass fibre reinforced gypsum,' in: *Proc. Int. Workshop 'High Performance Fiber Reinforced Cement Composites,' RILEM/ACI*, W. H. Reinhardt and A. E. Naaman eds, London; Spon/Chapman and Hall: pp. 271–87.

Brandt, A. M. (1965) 'Testing of concrete creep in non-reinforced beams subjected to bending' (in Polish), *Archives of Civil Engineering*, 11(1): 87–93.

Brandt, A. M., Hebda, L. (1989) 'Creep in SFRC element under long-term eccentric compressive loading,' in: *Proc. Int. Conf. 'Composite Structures,'* I. H. Marshall ed., London; Elsevier Applied Science: pp. 743–54.

Brandt, A. M., Kucharska, L. (2001) 'New trends in designing the durability of concrete.' in: *3rd Int. Conf. 'Concrete under Severe Conditions,"* N. Banthia ed., Vancouver: pp. 797–810.

Brandt, A. M., Glinicki, M. A., Potrzebowski, J. (1996) 'Application of FRC in construction of the underground railway track,' *Cement and Concrete Composites*, 18: 305–12.

Brara, A., Klepaczko, J. R. (2007) 'Fracture energy of concrete at high loading rates in tension,' *International Journal of Impact Engineering*, 34: 424–35.

Brooks, J. J., Megat Johari, M. A. (2001) 'Effect of metakaolin on creep and shrinkage of concrete,' *Cement and Concrete Composites*, 23: 495–502.

Burakiewicz, A., Marshall, A. L. (1987) 'Compressive strength and the test methods for concrete at cryogenic temperatures,' *Archives of Civil Engineering*, 33(1): 9–22.

Buyle-Bodin, F., Madhkhan, M. (2002) 'Performance and modelling of steel fibre reinforced piles under seismic loading,' *Engineering Structures*, 24(8): 1049–56.

Cachim, P. B., Figueiras, J. A., Pereira, P. A. A. (2002) 'Fatigue behavior of fiber-reinforced concrete in compression,' *Cement and Concrete Composites*, 24: 211–17.

Cheng, M.-Y., Parra-Montesinos, G. J. (2007) 'Punching shear resistance and deformation capacity of fiber reinforced concrete slab-column connections subjected to monotonic and reversed cyclic displacements,' in: *International Workshop HPFRCC 5 in Mainz, RILEM, Proc. 53*, W. H. Reinhardt and A. E. Naaman eds: pp. 489–96.

Chung, D.D.L. (2003) 'Damage in cement-based materials, studied by electrical resistance measurement,' *Materials, Science and Engineering*, R52(1): pp. 1–40.

Collepardi, M., Borsoi, A., Collepardi, S., Olagot, J. J. O., Troli, R. (2005) 'Effects of shrinkage reducing admixture in shrinkage compensating concrete under non-wet curing conditions,' *Cement and Concrete Composites*, 27: 704–8.

Çopuroğlu, O., Schlangen, E. (2008) 'Modeling of frost salt scaling,' *Cement and Concrete Research*, 38: 27–39.

Dougill, J. W. (1968) 'Some effects of thermal volume changes on the properties and behaviour of concrete,' in: *Proc. Int Conf. 'Structure of Concrete,'* London; Cement and Concrete Association.

Ferreira, M., Jalali, S. (2006) 'Quality control based on electrical resistivity measurement,' in: *Proc. ESCS: Europ. Symp. on Service Life and Serviceability of Concrete Struct.*, Helsinki: pp. 325–32.

Folliard, K. J. (1997) 'Properties of high-performance concrete containing shrinkage-reducing admixture,' *Cement and Concrete Research*, 27(9): 1357–64.

Fu, Y. F., Wong, Y. L., Poon, C. S., Tang, C. A. (2007) 'Numerical tests of thermal cracking induced by temperature gradient in cement-based composites under thermal loads,' *Cement and Concrete Composites*, 29: 103–16.

Gagné, R., Pigeon, M., Aïtcin, P. C. (1990) 'Durabilité au gel des bétons de haute performances mécaniques,' *Materials and Structures*, 23(134): 103–9.

Genconglu, M., Mobasher, B. (2007) 'Static and impact behaviour of fabric reinforced cement composites in flexure,' in: *Int. Workshop HPFRCC 5, RILEM, Proc. 53*, W. H. Reinhardt and A. E. Naaman eds: pp. 463–70.

Georgali, B., Tsakiridis, P. E. (2005) 'Microstructure of fire-damaged concrete: a case study,' *Cement and Concrete Composites*, 27: 255–9.

Georgin, J. F., Reynouard, J. M. (2003) 'Modeling of structures subjected to impact: concrete behaviour under high strain rate,' *Cement and Concrete Composites*, 25: 131–43.

Glanville, W. H. (1933) 'Creep of concrete under load,' *The Structural Engineer*, 11(2): 57–73.

Glinicki, M. A. (1991) 'Influence of the rate of loading on the strength and deformation of cement matrix composites' (in Polish), *PhD Thesis, IFTR*, Warsaw: Rep.1.

Gram, H. E., Persson, H., Skarendahl, Å. (1984) *Natural Fibre Concrete, SAREC Report*, Stockholm: pp. 139.

Grzybowski, M., Shah, S. P. (1990) 'Shrinkage cracking of fiber reinforced concrete,' *ACI Materials Journal*, 87(2): 138–48.

Grzybowski, M., Meyer, C. (1993) 'Damage accumulation in concrete with and without fiber reinforcement,' *ACI Materials Journal*, Nov–Dec: pp. 594–604.

Guyon, Y. (1959) 'Béton précontraint, 3rd ed., Eyrolles, Paris.

Hager, I., Pimienta, P. (2004) 'Impact of the polypropylene fibers on the mechanical properties of HPC concrete,' in: *Proc. 39 6th Int. RILEM Symp. BEFIB*, vol. 1.

Hertz, K. D. (2003) 'Limits of spalling of fire–exposed concrete,' *Fire Safety Journal*, 38: 103–16.

Holt, E. (2005) 'Contribution of mixture design to chemical and autogenous shrinkage of concrete at early ages,' *Cement and Concrete Research*, 35: 464–72.

Jianyong, L., Yan, Y. (2001) 'A study of creep and drying shrinkage of high performance concrete,' *Cement and Concrete Research*, 31: 1203–6.

Jóźwiak-Niedźwiedzka, D. (2005) 'Scaling resistance of high performance concretes containing a small portion of pre-wetted lightweight fine aggregate,' *Cement and Concrete Composites*, 27: 709–15.

Kalifa, P., Chéné, G., Gallé, C. (2001) 'High-temperature behaviour of HPC with polypropylene fibres: from spalling to microstructure,' *Cement and Concrete Research*, 31: 1487–99.

Kimura, H., Ishikawa, Y., Kambayashi, A., Takatsu, H. (2007) 'Seismic behaviour of the 200 MPa ultra-high-strength steel-fiber reinforced concrete columns under varying axial load,' *Journal of Advanced Concrete Technology*, 5(2): 193–200.

Kleen, E., Niepmann, D. (2007) 'Influences of pozzolans and slag on the chloride penetration in concrete,' in: *9th CANMET/ACI Int. Conf. on the Use of Fly Ash, Slag, Silica Fume and Other Supplementary Cementing Materials in Concrete*. May, Warsaw; Suppl.Vol.: pp. 201–17.

Kovler, K., Zhutowsky, S. (2006) 'Overview and future trends of shrinkage research,' *Materials and Structures, RILEM*, 39: 827–47.

Kovler, K. ed. (2008) *Internal Curing of Concrete*, RILEM TC 196-ICC STAR Report.

Krstulovic-Opara, N. (2007) 'Liquefied natural gas storage: material behavior of concrete at cryogenic temperatures,' *ACI Materials Journal*. May–June: pp. 297–306.

Kyung, K. H., Meyer, C. (2007) 'Aramid fiber mesh-reinforced thin sheet response to impact loads, in: *International Workshop HPFRCC 5, RILEM, Proc.53*, W. H. Reinhardt and A. E. Naaman eds: pp. 447–53.

Lau, A., Anson, M. (2006) 'Effect of high temperatures on high performance steel fibre reinforced concrete,' *Cement and Concrete Research*, 36: 1698–707.

Le Camus, B. (1947) 'Recherche expérimentale sur la deformation du bééton et du béton armé', *Ann. ITBTP*, nos 32, 33, 34.

Lee, M. K., Barr, B. I. G. (2004) 'An overview of the fatigue behaviour of plain and fibre reinforced concrete,' *Cement and Concrete Composites*, 26: 299–305.

Leung, C. K. Y., Cheung, Y. N., Zhang, J.(2007) 'Fatigue enhancement of concrete beam with ECC layer,' *Cement and Concrete Research*, 37: 743–50.

Lévy, C. (1992) 'La carbonation: comparaison BO/BHP du Pont de Joigny,' in: *Les Bétons à Hautes Performances*, Y. Malier ed., Paris; Presses de LCPC: pp. 359–69.

L'Hermite, R. (1955) 'Idées actuelles sur la technologie du béton,' *Docum. Tech. Bât. Trav. Publ.*, Paris.

——.(1957) 'Que savons-nous de la déformation plastique et du fluage du béton,' *Ann. ITBTP*, 10(117): 777–810.

Li, M., Zhang, M., Ou, J. (2007) 'Flexural fatigue performance of concrete containing nano-particles for pavements,' *International Journal of Fatigue*, 29: 1292–301.

Li, V. C., Matsumoto, T. (1998) 'Fatigue crack growth analysis of fiber reinforced concrete with effect of interfacial bond degradation,' *Cement and Concrete Composites*, 20: 339–51.

Luo, X., Sun, W., Chan S. Y. N. (2000) 'Characteristics of high-performance steel fibre-reinforced concrete subject to high velocity impact,' *Cement and Concrete Research*, 30: 907–14.

Lura, P., Jensen O.-M., Igarashi S.-I. (2006) 'Measurements of internal water curing and of its consequences,' in: *RILEM, TC-196 ICC Internal Curing of Concrete, part 6*.

Mahoutian, M., Bakhshi, M., Shekarachi, M.(2007) 'Study on gas permeability of air-entrained concrete,' in: *9th CANMET/ACI Int. Conf. on the Use of Fly Ash, Slag, Silica Fume and Other Supplementary Cementing Materials in Concrete*. May, Warsaw; Suppl.Vol.: pp. 441–53.

Mandelbrot, B. B. (1982) *The Fractal Geometry of Nature*, San Francisco, CA; Freeman.

Marshall, A. L. (1982) 'Cryogenic concrete,' *Cryogenics*, 21(11): 555–65.

——.(1990) *Marine Concrete*, Glasgow and London; Blakie.

Mehta, P. K. (1997) 'Durability – critical issues for the future,' *Concrete Int. ACI*, July: pp. 27–33.

——.(1999) 'Advances in concrete technology,' *Concrete Int. ACI*, June: pp. 69–76.

Meng, B. (1994) 'Calculation of moisture transport coefficients on the basis of relevant pore structure parameters,' *Materials and Structures RILEM*, 27: 125–34.

Mihashi, H., Izumi, M. (1977) 'A stochastic theory for conrete fracture', *Cement and Concrete Research*, 7: 411–41.

Mindess, S., Young, J. F., Darwin, D. (2003) *Concrete*, 2nd edition, Upper Saddle River, NJ; Prentice-Hall, Pearson Education.

Mitchell, I.J. (1953) 'Thermal expansion tests on aggregate, neat cements and concretes', *Proc. ASTM*, 53, pp. 963–77.

Miura, T. (1989) 'The properties of concrete at very low temperatures,' *Materials and Structures*, 22: 243–54.

Naaman, A. E., Hammoud, H. (1998) 'Fatigue characteristics of high performance fibre-reinforced concrete,' *Cement and Concrete Composites*, 20: 353–63.

Naus, D. J., ed. (2003) 'Life predition and aging management of concrete structures,' *2nd RILEM Workshop*, PRO 29.

Neville, A. M. (1997) *Properties of Concrete*, 4th edition, New York; John Wiley.

Nischer, P. (1984) 'Influence of concrete quality and environment on carbonation,' in: *Proc. of RILEM Sem. 'Durability of Concrete Structures under Normal Outdoor Exposure,'* 26–29 March, Hannover: pp. 231–8.

Noumowé, A. (2005) 'Mechanical properties and microstructure of high strength concrete containing polypropylene fibres exposed to temperatures up to 200°C,' *Cement and Concrete Research*, 35: 2192–8.

Ong, K. C. G., Basheerkhan, M., Paramasivam P. (1999) 'Resistance of fibre concrete slabs to low velocity projectile impact,' *Cement and Concrete Composites*, 21: 391–401.

Ožbolt, J., Reinhardt, H. W. (2001) 'Sustained loading strength of concrete modelled by creep-cracking interaction,' *Otto-Graf-Journal*, 12: 9–19.

Pade, C., Guimaraes, M. (2007) 'The CO_2 uptake of concrete in a 100 year perspective,' *Cement and Concrete Research*, 37: 1348–56.

Page, C.L. (2007) 'Durability of Concrete and Cement Composites', Cambridge, UK, Woodhead Publishing.

Parant, E., Rossi, P., Boulay, C. (2007) 'Fatigue behaviour of a multi-scale cement composite', *Cement and Concrete Research*, 37, pp. 264–269.

Parra-Montesinos, G. J. (2006) 'High-performance fiber-reinforced cement composites: a new alternative for seismic design of structures,' *ACI Structural Journal*, 102(5): 668–75.

Paris, P. C., Erdogan, A. (1963) 'A critical analysis of crack propagation laws,' *Journal of Basic Engineering ASME*, 85: 528–34.

Persson, B. (1998) 'Experimental studies on shrinkage of high-performance concrete,' *Cement and Concrete Research*, 28(7): 1023–36.

Polder, R. B. (2001) 'Test methods for on site measurement of resistivity of concrete – a RILEM TC-154 technical recommendation,' *Construction and Building Materials*, 15: 125–31.

Poon, C. S., Shui, Z. H., Lam, L. (2004) 'Compressive behavior of fiber reinforced high-performance concrete subjected to elevated temperatures,' *Cement and Concrete Research*, 34: 2215–22.

Powers, T.C., Copeland, L.E., Mann, H.M. (1959) 'Capillary Continuity or Discontinuity in Cement Paste', *Journal of PCA Research and Development Laboratories*, 1(2), pp. 38–48.

Purkiss, J. A. (1985) 'Some mechanical properties of glass reinforced concrete at elevated temperatures,' in: *Proc. 3rd Int. Conf. on Composite Structures*, Paisley College, I.H. Marshall ed., Elsevier Applied Science: pp. 230–41.

——.(1988) 'Toughness measurements on steel fibre concrete at elevated temperatures,' *International Journal of Cement Composites and Lightweight Concrete*, 10: 39–47.

——.(1984) 'Steel-fibre-reinforced concrete at elevated temperatures', *International Journal of Cement Composites and Lightweight Concrete*, 6(3): 179–84.

Purnell, P., Beddows, J. (2005) 'Durability and simulated ageing of new matrix glass fibre reinforced concrete,' *Cement and Concrete Research*, 27: 875–84.

Radomski, W. (1981) 'Application of the rotating impact machine for testing fibre-reinforced concrete,' *International Journal of Cement Composites and Lightweight Concrete*, 3(1): 3–12.

Ramakrishna, G., Sundararajan, T. (2005) 'Impact strength of a few natural fibre reinforced cement mortar slabs: a comparative study,' *Cement and Concrete Composites*, 27: 547–53.

Ramakrishnan, V., Lokvik, B. J. (1992) 'Flexural fatigue strength of fiber reinforced concretes, in: *Proc. Int. RILEM/ACI Workshop HPFRCC*, W. H. Reinhardt and A. E. Naaman eds, London; Spon/Chapman and Hall: pp. 271–87.

Reinhardt, W. H. (1982) Concrete under impact loading. Tensile strength and bond, *Heron*, 27(3): TU Delft, The Netherlands, 1–48.

Rongbing, B., Jian, S. (2005) 'Synthesis and evaluation of shrinkage-reducing admixture for cementitious materials,' *Cement and Concrete Research*, 35: 445–8.

Rostasy, F. R., Pusch, U. (1987) 'Strength and deformation of lightweight concrete of variable moisture content at very low temperatures,' *International Journal of Cement Composites and Lightweight Concrete*, 9(1): 3–17.

Rostasy, F. S., Sprenger, K. H. (1984) 'Strength and deformation of steel fibre reinforced concrete at very low temperature,' *International Journal of Cement Composites and Lightweight Concrete*, 6: 47–51.

Rüsch, H. (1960) 'Researches toward a general flexural theory for structural concrete,' *Journal of the ACI*, 57: 1–28.

Sahmaran, M., Li, V.C. (2007) 'De-icing salt scaling resistance of mechanically loaded engineered cementitious composites', *Cem. and Concr. Res.*, 37: 1035–46.

Sarja, A., ed. (2000) 'Durability design of concrete structures. Committee report 130-CSL,' *Materials and Structures RILEM*, 33: 14–20.

Sąsiadek, S. (1980) 'Fatigue strength of concrete with small aggregate grains and steel-fibre-reinforcement' (in Polish), *PhD Thesis*, Cracow Technological University, Cracow.

Sąsiadek, S. (1991) 'Fatigue strength of concrete with limestone aggregate,' in: *Proc. Int. Symp. 'Brittle Matrix Composites 3'*, *Warsaw*, Brandt, A.M., Marshall, I. H., eds, Elsevier Applies Science, London, pp. 148–53.

Savva, A, Manita, P., Sideris, K.K. (2005) Influence of elevated temperatures on the mechanical properties of blended cement concretes prepared with limestone and siliceous aggregates. *Cement and Concrete Composites*, 27: 239–48.

Schiessl, P. (1983) 'Carbonation rate as function of w/c ratio,' in: *Proc. CEB-RILEM Int. Workshop 'Durability of Concrete Structures,'* Copenhagen; Bull.d'Inf.: CEB no 152.

Shah, H. R., Weiss, J. (2006) 'Quantifying shrinkage cracking in fiber reinforced concrete using ring test,' *Materials and Structures RILEM*, 39: 887–99.

Shayan, A., Morris, H. (2001) 'A comparison of RTA T363 and ASTM C1260 accelerated mortar bar test methods for detecting reactive aggregates', *Cement and Concrete Research*, 31: 655–663.

Singh, S. P., Kaushik, S. K. (2003) 'Fatigue strength of steel fibre reinforced concrete in flexure,' *Cement and Concrete Composites*, 25: 779–86.

Sisomphon, K., Franke, L. (2007) 'Carbonation rates of concretes containing high volume of pozzolanic materials,' *Cement and Concrete Research*, 37: 1647–53.

Skinner, B. J. (1966) *Thermal expansion, Handbook of Physical Constants*, sec.6, Geological Society of America, vol. 97.

Steffens, A., Dinkler, D., Ahrens, H. (2002) 'Modeling carbonation for corrosion risk prediction of concrete structures,' *Cement and Concrete Research*, 32: 935–41.

Suaris, W., Shah, S. P. (1983) 'Properties of concrete subjected to impact,' *Journal of Structural Engineering*, 109: 1727–41.

Subramaniam, K. V., Shah, S. P. (2003) 'Biaxial tension fatigue response of concrete,' *Cement and Concrete Composites*, 25: 617–23.

Sukontasukkul, P., Nimityongskul, P, Mindess, S. (2004) Effect of loading rate on damage of concrete. *Cement and Concrete Research*, 34: 2127–34.

Sukontasukkul, P., Mindess, S., Banthia, N. (2005) Properties of confined fibre-reinforced concrete under uniaxial compressive impact. *Cement and Concrete Research*, 35: 11–18.

Swamy, R. N., Theodorakopoulos, D. D., Stavrides, H. (1977) 'Shrinkage and creep characteristics of glass fibre reinforced cement composites,' in: *Proc. Int. Congress on Glass Fibre Reinforced Cement*, Brighton: pp. 75–96.

Tandon, S., Faber, K. T. (1999) 'Effects of loading rate on the fracture of cementitious materials,' *Cement and Concrete Research*, 29: 397–406.

Torrent, R. J. (1999) 'Gas permeability of high-performance concrete – site and laboratory tests,' in: *ACI SP 186 High-Performance Concrete: Performance and Quality of Concrete Structures: pp. 293–308.*

Valenza II, J. J., Scherer, G. W. (2007) 'A review of salt scaling, 1. Phenomenology, 2. Mechanisms,' *Cement and Concrete Research*, 37: 1007–21; 1022–34.

Vandewalle, L. (2000) 'Concrete creep and shrinkage at cyclic ambient conditions,' *Cement and Concrete Composites*, 22: 201–8.

Wang, N., Mindess, S., Ko, K. (1996) 'Fibre reinforced concrete beams under impact loading,' *Cement and Concrete Research*, 26: 363–76.

Wang. W., Wu, S., Dai, H. (2006) 'Fatigue behavior and life prediction of carbon fiber reinforced concrete under cyclic flexural loading,' *Materials, Science & Engineering*, A 434: 347–51.

Watstein, D. (1953) 'Effect of straining rate on the compressive strength and elastic properties of concrete,' *Journal of the ACI*, 49: 729–44.

Weber, S., Reinhardt, H.W. (1997) 'A new generation of high performance concrete: concrete with autogenous curing,' *Advanced Cement Based Materials*, 6: 59–68.

Wight, J. K., Parra-Montesinos, G. J., Lequesne, R. D. (2007) 'High-performance fiber reinforced concrete for earthquake-resistant design of coupled wall systems,' in: *Int. Workshop HPFRCC 5, RILEM, Proc.53*, W. H. Reinhardt and A. E. Naaman eds: pp. 481–8.

Wittmann, F. H. (1984) 'Influence of time on crack formation and failure of concrete,' *NATO Adv. Res. Workshop*, S. P. Shah, ed., Evanston, IL; Northwestern University: 593–616.

Yan, D., Lin, G. (2006) 'Dynamic properties of concrete in direct tension,' *Cement and Concrete Research*, 36: 1371–8.

Yan, H., Sun, W., Chen, H. (1999) 'The effect of silica fume and steel fiber on the dynamic mechanical performance of high-strength concrete,' *Cement and Concrete Research*, 29: 423–6.

Ye, G. (2005) 'Percolation of capillary pores in hardening cement pastes,' *Cement and Concrete Research*, 35(1): 167–76.

Zhang, J., Li, V. C. (2002) 'Monotonic and fatigue performance in bending of fiber-reinforced engineered cementitious composite in overlay system,' *Cement and Concrete Research*, 32: 415–23.

Zhang, J., Stang, H., Li, V. C. (1999) 'Fatigue life prediction of fiber reinforced concrete under flexural load,' *International Journal of Fatigue*, 2: 1033–49.

Zielinski, A. J. (1982) *Fracture of Concrete and Mortar under Uniaxial Impact Tensile Loading*, Delft, The Netherlands; Delft University Press.

Standards

ACI 209R-92 (reapproved 2008) *Prediction of Creep, Shrinkage and Temperature Effects in Concrete Structures.*

ACI 209.1R-05 *Guide to Factors Affecting Shrinkage and Creep of Hardened Concrete.*

ACI 216.1-97 *Standard Method for Determining the Fire Resistance of Concrete and Masonry Construction Assemblies.*

ACI 544.2R-89 (reapproved 1999) *Measurement of Properties of Fiber Reinforced Concrete.*

ACI (1999) 'Prediction of creep, shrinkage and temperature effects in concrete structures,' *ACI Manual of Concrete Practice, Com. 209, Part 1.*

ASTM C672/C672M-03 (2003) 'Standard test method for scaling resistance of concrete surfaces exposed to de-icing chemicals,' *Annual Book of ASTM Standards 2004*, Volume 04.02, ASTM International, PA.

ASTM C 666/C666M –03 (2003) 'Standard test method for resistance of concrete to rapid freezing and thawing,' *Annual Book of ASTM Standards 2004*, Volume 04.02, ASTM International.

ASTM C684-99R03 (2003) *Test Method for Making, Accelerated Curing, and Testing Concrete Compression Test Specimens.*

ASTM C457-06 (2006) *Standard Test Method for Microscopical Determination of Parameters of the Air-Void System in Hardened Concrete.*

BSI 7543 (1992) *British Standard: Guide to Durability of Buildings and Building Elements, Products and Components.*

CEB-FIP (1990) *Model Code for Concrete Structures*, Lausanne; Comité Euro-International du Béton (CEB).

CEB (1992) 'Durable concrete structures,' *Design Guide*. Bull. d'Inf. No 182 and 183, London; Thomas Telford Publishing.

EN 1992-1-1 Eurocode 2, (2004) 'Design of concrete structures,' *Part 1: General Rules and Rules for Buildings*, 1-2, Structural fire design.

EN 12390 (2007) 'Testing hardened concrete,' *Part 9: Freeze-Thaw Resistance – Scaling.*

RILEM (1999) 'Technical recommendation: tests for gas permeability of concrete, TC 116-PCD: permeability of concrete as criterion of its durability, *Materials and Structures*, 32: 174–9.

RILEM (2004) 'Report of TC 176-IDC: internal damage of concrete due to frost action,' *Materials and Structures*, 37: 740–2.

12 Design and optimization of cement-based composites

12.1 Methodology of design

The selection and design of a material is quite similar to the design of a structure or any other product. Both are more-or-less iterative processes in which an acceptable design is obtained starting from some theoretical basis then by successive trial and verification. This should reach a well-defined aim: all design parameters should be specified to enable execution in an unambiguous way. The final result should satisfy all imposed conditions during exploitation as to the strength, durability and aspect of the material. Additionally, the final product should correspond to the conditions of execution and of economy and ecology during its service life, including its construction, demolition and possible reuse.

However, there is an important feature in designing cement-based materials: it is carried on with numerous uncertainties in relation to the properties of the components, their effective volume fractions, future conditions of execution and cure, etc. The result is subject to random variations of all properties. That is why corrections and adjustments of composition, selection of components and technologies are necessary before the final result is obtained. Then the realised design of a material is again verified in local conditions. The personal knowledge and experience of the designer are considerably valued in that process. The verifications are executed by calculation and testing and more often by a combination of both.

In the design of structures, the randomness of the dimensions of elements and of mechanical properties of materials are covered by a system of safety coefficients. Trial assembling to allow for corrections of dimensions is rare and limited to very special structures. Also, acceptance tests of structures are not necessarily followed by corrections; in the case of materials the procedures are different.

Strong variability of the material's components and their properties is important and leads to a situation in which using the same nominal technologies and components means the final results are always somewhat different. Uncertainty of the final material's properties and its behaviour in service conditions is also related to the variability of the conditions of execution, curing and ageing. All these data are assessed quantitatively, bearing in mind that the random distribution of final results is unavoidable. As full information about the statistical distribution of all parameters is not available, experimental verification and appropriate safety coefficients are therefore necessary. The strength and all other material properties are considered as random variables. The statistical distribution of the strength is particularly important and not only its mean value, because randomly distributed weak regions may determine the safety of the entire structure. One of the objectives of the mix design and the execution technology for concrete-like composites is to achieve the smallest possible scatter of strength and other properties.

The design may be accompanied and preceded by an optimization approach, which is described briefly in Section 12.7. The design methodology is presented in a schematic form in Figure 12.1.

The preliminary selection of materials is the first step. Here the decision is based on initial input data: functions, requirements, environmental condition, etc., which are used in a somewhat vague and intuitive way with personal experience and the rational belief of the designer. This decision is derived from apparently obvious arguments, which include basic mechanical properties, service requirements, total cost, etc. In various cases the choice between steel, concrete or wooden structures is more or less evident; sometimes it requires serious studies and analyses. If the cement-based composite materials are selected, then the material's design is initiated.

In the second step, the data about components and possible technologies are collected, together with requirements on strength, durability, etc.

The third step is again preceded by a decision: is the mix design to be performed with or without application of an optimization approach? In both cases, iterative procedure is necessary. However, it is believed that the left-hand path in Figure 12.1 is shorter and leads to better results, because objective indications from the optimization approach are used. The tentative design may in many cases be started with a very distant solution from the final one. The end of the design is presented in the form of a list of components with their volume or mass fractions, computed for one cubic meter or for one batch.

The fourth step corresponds to multiple checking on the site during execution and making possible corrections. For practical reasons, the design procedures are presented in a few stages corresponding to separate operations.

In this chapter, the standard mix design methods for concretes are only briefly described as they are presented in full detail in many manuals, e.g. de Larrard (1998), Day (1999), Dewar (1999), and Kett (1999). Design and detailing of structural elements with ordinary concrete, high performance concrete or fibre-reinforced concrete is not considered here.

12.2 Requirements imposed by the conditions of execution

In the execution of concrete structures, several processes are involved that should be carried on and lead to the results expected in design. These

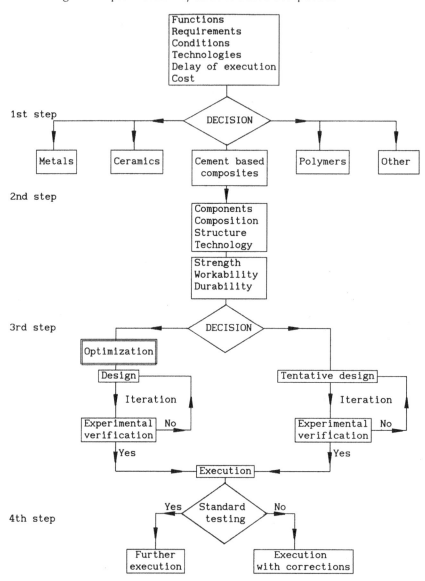

Figure 12.1 Schematic of the design methodology for cement based materials.

processes are: transporting, pumping, pouring, casting, spreading and compacting, etc. Additionally, for certain cases there is injection, spraying,
self-levelling, trowelling and moulding, because the consolidation and all
finishing operations should not introduce any detrimental segregation. In
the initial stage the cement-based composites behave as more-or-less viscous
fluids and this stage usually lasts up to two hours approximately. During that

time (and later) as a result of the hydration processes the fresh mix is continually transformed into a solid. The properties in liquid state are essential for the above processes. The expected results depend on the rheology of the fresh concrete. The following properties of the fresh concrete should reach appropriate values, Banfill (2006):

- flow and frictional resistance against surfaces;
- adhesion;
- resistance to segregation, settlement and bleeding;
- low water content to ensure required strength and density;
- resistance to sagging under self weight on vertical or inclined surfaces;
- low pressure on the formworks.

The material behaviour in the fresh stage may be represented by different rheological models in which shear stress and displacements are related. Two of these models are described below. According to Newton's model, shear stress τ is related to coefficient of viscosity η and to the rate of shear γ by the following equation: $\tau = \eta\dot{\gamma}$. This is the model of an ideal fluid and even fresh cement paste behaves in a different way than that due to many factors; for example, attraction between cement particles in a relatively dense suspension.

The Bingham model (two-point model) allows for an initial shear stress τ_o which represents a threshold before any displacement may occur: $\tau = \tau_o + \mu\dot{\gamma}$. Two constants τ_o – yield value and μ – plastic viscosity, characterize the viscous fluid considered. The Bingham model is used to describe cement matrix behaviour with satisfactory results. It is combined with thixotropic effects in which the apparent viscosity decreases with shear stress application. Both the Newton and Bingham models are presented in Figure 12.2. Several more complicated models exist in which other factors are considered; for examples see Tattersall and Banfill (1983) and Banfill (2006).

The selection of an appropriate model is made according to its ability to represent experimental results and also its simplicity. The properties of cement paste are not the only factors determining the behaviour of cement-based composites, concretes, mortars, fibre concretes, etc. Other components of the fresh mix, their volume fractions and properties are also important. That is why, for practical considerations, it is not necessary to use very complicated models for the cement paste.

The appropriate behaviour of the fresh mix is obtained in the material's design by selection of its composition and structure and particularly by the application of admixtures (cf. Section 4.3), then the workability of the fresh mix is controlled and possibly corrected in situ.

The general term 'workability' is used to express these overall properties of the fresh mix and is applied to all kinds of concrete. Special concretes that do not require any additional operations for placing and compacting, so-called self-compacting concretes (SCC) and self-levelling concretes (SLC)

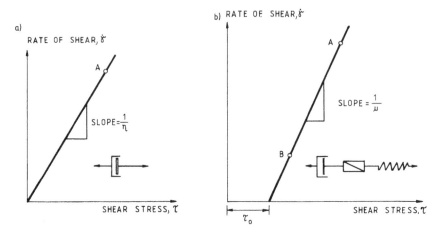

Figure 12.2 Two simple rheological models used for fresh cement paste:
  (a) Newton model
  (b) Bingham model.

are considered in Chapter 13. The workability may be also defined by the amount of energy necessary for its compaction. Workability is considered in the mix design and also takes into account the methods of execution: concrete may be batched and mixed in situ or in a pre-cast plant; in most cases it is delivered on site as a ready-mixed concrete in special trucks. In this case, the duration of the transport should be considered in designing the concrete's workability.

As measures of workability of the fresh mix, several different tests are proposed, but each of them gives a certain indication which is not comparable with the others. The slump test (Abrams cone) and the Vebe test give only single values by which the mix is characterized, in millimeters and seconds, respectively. Therefore, nothing more complicated than the Newton model is used and the test results do not characterize the mix without ambiguity. However, these and other similar one-point tests are used frequently, mainly for their simplicity and, in most cases, acceptable repeatability of results. The required workability is expressed as minimum slump, maximum Vebe time, or any other measure, and in many cases such a simple result is sufficient; however, it is necessary to understand its limitations. In certain compositions the results are ambiguous or even impossible to execute: very stiff mixes cannot be tested simply and compared with valid results; the same concerns very fluid mixes with superplasticizers.

Special measures of workability may be necessary when improved resistance to segregation is required for special reasons. Estimation 'by eye' given by an experienced practitioner is also valid. The relation between Bingham model parameters and slump test results has been studied by Wallevik (2006) who has shown that the slump may be, to some extent, predicted from the yield stress τ_o, but this dependence becomes weaker with increasing workability of the mix. However, there was only a low correlation observed between the slump and plastic viscosity μ. Therefore, various methods are applicable but in different circumstances: the more sophisticated and difficult – in a laboratory, simple tests – in situ. The main and non-negligible advantage of simple tests is that their result is expressed by one single value.

On the other hand, more complex tests are suitable for advanced laboratories and even then only approximate results may be expected. There are serious shortcomings for every test of that kind related to different simplifications and limits of applicability.

The two-point tests are executed on various types of coaxial cylinder viscometers (rheometers), which are basically composed of two cylinders. The space between them is filled with the fresh mix under examination (Figure 12.3). The external cylinder is rotated and the movement is transferred through the liquid to the internal cylinder, where the torque is measured. The dimension of the maximal aggregate grain is decisive on the selection of a viscometer.

Another type of rheometer with parallel plates was designed and tested by Kuder *et al.* (2007) to evaluate the rheological behaviour of the fibre-reinforced cement matrices that are too stiff for testing with ordinary rheometers as shown in Figure 12.3. It appeared from the tests of fibre-reinforced mortars that at lower fibre amounts the properties of the mix were decided by the matrix and the fibres only reduced its yield stress and viscosity. At higher fibre contents, the mechanical interlocking and entangling of fibres determined the mix behaviour. Similar influence of the fibres may be expected for concrete mixes also.

The analysis of the results allows both constants in the Bingham equation to be determined. Different types of such computer-assisted equipment are available, but still the measurement techniques are not perfect and the results are not entirely independent of the measurement method.

In the case of fibre reinforced composites a special measure of workability was proposed by ACI 544.2R-89 called inverted slump cone. A standard Abrams cone is suspended upside down and filled with tested material without any compaction. The internal vibrator is activated and the bottom cover of the cone is removed, then the time of flow out of the material is recorded, which is usually between 10 and 30 seconds. This method is still in use, but is subject to some criticisms. Other methods of the flowability of the fresh mix with dispersed fibres are proposed by Laboratoire des Ponts et Chaussées in France (cf. Section 12.5).

It has been observed that with high fibre content both τ_o and μ are increased, while the length of the fibres influences only the plastic viscosity μ. However, the main factor is the amount of dispersed fibres (cf. Chapter 5) when ordinary technology is applied. For higher fibre content non-conventional methods of placing the fresh mix in the forms like SIFCON are necessary (cf. Section 13.5).

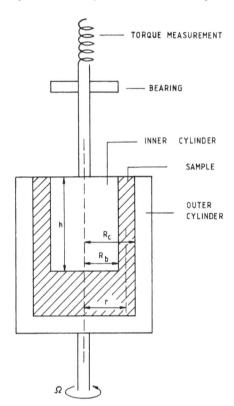

Figure 12.3 Schematic of a coaxial cylinder rheometer, after Tattersall and Banfill (1983).

Appropriate workability should ensure easy and correct placing of the fresh mix in the forms, without excessive bleeding and segregation. The workability should be maintained during a certain timeframe, for example one hour or more, which is needed for necessary transportation and operations on the site.

All mix components contribute to the final workability. Large grains reduce the volume of voids and consumption of the cement paste and consequently improve the economy of the mix's composition. In many cases the continuous sieve curve is the best solution, but the standard regulations may also be satisfied with a step sieve curve, that is, composed only of a few selected fractions of the coarse aggregate and without the others. The maximum size of aggregate grains must be related to the dimensions of elements to be cast and to the density of reinforcement.

The voids between coarse aggregate grains are reduced by the spherical shape of grains with different dimensions, which are also more easily compacted. The general condition imposed on the aggregate grading is the minimum of voids, that is, the maximum density: small particles in the aggregate skeleton fill the voids between larger ones. The smallest particles of microfillers like SF or fly ash require more water in the mix.

The simplest way to improve workability, by the addition of water, precludes obtaining designed strength and high performance in general. It is therefore necessary to maintain the amount of water as low as possible and there is a certain conflict between the amount of water, which should be kept as prescribed for the required strength and the spontaneous trend on the site to add water for easier placing. The designed composition must be maintained and carefully controlled.

The behaviour of the fresh mix is improved by the application of air-entraining agents and water-reducing admixtures; the latter exist in consecutive generations known as plasticizers or superplasticizers. The mechanisms of increasing the flow of the fresh mix are based on the formation of a kind of double layer around each cement particle that facilitates reciprocal displacements and the decreasing of interparticle attraction forces transforming them into repulsion. There are superplasticizers containing reactive polymers, which allows the slump of the fresh mix to be maintained at the range of approximately 200 mm during the time necessary for casting and compaction. Here, not only the properties of the cement paste are of importance, but also the amount of the paste. The obtained effects may be enhanced in a classic way by mechanical vibration, but also flowing concrete is used (SCC and SLC mentioned above) that do not require any vibration.

Considering the behaviour of the fresh mix in the context of the Bingham model it should be mentioned that the superplasticizers mostly decrease the yield value τ_o and air-entrainers decrease the plastic viscosity μ.

12.3 Requirements imposed on hardened material

Requirements imposed by structural applications and related to the properties of the hardened material concern basically compressive and tensile strength, frost resistance, impact strength and durability. The mechanical properties determine the dimensions of the structural elements and a conflict may be specified between the requirement of high strength and the difficulties of obtaining it with given components and in given conditions in situ or in a pre-cast factory. The requirement of high strength and other properties of hardened material should be accompanied by a system of frequent controls at all consecutive stages: quality of components, composition of the mix, methods of placing and compaction and cure of hardening elements.

The strength at an early age is often specified for reasons of serviceability or economy. Special requirements formulated for high performance materials are described in more detail in Section 13.4.

Durability is another requirement which is imposed on the hardened material, particularly for application in outdoor structures. This means that its behaviour during the service life and in given conditions should satisfy imposed limits. Durability is related to the maintenance cost and for many structural applications it is as important as the strength; both requirements often can be satisfied simultaneously (cf. Chapter 11.5). In the mix design the condition of durability means requirements of high density and impermeability of the hardened composite and of its high resistance against cracking.

The compressive strength which is considered in this chapter and presents a basis for mechanical characterization is so-called characteristic strength. In the structural design another notion is applied, termed 'design strength,' which is calculated from the previous one taking into account assumptions concerning the general format of structural design. For all problems related to the design strength the reader is referred to manuals and standards for structural design and verification of reliability.

Partial replacement of Portland cement by natural or artificial pozzolans reduces the heat of hydration and also the unit price of concrete; it slows down the process of hardening and the early strength is lower. Nevertheless, the majority of the volume of structural concrete used in technically advanced regions of the world are based on rational compositions of Portland cement and so called secondary binding materials: fly ash, ground granulated blast-furnace slag, metakaolin, SF, and others.

12.4 Other requirements

Economic conditions, understood as design requirements, concern the specific cost of the components and labour, all foreseeable maintenance and repair costs, as well as the possible cost of destruction and demolition of concrete structures at the end of their service life. The technical requirements should be satisfied in an economical way and this condition is understood in a general sense. Therefore, it is necessary to take into account that better quality of components and higher level of execution will increase the specific cost of the material, but probably also the quality of the product; this means the strength and durability of the structure will also be higher. The cost of the components is certainly the easiest to calculate and there is often a trend towards making economies on the consumption of the most expensive component, which may be Portland cement or fibres. In a correct calculation of the total cost it usually appears that the expenses for maintenance of the structure in its lifetime may be of similar importance.

Other more complex requirements, such as the outward aspect of the structure or resistance against particular external factors may be derived from these basic ones and are taken into account in a direct way, for example, by an imposed method of finishing of the external surfaces.

Several requirements imposed on cement-based composites conflict. For obvious reasons high strength and improved durability increase certain components of the cost. Materials for outstanding structures or for elements in a particularly corrosive environment are more expensive. Therefore, in the material's design some kind of compromise is always looked for to accommodate contradictory conditions, either by trial and error methods or, better, by the solution of an optimization problem.

The restrictions on the mix's composition of concrete-like composites are given by standards and recommendations in various countries and regions, which are appropriate for the given conditions. In some cases they are imposed in the form of laws that must be complied with, in others they are regulations that may be compulsory for other reasons, for example, the insurance of buildings. Standard requirements may concern the lowest admissible qualities of cement for certain categories of structures, minimum and maximum contents of cement or recommended sieve distribution of aggregate grains. These restrictions limit the freedom of the designer and present constraints in the design process, nevertheless they are intended to ensure that the minimum acceptable quality of the product is maintained.

Two basic principles are accepted for nearly all methods of mix design:

1   The relation between w/c ratio and compressive strength of hardened material in the form of an equation as proposed by Féret, Abrams or Bolomey (cf. Section 8.2).
2   Aggregate grading ensuring the best packing, which means leaving minimum voids to be filled by cement paste (cf. Section 4.2.5).

The first principle may be modified when admixtures, improving strength, or reinforcement with dispersed fibres are used. However, it does give a basic relationship which is generally valid for all cement-based materials after minor adjustments allowing for local conditions. It is assumed that plasticizers and superplasticizers do not directly influence the strength and other mechanical parameters of the hardened material – they only improve workability during the initial time necessary for placing. In other words, their role is to make low values of w/c ratio acceptable from the viewpoint of the workability of the fresh mix, although it should be taken into account that certain admixtures, such as air-entraining agents, influence both the fresh mix and the hardened composite. In fact, high quality superplasticizers improve the final strength, often as much as by 15–20%, due to better packing of the particles. This is a supplementary increase besides that due to lower w/c ratio.

The second principle was initially presented in the form of an optimum grading curve according to Fuller and Thompson (1903) and translated into grading limits imposed by various standards and recommendations (cf. Chapter 4.2). Such an ideal or optimal aggregate grading is usually impossible to apply for various reasons:

•   for good workability more fine particles are needed;
•   ideal grading is rare in natural aggregate resources and in normal

production, and it would be very expensive to modify the available aggregate to obtain the optimum grading. It is believed that excessive differences with grading would be impractical and uneconomical and some compromise between an optimal grading and that as obtained from the quarry is necessary. Moreover, the great range of size fractions may excessively increase the cost of stockpiling.

When more fine aggregate is used, it means that a lower fineness modulus of the aggregate is accepted, so lower values of the Young's modulus may be expected with all other parameters remaining constant. Also, increased cement content may be necessary in that case to maintain the required strength, and higher shrinkage is then very probable. However, these disadvantages may be balanced by the lower cost of the aggregate.

12.5 Review of concrete mix design methods

There are many methods of concrete mix design and all of them are computer-assisted, for example, FirstMix, Mix III, MixSim, Betonlab Pro2 (2000), etc. In these methods, which are available on the market but also free of charge through the internet, various computer programs for concrete design are proposed. After the selection of the types of components, limits of their volume or mass fractions and various requirements, the solutions are obtained in the form of proportions of components. In some programs, optimization procedures are included with the total cost as the optimization criterion and other variables (strength, density, etc.) are imposed. In the multi-criteria approach, several variables may be used to formulate criteria for optimization. If the optimization approach is not included, then repeating computation with different values of selected variables, different solutions can be obtained to choose the most appropriate components, using the designer's experience and rational belief.

Brief remarks are given below that relate to the original bases of the mix design methods, also called 'selection of the mixture proportions.' These remarks are mainly of historical importance, but they are still interesting because they help the basis of the computerized programs that are exclusively used at present to be understood.

The designations of methods 'French' or 'Polish' have no general meaning and are used here only to indicate the origin of their written presentations. It is obvious that in various countries different methods were traditionally developed and modified according to available components, local requirements and applications. If properly used, all methods led to good results even if they were developed from slightly different assumptions. A large input of experimental data was a factor that ensured convergent results even when starting points were different.

The methods discussed here are called 'analytical' as they are based on certain analytical relations which have been derived from practical observations and experience. It was always assumed that the results obtained were verified experimentally and some corrections were introduced – usually first in a laboratory, and then on the site.

(a) American Concrete Institute (USA) method

The ACI method was largely used in North America. The design was derived from a set of tables where basic assumptions were presented in the form of simple values for application. First, the maximum aggregate size and the workability expressed by slump had to be selected according to the type of structure. Then, for that data the necessary amount of water was estimated for two cases: air-entrained and non-air-entrained concretes. In the same table, the approximate volume of air was given; air that was entrapped during mixing and also air entrained if it was provided, according to exposure to freeze-thaw cycles. The *w/c* ratio for the required compressive strength was selected from the next table, based on the Abrams' formula (cf. Section 8.2). The volume of coarse aggregate was again given in a table according to two parameters: maximum grain diameter and the fineness modulus of sand. The last component – sand – was determined as the difference between a unit volume and the sum of the volumes of all other components: cement, water, coarse aggregate and air.

   The ACI method was based entirely on experience over a long period and its application, with the tables, was easy. The results had to be corrected according to particular requirements and local conditions concerning workability, strength, durability, aggregate moisture, etc., and to possible application of admixtures (ACI 211.1-91 2002).

(b) United Kingdom method

This method was based on the Road Note No 4 (1950) and later a version was published by Teychenne *et al.* (1975), which was a further development.

   At the beginning of the design procedure (divided into five stages) the grading of the aggregate was determined and it had to be within the imposed limit curves. The minimum amount of cement was imposed as a function of the kind of concrete structure (plain, reinforced or prestressed) and aggressiveness of environment. For a given aggregate and *w/c* ratio = 0.5 the compressive strength at age of 28 days was found from a table and derived from previous experience. These two values determined a point that was placed between a set of curves based on Abrams' formula. Then, an appropriate curve was interpolated as belonging to the same family, passing through the point mentioned above. It was used to determine the *w/c* ratio for the required strength (cf. Figure 12.4). The water content for a given quality of coarse aggregate and for required slump was taken from another table, and therefore the amount of cement was easily calculated. The total amount of aggregate was established from tables for four different workabilities. The

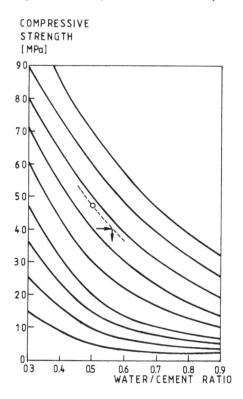

COMPRESSIVE
STRENGTH
[MPa]

WATER/CEMENT RATIO

Figure 12.4 Set of curves representing relation between compressive strength and *w/c* ratio for the UK mix design method, after Teychenne *et al.* (1975).

amount of sand was established, again using a series of tables for various values of slump, etc.

(c) Baron-Lessage (French) method

This method was proposed by R. Lesage and later developed among the others by J. Baron and R. Lesage (1976). It was based on two traditional approaches: by J. Faury and A. Joisel, and by R. Vallette. Both methods together with third called G. Dreux's method were in use in France. The authors of the Baron-Lesage method proposed to start with a selection of the ratio between fine and coarse aggregate to obtain the best workability, which is the most important criterion of the mix quality. The workability for that method is determined on a special LCL Workabilimeter shown in Figure 12.5, on which a result is expressed as flowing time.

The design was initiated by selecting the *w/c* ratio and the amount of cement that followed the general indications for the kind of work intended. The initial data selected do not have much influence on the final result and

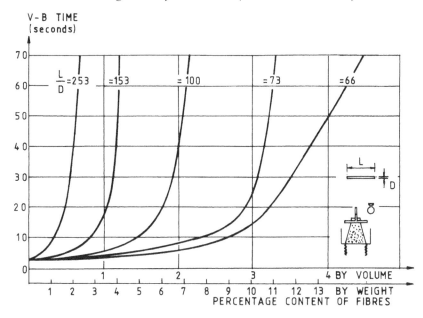

Figure 12.5 Influence of steel fibre volume fraction on the workability of cement-based materials for different fibre aspect ratios ℓ/d, after Edgington *et al.* (1974).

should only ensure that the flow time was contained within the limits of 15 and 40 seconds. Then, the minimum flow time was determined experimentally for various values of fine to coarse aggregate ratio (s/g). Usually not more than five experimental mixes were needed and it depended how many different sizes of aggregate were distinguished for design purposes. In a more-or-less complicated way the optimum aggregate size distribution was determined for the minimum flow time.

The next stage of the design was establishing the actual *w/c* ratio as a function of the required flow time and related to the kind of structural elements to be cast, difficulty of placing and vibrating, etc. For that aim, a special table with recommendations was available, and again a few mixes were needed with only one variable: volume of water. The last step of design was calculation of mix composition for 1 m³ and final verification together with possible corrections for all additional factors and effects, which were not considered in the design procedure.

In the Baron-Lesage method, the application of superplasticizers modified only the flow time. When fibres were added, then accordingly the amount of coarse aggregate had to be decreased to compensate for their influence on the flow time.

(d) Polish method

The method used in Poland was based on the solution of three linear equations, which expressed strength, density and water requirements of particular fractions of aggregate.

The compressive strength was calculated after the so-called Bolomey's formula (cf. Section 8.2) from equations of the following form:

$$f_{28} = A_1\left(\frac{1}{w/c} - 0.5\right) \text{ for } w/c > 0.4$$

$$f_{28} = A_2\left(\frac{1}{w/c} + 0.5\right) \text{ for } w/c \leqslant 0.4$$

Both constants A_1 and A_2 were considered after quality of cement and aggregate.

The second equation, expressing density, reflected the fact that the sum of volumes of all components should be to 1 m³, and had the following form:

$$\frac{C}{\rho_c} + \frac{K}{\rho_k} + W = 1.0 \text{ [m}^3\text{]}$$

where C and K are masses of cement and aggregate and ρ_c and ρ_k are specific densities of cement and aggregate, respectively.

The third equation expressed the amount of water W required for cement and for aggregate:

$$W = C\,w_c + K\,w_k$$

where, according to Bolomey, $w_c = 0.23$ and according to Stern w_k may take values from 0.23 up to 0.27 for concretes of low and high workability, respectively.

In these three equations the unknown values were: w/c, C and K. As in other methods minimum amount of cement and aggregate grading were imposed by standard regulations.

In all four methods presented above it was assumed that after the calculation of masses of all components, a trial batch should be executed and further corrections were necessary to adjust standard coefficients and data from tables to local conditions and actual components, their moisture, etc.

The design methods for ordinary concretes are proposed by many authors and firms, also in the form of user friendly computer programs, which provide quick answers to basic problems concerning the mix design for required properties. The programs are built after a few available analytic formulae and using important collections of experimental results. The computer-aided design of materials is aimed at easy exploitation of all previous experience in that field. Moreover, the final results are less dependent on the personal ability of the designer. That is valid in most ordinary problems where requirements and components correspond to normal practice. These programs, as well as the abovementioned design methods do not cover all varieties of admixtures and types of dispersed reinforcement which are available to designers for special concretes: high or very high performance, with increased resistance against corrosion, improved permeability, etc. In such cases, again a method of trial and error is to be applied.

12.6 Recommendations for design of HPCs, FRC and polymer-modified concretes

The design of high performance concrete (HPC), very high performance concrete (VHPC) and ultra high performance concrete (UHPC) is similar to that of ordinary concretes and selection of proportions of components is based on the influence of the *w/c* ratio on compressive strength and on optimum packing of aggregate grains to minimise voids. However, the process of mix design is more complicated because various new components should be used. These are different supplementary cementitious materials like fly ash, GGBS, SF, etc. and also short dispersed fibres of different origin. An interesting approach was published by de Larrard (1998), based on the development of the traditional relation proposed by Féret (cf. Section 8.2). Moreover, not only strength is considered, but also rheological properties of the fresh mix. That is why various computer-assisted programs for mix design are developing.

Design of high performance concretes includes application a few special measures:

- selection of the best quality of all components (aggregate, cement, secondary binders, admixtures) and optimal sieve distribution of aggregate grains;
- reduction of maximum grain of aggregate and increase of volume fraction of fine aggregate;
- selecting aggregate grains of the best shape and strength, with a good adherence to cement paste;
- increase of the cement content;
- replacement of a part of Portland cement by other pozzolanic materials, secondary binders and microfillers (fly ash, SF, limestone powder, etc.), particularly as ultrafine particles;
- verification of the mix composition in the actual technology of mixing, transportation, placing, vibration and curing in view of obtaining excellent workability;
- application of various types of chemical admixtures, particularly superplasticizers, and in that way reduction of *w/c* ratio;

- verification of whether there is no incompatibility between different components (cement and admixtures, aggregate and cement, cement and fibres, etc.).

The first requirement imposed on high performance concretes (HPC) of any kind is their good workability, and this property should be obtained with minimum water content to avoid all possible negative effects of additional water over that necessary for cement hydration, such as bleeding, segregation, difficulties in correct placing in the moulds, etc. On the other side, excess cement is also unacceptable to control shrinkage and creep.

In all cases of the material design the iterative process is used, by which the possible differences between nominal and actual component properties and conditions of execution may be corrected (cf. Figure 12.1). For the design methods, it is assumed that in the execution of concrete mix strict realization of the designed composition is ensured, without additional water to facilitate casting; this includes transportation and placing. Immediately after placing the fresh mix, a well-prepared procedure of curing should be initiated and continued during the period of time indicated, dependent on local conditions in situ. Because of the low content of water, these concretes are particularly vulnerable to desiccation. Several examples of the successful design of HPC and VHPC are given by Malier (1992), among others, together with a description of a few failures.

Nehdi *et al.* (1998) indicated the importance of ultrafine particles, in most cases made by grinding various kinds of waste materials. These particles of dimensions down to 0.3 μm replace part of the cement grains, decrease bleeding and densify the structures; the use of waste materials is also beneficial.

Recommendations for design of UHPC type concrete called Ductal® were suggested by Chanvillard and Rigaud (2003):

- *w/c* ratio around 0.2; i.e. below the quantity of water needed for full hydration of cement;
- fine grading of sand with grains # 0.6 mm;
- addition of SF and appropriate selection of chemical admixtures as a necessity;
- dispersed reinforcement with steel fibres, e.g. 2% vol of 13–15 mm long fibres.

Alves *et al.* (2004) have compared four different experimental methods of calculating the mix proportions for high strength concretes and the results have been verified experimentally. It appeared that different methods lead to concretes of minimum cost depending on their compressive strength from 45 to 90 MPa; that is, no single method gave the best results for all concretes within that scope of strength. Using basic principles and rheological models for designing high performance concrete, Sobolev (2004) proposed

Table 12.1 Examples of four HPC mixture proportions

	A	B	C	D
Mixture proportions, kg/m³				
Portland cement	426	449	468	478
silica fume	22	50	83	120
superplasticizer	2.2	5.0	8.3	12
coarse aggregate	1169	1155	1136	1119
fine aggregate	630	622	612	603
water	153	142	132	121
Mixture parameters				
silica fume (% of cement)	5	10	15	20
w/c	0.342	0.284	0.239	0.203
cement paste volume, dm³	298	307	318	327
Compressive strength, MPa				
1 day	16.8	24.1	34.4	45.1
3 days	28.6	42.2	63.0	84.9
7 days	50.1	67.2	84.8	102.5
28 days	60.0	80.0	100.0	120.0

Source: Sobolev (2004).

four different mixture proportions presented in Table 12.1; they cover concretes characterized by 28-day compressive strength from 60 to 120 MPa and may serve as examples for practical applications, perhaps with some modifications. For all four mixture proportions, certain characteristics were the same: slump 100 mm; fine to total aggregate ratio 0.35; fresh mix unit weight appr. 2400 kg/m³; and air content 2.5%. As coarse aggregate granite grains below 20 mm were used and nearly 97% of sand grains were below 2.5 mm. It is easy to observe how with increasing strength, volume fraction of SF and Portland cement increase while water content and values of *w/c* ratio are reduced. These are general indications for the mix design of HPC.

An example of HPC mixture proportions, specially proposed for pavements, is shown in Table 12.2. The application of limestone filler allows the reduction of cement content in comparison to concrete B proposed in Table 12.1 for the same strength.

The application of artificial neural networks (ANN) to the mix proportion design, using the least paste content and a mathematical model to express the relation between main variables (*w/c* ratio, fly ash/binder ratio, aggregate grain distribution) was proposed by many authors, for example, Ji *et al.* (2006) who declared better mechanical and economical properties of obtained concretes.

All stages of production of high performance concrete and similar composites should be controlled to exactly follow the design proportions

Table 12.2 Example of special high performance concrete mixture proportions for pavement

	Specifications	Theoretical	Actual
Mixture proportions, kg/m³			
Portland cement		406	408
silica fume		40.6	39
superplasticizer		4.35	5.62
coarse aggregate	60% D_{max} ≤6mm	935	912
fine aggregate	0–4 mm	623	608
water		185	190
limestone filler		101	139
Mixture parameters			
silica fume (% of cement)		10	9.5
w/c		0.46	0.47
slump (mm)		150	160
Compressive strength, MPa	28 days	80	78.1

Source: de Larrard and Sedran (2000).

and technology. In the case of difficulties, consultation from a specialized laboratory should be ensured.

There exist many recommendations for design of fibre reinforced materials together with examples of compositions applied at various occasions; also producers of fibres propose ready-to-use mix proportions. The first question to be answered in design is what type of fibres should be selected; that is, what material, dimensions and shape? The objective of fibre reinforcement should be clearly specified and the following cases may be considered:

1 Fibres are added to improve tensile strength and therefore to increase the load corresponding to the first crack in elements subjected to bending or tension, and when increased impermeability of hardened composite material is required.
2 Fibres have to increase ductility and the fracture toughness of the material by control of the cracking process; this ability may be checked by one of available fracture indices (cf. Section 10.3). The higher the concrete strength, the lower its ductility and the more fibres may be needed to control excessive brittleness.
3 Fibres are expected to improve particular mechanical parameters, such as abrasion or impact resistance, and are added for better durability or for any other reason.

In the majority of conventional structures, the fibre-reinforcement is not used to replace classic steel reinforcement in the form of steel bars, which are situated in the specified places in structural elements to carry principal tensions. Fibres are used in parallel with steel bars to improve bonding, to carry a part of the shearing stresses, to control local cracking, etc. On the other hand, different types of fibres (steel, glass, polypropylene) are used as main reinforcement in such elements as floors, claddings, and tiles. Fibres are also applied where the reinforcement is introduced for temporary loads during transportation or against shrinkage cracking. For obvious reasons the design of reinforcement in non-structural elements is mostly based on the designer's experience and on experimental verifications.

In recent years, structural elements like bridge main girders may be reinforced exclusively with steel fibres (cf. Chapter 14).

In particularly complex problems of heavily loaded elements, the application of hybrid reinforcement composed of two or more kinds of different fibres may be considered. In the selection of reinforcement the total cost (materials and execution) should also be taken into account (cf. Chapter 5). The volume of fibres is usually limited by their cost and by the workability of the fresh mix. Also the efficiency of fibres decreases when a high percentage is applied. It is rare in practical cases that a volume fraction of steel fibres higher than 2.5% is used and composites with reinforcement lower than 1% are often applied. This remark does not concern particular cases where higher volume fractions of fibres are purposefully used and where other technologies of placing are applied, for example, SIFCON (cf. Section 13.5). The application of polymer admixtures with fibres may also be foreseen, if the exploitation requirements cannot be satisfied in a less expensive way.

In most cases the random distribution (3D) of the fibres is considered in the design of fibre-reinforced concrete. If for any reason other distribution of fibres is envisaged, for example linearized fibres, 1D or 2D, or continuous fibres in the form of mats or meshes, then the design of the material's composition should be modified. For that purpose, experience from other results or a trial and error method must be employed.

The addition of fibres to a given composition of the matrix is not a correct procedure. Because worse workability always results from the addition of fibres, this should be compensated for in the design by the appropriate reduction of aggregate, increasing the water content or increasing the dosage of a superplasticizer. The fibres completely transform the behaviour of the fresh mix and also considerably change the properties of the hardened material. It is then necessary to design the composite material from the beginning either as a plain matrix, that is, no-fibre composite, or as a composite material with dispersed reinforcement, uniform distribution of fibres being ensured. For that aim there are various concepts of how to consider fibres in the modelling of the fresh mix in the design, for example, including fibres in the optimization of the solid skeleton (Ferrara *et al.* 2007).

Fibres modify the workability of the fresh mix in a similar way to coarse aggregate. It is therefore possible to counterbalance the addition of fibre reinforcement by increasing the fine-to-coarse aggregate ratio (Rossi *et al.*

1989). The influence of steel fibres on the workability of the fresh mix was studied already by Edgington *et al.* (1974) and is shown in Figure 12.5. Yu *et al.* (1992) proposed an 'equivalent packing model' relating the dimensions of a fibre as a non-spherical particle to a replacing sphere with a given diameter. Also de Larrard (1998) represented fibres as a kind of 'perturbation volume' for his compressible packing mode. Grünewald (2004) defined the 'maximum fibre factor' using the ratio of the fibre length to the maximum aggregate diameter. Bui *et al.* (2003) tried to include the fibers' contribution to the rheological model of paste through the optimization of the packing density of the fibre-reinforced solid skeleton. Ferrara *et al.* (2007) proposed to design SCC with fibres by including the fibres into the particle size distribution using the concept of an equivalent diameter, determined on the basis of the specific surface area of the fibres. Through the rheology tests on cement pastes and on plain and fibre reinforced concrete, the overall influence of fibres has been evaluated; however the distribution and orientation of the fibres was neglected.

The mix compositions with fibres are therefore characterized by higher cement content and by higher values of fine/coarse aggregate ratio to increase the amount of the cement paste. These modifications are called 'reproportioning'; some recommendations are given by Nataraja *et al.* (2005) in view of maintaining the workability by the trial-and-error method. It has been proved that with appropriate mixture proportions and efficient superplasticizer it is possible to obtain SCC with uniformly distributed short fibres (Ferrara *et al.* 2007).

The matrix itself exhibits higher porosity than in non-reinforced materials because of the air entrapped with fibres. Steel fibres are used in compositions with smaller aggregate grains: the maximum size is usually about 10 mm. The workability of the fresh mix with fibres is usually improved by superplasticizers.

General recommendations related to the mix proportions of SFRC issued by ACI Committee 544 (2002) are shown in Table 12.3. In the guidance published in Japan, the importance of high quality of all components and inspection at all stages of execution is stressed (Rokugo *et al.* 2007). To establish these values different arguments are taken into account: mechanical efficiency, workability and economy. As examples, a few mix design proportions are shown in Table 12.4. It should be reiterated here that in cement-based composite materials the fibres are by far the most expensive component, which considerably influences the total cost of materials. That aspect is considered also in Chapter 13.

The design of polymer modified concretes is based on similar reasoning as for Portland cement concretes, but there are no published programs. In PCC, the binder is composed with two phases: Portland cement and polymer phase that may be introduced to the mix as polymer latexes, and water-soluble polymers or liquid polymers. There is a continuous relation between PCC and PC: when the polymer fraction (starting from 10% mass) is increased

Table 12.3 Recommendations for mix proportions for ordinary steel-fibre-reinforced concrete

Description of mix	9.5 mm maximum-size aggregate	19 mm maximum-size aggregate	38 mm maximum-size aggregate
Cement (kgs/m^3)	356–593	297–534	279–415
w/c ratio	0.35–0.45	0.35–0.50	0.35–0.55
Fine-to-coarse aggregate, %	45–60	45–50	40–55
Entrained air, %	4–8	4–6	4–5
Steel fibre volume content, %			
deformed fibre	0.4–1.0	0.3–0.8	0.2–0.7
smooth fibre	0.8–2.0	0.6–1.6	0.4–1.4

Source: ACI 544.1R-96 (2002).

then PCC is transformed into PC where there is no Portland cement. The polymers in the form of a dispersion or polymer powders are added to the fresh mix; the following materials are mentioned: Ohama (1996), Ohama *et al.*(1997): styrene-butadiene rubber (SBR), polyethylene-vinyl acetate (EVA) and polyacrylic ester (PAE). The existence of polymer particles and their hardening influence the hydration process of the cement phase. The coexistence of both phases may have some effect on their processes of hardening and it depends on the conditions of curing. For composition with an increased polymer content, the process of cement hydration is slowed, but this is compensated for in final strength by the influence of the polymer film around cement particles. It is essential to limit any adverse influence of polymer addition on cement hydration (ACI Com.548 2003). The properties of PCC are improved with the addition of SF and short steel or polypropylene fibres; such composites are used for repair of concrete structures.

Polymer impregnated concretes (PIC) are produced on the basis of ordinary concretes that are impregnated when hardened (cf. Section 13.3). The high cost of polymer admixture in PCC and PC or of impregnation in the case of PIC should be justified by clearly established goals, for example:

• repair works and patching, where durability and good bond to old concrete are essential;
• impermeability of containers for gases or liquids or of the external cover layers in the case of a corrosive environment;
• improved durability of elements subjected to abrasion (industrial floors, pavements);
• high strength and toughness of elements exposed to impacts (shields, special walls).

Table 12.4 Examples of steel-fibre-reinforced concrete mixture proportions

	1	2	3	4	5	6
Mixture proportions, kg/m³						
Portland cement	500	500	349	400		
blast-furnace slag cement					382	382
fly ash					179	179
silica fume		40				
superplasticizer	7.5	12.5	7.0	4.8	3.6	3.6
coarse aggregate 2–16 mm	695		1212	1075	489	489
fine aggregate # 2 mm	800	1480	584			
fine aggregate # 4 mm				720	1044	1044
water	177	262	185	180	183	183
steel fibres with hooks 0.45 × 30mm		156	30		60	
steel fibres with hooks 0.8 × 60mm				39		60
steel fibres straight plain 0.4 × 27mm	79					
Mixture parameters						
silica fume (% of cement)		8				
w/c ratio	0.35	0.51	0.53	0.45	0.33	0.33
slump, mm	65	90	150			
Compressive strength, 28 days MPa	60.7	36.0	37.6	47.5	66.45	67.5
Tensile strength, MPa						
modulus of rupture		6.0	3.2			
splitting strength	3.4			3.9	5.7	5.6

Example 1 was for ordinary reinforced structural elements.
Example 2 was designed for shotcreting.
Example 3 was designed for railway subbase.
Example 4 is after CEB Bulletin.
Examples 5 and 6 are self-compacting concretes, from Grünewald *et al.* (2003).

Also for other reasons, for example weak performance in elevated temperature, flammability and possibly toxicity, the application of all kinds of polymer-modified concretes is limited.

The general conclusion is that while there are many well-established procedures to design ordinary concretes, which may be extended also to cover high performance concrete and ultra high performance concrete, there are no such methods for polymer modified concretes and for fibre-reinforced concretes. Also for fibres other than short steel ones (glass, polypropylene, long steel fibres, etc.) there are no general recommendations.

In each case, only previous experience in the form of approximate formulae and application of general principles may help in design, which has to be completed using producers' guidance, trial mixes and the results of tests of hardened material.

In all ready-mix concrete plants the material data and compositions for a variety of needs are stored and may be used again in various combinations, together with obtained results: strength, its development in time, rheological properties, etc. Such databases enable easy design of any kind of concrete in an economic way, with some modifications for actual demand. Also, necessary trial batches should always be executed.

12.7 Remarks on optimization of cement-based composites

(a) General introduction to optimization

The optimization approach to any activity in technology, economics or other fields is aimed at finding methods of determining the best solutions in an objective way. In particular, optimization of composite materials deals with problems of the selection of the values of several variables, which determine their composition and internal structure. Other variables that describe the conditions of mixing, vibration, placing and curing may also be included.

This section is limited to the application of methods of mathematical optimization to materials with cement-based matrices, like ordinary concretes, polymer concretes, fibre-reinforced concretes, etc. Other methods of improving the material design by separate analyses of particular aspects and criteria, or by computer assisted comparisons of a number of designs to select the best one, are mentioned only briefly.

Optimization methods are aimed at furnishing certain inputs of rational indicators for design procedures. It is considered useful, and in many cases even necessary, to support traditional design methods by an objective approach. This seems particularly helpful in the design of composite materials where, due to a large number of design criteria and variables, any intuitive approach to design is difficult. The optimization of structures had been studied already by Galileo in 1634 and since then many valuable papers have been published in this field. Among others, in two books by Brandt (1984, 1989) the criteria and methods of structural optimization were presented. The optimization of materials was started not so long ago and the results obtained are not numerous. For a review of the optimization methods applied particularly to cement-based materials design the reader is referred to the book edited by Brandt (1998).

Concrete mix design covers requirements related to the mechanical properties, technological conditions and economical restraints within an area created by multiple limitations. In such a problem, there are several variables (e.g. proportions of components), certain boundary conditions (e.g. minimum strength) and optimization criteria, also called objective functions (e.g. minimum cost). In general, all this forms a problem of multi-criteria optimization. Its solution may be determined by more-or-less straightforward approaches from a simple selection from sets of arbitrarily selected values of the variables up to a precise solution of an optimization problem. The solution should be looked for within the imposed limits (constraints).

After mix design methods of historical importance mentioned in Section 12.5, there is a trend to base all methods on rational bases, which obviously include optimization aimed at the most economical mixture composition, satisfying all imposed requirements, but reducing the number of consecutive trials. Here various approaches are noted from the analytical to the application of the ANN (artificial neural networks). Mathematical programming techniques were successfully applied by Karihaloo (2000) for optimization of HPS fibre-reinforced concrete for high tensile strength and high ductility.

A multi-criteria (multi-objective) optimization method was presented by Bayramov *et al.* (2004). The aim was to obtain concrete reinforced with dispersed fibres that behaves as less brittle than plain concrete. The selected strength and fracture properties of the concrete were: compressive f_c, splitting f_{st} and flexural f_{flex} strength, Young's modulus E, fracture energy G_F and characteristic length ℓ_{ch}. These properties were obtained by the application of three-level full factorial experimental design and the test results, after regression analysis, were represented as response surfaces in a three-dimensional space. Then, two optimization problems have been solved using the response surface method and solutions were obtained again in the form of surfaces in three-dimensional space.

In the first solution, the highest values of f_{st}, ℓ_{ch} and f_{flex} have been for steel fibres of volume fraction V_f and fibre aspect ratio L/d. In the second problem, the optimization of cost, which means minimum volume of fibres V_f, was looked for. In both problems the importance of variables was the same, that is, all variables were considered with the same weights. Numerical solutions to both problems show the utility of the proposed method for practical applications in different formulations.

The optimization procedure may be based on a sufficient number of experimental examples covering the feasible region of expected results. Such a solution to the problem was proposed by Yeh (2007). He described an optimization method based on the minimum cost criterion as an objective function, and five sets of constraints covering strength, workability, contents of components and their proportions, and finally, the unit total volume. Using a Computer-Aided Design (CAD) tool, five different workabilities (slump from 50 mm to 250 mm) and five values of compressive strength (from 25 MPa to 55 MPa) have been computed. As a result, 25 mixture proportions were considered, each being optimal in its category. This set of solutions may be presented as a basis for other mixture proportions with additional components and satisfying other constraints.

(b) Basic concepts of material's optimization

In the optimization of cement-based composites, the parameters and methods are taken from three well-advanced fields: concrete technology, mechanics of composites and mathematical optimization.

The general formulation of the problem of optimization of a material is the same as in the structural optimization. An optimal material is described by a set of decisive variables x (i = 1,2,..., n) which minimize or maximize an optimization criterion or multiple criteria. The solutions should be within or on the border of the feasible domain.

Variables x are considered as independent and together with arbitrary selected parameters they completely determine the object of optimization – a cement-based material.

In the case of material optimization, the variables are: volume fractions, kinds and qualities of components, their distribution in space, their reciprocal relations like adherence, etc. The variables may be defined as continuous or discontinuous (discrete) ones. Quantities (e.g. volume contents) characterizing particular components are the continuous variables. A few discrete kinds of components represent discontinuous variables, that is, only a discrete list of available types of Portland cement and different kinds of fibres may be used in the composition. Methods of production may also be considered as discrete variables provided that they determine final or transitory material properties.

The decisive variables should belong to a feasible set. This means that their values, beyond imposed limits, cannot be accepted in the problem for constructional, functional or other reasons. The constraints may have the form of equalities or inequalities:

$$g_p(x_i) = 0, p = 1, 2,..., r \qquad (12.1)$$

$$h_s(x_i) \leqslant 0, s = 1, 2,..., t$$

or simply limit values may be imposed

$$\underline{x_i} \leqslant x_i \leqslant \overline{x_i}, i = 1, 2,..., n,$$

where lower and upper bars indicate imposed lower and upper limit values, respectively. For example, the variables which describe the components are limited to available materials (cement, sand, gravel, etc.) and their possible properties. The constraints determine the feasible domain.

The volume fractions and properties of the components, as well as their effective distribution in a composite material are random variables. Their final values and their nominal values, determined by testing, are subjected to unavoidable scatter. When in an optimization problem only design and nominal values are considered, then that is a deterministic approach. In the opposite case, when the distribution functions are taken into account, the stochastic problem of optimization is formulated.

Optimization criteria describe basic properties of materials. They are also called 'objective functions.' In material optimization the objective functions describe selected properties that were considered as important and decisive for the material's quality and applicability. The solution consists of the determination of those values of design variables that extremize these properties. All physical, chemical and other properties may be treated as material properties. Particularly important for engineering materials are mechanical properties such as strength, Young's modulus, specific fracture energy, durability and specific cost.

When a set of independent variables x_i, ($i = 1, 2,..., k$) is considered, the optimization criterion is expressed by those variables as a function $F(x_i)$ subject to constraints of different kinds and forms; for example, the limited volume of material may be presented as integral.

If the problem is formulated with one single criterion $F(x_i)$ with the constraints (12.1), then the necessary conditions for a maximum are derived from the Kuhn-Tucker theorem (Kuhn and Tucker (1951) and have the following form:

$$x_i \frac{\partial F^*}{\partial x_i} = 0; \quad \frac{\partial F^*}{\partial x_i} \leq 0; \quad \frac{\partial F^*}{\partial \mu_p} = 0$$

$$\mu_s \frac{\partial F^*}{\partial \mu_s} = 0; \quad \frac{\partial F^*}{\partial \mu_s} \leq 0; \quad x_i \geq 0, \, i = 1, 2,..., n \tag{12.2}$$

here $\mu_s \geq 0$, $s = 1, 2,..., t$, and μ_p, $p = 1, 2,..., r$ are so called Lagrange multipliers, and

$$F^*(x_i, \mu_p, \mu_s) = F(x_i) + \sum_{p=1}^{r} \mu_p g_p(x_i) + \sum_{s=1}^{t} \mu_s h_s(x_i) \tag{12.3}$$

In structural and material optimization problems there are several common features. These problems are correctly formulated when criteria, constraints and variables are defined. Sometimes the design variants are imprecisely called 'optimal solutions'. Calculation of a few cases and selection of the best one is not an optimization approach. In such a simplified formulation there is no precise definition of what criteria are used for selection and neither from which feasible region this solution is found out.

It should also be emphasized that the optimization problem is solved not on a real structure or material, but on their approximate models. The results of an optimization procedure are dependent on simplifying assumptions and approximations admitted for those models, for example, real cement-based composite is assumed as elastic and homogeneous material.

Material optimization does not entirely replace the material's design, because it may not cover certain aspects and requirements, which make up the complete material. The omission of some secondary aspects is justified by the necessary simplification of the optimization problem. In the next step of the material's design, in contrast, all conditions and requirements concerning safety, serviceability, economy, etc., should be satisfied and necessary modifications should be introduced to the optimized material's composition and structure. That is why the material's optimization, like structural optimization, does not replace design, but is that part of it in which some intuitive procedures are replaced by objective calculations.

The sensitivity of the objective function with respect to variables is a separate problem. When the objective function does not really depend on the decisive variables, then probably the variables are incorrectly selected.

Determination of constraints and objective functions is important in the formulation of the problem. It is based on given conditions, but sometimes an objective function may be replaced by a constraint or vice versa. It sometimes also occurs that the small modification of a constraint may considerably influence the objective function. In such a case the resignation of preliminary assumptions concerning constraints may be justified.

In rare problems it is admissible to limit the optimization to one single criterion, for example, a structure of minimum cost or a material of maximum strength may be considered as an appropriate solution. In general, such a formulation has somewhat academic character and may be used only as a simplified example for preliminary explanation of the problem. In most cases the existence and necessity of several criteria is obvious, though often they are considered in an indirect way, which means by appropriate constraints. Multi-objective or multi-criteria optimization is the next step, presented below, in which several criteria are directly considered.

(c) Multi-criteria optimization of materials

Let us consider the n-dimensional space of variables x_i in which objective functions $F_j(x_i)$, $j = 1, 2,..., k$ are determined, where k is the number of objective functions or functionals. It means that a solution to the problem should satisfy k objective functions and the problem may be formulated as follows:

'determine *n*-dimensional vector in the space of decisive functions which satisfies all constraints and ensures that the functions F_j have their extrema.'

The decisive functions are the components of the vector X^N in the *n*-dimensional space. Every point of that space indicates one particular material defined by *n* decisive variables. The feasible region Q is a part of the *n*-dimensional space and is defined by the constraints (equation 12.1).

The space of the objective functions R^k has k dimensions. Every point of that space corresponds to one vector of the objective function $F_j(x_i)$. In that space the feasible region Q is represented by a region $F(Q)$. Without entering in all mathematical formulations, which may be found in manuals of optimization, it may be proved that the points of feasible region Q are represented by the points in the region $F(Q)$. An example of regions Q and $F(Q)$ are shown in Figure 12.6 in the case of $k = 2$ and 2D region. In real problems of the material's optimization the number of dimensions is usually higher.

The optimization problem formulated in this way may have several solutions and one single solution appropriate for given conditions should be selected using other arguments. To present that procedure a few definitions are necessary.

a)

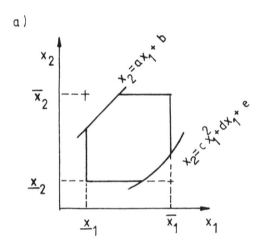

b)

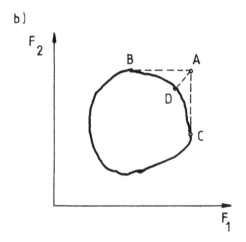

Figure 12.6 Feasible regions:
    (a) in variables space
    (b) in objective functions space, after Brandt ed. (1989).

The ideal solution is an extremum for all objective functions. Because a characteristic feature in multi-objective optimization is that the criteria are in conflict, then the ideal solution is outside the feasible region. That is indicated by point A in Figure 12.6, in which both objective functions F_1 and F_2 have their maximal values at this point, but outside region $F(Q)$.

The solutions of the problem are situated on section BDC of the boundary of the feasible region. These are called non-dominated solutions, Pareto ideal solutions or the compromise set; that is, no single objective function can be increased without causing the simultaneous decrease of at least one other function. In effect, the difficulty is that there are a great number of Pareto solutions. It is therefore necessary to apply other procedures for the selection of one solution, called 'the preferable solution.'

The compromise set may be determined from the Kuhn-Tucker theorem (equation 12.2) in which function F from equation 13.3 is expressed by

$$F = \sum_{j=1}^{k} \lambda_j F_j, \text{ where } \sum_{j=1}^{k} \lambda_j = 1, \lambda_j \geq 0 \qquad (12.4)$$

For different λ_j the particular points belonging to the compromise set may be found from equations 12.2.

The compromise set is a rigorous and strict solution to the problem, but the next step, the selection of the preferable solution, is based on a subjective decision. It may be found by using different assumptions or additional hypotheses. For example, when strength and cost are two conflicting objective functions, then the compromise set contains all possible solutions. For one application the cheapest solution may be selected, and for the other the strongest one. In general, a system of arbitrary weights may be assumed for each objective function, that is, some functions are considered more important than others. Another development of that method is by creation of a utility function, by which the weights of various criteria are introduced. The weights are selected, taking into account other arguments, from experience or from practical requirements, which have not been applied in the formulation of the problem.

Another method is based on the selection of a point on the curve of Pareto solutions, which is closest to the ideal solution in the space of normalized functions. In that second method the definition of the closest point needs to be justified, again using some new arguments, for example, from the conditions of execution in practice.

The application of the optimization approach to cement-based composite materials requires competent and well-founded selection of appropriate decisive functions (variables) and objective functions (criteria). Effective solutions to such problems are possible only when the relations between these functions may be presented in an analytical form. These relations should be based either on verified models of materials or on experimental results, presented in the form of approximate functions.

The principal difficulty in the problems of material optimization is in establishing their correct formulation, from which effective and useful solutions may be derived. The sensitivity of the objective functions, with respect to the variables, is a separate question related to the quality of the general formulation of a given problem.

The next difficulty is the determination of the analytical relationships between the objective functions and the variables. Such relations may be established from various test results available from publications. It is unnecessary to carry out experimental research in each case when an optimization problem is formulated. Often, however, the relations between all variables and objective functions are not given explicitly and special methods should be applied to obtain approximate solutions from incomplete test results.

The last difficulty in these problems is the effective solution of the equations obtained and various numerical methods are developed.

All values characterizing the properties of the material's components are subject to random and systematic variations. The same variations concern their effective volume fractions and distribution in space. That is why the experimental verification with actual materials and in local conditions of implementation is necessary to test the final result, and in most cases to introduce certain modifications into the set of values of the design variables as established in the optimization procedure. The requirement for tests before any full-scale application is also accepted in simple mix design methods for ordinary concretes. It is also compulsory when the optimization is applied and is even more important in the case of high performance concretes (cf. Chapter 13).

Further development of the optimization approach for cement-based composites should be directed at various realistic objective functions and variables and at better expressions for objective functions. The generalization for random variables and the improvement of mechanical models may also be introduced to the optimization approach. Optimization for the least cost seems to be of particular importance.

For more detailed description of problems related to the optimization of concrete-like materials, the reader is referred to specialized books, for example that edited by Brandt (1998).

References

Alves, M. F., Cremonini, R. A., Dal Molin, D. C. C. (2004) 'A comparison of mix proportioning methods for high-strength concrete,' *Cement and Concrete Composites*, 26: 613–21.

Banfill, P. F. G. (2006) 'Rheology of fresh cement and concrete,' *Rheology Reviews*, London; The British Society of Rheology: pp. 61–130.

Baron, J., Lesage, R. (1976) 'La composition de béton hydraulique du laboratoire au chantier,' *Rapp.Tech. LCPC*, no. 64.

Bayramov, F., Taşdemir, C., Taşdemir, M. A. (2004) 'Optimisation of steel fibre reinforced concretes by means of statistical response surface method,' *Cement and Concrete Composites*, 26: 665–75.

Brandt, A. M. ed. (1984) *Criteria and Methods of Structural Optimization*, The Netherlands; Martinus Nijhoff

——.(1989) *Foundations of Optimum Design in Civil Engineering*, The Netherlands; Martinus Nijhoff.

——.(1998) *Optimization Methods for Material Design of Cement-Based Composites*, London and New York; E. & F. N. Spon.

Bui, V. K., Geiker, M. R., Shah, S. P. (2003) 'Rheology of fiber-reinforced cementitious materials,' in: *Int.Workshop HPFRCC 4, RILEM, Proc.53*, W. H. Reinhardt and A. E. Naaman eds: pp. 221–31.

Chanvillard, G., Rigaud, S. (2003) 'Complete characterisation of tensile properties of Ductal® UHP fibre-reinforced concrete according to the French recommendations,' in: *Int. Workshop HPFRCC 4, RILEM, Proc. 53*, W. H. Reinhardt and A. E. Naaman eds: pp. 21–34.

Day, W. K. (1999) *Concrete mix design, quality control and specification*, 2nd edition. London: Taylor & Francis.

de Larrard, F. (1998) *Concrete mixture-proportioning*, London; E. & F. N. Spon.

de Larrard, F., Sedran, T. (2002) 'Mixture-proportioning of high-peformance concrete,' *Cement and Concrete Research*, 32(11): 1699–1704.

Dewar, J. D. (1999) *Computer modelling of concrete mixtures*, London and New York; E. & F. N. Spon.

Edgington, J., Hannant, D. J., Williams, R. I. T. (1974) *Steel fibre reinforced concrete*. CP 69/74, Building Research Establishment, Watford, UK.

Ferrara, L., Park, Y.-D., Shah, S. P. (2007) 'A method for mix-design of fiber-reinforced self-compacting concrete,' *Cement Concrete Research*, 3: 957–71.

Fuller, W. B., Thompson, S. E. (1903) 'The laws of proportioning concrete,' *ASCE*, 59: 67–143.

Grünewald, S., Walraven, J. C., Obladen, B., Zegwaard, J. W., Langbroek, M., Nemegeer, D. (2003) 'Tunnel segments of self-compacting steel fibre reinforced concrete,' in: *Int. RILEM Symp. on Self-Compacting Concrete*, O. Wallevik and I. Nielsson, eds, Bagneux, France; RILEM Publications S.a.r.l.

Grünewald, S. (2004) 'Performance based design of self compacting steel fiber reinforced concrete,' *PhD Thesis*, Delft, The Netherlands; Delft University of Technology.

Ji, T., Lin, T., Lin, X. (2006) 'A concrete mix proportion design algorithm based on artificial neural networks,' *Cement and Concrete Research*, 36: 1399–1408.

Karihaloo, B. L. (2000) 'Optimum design of high-performance steel fibre-reinforced concrete mixes,' in: *Proc. Int. Symp. Brittle Matrix Composites 6*, A. M. Brandt, V. C. Li and I. H. Marshall eds, Warsaw; ZTurek and Woodhead Publishing: pp. 3–16.

Kett, I. (1999) *Engineered Concrete Mix Design and Test Methods*, London and New York; CRC Press.

Kuder, K. G., Ozyurt, N., Mu, E. B., Shah, S. P. (2007) 'Rheology of fiber-reinforced cementitious materials,' *Cement and Concrete Research*, 37: 191–9.

Kuhn, H. W., Tucker, A. W. (1951) 'Nonlinear programming,' in: *Proc. of 2nd Berkeley Symposium on Mathematical Statistics and Probablity*, Berkeley, CA; University of California Press: pp. 481–92.

Malier, Y. ed., (1992) *Les Bétons à Hautes Performances*, Paris; Presses de LCPC.

Nataraja, M. C., Nagaraj, T. S., Basavaraja, S. B. (2005) 'Reproportioning of steel fibre reinforced concrete mixes and their impact resistance,' *Cement and Concrete Research*, 35: 2350–9.

Nehdi, M., Mindess, S., Aïtcin, P. C. (1998) 'Rheology of high-performance concrete: effect of ultrafine particles,' *Cement and Concrete Research*, 28: 687–97.

Ohama, Y. (1996) Polymer-based materials for repair and improved durability: Japanese experience. *Construction and Building Materials*, 10(1): 77–82.

Ohama, Y., Kawakami, M., Fukuzawa, K. eds (1997) *Polymer in Concrete*. London; E. & F. N. Spon.

Rokugo, K., Kanda, T., Yokota, H., Sakata, N. (2007) 'Outline of JSCE Recommendation for Design and Construction of Multiple Fine Cracking Type Fiber Reinforced Cementitious Composite (HPC),' in: *Int. Workshop HPFRCC 5, RILEM, Proc.53,* W. H. Reinhardt and A. E. Naaman eds: pp. 203–12.

Rossi, P., Harrouche, N., de Larrard, F. (1989) 'Method for optimizing the composition of metal-fibre-reinforced-concrete,' in: Proc. *Int. Conf. Fibre Reinforced Cements and Concretes: Recent Developments,* Cardiff, R. N. Swamy and B. Barr, eds, London; Elsevier Applied Science: pp. 3–10.

Sedran, T., de Larrard, F. (2000) BetonlabPro 2. *Computer-Aided Mix design Software.* Paris; Presses de l'Ecole Nationale des Ponts et Chaussées.

Sobolev, K. (2004) 'The development of a new method for the proportioning of high-performance concrete mixtures,' *Cement and Concrete Composites* 26: 901–7.

Tattersall, G. H., Banfill, P. F. G. (1983) *The Rheology of Fresh Concrete,* New York; Pitman Publishing.

Teychenne, D. C., Franklin, R. E., Erntroy, H. C. (1975) *Design of Normal Concrete Mixes,* Building Research Establishment, Watford, UK.

Wallevik, J. E. (2006) 'Relationships between Bingham parameters and slump,' *Cement and Concrete Research,* 36(11): 1214–21.

Yeh, I-C. (2007) 'Computer-aided design for optimum concrete mixtures,' *Cement and Concrete Composites,* 29: 193–202.

Yu, A. B., Zou, R. P., Standish, N. (1992) Packing of ternary mixtures of non-spherical particles. *J. of the Amer. Cer. Soc.,* 75(10): 265–72.

Standards

ACI 211.1-91 (reapproved 2002) *Standard Practice for Selecting Proportions for Normal, Heavyweight and Mass Concrete.*

Road Note No 4 (1950) *Design of concrete mixes,* Road Res. Lab., 2nd edition. London; HMSO.

ACI 544.2R-89 (reapproved 1999) *Measurement of Properties of Fiber Reinforced Concrete.*

ACI 544.1R-96 (reapproved 2002) *State-of-the-Art Report on Fiber reinforced Concrete.* ACI Committee 544.

ACI 548.3R-03 (2003) *Polymer-Modified Concrete.*

13 High strength and high performance concretes

13.1 High performance materials

Not long ago, the terms 'high performance composites' and 'advanced composites' were applied only to such composite materials as carbon fibre reinforced plastics and metal matrix composites, used in the construction of aircraft, rockets and satellites. These terms are now used for concrete-like materials and have similar meaning: they are materials of improved selected properties, designed and produced for special applications. The concept of high performance is expressed by the following technical terms:

- better consistency of the fresh mix, i.e. its workability, mobility, compactability, pumpability and finishability, to ensure good result of execution without much energy expense or much effort from workers, often resulting in excessive scatter of local properties;
- good behaviour of materials in their hardened state, i.e. strength and deformations fulfilling standard requirements with a sufficient safety margin for certain unpredictable situations, and without weak regions;
- relatively high strength, also at early ages, i.e. after one, three or seven days;
- acceptable behaviour in the long-term; therefore improved durability adequate to requirements during the forecast life of the structure, low maintenance costs and relative facility of repair works;
- good aspect of the structure during service life, i.e. without visible cracks and voids, excessive deflections, spallings etc.

There may be other special properties that are needed in certain applications, for example, low permeability, corrosion protection or abrasion resistance.

The above characteristics are described from the viewpoint of the user, who is considered to be a contractor who should produce, transport and cast the fresh mix as a first user. Other users are the investor, proprietor of the structure and the general public. All of them are interested in low general cost of the structure, its long-term serviceability and safety. The users are less interested in various physical and chemical properties, which are necessary to ensure high overall performance of the materials. It means that the performance approach to material and structure properties becomes of primary importance (cf. Section 14.4).

The term 'high performance' does not necessarily cover all the material's properties, that is, it does not mean that a high performance concrete is not only very strong at the traditional age of 28 days, but also has high early strength after one, three or seven days, generates low heat during hardening and is frost-resistant. In most cases not all properties are required for a given material, but only those rationally selected for given application. For example, not all high strength concretes have improved durability and vice versa: it is easy to design very durable concrete with medium compressive strength. Therefore, the term 'high performance concrete' concerns a concrete that is particularly well adapted to its function or 'tailored' for requirements. Without any doubt, the first and foremost characteristic property of high performance concrete is its high compressive strength, but some other requirements have to be satisfied, otherwise a more narrow term would be correct: 'high strength concrete.'

In this chapter, the following concrete-like composites are considered, which are characterized by particularly high values of certain properties, for example, strength or toughness:

- high strength matrices (HSM) that are closely related to the initiation and development of other kinds of high performance concretes;
- polymer-modified concretes (PMC), which are characterized by a few properties at a high level;
- high performance concretes (HPC), plain and with dispersed reinforcement;
- slurry infiltrated fibre concretes (SIFCON) and slurry infiltrated mat concretes (SIMCON), that also belong to the high performance concrete family.

13.2 High strength matrices

Very high compressive strength of cement matrices was obtained over 70 years ago. In the laboratory of the Lone Star Cement Corporation in the USA under the direction of D. A. Abrams, the compressive strength of cement paste equal to 276 MPa at an age of 28 days was achieved in 1930 (Powers 1947). The paste was of very low w/c ratio equal to 0.08. This record achieved in a laboratory had no particular influence on contemporary practice and up to the 1970s the strength of concretes rarely exceeded 30 MPa or 40 MPa.

The high strength of cement based materials is directly related to low porosity and closed capillary pores in the matrix, which may be achieved in various ways: low water content, dense packing of grains and particles, and the addition of micro-fillers, improved mixing and compressing of the fresh mix, compacting by efficient vibration, increasing density by long and complete hydration of cement, additional impregnation of pore and void systems and autoclaving. All these methods are used and simultaneously combined to decrease the total porosity and to control pore dimension distribution.

A few well known and traditional formulae for compressive strength based on classical law proposed by Féret in 1892 and others are given in Section 8.2. The excess of water creates additional capillary pores, which considerably decrease the matrix strength (cf. Section 6.5):

- as the weakest component in the hardened composite material, (law of mixtures);
- as initiators of cracking due to local stress concentration.

That is why the mix compositions of high strength concrete contain as small an amount of water as necessary for hydration or even less. The workability of the fresh mix is ensured by admixtures and the application of plasticizers and superplasticizers (cf. Section 4.3) leads to appropriate fluidity and workability of the fresh mix with low values of *w/c* ratio. It is generally required that the slump, as a simple measure of the workability, is between 120 mm and 200 mm corresponding to the technology that is applied. Such workability should be maintained during about one or two hours, that is, time sufficient for transportation and casting of the fresh mix.

Further improvements of packing are achieved by microfillers and secondary cementitious materials, for example, the application of condensed SF for high performance concretes is considered necessary. The properties of SF are described in Section 4.2.5.

A special procedure for the introduction of small particles has been known since 1981 as DSP: 'densified with small particles' (Bache 1981). The ratio of 1:30 was considered as the optimum between smaller and bigger particle dimensions (Jennings 1992; Tan *et al.* 1987). The smallest particles are not available in the natural sand composition and should be admixed with fly ash, SF or microsilica. SF enters into interstitial spaces between Portland cement grains (30–100 μm) and satisfies the abovementioned ratio. The second effect of SF particles is their pozzolanity, that is, ability to react with $Ca(OH)_2$ during the hardening of the material. The products of the reaction fill the pores and increase toughness and strength. The percentage of non-hydrated cement grains is relatively high because of low water content in the fresh mix, and hard cement grains also strengthen the microstructure. DSP materials are characterized by a high compressive strength up to 250 MPa, but rather low tensile strength and increased brittleness.

To improve toughness, fibres as a dispersed reinforcement are applied, which may be higher stressed than in ordinary FRC due to the higher strength of the matrix. The difference in behaviour of concretes and DSP matrices is presented in Figure 13.1.

The application of DSP materials extends beyond building construction to heavy duty industrial floors and mechanical parts. The service life of elements

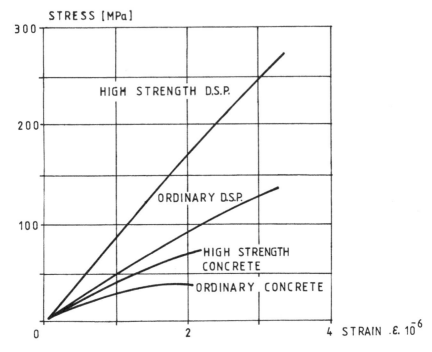

Figure 13.1 Stress-strain curves for concretes and DSP matrices, after Hjorth (1983).

subjected to intensive abrasion was five times longer than those made with steel or cast basalt (Hjorth 1983).

Another way to increase packing and to reduce the thickness of the interface is to push particles closer together by initial pressure exerted on the fresh mix. According to tests by Birchall *et al.* (1981) and Tan *et al.* (1987) a compaction pressure of 345 MPa enabled a very low porosity of 2% to be reached, along with improved tensile strength up to 64 MPa and compressive strength up to 655 MPa. That example is probably essentially a laboratory record and cannot easily be transferred into building practice for larger elements or structures.

It was discovered later that a much lower pressure of about 5 MPa is sufficient if accompanied by the addition of a water soluble polymer, for example, hydroxypropylomethyl cellulose or hydrolyzed polyvinylacetate, and the application of high energy shear mixing. The mass proportions of cement, polymer and water are approximately 100:7:10. The addition of the polymer reduces interparticle friction and together with intensive mixing enables cement particles to be rearranged and their closer packing obtained. Not only is porosity considerably lower and may reach 1%, but the microstructure is also significantly improved and appears more amorphous than the ordinary Portland cement pastes. The bond to aggregate is increased and transgranular fractures are observed. The material obtained in such a way is called 'macro-defect-free (MDF) paste.'

It has been shown by Kendall and Birchall (1985) that by decreasing porosity, very high values for fracture energy may be obtained. In these investigations, the notch sensitivity of MDF paste was also tested on specimens with different notches subjected to bending. The relation between modulus of rupture and notch depth for two different porosities is shown in Figure 13.2. The equation proposed by Kendall *et al.* (1983) is verified and it is proved that these brittle materials considerably increase their strength and fracture energy with decreasing porosity (cf. Section 6.5). However, they exhibit high notch sensitivity and this may be correctly determined from the fracture mechanics approach. The conclusion concerning high strength matrices is that the reduction of the volume of pores from 30% to 0% may double the strength, while the reduction of microcrack length from 1 mm to 0.01 mm may increase the strength by a factor of 10. However, both procedures are applied with independent effects.

The MDF pastes are used to produce elements of complicated shapes by various technologies: press-moulding, extruding, calendering, roll-milling, etc. The final product has no macropores, exhibits high compressive and also tensile strength and improved ductility thanks to the polymer component creating a film in the interface between cement grains. The compressive strength may reach 400 MPa and modulus of rupture up to 200 MPa. Also, Young's modulus of 40–50 GPa is considerably increased if compared with ordinary Portland cement concretes.

The MDF pastes are also filled with metallic powders, silicon carbide, fine sand and with other kinds of hard materials to obtain particular properties,

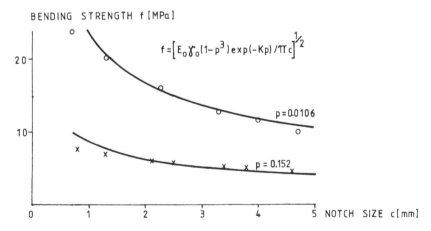

Figure 13.2 Bending strength for notch size c and for two sample porosities *p*, where *E* and *g* are Young's modulus and fracture energy for *p* = 0, after Kendall and Birchall (1985).

464   *High strength and high performance concretes*

for example, improved abrasion resistance. When polymers are not added, then the increase of compressive strength and low porosity are accompanied by high brittleness. In MDF pastes with high alumina cements the process of conversion (cf. Section 4.1.1) was never observed, certainly because of very low water content and w/c ratio values.

Other combinations of special components and technologies were applied by Ohama *et al.* (1990) to obtain so-called super-high-strength mortars characterized by compressive strength over 200 MPa. As fine aggregate, silica sand was used of two sizes: 0.05–0.21 mm and 0.70–1.70 mm, together with stainless steel particles of 0.40 and 0.80 mm. Two kinds of microfillers were added: high purity silica of size up to 4.4 µm and SF (silica fume) of 0.1–0.3 µm. High range superplasticizers of different kinds were tested on that occasion but ordinary Portland cement was used. Examples of the mix proportions are given in Table 13.1.

The specimens were subjected to different kinds of cure: ordinary cure in water, six-day cure in hot water, autoclave cure alone and combined with hot water. In Figure 13.3, the compressive strength is shown as a function of the high purity silica content. A compressive strength of over 300 MPa was observed for compositions with steel particles.

Impregnation of pore systems in hardened cement-based materials is achieved by various kinds of organic monomers (cf. Chapters 3.2.5 and 13.3), which are later polymerized.

The basic mechanical properties of the high performance materials are shown in Table 13.2, where two special materials are considered as well as cement paste and concrete. High strength materials are also used as matrices for composites in which reinforcement by dispersed fibre increases their toughness and ensure ductility (cf. Tables 13.5 and 13.6).

Table 13.1 Composition of super-high-strength mortars

Cementitious materials:fine aggregate ratio (by mass)	Water: cementitious materials	High-purity silica: cementitious materials	Silica fume: cementitious materials	Superplasticizer: cementitious materials
		0		
	0.17	0.10	0.10	
		0.20		
1:0.5		0.30		0.03
		0		
	0.15	0.10	0.20	
		0.20		

Source: Ohama *et al.* (1990).

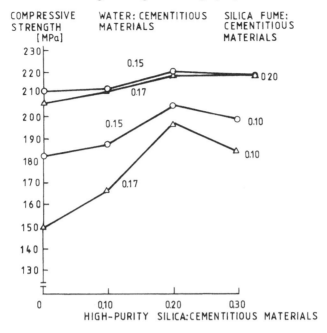

Figure 13.3 Compressive strength of super-high-strength mortars as a function of high-purity silica content, after Ohama *et al.* (1990).

Table 13.2 Mechanical properties of high strength cement-based materials

Materials	Compressive strength f_{c28} (MPa)	Tensile strength f_{t28} (MPa)	f_{c28}/f_{t28}	Young's modulus (GPa)	Stress intensity factor K_{Ic} (MPa.m$^{1/2}$)	Specific fracture energy γ (J/m)
Cement paste	150	6	25	18	0.2	15
Concrete	60–120	5–12	10–12	25	1.2	80
DSP	250	15–25	10–16	80	2.0	1.0
MDF	300	150	2	50	3.0	200
Steel	>500	>500–700	1.0–1.4	210		10^5
Sintered alumina	3000	>500	6	350		30

13.3 Polymer modified concretes

Three kinds of concretes with polymers and resins may be distinguished according to definitions adopted by the ACI (2003):

1   Polymer Cement Concrete (PCC), in which during mixing monomers are added to concrete or mortar composition. After mixing, the added

monomers simultaneously polymerize with hardening of the cement or are used in the form of latexes.

2   Polymer Impregnated Concrete (PIC) obtained with hardened ordinary concretes, which are dried and then permeated with monomers of sufficient viscosity by vacuum or monomer displacement and pressure. Then, in-situ polymerization of the monomer is carried on by radiation, heating or chemical initiation. As an effect, a large part of the voids and pores is filled with polymer, which forms a continuous reinforcing network.

3   Polymer Concretes (PC) composed of polymers and fillers or aggregate, in which Portland cement is completely eliminated. The particulate constituent is added to pure monomer and the polymerization takes place after mixing.

Polymers and resins used in these kinds of modified concretes are briefly presented in Section 4.1.

In ordinary concretes the capillary pores are filled with water or air. In PCC, pores are filled in a similar way, but dispersed polymer particles are distributed in the matrix. In PIC, capillary pores are nearly completely filled with polymer phase. In PC there are practically no pores and only the polymer matrix represents the continuous phase without cement. These materials are not covered by the general term 'cement-based composites.' That is why they are only briefly mentioned below.

The role of polymers in PCC, PIC and PC is entirely different mainly because of the fact that additive and synergistic mechanisms are different. Additive mechanisms depend on the properties of the components and on their fractional volumes. The rule of mixture is decisive at the macro-structural level and it determines the final properties of the composite. The synergistic mechanisms depend rather on the phenomena occurring in the micro-structure of the interfacial layers on the surface of the components, and certain small quantities of admixture may play a considerable role.

In PC, both groups of resins – those hardened by polymerization and polycondensation – are applied. Polymers are more expensive, but there are no additional products emitted during hardening. Polycondensates are cheaper and better support elevated temperature during exploitation, but water from the condensation process increases the porosity of the final composite and consequently lower strength, frost resistance and impermeability are obtained. Large modifications to a material may be obtained simply by using different kinds of polymers. Polymers and resins are from 5 to 40 times more expensive than Portland cement and the optimum design of a material composition is of particular importance because in that way the minimum consumption of polymers is often required.

The application of PC has developed quickly since the 1950s when it was initially used to produce synthetic marble. Since the 1970s, PC has been widely known as a material for the repair of reinforced concrete structures, mainly highway overlays and bridges, but also for production of machine bases, building panels and utility boxes (Ohama 1995). Because of its excellent bonding to ordinary concrete and high mechanical properties, PC is used for resurfacing overlays on bridges or industrial floors, when a short time of execution and a low dead load, thanks to thin layers (about 13 mm), are valued. The quality of overlays made with PC has been verified during many years of service on several structures and it has proved to have good serviceability. Other important applications of PC are in pre-cast elements: internal lining of steel pipes, skins for lightweight sandwich panels for buildings, ballistic panels and other structures for military purposes, etc., and only high materials costs and sensitivity to elevated temperature present certain limitations. PC may be also reinforced with steel or glass fibres and steel bars (Fowler 1999).

In PCC, natural and synthetic caoutchouc, polyvinylacetate, acrylics, vinyls and others are added mostly in the form of latexes, that is, as colloidal suspensions in water. Latexes were used as modifying admixtures for concretes as early as in the 1920s. In PCC, synthetic materials are added to the Portland cement or to water before mixing or eventually all components are mixed together. The selection of the best technology is essential in each particular case to obtain an appropriate distribution of the polymer in the form of a film around all particles and in pores. However, the processes of polymerization and cement hydration are conflicting. Polymerization may be disturbed by the presence of water in the cement slurry and free water from polymerization may also modify, in a negative sense, the *w/c* ratio, which determines the properties of the hardened cement.

Polymer composition has a major influence on the final properties of the hardened composites and also on the workability of the fresh mix. PCC is used when the following properties need to be improved:

*   compressive and tensile strength of thin layers;
*   resistance against various chemical agents and bacteria;
*   durability in freeze-thaw cycles;
*   resistance against ageing and the influence of sunlight.

The resultant properties of the PCC elements depend, to a considerable extent, upon the quality of latex and the additional ingredients, and on the appropriate execution procedures as well as on the later cure.

Styrene-butadiene latex is used as an admixture for materials applied to the repair of concrete structures, for protective overlays and for industrial floors. A typical application would be for the repair of bridge decks damaged by de-icing agents and freeze-thaw cycles. The PCC is also applied for special purposes, such as protective spray-on coatings for various kinds of structures or decorative elements. The latex is added as 5% and 20% of cement mass during mixing. If the latex content is below 5% of cement mass the term PMC is used instead of PCC.

In PMC, the polymers are added as polymer latexes, redispersible polymer powder, water soluble polymers and liquid polymers. The cement hydration process rather precedes the polymer film formation around aggregate grains and a kind of co-matrix is formed in which cement phase and polymer phase are interpenetrated (Ohama 1998). The formation of such a co-matrix is shown schematically in Figure 13.4.

If compared with a basic material without latex admixture the following improvements are observed:

- several times (5–10) increased bond to concrete surface;
- diminished water absorption;
- considerably decreased carbonation depth and chloride ion penetration.

Also, the creep strain and creep coefficient are significantly smaller than for unmodified concretes, and special expansive additives are used to reduce drying shrinkage (Ohama and Demura 1991). The bonding of PMC to ordinary concretes is excellent over long periods of time, if not destroyed

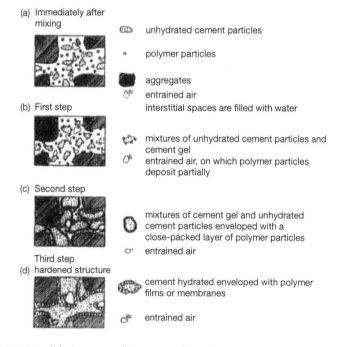

Figure 13.4 Simplified images of structure of a polymer cement concrete: consecutive stages of formation of a polymer cement co-matrix. Reprinted from *Cement and Concrete Composites*, Vol 20, Ohama, Y., 'Polymer-based admixtures', pp. 24., Copyright (1998), with permission from Elsevier.

by high temperature. That property is of considerable importance for repair works to concrete structures.

Latex modified concretes should not be placed under water and exposure to highly concentrated chemicals is not permitted. A temperature above +10°C is recommended for placement in outdoor structures. Because of the large variety of products commercially available for use in PC it is impossible to show here a complete image of their mechanical properties and applications. The results of tests executed by Krüger (1991) are presented in Figure 13.5 as an example. The improvement of mechanical properties was different for the three kinds of admixture examined, but tensile strength was increased for all of them. The influence of polymers did not give such a clear image for compression. When certain polymers are applied, for example, an acrylic latex emulsion, the increase of creep strain was also observed (Mangat *et al.* 1981).

After many years of research and numerous successful applications there are still a few unexplained behaviours, namely:

* Usually only the large pores in PCC are filled by polymer and disappear, while smaller ones may become even larger.
* It is believed that to obtain the real effect of polymer admixtures a minimum amount of 15% of the Portland cement mass is needed and only with that amount can a kind of continuous film around solid particles in the mix be expected.

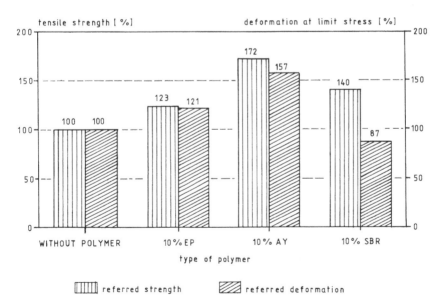

Figure 13.5 Influence of polymer type on the mechanical properties under tensile stress, EP – epoxy, AY – acrylate, SBR – styrene-butadiene, after Krüger (1991).

- High alkalinity of Portland cement paste may degrade some kinds of monomers, like methyl methacrylate (MMA) and styrenes.

The execution of PCC is more difficult than that of ordinary concretes and rigorous procedures should be applied for mixing and curing. In PIC, the system of pores in hardened cement is filled by liquid monomers. The effects are multi-fold:

- decrease of porosity;
- formation of a reinforcing system in the porous hardened matrix;
- increase of matrix-aggregate and matrix-reinforcement bond;
- modification of the internal stress state by decreasing stress concentrations.

The examples of monomers used in impregnation are shown in Table 4.4. The composite properties may be tailored to suit imposed conditions by the application of different monomers. Hardened polymers considerably increase the strength, toughness and durability of final products. The best results are obtained with MMA combined with trimethylolpropane trimethacrylate (TMPTMA) as a cross-linking agent, which builds a polymer structure; both components are used in various proportions according to requirements for strength and other mechanical properties. Examples of test results of PIC specimens under compression and tension are presented in Section 10.2 and prove how behaviour and strength may be modified by proper material design. The composition is usually completed by addition of inhibitors, catalysts and promoters in small volume fractions. Efficient processing systems were designed and best rates of impregnation and polymerization were obtained. Also, the performance of the final product – pore sealing, mechanical properties and durability – were found satisfactory by several authors and were recommended by ACI (2003). The tightness of the concrete is increased considerably by penetration of a small percentage of synthetic materials and that improvement was also observed for relatively high quality concretes with a *w/c* ratio < 0.5 (Ohama 1997).

Even though the extensive research programs have been realized the full-depth of PIC, it did not become a commercially applied material because of technological difficulties in situ. Its use is limited to partial-depth impregnation and to relatively small elements produced in pre-cast plants, for example, power transformer boxes are made of PIC thanks to its low dielectric constant and low creep. Fibre reinforcement is also used.

13.4 High performance concretes (HPC)

13.4.1 Definitions and general information

The concept of high performance concrete was created on the basis of high strength concretes by a kind of generalization of that notion: not only high strength, but also certain selected properties, as explained in Section 13.1 (Mailer 1992).

The meaning of the designation 'high performance concrete' has varied since the beginning of its application in building construction and this evolution is presented in an abbreviated way in Table 13.3. The compressive strength of ordinary concrete increased from about 15 MPa in 1910, to 60 MPa in 1991 in the relevant recommendations and to 200–800 MPa for ultra high performance concrete. There is a conventional distinction between high performance (high strength) concretes that are characterized by compressive strength $f_{28} \geq 60$ MPa and very high performance concretes (VHPC) with $f_{28} \geq 120$ MPa. Other authors also proposed 50 MPa and 100 MPa as respective limits and consider them as mean or characteristic values. Concretes over 200 MPa are called ultra high performance concrete.

The term 'high performance concrete' is used here with general meaning, i.e. it covers various types of concrete with high performances of different kinds. Some are described here more in detail and their particular names and acronyms are used.

13.4.2 Components

The composition of high performance concrete is basically characterized with respect to ordinary concretes by:

- lower w/c ratio and use of superplasticizers, in this context also called 'high-range water-reducing admixtures' (HRWRA);
- increased fractions of fine and very fine grains of different origin (microsilica, SF (silica fume), fly ash, metakaolin);
- reduced volume fraction of coarse aggregate and smaller maximum grain dimension.

In the design of high performance concrete composition two main groups of problems need to be solved in a well-coordinated way. The first concerns the chemical and physical properties of all components, their compatibility

Table 13.3 Evolution of cement concrete compressive strength f_{28}

Year:	1850	1910	1950	1979	1991	2007	Remarks
f_{28} (MPa)	8 – 10	12 –15	20–30	$\leqslant$40	$\leqslant$60	$\leqslant$60	Ordinary concrete
				$\geqslant$40	$\geqslant$60	$\geqslant$60	High performance concrete
					$\geqslant$120	$\geqslant$120	Very high performance concrete
						200 – 800*	Ultra high performance concrete

*   with heat curing.

and synergisms, together with their particular roles in the material structures. Selection of these components is closely related to the properties of the final composite material. The second group of problems concerns the feasibility of obtaining the designed material in local conditions and with a reasonable cost.

The examples of composition of high performance concrete and very high performance concrete presented in Table 13.4 show the features listed above.

(a) Admixtures and additives

Two main groups of admixtures are considered to be very important components of high performance concrete: (i) superplasticizers, which improve the workability of the fresh mix with low values of w/c ratio; and (ii) micro-fillers, which increase the density of the hardened material.

There are several kinds of superplasticizers available and every year many others appear on the market. Two main groups may be mentioned: sulphonated melamine-formaldehyde condensates and sulphonated naphthalene-formaldehyde condensates. Their action is explained by the adsorption of polyanions on the surface of cement grains and by the generation of negative potential, which eliminates the attraction and coagulation of the grains. As a result, a decrease of internal friction is obtained and the workability expressed, for example, by slump of 180–250 mm ensured during 1–1.5 hours. The correct selection of a superplasticizer for other mix components is important. The superplasticizers are added to water, according to the producer's prescription. The dosage of superplasticizers is verified by successive steps; for example, adding 0.5%, 1.0%, 1.5% by mass of cement; 2% is rarely exceeded. With higher dosages some delay may occur in hydration and hardening together with the apparent early setting of the fresh mix. When Portland cement is partly replaced by fly ash or blast-furnace slag, then better flowability of the fresh mix is usually obtained and lower dosages of superplasticizers are possible. This is quite profitable from an economic viewpoint: superplasticizer is an important part of the cost of concrete components.

The compatibility of superplasticizers with cement as well as their efficiency, duration of the fresh mix fluidity, sensitivity to ambient temperature and other factors should be verified by experiments executed in local conditions (Kucharska and Moczko 1993). The composition and procedures established in the laboratory and checked in field conditions should be carefully applied on the site.

Fly ash (FA), silica fume (SF), metakaolin and other micro-fillers are furnished as slurry or powders. Silica fume is considered as the most efficient micro-filler for HPC. Its role is based on:

1   reduction of water/cement (w/c), or water/binder ($w/c+m$) ratio when silica fume or other micro-filler is used jointly with superplasticizers;

2   increase in the strengh of hardened composite with respect to a concrete
    with the same *w/c* ratio but without silica fume.

Micro-fillers enter into the voids between cement grains (Figure 13.6) and
by acting as water-reductors enhance the efficiency of superplasticizers. By
their addition, the contacts between aggregate grains and cement paste are
increased, the intergranular friction is reduced and hydration processes are
modified. Fine spheroidal grains of micro-fillers improve material cohesion,
and decrease the possibility of bleeding and segregation of the fresh mix
during transportation and casting.

   The pozzolanic properties of SF result in a slow hydration process and in
more efficient gel development. SF considerably improves the performance
of the binder phase and increases its bonding action with the aggregate and
reinforcement. The highly porous interface is the weakest element in the
structure of an ordinary concrete and its strengthening is decisive for high
performance concrete. This is confirmed by the curves in Figure 13.7; the
strength of cement paste is not improved by SF, but the concrete strength is
greatly increased.

   High quality SF is composed mostly of SiO_2 and has very fine grains, that
is, one or two orders smaller than Portland cement (cf. Section 4.1.3). The
application of SF is not necessary for concretes up to 60 MPa of compressive
strength (cf. Table 13.4), but it improves their density and durability; it is
considered as a compulsory component of very high performance concrete.
The content of SF is 7–15% of cement mass – usually 8–10%. A higher
dosage may increase brittleness and unfavourably influence the total cost
of the final composite material. Another siliceous additive is so-called 'high
purity silica' with grains of about 4 µm of diameter (Ohama *et al.* 1990).

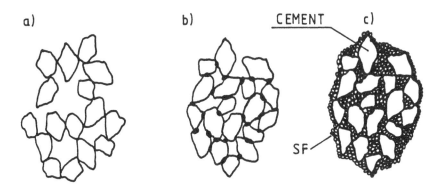

Figure 13.6 Packing of cement grains:
  (a) in a coagulated cement paste
  (b) in cement paste with superplasticizer
  (c) in cement paste with superplasticizer and SF.
  after Roy (1987)

Because of the limited availability of SF on the world market and its increasing price, it is difficult to predict its future application for high performance concrete. It is probable that various kinds of fly ash, blast-furnace slag or metakaolin will play a similar role in high performance concrete composition, certainly after necessary verification of their quality and compatibility with other components.

The influence of fly ash, ground blast-furnace slag and other micro-fillers on the properties of high performance concrete is positive. As for ordinary concretes, these mix components densify the structure, and because of their pozzolanic properties they take part in hydration processes. Partial replacement of Portland cement by fly ash and ground slag enables a decrease in the cost of materials, improves the workability and reduces the heat of hydration. In practice, the majority of concrete structures are made with binary or ternary blended cements, which means that more than one additional binder is used with Portland cement.

Other admixtures are used for high performance concrete as for ordinary concretes. However, their necessity should be verified in terms of improved impermeability and early strength. It has been observed that air-entraining agents are not needed for high performance concrete in most cases and certainly not for very high performance concrete, because pores are already well distributed and total porosity is low. Furthermore, it can be difficult to produce an additional pore system in a dense and coherent fresh mix. However, this question is still subject to tests and discussions.

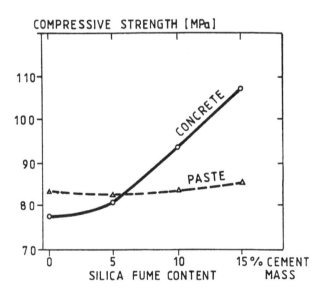

Figure 13.7 Effect of silica fume on the strength of 28-day-old paste and concrete of the same *w/c* or w/(cement + sf) ratio equal 0.33, after CEB/FIP Bulletin (1990).

Table 13.4 Examples of mixture proportions of high performance concrete and very high performance concrete, after different published papers

	high performance concrete					very high performance concrete		
	kg/m³		kg/m³		kg/m³		kg/m³	kg/m³
gravel 12.5/20	852		698		955	2.5/10	1075.6	1093.6
gravel 5/12.5	267	10/20	465	5/16	217	0/5	753.9	772.6
sand 0/5	765	6/14	738	2.5/6.3	934	cement	502.6	506.6
cement	425	0.1/2.5	425	0/2.5	425	water	115.6	115.4
water	150		160		160	superpl.	12.1	21.1
superplasticizer	6.4		8.5		12.8	silica f.	50.2	50.7
retarder	1.7		1.7		1.7	slump	230	230
slump (mm)	180–250		110–190		200–210			
	MPa		MPa		MPa		MPa	MPa
f_{c1}	17.8		36.2		57.7	f_{c28}	120.7	120.0
f_{c7}	60.8		68.3		62.7			
f_{c28}	74.0		75.9		67.2	d (kg/m³)	2510	2560
f_{c90}	82.5		81.5			E (GPa)	49.7	50.3
ft_{28} splitting	5.3		4.5			w/c	0.21	0.21
water/cement	0.35		0.38		0.38			

(b) Portland cement

Ordinary Portland cement of good quality may be used for high performance concrete. A high content of tricalcium silicate C_3S and bicalcium silicate C_2S (alite and belite) is favoured together with a low content of tricalcium aluminate C_3A (cf. Section 4.1.1). Because of the low values of w/c ratio required, a relatively high amount of Portland cement is used: 400 kg/m^3 and more. The amount of cement may be reduced by blending with other supplementary binders.

Portland cement of rather fine grain is preferred, but not too fine, so as to avoid excessive acceleration of all processes of hydration. Ultra-fine cement (cf. Section 4.1) is used only for special purposes. The compatibility of cement with other components should be verified as well as low shrinkage and heat of hydration.

(c) Water/cement (w/c) or water/binder (w/b) ratio

As mentioned above, the reduced amount of water and low value of the w/c ratio are necessary for the high strength and low porosity which characterize high performance concrete. The excellent workability of fresh mix required is ensured by admixtures. When the w/c is equal to 0.22, it means that the amount of water is limited to that necessary for cement hydration. However, in such a case complete hydration is impossible. On the other hand, any additional amount of water, if not bound to cement, may decrease the strength and increase the porosity of the hardened matrix.

Several authors propose the value of 0.22 as an optimum w/c ratio (ACI 1998). In many published compositions of very high performance concrete (VHPC), w/c ratio remains between 0.25 and 0.30 and between 0.30 and 0.35 for high performance concrete (HPC), probably for higher percentage of hydration of Portland cement. Also, lower dosages of superplasticizer might be possible for required workability, and consequently, lower cost.

Because of the pozzolanic properties of micro-fillers, it is common to also calculate the water to cement and micro-filler ratio: $w/(c + m)$. In many published compositions, both these ratios are exposed and discussed.

(d) Aggregate

The properties of aggregate are extremely important and for high performance concrete they should satisfy several different requirements. First of all, good workability is to be ensured and for that reason continuous sieve distributions are preferred.

Maximum grain diameter should be limited to improve workability and to reduce discontinuities and stress concentrations. According to de Larrard and Malier (1992), the grains should not exceed 25 mm. In CEB-FIP (1990) 20 mm is indicated and certain authors propose 10 mm as a maximum grain diameter or even smaller for VHPC.

Spheroidal shape of the grains is considered as an advantage. Natural gravel is better for workability and lower water requirement, but crushed hard aggregate is characterized by stronger bonding to the cement matrix and is preferred. In terms of concrete strength, the following conditions are formulated:

- Young's modulus and strength should be close to those of hardened cement matrix to avoid stress concentrations;
- good bonding to cement paste;
- water absorption below 3%;
- in outdoor application frost resistance is required.

Low difference in strength and deformability of aggregate grains and cement matrix improve composite strength. The tests published by Baalbaki *et al.* (1991) show that the strongest concrete was obtained with coarse sandstone aggregate, which was characterized by the highest strength and lowest Young's modulus (Table 13.5). The authors concluded that the better deformability of sandstone grains, compared to quarzite and limestone, contributed to reduced stress concentrations. The high porosity of sandstone improved its bond to cement paste. The $\sigma - \varepsilon$ curves for three kinds of concretes are shown in Figure 13.8.

The appropriate sieve distribution and packing of aggregate grains, together with other particles in the mix, is the direct and classic way to ensure the high density of the hardened concrete (Figure 13.9). In practice, the composition of grains as obtained from quarries should be improved, but such modifications are expensive. In other situations, the correct proportions of grains of different diameters are furnished at request. In most instances, the production of high performance concrete should be based on a local source of good aggregate, taking into consideration the economic side of its improvement and transportation.

In certain cases of bridge girders working as cantilevers during construction, the dead load is so important that for the part of these girders and also for

Table 13.5 Mechanical properties of rocks compared with mortars and concretes at age of 91 days

	Quartzite			Limestone			Sandstone	
	rock	mortar	concrete	rock	mortar	concrete	rock	concrete
Compressive strength (MPa)	87	108	99	115	106	106	147	107
Young's modulus (GPa)	42	38	45	49	36	44	40	31
Porosity (%)	1.0			2.7			6.4	

Source: after Baalbaki *et al.* (1991).

STRESS 6

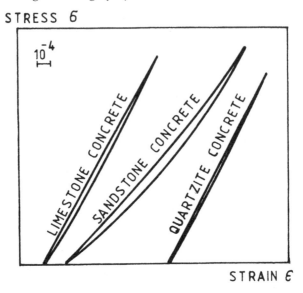

Figure 13.8 Comparison of histeresis for concretes made with different coarse aggregates, after Baalbaki *et al.* (1991).

bridge decks lightweight high strength concrete is used (cf. Section 4.2). Using good quality lightweight aggregate concrete strength over 60 MPa and higher may be obtained. Frost resistance of such concretes was tested with results suggesting the need for further studies (Gao *et al.* 2002).

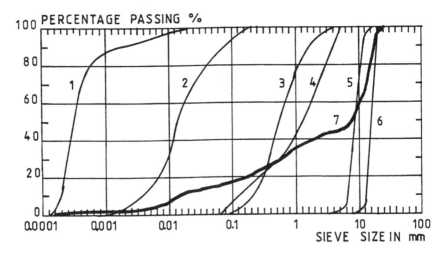

Figure 13.9 Grading of all components of a very high performance concrete, after Lecomte and Thomas (1992).

The quality of sand seems to play a smaller role and requirements as for ordinary concrete apply. Continuous sieve distribution is also favourable and mineralogical composition should be similar to that of the coarse aggregate.

13.4.3 Microstructure

The microstructure of high performance concrete is characterized by the following features in comparison to ordinary concretes:

- closer packing of grains of variable dimensions (aggregate, non-hydrated cement, micro-fillers);
- system of different pores, but their total porosity is smaller and larger pores are eliminated;
- better homogeneity by uniform distribution of hydration products;
- reduction of the least useful calcium hydroxide crystals.

The first feature is obtained by lower w/c ratio after the complex action of superplasticizer, SF and other micro-fillers as is schematically shown in Figure 13.6. The second feature is the simple consequence of the better packing of grains and of the additional pozzolanic reactions. The modification of the microstructure is particularly important in the interface between aggregate grains and cement matrix in high performance concrete. All differences between the interface and the bulk matrix are reduced considerably (Figure 13.10) when both superplasticizer and micro-fillers are used. Therefore, the interface is no longer the weakest zone in the composite material and that is why the transgranular fracture is characteristic for high performance concrete. For the same reasons the penetration of gases and fluids through the interface is reduced in high performance concrete compared with ordinary concretes and hence the overall durability is improved.

When Portland cement is partly replaced by fly ash, SF or blast-furnace slag, then usually better flowability of the fresh mix is obtained and lower dosages of superplasticizers are possible. This is quite profitable from an economic viewpoint, because superplasticizer is an important part of the cost of concrete components.

The distribution of pores in cement paste according to their diameter is shown in Figure 13.11; pores larger than 0.1 μm practically do not exist in pastes with SF.

In high performance concrete, dispersed fibres are used frequently because of increased brittleness if compared with ordinary concretes and two aims should be addressed: strengthening and increase of fracture toughness, and control of cracks. Various fibres (steel, carbon, synthetic) are used, preferably of small dimensions.

All design methods reviewed in Chapter 12 are applicable and the optimization approach may be useful. However, because of severe requirements

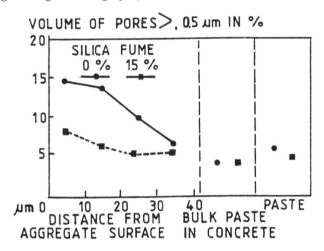

Figure 13.10 Relative volume of pores larger than 0.5 μm in the interface, in bulk cement paste in concrete and in neat cement paste at age of 180 days, after Scrivener *et al.* (1988).

the solution is much more difficult and very sensitive to variations in content and quality of the components. Insufficient results in strength of hardened concrete were obtained when incompatibility of components was encountered or when the aggregate proved to be the weakest element – either because of insufficient strength or by excessive stress concentration in the interface. Because of the low *w/c* ratio and high strength of the cement matrix, the properties of the aggregate determine the composite strength. Other effects

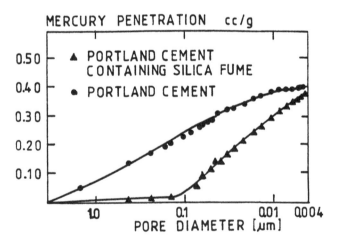

Figure 13.11 Pore distribution in cement pastes with and without SF, after Mehta and Gjørv (1982).

which may appear in high strength concretes are related to the possibility of increased shrinkage and higher Young's modulus – both factors may cause a danger of early cracking. Careful cure of the fresh concrete and long moisturizing are important in order to obtain a correct microstructure of high performance concrete (HPC).

13.4.4 Technology

The methods of production of high performance concrete are basically the same as those for ordinary concretes. A high quality of components and their prescribed proportions should be maintained and controlled at all stages of execution. High performance concretes are more often prepared using ordinary equipment on the site and in ready-mix plants. The special technologies, such as autoclaving, are used less in the main trend of high performance concretes, notwithstanding their efficiency in increasing strength, but perhaps because of the high consumption of energy. Five examples of high performance concrete and very high performance concrete compositions and properties of materials used in France are presented in Table 13.4.

The main reason why high performance concrete is accepted and fully supported by contractors and investors is the excellent workability of the fresh mix obtained by the appropriate application of superplasticizers, which are either added to the water before mixing or, when the mix is transported over a longer distance, is added in two portions: before mixing and after transportation, directly before casting. The facility to fill the moulds or forms by the fresh mix and its pumpability is ensured when sufficient slump of the Abrams cone and the coherence of the mix are maintained during all the time required for both transportation and casting.

The slump of the Abrams cone is not the best measure for the consistency of high performance concrete and other methods are available (cf. Section 12.2). However, these methods are not generally approved and standardized and special equipment varies in different countries, which is why the Abrams cone is still used. It seems that the degree of compactability as indicated by EN 206 (2000) is better adapted to high performance concrete, and its values close to 1.0 indicate high fluidity of the mix when the need for vibration is highly reduced. By that measure it is possible to determine slight differences between mixes of apparently similar high fluidity.

If vibration of the fresh mix is necessary, then the characteristics of the equipment used should be adapted with regard to its frequency and amplitude.

When micro-fillers like silica fume (SF) or fly ash (FA) are used for high performance concrete, the method of adding them to the mix should be elaborated and carefully executed; the same concerns other components. The good dispersion of SF is particularly important because its effects result from combined physical and chemical mechanisms.

The cure of high performance concrete after casting is different than in the case of ordinary concretes. There is no need to maintain high humidity over a long period of time because of the high rate of increase of strength. In order to avoid micro-cracking on the surface, it is advised that any evaporation of water during the first few hours should be prevented by perfect sealing or intense moisturizing. Furthermore, to prevent stress induction due to the characteristic ability of high performance concrete for self-desiccation, curing in water is not advised, because non-uniform swelling may occur due to the low permeability of high performance concrete. The absence of bleeding in high performance concrete is the reason why the humidity of the fresh mix should be maintained immediately following casting. Sufficient curing should allow the chemical mechanism to develop its full effect.

The high rate of hardening and relatively high early strength in high performance concrete after one, three or seven days, enables moulds and scaffolding to be removed earlier than with ordinary concrete. That is an advantage that is used for quick re-use of moulds or for the reduction of traffic closures on structures being repaired. The fresh mix is also only exposed for a short time to possible adverse ambient factors.

In view of a large variety of possible compositions of high performance concrete and of ambient conditions, the hardening process may develop differently depending on circumstances. That is why a detailed method of cure and the time delay for demoulding should be established after verification and observations in situ.

13.4.5 Mechanical properties

Analysis of the mechanical properties of high performance concrete is partly based on notions and methods developed not only for ordinary concretes, but also for advanced composite materials. The results obtained from the application of fracture mechanics and general damage theory appeared to be effective for examining crack propagation and energy accumulation in concrete elements, and also for analyzing their strength and deformability.

The mechanical properties of high performance concrete are related to their composition and structure, are characterized by a dense and strong matrix with good bonding to the aggregate grains and by an absence of excessive pores and other inhomogeneities. In ordinary concretes with normal weight aggregate the following inequalities between strength of aggregate, mortar and concrete are satisfied: $f_{agg} > f_m > f_c$, which indicates that fracture is determined as much by the strength of the interface zone as by the weakest element (cf. Table 13.5).

Thanks to the full transmission of stress between the mortar and aggregate grains, their contribution to composite strength of high performance concrete is increased and that effect may be observed on stress-strain curves. The different behaviour of specimens made of ordinary concrete and of high performance concrete is shown in Figure 13.12. The stress-strain curves for high performance concrete show their linear and elastic part to be longer and plastic deformation smaller or even non-existent before rupture. These features may enhance the danger of cracking if appropriate measures are not taken to prevent excessive shrinkage and local stresses, particularly at early ages.

If high performance concrete may be treated as a two-phase composite, both phases – hardened cement matrix and aggregate grains – behave as linear elastic bodies up to the point of highly brittle fracture. Then, the overall behaviour of the composite material under load is non-linear and quasi-plastic due to micro-cracking. The difference in Young's moduli (cf. Figure 8.7) is the reason for stress concentrations in the interface where a system of micro-cracks appear under relatively low external load. These concentrations are reduced in high performance concrete where high modulus matrix and low modulus aggregate are used. As a result, micro-cracks usually open in high performance concrete under stress corresponding to 70–90% of strength, while in ordinary concretes this already occurs under 40–50%. There are two consequences: high performance concrete may behave in a more brittle manner, and the application of linear elastic fracture mechanics (LEFM) corresponds better to their real behaviour than for ordinary concretes.

The results obtained from both calculations and measurements prove that fracture energy G_f and critical value of stress intensity factor K_{Ic} increase

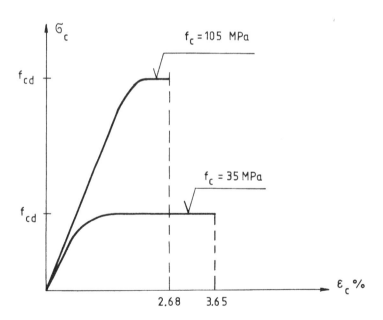

Figure 13.12 Stress-strain curves for ordinary and high performance concretes, after CEB/FIP (1990).

with the compressive strength, but these parameters are significantly less sensitive than compressive and tensile strength (cf. Figure 13.13). It means, for example, that when compressive strength f_c is increased by 2.5 times, the increase of G_f and K_{Ic} is only of 15% and 30%, respectively. Therefore, they are less effective for analysis of test results or other use. The values of parameters c_f – length of the process fracture zone, and l_o – length parameter, describing dimensions of a micro-cracked zone ahead of the tip of an advancing crack decrease for stronger concretes and consequently crack propagation is relatively less controlled than in ordinary concretes (Gettu *et al.* 1990).

The brittleness of high performance concrete (HPC) and very high performance concrete (VHPC) deserves more attention. Experimental results obtained by de Larrard and Malier (1992) on specimens made with three different concretes are given below as an example:

compressive strength	$f_c =$	54	76	105	[MPa]
critical value of stress intensity factor	$K_{Ic} =$	2.16	2.55	2.85	[MPa · m$^{1/2}$]
critical value of elastic energy release rate	$G_{Ic} =$	131	135	152	[J/m^2]

The smaller increase of fracture toughness than of compressive strength is related to the difference in values of tensile to compressive strength ratio f_t/f_c, which varies between 1/10 and 1/12 for ordinary concretes and may be as low as 1/20 for high performance concrete. In all three concretes compared above, the tensile strength f_t was the same and equal to 5 MPa if these relations applied. Therefore, even if higher compressive strength is admitted, there is no reason to expect that higher tensile stress may also be safely supported.

Critical strain also increases less than compressive strength and for high performance concrete values of 200 µm and 350 µm were observed under high and low rates of loading respectively. The observed values of Poisson ratio were lower for high performance concrete than for ordinary concretes and range from 0.13 to 0.16 (Persson 1999).

An increase of strength and of fracture toughness is related to two phenomena: decreased total porosity due to the application of superplasticizers and micro-fillers and reduction of micro-cracks and other discontinuities, which enhance stress concentrations and crack propagation according to basic relations proposed by Griffith. The second phenomenon was described by Kendall and Birchall (1985), among others.

The increased brittleness of high performance concrete compared to ordinary concretes may have no importance in structural applications. Cracks in concrete elements depend not only on the brittleness of the material and its tensile strength, but mostly on tensile forces induced by loads and by constraints imposed on displacements. The main tensions are supported by adequate reinforcement systems and not by the concrete itself. Also, local tensions in external zones of concrete elements due to shrinkage and high rate of drying, thermal stresses and freeze-thaw cycles are reduced in high

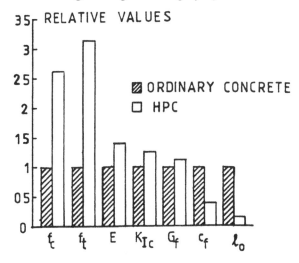

Figure 13.13 Comparison of mechanical properties of ordinary concretes and high performance concrete (relative values), after Gettu *et al.* (1990).

performance concrete by decreased permeability and absorption. Therefore, there is no reason to fear that structures made of high performance concrete may exhibit lower crack resistance or increased brittleness.

The most important property of high performance concrete is its improved durability thanks to increased impermeability and homogeneity of material structure. Hydration products are more amorphous than in ordinary concretes and capillary pores of 0.1–1.0 µm are considerably reduced or eliminated (cf. Figure 13.10). Thus the material's resistance to different climatic actions and corrosive factors is enhanced. Increased capability to adopt alien ions (Cl-, Na+, K+) and improve impermeability decrease diffusion of chlorides, improved resistance against long-term action of sulphates, and against alkali-aggregate reaction and related swelling.

The carbonation process in high performance concrete is basically reduced because CO_2 cannot penetrate its dense structure. Accelerated tests executed by Lévy (1992) showed practically no traces of carbonation. When SF is added as admixture, e.g. for very high performance concrete, the pH of concrete may decrease, thus creating conditions more favorable for corrosion of steel reinforcement. However, in that case the electrical resistance of the concrete is enhanced and consequently it will slow down the corrosion process.

The shrinkage of high performance concrete develops differently than in ordinary concretes. A higher rate of autogenous shrinkage may induce additional stresses when free displacements at an early age are constrained. In contrast, drying shrinkage is less important because of the smaller amount of free water and the lower permeability of the matrix. The total strain due to shrinkage depends on the size of the elements and on the curing conditions;

however, in most cases, smaller values than for ordinary concretes were observed.

So-called plastic shrinkage may cause cracks in external layers of elements, which appear immediately after casting. Such effects are related to local desiccation because of the dense structure of high performance concrete, plus the low w/c ratio pore water does not migrate from the internal core of elements and the phenomenon of bleeding is absent. The main remedy is abundant moisturizing of high performance concrete at a very young age and the application of PP fibres.

The evolution of heat during the hydration of high performance concrete should be considered carefully. The amount of heat depends on cement content and on the degree and rate of its hydration. The admixture of SF and low w/c ratio, result in a smaller degree of hydration and lower heat evolution. Therefore it is expected that a smaller amount of heat is produced in high performance concrete. However, appropriate measures should be prepared if necessary, to lower the temperature of components, to evacuate heat and to allow for possible displacements, etc. Furthermore, it should be borne in mind that for the appropriate action of SF a lower temperature is favourable. More research is needed in that area to quantify the effects of different material composition, methods of execution and ambient conditions.

The high performance concrete composites are obtained with conventional components, but used in a non-conventional way and also by quite new components and technologies (Navy 1996). Therefore two main groups of problems are considered at different levels. The first group is related to the design of materials, taking into account not only the chemical and physical properties of the components, their compatibility and synergisms, but also the material structure, which means the distribution of particular components in the space. The second group of problems covers the feasibility of execution of the material at a required scale and with reasonable cost. Also some negative effects of the application of a high amount of superplasticizers and of other admixtures and additives sometimes lead to mixture proportions that are very sensitive to minor differences in dosages and technologial conditions. These effects are: excessive bleeding or no bleeding at all, surface dessication, segregation of components, rapid loss of workability, high hydration heat, etc. All these negative effects may be avoided by highly developed technological methods carried out with care and competence.

Various kinds of discontinuities and inhomogeneities play an important role. They are related to the pore system, to the distribution of all components and to interfaces between them. The transfer of stress between particular phases and stress concentrations in the transmission zones become extremely important. That is why higher strength is observed for compositions with smaller aggregate grains, for example 10mm, than for 20mm grains, with all other parameters constant. The high porosity and weakness of the interface is considerably reduced by application of the SF.

13.4.6 *Special kinds of very high performance concrete and ultra high performance concrete*

There are several kinds of high performance concrete and very high performance concrete proposed by different authors and companies. Some of them are covered by patents and others are presented openly in published papers. A few of them are briefly described below in order to present a large variety of cement-based composites of the highest quality. Not all are further developed and used after successful laboratory investigations, but all have some particularities that deserve attention. These achievements perhaps indicate directions of future development of cement-based composites.

The common features of these composites in contrast to high performance concrete are:

- smaller grains of all components (below 0.3 or 0.5 mm);
- higher dosage of Portland cement (over 700 or 1000 kgs/m^3) and other supplementary cementitious materials (SF, micro silica, metakaolin);
- extensive use of superplasticizers.

In many cases, external pressure and/or heat treatment is applied during the hydration period. As a result, very high compressive and tensile strength is obtained, together with high brittleness that should be counterbalanced by fibres.

High performance fibre reinforced cement composites (HPFRCC) are materials reinforced in such a way that the effects of strain hardening and multiple cracking are obtained. The definition was proposed by Naaman and Reinhardt (2003) and explained in Figure 13.14. The ascending curve between points A and B indicates that effect. The area under the curve represents the energy accumulated, and the process of multi-cracking is always observed after the first crack at point A. In contrast to UHPFRC these composites are not characterized by very high compressive strength and aggregate grains may even reach 20 mm. HPFRCC are cured in normal conditions.

Ultra high performance fiber reinforced concrete (UHPFRC) are concretes of higher compressive strength than 120 MPa and based on various compositions of fine aggregate with a high dosage of cement and the use of supplementary cementitious materials (SCM) (Habel *et al.* 2006, cf. Section 4.1.2).

Engineered cementitious composites (ECC) is another name used for high performance composites elaborated by Li *et al.* (2007). Composed with very fine aggregate and PVA fibres at high percentage (also hybrid fibres), these materials are characterized not only by high compressive strength, but also by high ductility (strain over 3%). At large imposed deformation, even under uniaxial tension, the width of multiple cracks does not exceed 70 μm, therefore quasi-plastic behaviour is observed as well as improved durability also in aggressive environments.

Ductal® (Lafarge), *CeraCem*® (Sika), *CEMTEC*$_{multiscale}$® and others are kinds of UHPFRC developed by different companies. These materials are produced on the basis of Portland cement, with fine aggregate, and various kinds of admixtures, additives and fibres. During cement hydration, different levels of pressures are applied as well as heat treatment. Tensile strength over 10 MPa and compressive strength over 200 MPa are reached with high fluidity and the possibility of self-placing. Such properties combined with high density, ductility and improved durability create materials for various special applications, parts of equipment, also in bridges, off-coast platforms, etc. (Rossi 2008). Prestressed bridge beams made with *Ductal*® do not require mild steel reinforcement (Behloul 2007) and are described briefly in Chapter 14.

Strain hardening fibre reinforced cementitious composites (SHFRCC) is another name for highly reinforced composites that are characterized by strain hardening on the stress-strain or load-deflection diagrams, that is, after the first crack there is an increase of stress or load, which is related to multiple cracking; respective diagrams are shown in Figures 10.18 and 13.14. High value of strain reaching 5% was observed and this quality was obtained by carefully designed composition that is characteristic for high and very high performance concretes, and by high volume fraction of dispersed fibres (Ahmed and Mihashi 2007; Boshoff and van Zijl 2007).

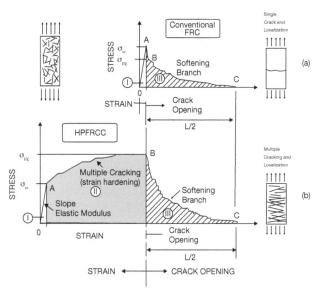

Figure 13.14 Comparison of typical stress-strain response in tension of HPFRCC with conventional FRCC. (Reproduced with permission from Naaman A. E. and Reinhardt H. W., Setting the stage: towards performance based classification of FRC composites; published by RILEM Publications S.a.r.l., 2003.)

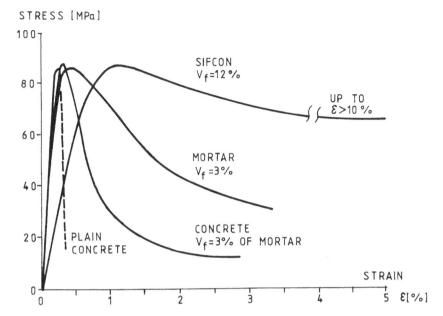

Figure 13.15 Stress-strain curves in compression of various composites as compared with SIFCON, after Naaman (1992).

Reactive powder concretes (RPC) were proposed by Richard and Cheyrezy (1995) as a new kind of ultra high performance concrete. Two different mix proportions and technologies have been elaborated with higher and lower strength (cf. Table 13.6). These composites reach compressive strength over 800 MPa and their properties exceed the usual requirements in building and civil engineering construction. To reach strength over 200 MPa, various curing technologies are necessary, including steam curing and high pressure curing. Beside special applications, they are considered as examples of the highest achievement in cement-based composites.

Special kinds of ultra high performance concrete have been considered by Xiao *et al.* (2004) where, in contrast to all composites of that kind, the use of coarse aggregate was proposed as a remedy against excessive brittleness. Mixtures with two proportions were compared: one with sand below 0.8 mm and one with crushed basalt of 2–5 mm grains, both with a compressive strength of 150 MPa. In a wedged splitting test, a significant increase of fracture parameters was obtained: in fracture energy G_f by 2.3 times, in characteristic length ℓ_{ch} by 2.3 times and in fracture toughness K_f by 2.0 times (cf. Section 10.3.4). Such materials may possibly be used without high fibre reinforcement.

Table 13.6 Typical mixture proportions and properties of Reactive Powder Concretes (RPC)

	RPC 200				RPC 800	
	Non fibred		Fibred		Silica aggregates	Steel aggregates
Portland cement	1	1	1	1	1	1
SF	0.25	0.23	0.25	0.23	0.23	0.23
sand 150–600 µm	1.1	1.1	1.1	1.1	0.5	–
crushed quartz d = 10 µm	–	0.39	–	0.39	0.39	0.39
superplasticizer	0.016	0.019	0.016	0.019	0.019	0.019
steel fibre L = 12 mm	–	–	0.175	0.175	–	–
steel fibre L = 3 mm	–	–	–	–	0.63	0.63
steel aggregates < 800 µm	–	–	–	–	–	1.49
water	0.15	0.17	0.17	0.19	0.19	0.19
compacting pressure (MPa)	–	–	–	–	50	50
heat treatment temperature (°C)	20	90	20	90	250–400	250–400
compressive strength (MPa)	170 – 230				490 – 680	650 – 810
flexural strength (MPa)	30 – 60				45 – 141	
Young's modulus (GPa)	50 – 60				65 – 75	

Source: Reprinted from *Cement and Concrete Research*, Vol 25, Richard, P.; Cheyrezy, M., 'Composition of reactive powder concretes', pp. 11., Copyright (1995), with permission from Elsevier.

13.5 Slurry infiltrated fibre concrete

Slurry infiltrated fibre concrete (SIFCON) is a highly reinforced material belonging to the group of high performance concretes. It is composed of a high volume of steel fibres, ranging between 4% and 20%, in a cement-based matrix. Fibres are pre-placed in a mould and the resulting fibre array is infiltrated by cement-based slurry. Steel fibres are straight and plain, often bigger than in ordinary FRC materials, and hooked fibres are also sometimes used. Polypropylene fibres are used in the form of mats and fabrics.

The cement slurry may be filled with fine sand, micro-aggregate and special additives like fly ash and SF. The mass proportions of the components are: Portland cement:fly ash:sand from 9:1:0 up to 3:2:5 with the *w/c* ratio varying from 0.20 up to 0.45, depending on the use of superplasticizers (Schneider 1992).

The high fluidity (low viscosity) of the slurry is necessary for adequate penetration of the dense fibre array in the mould. By the addition of fly ash and fine sand, the Portland cement content may be diminished and possible shrinkage cracking reduced. Vibration of pre-placed fibres before and after pouring of cement slurry is often advisable in order to improve the density and distribution of reinforcement. External pressure should also be applied.

Research into the properties and technology of SIFCON is in progress in several laboratories. One of the first published reports was by Lankard (1985). A comprehensive review of mechanical properties and test results may be found in Naaman (1992). The behaviour of SIFCON elements under compression and bending compared with classic fibre-reinforced mortar is presented in Figure 13.15 and 13.16.

The SIFCON elements exhibit a considerable increase in strength and ductility compared to ordinary fibre reinforced concrete. The fracture toughness, calculated as proportional to the area under load-deflection curves, was also increased and the curves were characterized by a post-peak plateau. After these tests, it was concluded that the quality of the matrix itself contributed significantly to the overall behaviour of the composite material. That effect was studied further by Shah (1991) who showed an increased strain and strength capacity for a matrix subjected to tension. Therefore, these results confirmed earlier achievements in SIFCON technology and also supported

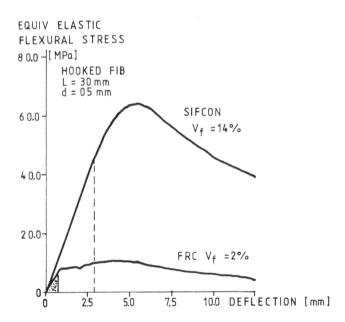

Figure 13.16 Stress-deflection curves in bending for SIFCON and FRC elements, after Naaman (1992).

theoretical predictions, which had been formulated by Aveston *et al.* (1972). It has also been shown that the matrix properties are improved only when fibre volume exceeds a certain threshold. Below that value the crack opening load is unaffected by reinforcement, as it is for ordinary fibre reinforced mortars and concretes. Beyond that threshold the tensile strength and fracture energy vary linearly with fibre content – at least within the scope of the fibre volumes used for SIFCON.

Multiple cracking under direct tension contributed to higher strength and increased energy absorption compared with FRC. The crack pattern as an external image showing the fracture mechanisms may be studied quantitatively and an approach using the concept of fractal dimension (cf. Section10.5) was proposed by Yan *et al.* (2002), who have shown that the main mechanical properties are linear functions of the fractal dimension increment $(D - 1)$, where D characterizes the crack pattern. The excellent ductility of SIFCON was also observed under cyclic compressive loads. Other mechanical properties like Young's modulus and bond strength to the matrix are also increased in SIFCON.

The majority of fibres are situated approximately parallel to the horizontal plane; therefore, the in-plane and out-of-plane properties are different and SIFCON exhibits clearly some orthotropic properties.

The considerably improved mechanical properties of SIFCON can be exploited in various structures or in their particular regions, where locally improved properties may modify overall structural behaviour, such as shields against projectiles or radiation, repair of outdoor structures, joint connections in structures exposed to seismic action or possible explosions, pre-cast impact resistant panels, walls of treasuries, etc. The practical applications of SIFCON are restricted to these special structures; probably there are some barriers on information about executed structures and details of their design, and such restrictions are compulsory for both designers and contractors. Also, the question as to whether the high additional costs of materials and construction are balanced by the performance of the product should be answered in every case.

A special type of this material is called SIFCA: slurry infiltrated fiber-reinforced castable. This is a composite material of ceramic matrix with calcium aluminate cement and with aggregate made of aluminum oxide, mullite, zircon and calcined fireclay. The matrix is reinforced with stainless-steel fibres. SIFCA is used to made pre-cast elements for refractory structures, where temperature can rise up to 1100°C. Heat curing is often used during the pre-casting. Another kind of similar material is called SIMCON: slurry infiltrated mat concrete in which arrays of single fibres are replaced by a system of steel mats for better and easier distribution of reinforcement (Murakami and Zeng 1998).

References

Ahmed, S. F. U., Mihashi, H. (2007) 'A review on durability properties of strain hardening fibre reinforced cementitious composites (SHFRCC),' *Cement and Concrete Composites*, 29: 365–76.

Aveston, J., Cooper, G. A., Kelly, A. (1971) 'Single and multiple fracture,' in: *Proc. Nat. Phys. Lab. Conf. The Properties of Fibre Compositesosites*, Guildford, UK; IPC Science and technology Press Ltd: pp. 15–24.

Baalbaki, W., Benmokrane, B., Chaallal, O., Aitcin, P. C. (1991) 'Influence of coarse aggregate on elastic properties of high performance concrete,' *ACI Materials Journal*, 88(5): 499–503.

Bache, H. H. (1981) 'Densified cement-ultrafine particle-base materials,' in: *Proc. 2nd Int. Conf. on Superplasticizers in Concrete*, Ottawa, ON: pp. 185–213.

Behloul, M. (2007) 'HPFRCC field applications: Ductal® recent experience,' in: *Proc. Int. Workshop on High Performance Fiber Reinforced Cement Composites HPFRCC-5*, H. W. Reinhardt and A. E. Naaman, eds, Mainz, Germany: pp. 213–22.

Birchall, J. D., Howard, A. J., Kendall, K. (1981) 'Flexural strength and porosity of cements,' *Nature*, 289: 388–9.

Boshoff, W. P., van Zijl, G. P. A. G. (2007) 'Tensile creep of SHCC,' in: *Proc. Int. Workshop on High Performance Fiber Reinforced Cement Composites HPFRCC-5*, H. W. Reinhardt and A. E. Naaman, eds, Mainz, Germany: pp. 87–95.

de Larrard, F., Mallier, Y. (1992) 'Proprietes constructives des betons a tres hautes performances: de la micro a la macrostructure,' in: *Les Bétons à Hautes Performances*, Malier, Y. ed., Paris; Presses de LCPC.

Fowler, D. W. (1999) 'Polymers in concrete: a vision for the 21st century,' *Cement and Concrete Composites*, 21: 449–52.

Gao, X. F., Lo, Y. T., Tam, C. M. (2002) 'Investigations of micro-cracks and microstructure of high performance lightweight aggregate concrete,' *Building and Environment*, 37: 485–9.

Gettu, R., Bažant, Z. P., Karr, M. E. (1990) 'Fracture properties and brittleness of high-strength concrete,' *ACI Materials Journal*, Nov–Dec: 608–18.

Habel, K., Viviani, M., Denarie, E., Bruhwiler, E. (2006) 'Development of the mechanical properties of an ultra-high performance fiber reinforced concrete (UHPFRC),' *Cement and Concrete Research*, 36: 1362–70.

Hjorth, B. L. (1983) 'Development and application of high-density cement-based materials,' in: *Phil. Trans. Royal Soc. London*, A310: pp. 167–73.

Jennings, H. M. (1992) 'Advanced cement-based matrices for composites,' in: *Proc. Int. HPFRCC Workshop RILEM/ACI*, H. W. Reinhardt and A. E. Naaman, eds, Mainz; Chapman and Hall/Spon: pp. 3–17.

Kendall, K., Birchall, J. G. (1985) 'Porosity and its relationship to the strength of hydraulic cement pastes,' in: *Mat. Res. Soc. Symp. Very High Strength Concrete-Based Materials*, 42: 153–8.

Kendall, K., Howard, A. J., Birchall, J. D. (1983) 'The relation between porosity, microstructure and strength, and the approach to advanced cement-based materials,' *Phil. Trans. Roy. Soc.*, A310: pp. 139–53.

Krüger, T. (1991) 'Mechanical behaviour of polymer modified cement mortars under complex stress states,' in: *Proc. Int. Symp. Concrete Polymer Composites*, H. Schorn and M. Middel, eds, Bochum, Germany: pp. 135–46.

Kucharska, L. (1992) 'Design of structure of high performance concretes: role of additives and admixtures' (in Polish), *Przegląd Budowlany*, 64(8/9): 351–4.

Kucharska, L., Moczko, M. (1993) 'Influence of cement chemical composition on its response to superplasticizer addition in the light of the rheological research,' in: *Proc. Interuniv. Sem. Tech*, The Netherlands; University of Eindhoven: pp. 135–43.

Lankard, D. R. (1985) 'Preparation, properties and application of cement-based composites containing 5 to 20 % steel fibres,' in: *US-Sweden Joint Seminar, Steel Fiber Concrete,* S. P. Shah and A. Skarendahl, eds, Stockholm; CBI: pp. 189–217.

Lecomte, A., Thomas, A. (1992) 'Caractère fractale des mélanges granulaires pour bétons de haute compacité,' Materials and Structures, RILEM, 25(149): 255–64.

Lévy, C. (1992) 'La carbonatation:comparaison BO/BHP du pont de Joigny,' in: *Les Bétons à Hautes Performances, caractérisation, durabilité, applications.* Paris; Presses de LCPC: pp. 359–76.

Li, M., Sahmaran, M., Li, V. C. (2007) 'Effect of cracking and healing on durability of engineered cementitious composites under marine environment,' in: *Proc. Int. HPFRCC5 Workshop RILEM,* H. W. Reinhardt and A. E. Naaman, eds, Mainz; RILEM PRO 53: pp. 313–22.

Malier, Y. ed., (1992) *Les Bétons à Hautes Performances, caractérisation, durabilité, applications.* Paris; Presses de LCPC.

Mangat, P. S., Baggot, R., Evans, D. A. (1981) 'Creep characteristics of polymer modified concrete under uniaxial compression,' in: *Proc. 3rd Int. Congr. Polymers in Concrete,* Koriyama, Japan: pp. 193–208.

Mehta, P. K., Gjørv, O. E. (1982) 'Properties of Portland cement concrete containing fly ash and condensed silica fume,' *Cement and Concrete Research,* 12(5): 587–95.

Murakami, H., Zeng, J. Y. (1998) 'Experimental and analytical study of SIMCON tension members.' *Mechanics of Materials,* 28: 181–95.

Naaman, A. E. (1992) 'SIFCON: tailored properties for structural performance,' in: *Proc. Int. RILEM Workshop, High Performance Fiber Reinforced Cement Composites, HPFRCC4,* RILEM Pro 30: pp. 18–38.

Naaman, A. E., Reinhardt, H. W. (2003) 'Setting the stage: towards performance based classification of FRC composites,' in: *Proc. Int. RILEM Workshop, High Performance Fiber Reinforced Cement Composites, HPFRCC4,* RILEM Pro 30: pp. 1–4.

Navy, E. G. (1996) *Fundamentals of High strength High performance Concrete,* London; Longman Group Ltd.

Ohama, Y. (1995) 'Polymer-based materials for repair and improved durability: Japanese experience,' *Construction and Building Materials,* 10(1): 77–82.

——.(1997) 'Recent progress in concrete-polymer composites,' *Advanced Cement Based Materials,* 5: 31–40.

——.(1998) 'Polymer-based admixtures,' *Cement and Concrete Composites,* 20: 189–212.

Ohama, Y., Demura, K. (1991) 'Properties of polymer-modified mortars with expansive additives,' *Int. Symp. 'Concrete Polymer Composites,'* H. Schorn and M. Middel, eds, Bochum, Germany: pp. 19–26.

Ohama, Y., Demura, K., Lin, Z. (1990) 'Development of super high-strength mortars,' in: *Proc. Int. Conf. Concrete for the Nineties,* W. B. Butler and I. Hinczak, eds, Concrete, Leura, Australia: pp. 1–12.

Persson, B. (1999) 'Poisson's ratio of high performance concrete,' *Cement and Concrete Research,* 29: 1647–53.

Powers, T. C. (1947) 'A discussion of cement hydration in relation to the curing of concrete,' in: *Proc. of Highway Research Board, 27th Ann. Meeting,* Washington, DC: pp.178–88.

Richard, P., Cheyrezy, M. (1995) 'Composition of reactive powder concretes,' *Cement and Concrete Research,* 25: 1501–11.

Rossi, P. (2008) 'Ultra high-performance concretes,' *Concrete International,* 30(2): 31–4.

Schneider, B. (1992); Development of SIFCON through applications,' in: *Proc. Int.*

RILEM Workshop, High Performance Fiber Reinforced Cement Composites, HPFRCC4, RILEM Pro 15, Mainz; Chapman and Hall/Spon: pp. 177–94.

Scrivener, K.L., Bentur, A., Pratt, P.L. (1988) 'Quantitative Characterization of the Transition Zone in High Strength Concretes', *Advances in Cement Research*, 1(4) pp. 230–37.

Shah, S. P. (1991) 'Do fibres improve the tensile strength of concrete?' in: *Proc. 1st Canadian Univ.-Ind. Workshop on Fibre Reinforced Concrete*, Quebec City, QC; Université Laval, pp. 10–30.

Tan, S. R., Howard, A. J., Birchall, J. D. (1987) 'Advanced materials from hydraulic cements,' *Phil. Trans. Roy. Soc. London*, A322: pp. 479–91.

Xiao, J., Schneider, H., Dönnecke, C., König, G. (2004) 'Wedge splitting test on fracture behaviour of ultra high strength concrete,' *Construction and Building Materials*, 18: 359–65.

Yan, A., Wu, K., Zhang, X. (2002) 'A quatitative study on the surface crack pattern of concrete with high content of steel fiber,' *Cement and Concrete Research*, 32: 1371–5.

Standards

ACI 548.3R-03 (2003) *Polymer Modified Concrete*

ACI 211.4R-93 (1998) *Guide for Selecting Proportions for High-Strength Concrete with Portland Cement and Fly Ash*

CEB-FIP (1990) 'Model Code for Concrete Structures,' *Comité Euro-International du Béton* (CEB), Lausanne

EN 206-1:2000 *Concrete – Specification, Performance, Production and Conformity*

14 Application and development of cement-based composites

14.1 Introduction

The great development of all kinds of cement-based materials was the result of their ability to satisfy new requirements in building and civil engineering structures, among other advantages described in previous chapters of this book.

Without entering into historical details, it may be observed that there were periods with steady continuous developments and periods with stepwise development related to outstanding innovations and discoveries. A few of the important steps are listed here:

- understanding of the influence of w/c ratio on the strength of concrete;
- discovery of plasticizers and superplasticizers;
- application of various kinds of mineral admixtures, additives and microfillers;
- application of dispersed reinforcement;
- concept of high performance cement composites.

After 100 years or so, the advanced cement-based composites appeared in the early 1980s and now represent a new generation of various composite materials in building and civil engineering (Brandt and Kucharska 1999). Without doubt their application will be increased in many kinds of structures to comply with new requirements. However, there are still several problems that should be investigated to ensure further development in building materials in general, and in cement-based composites in particular. The most important are:

- improved quality to satisfy the increasing needs for safe and durable structures of various kinds (housing, transportation, industrial buildings);
- requirement of sustainable development (economy of raw materials, energy and space, limitations on emission of gases and wastes, etc.).

Furthermore, the concrete structures should answer the requirements of new challenges due to an increasing population, intensification of natural disasters and the development of terrorism. On the other hand, there are new possibilities supplied from scientific research (physics, chemistry), new experimental and computation methods and an increase of international cooperation in the field of building and civil engineering.

This chapter is devoted primarily to the consideration of the present and future areas of application for advanced cement-based composites, and to the new directions of research and development imposed by these applications.

14.2 Improvement of quality

Improvements in the quality of structural materials will follow the requirement of sustainable development that is imposed in the world. Concrete-like materials have an important role to play. To fulfil that task a considerable development in our knowledge and technology is necessary, because:

- civil engineering and building structures are characterized by ill-defined duration of the service life; the random character of loads and other actions to be supported is unavoidable;
- present knowledge of main chemical and physical processes that are going on in materials during years of their service in variable conditions is far from satisfactory;
- new complex materials and their components provide unknown and often unpredictable synergistic effects on various scales – from macro to nano.

According to John Ruskin 'Quality is never an accident. It is always the result of intelligent effort,' and such an effort should be well organized.

The properties of advanced cement composites, which decide on future applications, not only comprise extended performance, but also these features that are related to improved durability in the first place:

- improved resistance against cracking and fracture toughness, and control of brittleness;
- increased compressive and tensile strength;
- water and gas tightness and improved resistance against various types of corrosive agents;
- improved resistance against local spallings and destruction caused by excessive mechanical and thermal actions;
- special features not directly related to high mechanical strength, but required for special applications, e.g. the aspect of concrete in architectural elements.

It is also important how properties of the advanced materials are presented to the designers and potential users, and how these properties are obtained at full scale and in industrial production.

14.3 Requirement of sustainable development

The concept of sustainable development of our civilization was discussed in the early 1970s, but finally in 1992 at the Earth Summit in Rio de Janeiro (Brazil) it was formally introduced to the world.

Cement-based composites, their production and application in concrete structures have a considerable impact on the environment because of a large consumption of natural resources and mass disposal of wastes. Every year in the world over 1 m^3 of concrete is produced per head and the cement industry is responsible for approximately 5–8% of man-made CO_2 emissions (Scrivener *et al.* 2008). It is therefore essential that the innovations in this field be directed towards environmentally-conscious materials, structures and technologies.

The concept called 'green concrete' was created in this respect analogically to another term used in the world: 'green building' (Malhotra 2002; Kashino and Ohama 2004; Mindess 2006). This means that the development of concrete-like materials should limit the exploitation of basic resources: raw materials, energy, water and space, and reduce all impact on the environment through better design, execution, maintenance and removal. As it is expected that the world consumption of Portland cement may soon reach 2.10^9 tons per year, it is essential to control that production while meeting the increasing demand for concrete. This goal may only be achieved by the extensive use of various additional materials, which may replace cement: granulated blast-furnace slag, fly ash, metakaolin and others. Using present knowledge and with developing technology, it is possible to multiply the durability of concrete structures by ten, and replace a large part of natural aggregate by recycled materials (Mehta 2001).

In the production of cement, the use of secondary cementitious materials (SCM) and supplementary fuels are completed by extensive application of all kinds of by-products and wastes, and significant progress in this field has been reported in last 20–30 years. The composition of cement-based materials and their properties, which are the main subject of the book, considerably influence the sustainability development of the world by:

- economic use of all resources for production of materials;
- limitation of all kinds of pollution in air and water;
- reliability and durability of structures made with these materials.

Therefore, the sustainability of concrete-like materials is a very complex problem covering different aspects of production, construction and exploitation, but the most relevant aspect is the durability of concrete structures (Swamy 2000; Mora 2007). The improved durability is closely related to the better quality of materials.

14.4 Life cycle and durability

The durability of structures is a subject frequently considered in investigations, reported in published papers and in contributions to many conferences. This is justified by poor durability of concrete structures in many countries. It is openly admitted that the structures require repair, rehabilitation and replacement often only a few years after completion.

The durability of a structure is considered as adequate if, throughout its designed life, it fulfils its intended functions related to serviceability, strength and stability without excessive and unforeseen maintenance (cf. Section 11.5).

Insufficient durability sometimes presents real danger to the safety of structures, but its main influence is on the economy, because it considerably increases the cost of the infrastructure. Besides, various inconveniences appear for society due to additional work during exploitation: disturbances for traffic, etc. These costs also should be taken into account when investments are planned.

It is not necessary to repeat here the detailed recommendations for durable concrete and to specify reliable technologies, because all the needed information may be easily found in academic manuals and books so that the state of knowledge may be considered as quite good and easily available. However, it is worth examining the problem of why that knowledge is not fully applied in building practice.

(a) Reasons of inadequate durability

Besides various factors related to the composition of concrete and to the detailing of structures, there are several non-technical factors that are only briefly mentioned here. They appear at all consecutive stages of the construction process:

1   Planning and preparation of the investment, including non-technical factors;
2   Collection of design data, technical and technological design, detailing;
3   Selection of materials, their components and internal structure;
4   Building process together with cure and hardening of concrete, etc.

Points 1 and 2 concern concrete structures, but points 3 and 4 are related to the cement-based materials and several reasons for damage are discussed in more detail in other chapters. In the near future, all these phenomena should be considered in the recommendations and standards from the viewpoint of durability in such a way that the designers of structures and materials take appropriate measures to avoid destruction after a relatively short time. The selection of raw materials for concretes, design of the mixture composition, technology of execution and care at an early age should be performed according to the best state of knowledge. Not all recommendations adequately cover the requirements of durability and even the European Standards are criticised in that respect.

It is necessary to extend the use of various SCM, which not only allow a decrease in the production of Portland cement clinker, but contribute to natural resource conservation and, paradoxically, yield more durable concrete products.

(b) State of knowledge in the technology of concretes

For many years the state of knowledge on the reasons of inadequate durability of concrete structures has been high and methods on how to design and execute durable structures are available. Already in the manual of the European Concrete Committee, CEB (1989) it is written that we know how to make durable concrete but we do not use that concrete. Neville (1963) observed that: 'Surprisingly, the ingredients of a good concrete are exactly the same [as in a bad one], and only the "know-how" ... is responsible for the difference.'

It is therefore necessary to distinguish the state of knowledge according to the textbooks, the competence of outstanding experts and the sophisticated methods used in specialised laboratories from the level of many other persons that participate in the process of planning investments, designing the structures and materials and executing the works in situ. The differences are decisive for frequent failures in the realization of durable structures, because incompetence and lack of adequate background of staff are often the main reasons for important mistakes. In many cases, there is also a lack of motivation, because the durability of the majority of structures is ill-defined and is neither particularly required nor remunerated.

The advance of knowledge is directed towards increasing durability. It may be observed that many (if not all) new kinds of high performance concretes like FRC, PMC, SCC, etc. are invented and developed mainly in view of their better durability with respect to traditional concretes.

(c) Standards and recommendations

It is necessary to introduce durability requirements to the standards related to concrete-like materials. Durability may be defined as a basic performance in a similar way as strength, stability and serviceability that are covered by present standards. Current practice shows that it is impossible to ensure durability by a system of multiple descriptive standards. The system of standards for durability should be developed and completed in the near future.

The application of advanced cement composites is still limited by a lack of standards in many countries. For example, there are only a few standards for concrete structures in which high performance concretes are admitted with compressive strength after 28 days exceeding 60 MPa or 100 MPa and the application of new methods of testing for advanced composites is not adequately presented.

Further development of standards and recommendations for all kinds of advanced cement-based composites is carried on through world and regional organizations like the ISO, EN, ACI, RILEM, ASTM and others. Recommendations published by specialized organizations are universally applied, though they have no formal validity as standard documents.

In the standards, so-called 'performance format' is needed. This means that the performance requirements define objectives in the form of properties in both fresh and hardened state, durability, aspect and other features of quality without prescribing how those are to be obtained (Mindess 2006). The quality of advanced materials is a result of appropriate organization and control. The quality measures comprise a definition of requirements, organizational methods and control at every stage of material design, execution and curing (Malier 1992). Here again specific standards should be developed, perhaps on the basis of ISO Standard 9000 and others belonging to this family.

14.5 Directions for further development

It may be foreseen that the difference between ordinary materials and advanced cement composites like high performance concretes and FRC will increase. While the mass production of concretes of low and medium quality will be developed to achieve minimum cost through maximum use of industrialization and automation, special outstanding structures and selected elements will require higher performance materials. To satisfy that demand several conditions should be fulfilled:

- transfer of information between research centres, design and construction firms, producers of component materials, etc.;
- implementation of new methods and research results in educational programmes;
- improvement of the system of standards and approvals for new products and technologies;
- improvement of life-costing of construction to fully take into account the better performance and improved durability.

The application of all kinds of high performance concrete and FRC in building and civil engineering structures is the object of many manuals, guidelines from producers and detailed reports.

(a) Main applications of advanced cement composites in structures

The increased demand for high performance materials in general, and of high performance cement-based composites in particular, is related to two groups of concerns:

1 the developed needs for structures exposed to high loads and external actions of different kinds, e.g. offshore platforms, underwater tunnels,

tall buildings and towers, long span bridges, heavy duty industrial floors, road and runway overlays, seismic buildings, etc.;

2   increased cost of maintenance of existing infrastructure, which exhibits insufficient durability even if built 5–10 years ago, and requires a very high percentage of national and regional budgets.

There are reasons to expect that, if correctly executed, structures made of high performance concretes will rationally fulfil new tasks for 100 years or more.

Increased interest in high performance concrete has been exhibited since the mid-1980s by the sponsoring of investigations in that direction. It appeared that for large applications of high performance concrete the coordinated support of the various parties involved was required: research institutes and high schools; producers of Portland cement, of other components and ready mixed concrete; constructors of outstanding structures; and governmental bodies at central and local levels. Successful projects in that field in France, Canada, USA, Japan and other countries were executed by multiple agencies working for a common benefit. Also at present, concentrated efforts are required to prepare appropriate design codes and recommendations and this is realized by large groups and consortia, among others, combining academic and industrial research centres with their facilities and highly qualified staff, for example, NANOCEM.

The use of high performance concretes allows steel structures to be replaced by concrete structures in tall buildings over 250 m. The first case where the concrete of mean characteristic strength f_{c28} = 120 MPa was used was the 62-storey Two Union Square Building of in Seattle, Washington, USA, built in 1988–1989. In all cases where the final selection was in favour of high performance concretes, not only material and construction costs were taken into account, but also the reduction in the consumption of materials, lower weight of structural elements, and increase in the rental space of buildings, etc. Improved durability as an additional and important factor in the economic analysis is usually not taken directly into account; thus, the increased durability was gratis.

High performance concrete is used for large pre-cast girders for bridges and for giant offshore platforms. Both types of structures are used in particularly difficult climatic conditions and under heavy loads. Improved durability and reduction of dimensions due to high strength are therefore important. Similar materials are also foreseen for submerged tube bridges or tunnels, which are likely to be scheduled for construction in the near future.

In road pavements, high performance concrete with carefully selected aggregate and compressive strength up to 135 MPa exhibited a level of abrasion resistance from studded tyres similar to massive granite. According to laboratory tests, this concrete may last four times longer than traditional concrete of 55 MPa. The addition of a small amount of steel and polypropylene fibres improves crack control in elements reinforced with traditional steel bars, for example, in deep beams where steel fibres partly replace stirrups. A considerable increase of steel-matrix bonding is brought about by fibre reinforcement.

High performance concrete is also used in special elements in which rather than large dimensions, non-conventional requirements are decisive. As an example, the containers used for radioactive waste may be quoted, which should ensure:

- extensive durability expressed in hundreds of years, corresponding to periods of decaying of radioactive compounds;
- high strength against accidental loads of unpredictable character;
- high resistance against all possible corrosive agents.

In the case of such containers, the specific cost of material is quite negligible because of its small volume. As an additional requirement, a project for accelerated testing should be proposed in relation to the design of the material's structure and composition. Such problems are often solved by a design of boxes made of fibre-reinforced resin concrete or other highly sophisticated composite.

Stainless steel fibres are added to concrete refractory elements to provide better resistance against cracking and spalling due to thermal cycles and thermal shocks. A considerable increase in service life abundantly covers the additional costs paid for fibres. Strongly reinforced cement-based elements are also applied as part of machinery equipment.

The development of high performance cement-based materials has another important advantage: application of advanced methods in research, testing and design of materials attracts competent and highly motivated staff. That factor completely changed the image of cement-based materials, which are now becoming 'hi-tech.' It is not necessary to stress all beneficial effects of such an upgrading for employment potential in the field of cement-based materials.

The main fields of present and future applications of cement-based composites besides ordinary concrete and traditional reinforced concrete structures are:

- heavy-duty pavements and industrial floors;
- airport overlays, runways, taxiways, etc;
- nuclear energy reactor buildings;
- water supply and industrial waste-retaining structures;
- structures of tall buildings, particularly in seismic zones;
- long span bridges, huge sea platforms and other large structures;
- refractory elements;
- 'hot spots' in structures, e.g. hinges, free exits, connections, etc;
- different kinds of external claddings for buildings and tunnels;
- tunnel and slope strengthening (shotcrete);
- repair of concrete structures in various situations.

(b) Application of FRC in civil engineering structures

In the last few years, dispersed steel fibres were successfully used in prestressed concrete bridge beams where they replaced mild steel reinforcement. As one of many examples, in Figure 14.1 the cross-section of the bridge deck for Saint-Pierre-la-Cour (France) is shown. Ten prestressed concrete beams form the main structure of the bridge deck. Additional plates are used as the lost shuttering for ordinary concrete cast in situ with ordinary reinforcement and covered with road pavement. The prestressed beams and plates are made with Ductal®. Thanks to 3% volume of dispersed steel microfibres in Ductal® matrix, there were neither stirrups nor other mild steel reinforcement needed against shearing and local stresses in the beams and considerable economy was obtained in time and the cost of labour. This is certainly a new and important direction for the future application of fibres in structural elements without any other reinforcement.

(c) Non-structural applications

Several elements made of cement composites have no structural functions. It means that their mechanical role is either limited to supporting their own weight in service life and minor local loads, or that the structural role is important mostly during short periods of manufacturing, transportation and erection of structures. Non-structural elements made with cement-based composites represent a significant field of application in the construction industry and account for a large percentage in mass and value. Some examples are various kinds of wall sandwich panels, claddings, plates for facades and architectural details.

Non-structural elements should satisfy several particular requirements and their rapid development is possible thanks to use of large variety of

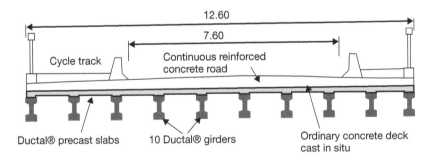

Figure 14.1 Cross-section of the bridge deck in Saint-Pierre-la-Cour (France) made with SFRC, V_f = 3%. (Reproduced with permission from Behloul M., HPFRCC field applications: Ductal® recent experience; published by RILEM Publications S.a.r.l., 2997.)

special cements, fibres, admixtures and technologies, for example, precast decorative elements.

(d) New components and technologies

In the mix composition of advanced cement-based materials a few admixtures for improving their special properties in fresh and in hardened states are normally included (cf. Section 4.3). In particular, different water reducers and plasticizers are used in a large majority of concrete production. This trend will intensify in the future, based on the results of investigations.

It is worth mentioning that in the last decades much attention has been paid to the application of mineral admixtures like ground blast-furnace slag, fly ash and various pozzolanic materials. That fact is justified by at least three reasons:

1   It appears that these admixtures, when carefully selected and correctly proportioned, impart considerable technical advantages to the resulting composites. Among these advantages are better workability, lower heat exhaustion during hydration and improved resistance against chemical aggression (cf. Section 4.3.9).
2   These materials are obtained from industrial wastes, which are otherwise stored in large volumes and present difficult problems for environmental protection.
3   Partial replacement of Portland cement with mineral admixtures enables appreciable savings in energy and cost to be made.

Increasing application of these waste materials in ordinary and advanced concretes requires further research aimed at standards and recommendations related to their quality and proper use. The reader is referred to specialized publications, for example, Swamy (1993) and Aïtcin (2000).

New technologies in the production of concrete should be considered from two viewpoints. The fresh mix may be prepared in different ways, improving the final properties, for example, special active mixing, the introduction of a superplasticizer to the mixing of concrete in two portions in view of its prolonged action, etc. On the other hand, in many cases special requirements are imposed, that of pumpability for longer distances or of self-levelling for industrial pavements. In such situations, selection of mixture components and their proportions are conditioned by the methods of execution. The application of multi-criterial optimization (cf. Section 12.7) would be a useful or even necessary approach.

(e) Elements for low-cost buildings

Over many years, ordinary, low-quality concrete will remain the main construction material for low cost houses in developing countries. The application of local raw materials and unqualified workers will be essential for solving the housing problem in large regions of the globe.

Low cost cement may be produced from natural zeolite, which is a rich resource and is globally spread. As a soft mineral it requires less energy for milling. It may be used as a replacement for the Portland cement for many applications where high strength is not required, or as cement admixture.

The problems related to safe and durable roofs for low cost houses are usually more difficult than those for walls and other elements of the house. For that reason, various kinds of low-cost fibre reinforcement are used in regions where cellulose pulp (in Nordic countries) and natural vegetal fibres (in tropical and subtropical countries) are available (cf. Section 5.7).

(f) Linings of tunnels and slopes

Lining of tunnels may be only partly considered as a non-structural element. The lining in coal mine tunnels was executed in Poland in the 1960s by shotcreting with concrete with 1.5% volume of short steel-fibres (Sikorski 1961). Results were excellent after many years of service. Similar applications in coal mine tunnels in Canada were reported by Morgan (1991) and Wood (1991). The lining for road and railway tunnels is also executed in many countries using a wet-process of shotcrete and steel-fibres as dispersed reinforcement. With a dosage of about 1–1.2% of Dramix or EE fibres, the following results were achieved:

- a reduction of construction time from three weeks, in the case of a cast lining, to 3–4 days;
- a reduction of the costs by half;
- improved ductility of the lining, which was applied directly to the rough surface of the rock.

The difference between standard shotcrete reinforcement and steel-fibre-reinforced shotcrete is that the mesh reinforcement should be fixed at a certain distance from the uneven surface of the rock, while the steel-fibre-reinforced shotcrete is laid directly on the rock surface. Hence, additional fixing operations are avoided and there are no void spaces behind the shotcreted layer. Tens of applications are reported: mines, road and railway tunnels, slope stabilisation, repair and rehabilitation works, etc., in Europe, North America, Southern Asia and Japan.

Thin claddings with or without fibre reinforcement, also curved in the form of shells, are pre-cast as:

- roofing sheets, panels and tiles for facades, etc.;
- curtain walls made of fibre-reinforced cement and of ferrocement;
- wave absorbers and anti-ballistic panels;
- permanent forms.

Claddings and curtain walls reinforced with carbon fibres were used in Japan with good results: reduction of thickness was possible because of higher strength, toughness and better fire resistance (Ohama *et al.* 1983).

Claddings reinforced with glass fibres and carbon fibres are used for facades, and pre-cast blocks for retaining walls. Cladding panels are usually produced by the spray-up process with dimensions up to a few meters and a depth of 5–15 mm. Alkali resistant glass fibres are used as a reinforcement even when their long-term durability is not perfectly ensured, because the reinforcement is needed only or mostly during execution, transportation and construction. Later, when fixed onto walls, these elements do not need to support loads and the poor durability of ordinary glass fibres in a cement-based matrix may be sufficient as it does not determine the durability of all structures. Glass fibres are applied also as reinforcement for matrices made with binders other than Portland cement: gypsum, high alumina cement, etc.

Extensive effort in research and technology led to positive results concerning the durability of glass fibre reinforced cement elements. A combination of modifications in the matrix composition and an improvement of the properties of the fibres enabled the use these composites for their long term durability (cf. also Section 5.3). For example, glass fibre reinforced cement thin plates are used as elements of partition walls, which may be subjected to accidental impact.

Polymeric low-modulus fibres are added for crack control during execution in claddings, tiles and decorations made of products known under various trade names like Caircrete and Faircrete in the UK. The matrix with these fibres in the fresh state may be easily formed in various shapes. In several cases, where strength and toughness of cladding is required, the hybrid combinations of fibres (carbon + steel, carbon + polypropylene, etc.) are applied, for example, for harbour facilities, channel linings or lining of oil-storage caverns. Polypropylene fibres are extensively used for control of micro-cracking due to thermal and shrinkage strain (cf. Section 5.6). Thus the large variety of available fibres gives the possibility of tailoring the cement-based composites to different applications.

(g) *Permanent formworks*

Cement-based composites are used to pre-cast permanent formworks for construction of large-scale bridges and industrial halls. Formwork in the form of plain or ribbed sheets is placed directly on main girders and the slab may be cast directly onto it. The ribs in the slab may be formed by profiling the sheets or by the inclusion of polystyrene void formers. Permanent formworks with glass fibre reinforcement are also used to produce pre-cast concrete blocks for retaining walls. In all these cases, long-term durability of glass fibres is not necessary.

(h) Repair and rehabilitation of concrete structures

Concrete structures deteriorate through climatic, physical, chemical and mechanical actions of different kinds. Outdoor structures are particularly exposed. The main single reason for the destruction of outdoor concrete structures is corrosion of reinforcement due to increasing intensity of the detrimental influence of polluted air and water, extensive use of de-icing salts on roads, frequent errors in the execution of concrete structures and so on. Cracking and spalling caused by internal pressure exerted by corrosion products are by far the most frequent reason for reconstruction and rehabilitation of structures. It is essential that in repair all possible sources of steel corrosion are consequently stopped; these may be due to depassivation of the concrete cover, to advanced carbonation, chloride attack, cracks with excessive width, etc. Long-term impermeability of repair material is the condition that may be satisfied by high-quality repair materials, for example, fibre-reinforced mortars and concretes, polymer modified concretes, etc.

Compatibility of repair materials and existing substrate is another important condition for a successful repair. Compatibility should be investigated analytically and then experimentally verified both in the laboratory and in situ. The interface between new and old material should ensure that all stresses that unavoidably appear due to mechanical and hygrothermal actions are safely transferred.

The problems related to repair and rehabilitation of structures are of great importance in many countries. For example, the highway infrastructure in USA is in a state of severe deterioration and thousands of bridges require repair and rehabilitation because of climatic-induced stresses, corrosion of reinforcing steel and repeated overloading. In Japan, the high amount of salt in the air and frequent alkali-aggregate reactions are the main reasons for degradation of structures. In Poland, the low quality of the construction of bridges in the 1950s and 1960s as well as severe climatic conditions characterized by a large number of freezing and thawing cycles in a year, are probably the most detrimental factors. Similar situations in all kinds of outdoor structures are observed in other countries, but for different reasons. As a result, the repair and rehabilitation of concrete structures requires large funds from local and central budgets. The appropriate methods for repair, which may ensure high durability together with technical and economical feasibility in various conditions, are developed and improved in many research centres around the world.

Modified cement-based matrices of various kinds were tested in view of the application of surface repair of concrete structures (Fowler 1999). Several compositions are used with:

- latexes based on styrene-butadiene, acrylic and other co-polymers, with the amount of latex solids to cement mass from 5% to 20%;
- dry polymers, e.g. vinyl acetate and vinyl versatic acid;

- microfillers, e.g. SF and fly ash.
- alkaline-resistent glass fibre and polymeric low modulus fibres.

Layers of composite mortars with latexes and SF varied from 6 mm to 250 mm in depth and were applied by a wet shotcrete method on walls and on overhead shotcreted slabs. The following mechanical properties were obtained (Ohama 1995):

- modulus of rupture 5–13 MPa;
- compressive strength 35–55 MPa;
- tensile strength 3.5–4.8 MPa

Good resistance to freeze-thaw cycles and to chemical attack due to high density and very low permeability was also obtained. Carbonation and chloride ions penetration are strongly decreased. Layers of thickness over 12 mm exhibited sufficient strength and ductility to bridge and waterproof the substrate cracks. Increased adhesion to substrate material, that is, old concrete, was excellent provided that the surface was cleaned and wetted. The adhesion was tested using tensile and shear bond specimens.

The application of latex modified concretes (LMC) for repair of outdoor structures is developing for the following structures:

- bridge desk overlays;
- pavements and stadiums;
- thin coatings of swimming pools;
- liners for pipelines, etc.

LMC appeared to be easy to place using various methods related to local conditions and dimensions of the projects: from pumping to mobile mixers and drum mixers for small size works.

Rapid set cements of various kinds are applied for repair work in order to decrease the duration of works and inconvenience for the users. Over 20 MPa of compressive strength is ensured already after three hours and over 40 MPa in 24 hours for such materials.

Steel fibres are used for tunnelling, mining and concrete repair, and glass fibres are rather used as reinforcement for cement mortar. Polymeric fibres are used mostly for control of the shrinkage cracking of external layers and as remedy in the case of fire (cf. Section 5.6).

Both main methods of shotcreting – dry and wet – are used in repair works with additional modifications. Dry shotcrete is based on mixing components with water in the nozzle. It results in high rebound and heavy dusting, which is detrimental for workers' health, particularly in closed spaces. Wet shotcrete is mixed with water before projecting. If used overhead, it requires a high amount of set accelerator, which in turn decreases the strength. A mortar modified with polymers and SF and reinforced with fibres is often applied by the wet process. The fibre-reinforced shotcrete is used extensively for different repair works, such as tunnel and channel linings, water storage and treatment facilities, bridge decks, piers and abutments, retaining walls, etc.

In many cases, the repair of a structure may be limited to casting a new external layer of highly impermeable material with properties carefully adjusted to those of the substrate. In other cases, however, it is essential that new material takes part in the load bearing. Then, active and durable connections are to be anticipated to ensure transfer of load on to new material. For example, increasing the load-carrying capacity of reinforced concrete beams may be realized by different methods:

- bonding steel plates to the beam;
- applying external post-tensioned tendons;
- providing a jacket of reinforced concrete over existing host member;
- applying non-metallic reinforcement.

Recommendations and guidelines on the economic and organizational sides of repair works, together with a binding procedure and assurance of quality, are given in many manuals (Czarnecki and Emmons 2002).

14.6 Economic aspects of development of cement-based composites

The economic questions concerning applications of cement-based composites have considerable importance for their development in various building and civil engineering structures. The advanced components and products are relatively expensive and their application requires special equipment, highly qualified staff and assistance from research centres. Consequently, the specific cost of such composites is higher when compared with medium-strength ordinary concretes, taking into account the so-called 'first cost.'

However, the overall cost of high performance concrete is not so high if that problem is considered on a larger basis. On the contrary, commercially available nearly 100 MPa concrete costs significantly less than 3.5 times the price of ordinary 30 MPa concrete. Thus, more strength is obtained per unit cost and per unit mass, together with more stiffness per unit cost and lower specific creep.

Furthermore, the specific cost of the material, calculated after the cost of components and execution, is only a small part of the total cost, which also comprises maintenance cost of various kinds, including repairs, possible replacement, breaks in normal exploitation, etc. Using high performance materials, smaller dimensions of structures are necessary, and lower maintenance cost may be expected. There are also other advantages: already Eiger (1933) has observed that high performance concrete elements may be subjected very early to partial service loads. All these factors decide upon the total cost of the entire structure. As a result, apparently expensive composite materials appear very interesting economically (Malier 1992; CEB 1989).

In the economic calculations, the cost of increased control of all components and of high performance materials in subsequent stages of execution should be at least partly compensated by gains due to the lower mean values of their required properties. When variations of the material's properties are reduced, then lower mean values may be designed and that aspect again allows a decrease in cost. Methods to evaluate the increased durability and other advantages of high performance concrete are needed. These should be based on rational provision of repair and maintenance expenses during the lifetime of the designed structure. Without such calculations the investor will not be convinced that in many situations the use of high performance concrete is the optimum solution.

Questions of economy determine the fields of application of cement matrix composites, which is the reason why ordinary concretes are still used and will continue to be used in the future for traditional plain and reinforced structures where high strength and improved durability are not necessarily required. Advanced composite materials are needed for special outstanding structures or for their particular regions, sometimes called 'hot-points' like joints and nodes, or for their elements particularly exposed to destruction.

A new generation of advanced cement-based composites is developing quickly. Increased interest for high performance concrete has been exhibited since the mid-1980s through the sponsoring of research and pilot construction projects up to the present situation of their common application.

14.7 Development of research and testing methods

In the past few decades, rapid development of methods of testing in the field of cement-based composites enabled the knowledge about processes and influences of different parameters to grow considerably. Certain methods are imported from the mechanics of high strength composites and ceramics; other methods are from basic research in physics and chemistry. There have been significant achievements in many fields.

Optical and electron microscopes allow the obtaining and recording of images at various levels, entering deeply into the materials' structure.

Extensive application of computer image analysis to various kinds of images obtained by different methods has led to quantitative results in several tests where, until last year, mostly qualitative conclusions were available.

Advanced modelling methods are used to a large extent. Simulating models are developed to represent various processes in microstructure, such as hydration of cement, progress of corrosion and of alkali-aggregate reaction, etc. Two- and three-dimensional simulations not only allow the verification of the influence of different conditions on examined processes in quick and inexpensive operations, but also provide the ability to demonstrate them to students in an extremely instructive way.

Advanced analysis of the acoustic emission effects, with the location of sources and analysis of energies at different levels, are able to distinguish the place and origin of phenomena. Thus matrix cracking, fibre debonding and other mechanisms of material damage may be separated and analysed.

These few examples concern the observation of the microstructure of cement-based composites, which is considered essential for understanding the basic processes that determine the final properties of these materials. It is therefore possible to improve the microstructure, for example, to obtain:

- segmentation of pores to avoid or at least to reduce permeability;
- densification of microstructure leading to higher strength;
- homogenisation of interfacial zones, which determine the overall properties of the material.

Using all these new methods, it will be possible to introduce new components to the technology of cement-based composites and to create new processes with full control of their results.

A large variety of new components, new methods of research and testing and advanced possibilities of modelling, together with improved methods of execution on site, create a real development of cement-based materials towards ceramics and high strength composites (Richard and Cheyrezy 1995).

If these achievements are so important why are there still difficulties in determining the behaviour of real structures in real service conditions?

There are large gaps in the present knowledge of cement-based composites, partly similar to those existing in other domains. Though it is possible to build theoretical models that are very sophisticated and apparently cover various situation and conditions, they do not correctly represent all the real-time dependent processes that determine the behaviour of these materials. External factors vary during the life cycle of the structures in such a way that their combined influence on these materials is outside these modelled predictions. The structure and composition of the materials is far more complex than is represented by the models.

The forms of damage at various levels and the fractal nature of the materials are other causes of complication. Though advanced methods are applied for experimental investigations, a large part of the present knowledge of the behaviour of cement-based materials is based on results obtained from carefully prepared specimens, tested in ideal and precisely controlled conditions and in limited periods of time. External influences are selected in order to obtain clear and unambiguous results, but frequently they have only a distant relation to the conditions in which real structures are expected to fulfil their service.

There are simple explanations why different results are obtained in the tests carried out intentionally on the same specimens and in the same conditions, and why the behaviour of small-sized specimens tested in ideal laboratory conditions are so different from the behaviour of the same material in real dimension structures that are ageing under the action of the natural environment. The extrapolations from tested specimens to structures are often doubtful and full of risk.

Without any doubt, a new approach is needed to develop existing knowledge towards real structures during their service life and to better use the available resources (Idorn 2005). There are reasons to expect new results with:

- development of the fuzzy logic concept, where randomness and non-linearity of relations are accepted to a large extent;
- application of artificial neural networks, where various data and results, apparently incoherent and incomplete, may be usefully exploited;
- use of fractal dimension of fractured surfaces, crack patterns and other effects to analyze real phenomena and processes at various levels;
- entrance to the nano level of matter, first to better understand various micro- and macro-processes, and next to modify them in a rational way, using nanotechnology (Scrivener *et al.* 2008).

There are several areas where new knowledge and better understanding are necessary. The following may be mentioned as important examples:

- rheology of fresh mix and relations of the rheological parameters to the properties of hardened materials;
- reactions between different minerals in Portland cement and various additives and admixtures, sometimes producing incompatibilities; cohesion and repulsion between particles in the fresh mix, etc.;
- delayed deformations in composites due to variation of temperature (heat of cement hydration) and influence of moisture diffusion in constrained conditions, in which structural elements exist at various levels – from micro- to macro-level;
- flow of external agents (ions, gases, fluids) into pore systems in concrete and their influence on maturity and durability of the composite.

Considerable effort in goal-oriented research is still needed in relation to applications of advanced cement-based materials. Notwithstanding recent achievements, the use of very high performance concrete and ultra high performance concrete is still limited to single special structures or so-called 'hot points' in the structures.

Selection of components and their proportions are, in most cases, based on theoretical models and consecutive trials. There is a place for considerable improvement in the material design by the application of the optimization approach, from which objective indications may be obtained (cf. Section 12.7). The optimization of concrete mixtures for multiple design objectives is a challenging and important area of research. Spatial gradation of the material composition and structure adds a new set of design parameters that must be accommodated within such an optimization framework.

The compatibility between particular components should be *a priori* determined together with their efficiency and not observed afterwards as a result of a success or a failure. This concerns cements, admixtures, microfillers and fibres, which introduced together to a mix, may cause positive or negative synergistic effects.

At present the design and execution methods of cement-based composites are strongly limited to a few given components and technological procedures. They are, in fact, reduced to certain generalizations of test results with numerical coefficients obtained from curve fitting. The proposed formulae for the forecast of strength and toughness need large test programs in which these coefficients can be determined. The extrapolation from laboratory test results to the conditions in which real structures should satisfy all requirements is still very difficult and should be based on great caution.

In big pre-casting factories and construction firms, the design formulae are restricted for local conditions and internal use, and are not published. Such a situation reduces, to some extent, the spread of design methods, their open discussion and further development. The proportioning of advanced and high performance materials cannot be entirely based on empiricism, because the application of a large variety of cements, different admixtures, dispersed reinforcement and various technologies for mixing, compaction and all other execution operations is not compatible with trial-and-error design methods. New methods based on expert systems, optimization approach and artificial neural networks are developing.

A breakthrough is also expected in material testing. Generally adopted test methods and acceptance requirements should open possibilities for international exchange and cooperation in the design and execution of elements and structures of cement-based composites. The testing recommendations proposed by regional (ASTM, CEN) and world (RILEM, FIB, ISO) organizations cover only selected types of materials and are not universally recognized. New kinds of materials in the family of high performance concretes require special methods and special equipment.

Such an important feature as material durability should be tested using accelerated methods to predict the effects of components and technologies in given conditions.

Execution methods of cement matrix composites were initially based exclusively on traditional operations developed decades ago for low- and medium-quality concretes. With the increasing variety of components and final products, those methods proved to be insufficient and inappropriate. That is why they should be continuously developed and improved. It is recognized that because of large volumes of produced materials and the types of components used and conditions of production, the methods should be relatively simple and as far as possible insensible to technological variations of composition, quality of components and conditions of execution and cure. Well-defined execution procedures are needed for materials that have to satisfy special requirements, for example, air-entrained concretes and internal curing that improve resistance of the outdoor structures against freezing and thawing cycles.

Special execution methods in which fresh mix is pumped at long distances or projected directly on the substrate need specialized test methods to establish their efficiency and influence on properties of the final product. Problems of their economical validity and possible effects on the health of workers are also not entirely solved. The progress of automatization is forecast in many operations in situ.

Notwithstanding present achievements, future research oriented to applications is still needed to support and develop the use of cement-based composites of various kinds in building and civil engineering (Table 14.1).

Table 14.1 Examples of goal-oriented research in cement-based materials

Research directions	*Goals to be achieved*
Materials science	New components
	Synergism of components
	Nano-, micro- and macro-structural effects
	Stress concentrations at inclusions
	Analysis of pore systems
Design of materials	Optimization of material design
	Computerization of design
	Exploitation of data with ANN
	Modellization of behaviour
	Extended use of recycled materials
Design of structures	Quantitative data
	Long-term behaviour of materials and durability
	Life cycle design
Execution and cure	Improved workability as a main feature
	Full application of available admixtures
	Low energy consumption
	Decrease of scatter of properties
Test methods	Standardization of test methods
	Relation between properties of specimens and of a material in a structure
	Testing of structures in natural scale
	Accelerated test methods and evaluation of durability
	New test methods: computer image analysis, tomography, micro-indentation, etc.
Economic methods	Evaluation of entire life cycle cost
	Optimization of maintenance operations
	Application of rules derived from sustainable development

14.8 Conclusions

The development of various kinds of high performance and ultra high performance concretes, also reinforced with dispersed fibres, has resulted in the creation of a group of very important building materials. At present, for many outstanding structures or for construction in special conditions, application of these materials is considered a necessity, and this situation will be extended in the future (Brandt 2008).

Successful use of various high performance materials based on the cement matrix has a considerable positive influence on the production of ordinary concretes. New components and technologies developed for special purposes are now, at least partly, applied in everyday production in ready-mix concrete plants. A large variety and better quality of admixtures, improved precision of execution and adequate curing are the bases for ordinary concretes that are becoming inexpensive, strong and ensure improved durability of buildings and civil infrastructure.

In general, concrete, and particularly concrete with dispersed fibre reinforcement, is becoming a high-tech material that provides excellent performance, but requires competent design and execution. Various experimental and theoretical methods that are successfully applied will certainly be used in further research and development.

References

Aïtcin, P. C. (2000) 'Cements of yesterday and today: concrete of tomorrow,' *Cement and Concrete Research*, 30: 1349–59.

Behloul, M. (2007) 'HPFRCC field applications: Ductal® recent experience,' in: *Proc. Int. Workshop on High Performance Fiber Reinforced Cement Composites HPFRCC-5*, H. W. Reinhardt and A. E. Naaman, eds, Mainz, Germany: pp. 213–22.

Brandt, A. M. (2008) 'Fibre reinforced cement-based (FRC) composites after over 40 years of development in building and civil engineering,' *Composite Structures*, 86, pp. 3–9.

Brandt, A. M., Kucharska, L. (1999) 'Developments in cement-based composites,' in: *Proc. Int. Conf. Extending Performance of Concrete Structures*, R. K. Dhir and P. A. J. Tittle, eds, Dundee; Thomas Telford: pp. 17–31.

Czarnecki, L., Emmons, P.H. (2002) 'Repair and Maintenance of Concrete Structures' (in Polish), *Cement Polski*, Cracow.

Eiger, A. (1933) 'High performance concrete' (in Polish), *Cement*, 4(1): 4–8.

Fowler, D. W. (1999) 'Polymers in concrete: a vision for the 21st century,' *Cement and Concrete Composites*, 21: 449–52.

Idorn, G. M. (2005) 'Innovation in concrete research: review and perspective,' *Cement and Concrete Research*, 35: 3–10.

Kashino, N., Ohama, Y. eds (2004) 'Environment-conscious materials and systems for sustainable development,' in: *Proc. RILEM Int. Symp.*, Koriyama, Japan; College of Engineering, Nihon University.

Malhotra, V. M. (2002) 'High-performance high-volume fly ash concrete,' *Concrete International*, 24: 45–9.

Malier, Y., ed., (1992), *Les Bétons à Hautes Performances: Caractérisation, Durabilité, Applications*. Paris; Presses de l'ENPC.

Mehta, P. K. (2001) 'Reducing the environmental impact of concrete,' *Concrete International*, 32(10): 61–6.

Mindess, S., (2006) 'High performance concrete: where do we go from here?' in: *Proc. Int. Symp. Brittle Matrix Composites 8*, A. M. Brandt, V. C. Li and I. H. Marshall, eds, Warsaw; ZTurek and Woodhead Publishing: pp. 15–23.

Mora, E. P. (2007) 'Life cycle, sustainability and the transcendent quality of building materials,' *Building and Environment*, 42: 1329–34.

Morgan, D. R. (1991) 'Use of steel fibre reinforced shotcrete in Canada,' in: *Proc. 1st Canadian University-Industry Workshop on Fibre Reinforced Concrete*, Quebec City, QC; Université Lava: pp. 164–82.

Neville, A. M. (1963) *Properties of Concrete*, London; Pitman & Sons.

Ohama, Y. (1995) 'Polymer-based materials for repair and improved durability: Japanese experience,' *Construction and Building Materials*, 10(1): 77–82.

Ohama, Y., Amano, M., Endo, M. (1983) 'Properties of carbon fiber reinforced cement with silica fume, *Concrete International*, 7(3): 58–62.

Richard, P., Cheyrezy, M. (1995) 'Composition of reactive powder concretes,' *Cement and Concrete Research*, 25: 1501–11.

Scrivener, K. L., Kirkpatrick, R. J. (2008) 'Innovations in use and research on cementitious materials,' *Cement and Concrete Research*, 38: 128–36.

Sikorski, C. (1961) *Steel Fibre Reinforced Concrete* (in Polish), Patent No. 58128, kl.80a, 51.

Swamy, R. N. (1993) 'Fly ash and slag: standards and specifications: help or hindrance?' *Materials and Structures*, RILEM, 26: 600–13.

Swamy, R. N. (2000) Editorial. *Cement and Concrete Composites*, 22(6): iii–iv.

Wood, D. F. (1991) 'Application of fibre reinforced shotcrete in tunnelling,' in: *Proc. 1st Canadian University-Industry Workshop on Fibre Reinforced Concrete*, Quebec City, QC; Université Laval: pp. 183–96.

Standards

CEB (1989) 'Durable concrete structures,' *Bulletin d'Information*, No 182.

Subject index

Subjects in the main text, tables, subscripts and figures are indicated by selected page numbers. Subjects appearing in references are not taken into account

eBooks – at www.eBookstore.tandf.co.uk

A library at your fingertips!

eBooks are electronic versions of printed books. You can store them on your PC/laptop or browse them online.

They have advantages for anyone needing rapid access to a wide variety of published, copyright information.

eBooks can help your research by enabling you to bookmark chapters, annotate text and use instant searches to find specific words or phrases. Several eBook files would fit on even a small laptop or PDA.

NEW: Save money by eSubscribing: cheap, online access to any eBook for as long as you need it.

Annual subscription packages

We now offer special low-cost bulk subscriptions to packages of eBooks in certain subject areas. These are available to libraries or to individuals.

For more information please contact webmaster.ebooks@tandf.co.uk

We're continually developing the eBook concept, so keep up to date by visiting the website.

www.eBookstore.tandf.co.uk